D1768577

Material Instabilities in
Continuum Mechanics

Material Instabilities in Continuum Mechanics
Related Mathematical Problems

The Proceedings of a Symposium Year on Material Instabilities in Continuum Mechanics organized by the Department of Mathematics, Heriot-Watt University, Edinburgh, 1985–1986

Edited by

J. M. BALL
Department of Mathematics
Heriot-Watt University

CLARENDON PRESS · OXFORD · 1988

Oxford University Press, Walton Street, Oxford OX2 6DP
Oxford New York Toronto
Delhi Bombay Calcutta Madras Karachi
Petaling Jaya Singapore Hong Kong Tokyo
Nairobi Dar es Salaam Cape Town
Melbourne Auckland
and associated companies in
Beirut Berlin Ibadan Nicosia

Oxford is a trade mark of Oxford University Press

Published in the United States
by Oxford University Press, New York

British Library Cataloguing in Publication Data
Material instabilities in continuum meachanics
and related mathematical problems: the
proceedings of a symposium year on material
instabilities in continuum mechanics,
organized by the Department of Mathematics,
Heriot-Watt University, Edinburgh, 1985–1986.
1. Continuum mechanics—Mathematics
I. Ball, J.M.
531 QA808.2
ISBN 0–19–853273–3

Library of Congress Cataloging-in-Publication Data
Material instabilities in continuum mechanics and related mathematical problems: the proceedings of a symposium year on material instabilities in continuum mechanics/organized by the Department of Mathematics, Heriot-Watt University, Edinburgh, 1985–1986: edited by J.M. Ball.
p. cm. Includes index.
1. Phase transformations (statistical physics)—Mathematical models—Congresses. 2. Mechanics—Mathematical models—Congresses. 3. Solid state physics—Mathematical models—Congresses. 5. Fluid mechanics—Mathematical models—Congresses. I. Ball, J. M. (John MacLeod), 1948– . II. Heriot-Watt University. Dept. of Mathematics.
QC176.8.P45M38 1987 531—dc 19 87-23297
ISBN 0–19–853273–3

Printed in Great Britain
by St Edmundsbury Press,
Bury St Edmunds, Suffolk

PREFACE

This volume contains the proceedings of the symposium year on 'Material Instabilities in Continuum Mechanics' which was held during 1985-86 in the Department of Mathematics of Heriot-Watt University, Edinburgh. The majority of the articles concern various aspects of the mathematical theory of phase transitions in solids and fluids, and reflect the revitalisation of this area by recent advances in the theory of systems of nonlinear partial differential equations and in the calculus of variations. These advances have led to new insights into the morphology of phase transitions and their dynamic behaviour. The remaining articles concern related problems and techniques in solid and fluid mechanics, mathematical physics and optimization.

The symposium year attracted over 80 visitors to the department: I would like to thank them all for the efforts they made to come, for the work they put into their lectures, and for their lively contributions to the discussions. Funding for the year was made available by the Science and Engineering Research Council. Invaluable advice and assistance in planning of the scientific programme was provided by Jack Carr and Robin Knops. That the visitors found accommodation waiting for them on arrival in Edinburgh was largely due to the heroic efforts of Patricia Hampton. Vital help with organization was given by Maureen Gardiner. The index to the proceedings was compiled by Denyse Lyon. Last but not least, the immaculate typing of the entire book, together with the related correspondence with authors, was undertaken by Ina Godwin.

J.M. Ball

CONTENTS

CONTRIBUTORS

E. Acerbi
Scuola Normale Superiore, Piazza dei Cavalieri 7, 56100 Pisa, Italy

G. Buttazzo
Scuola Normale Superiore, Piazza dei Cavalieri 7, 56100 Pisa, Italy

E. Cabib
Istituto di Meccanica Teorica ed Applicata, Universita' di Udine, Viale Ungheria 43, 33100 Udine, Italy

G. Caginalp
Department of Mathematics, University of Pittsburgh, Pittsburgh, PA 15260, U.S.A.

G. Capriz
Dipartimento di Matematica, Universita di Pisa, Pisa, Italy

A.V. Cherkaev
A.F. Ioffe Physical Technical Institute, Leningrad, USSR

D. Cioranescu
CNRS, Université Pierre et Marie Curie, Laboratoire d'Analyse Numérique, 4 Place Jussieu, 75230 Paris Cedex 05, France

B. Dacorogna
Département de Mathématiques, École Polytechnique Fédérale de Lausanne, MA (Ecublens), CH-1015, Lausanne, Switzerland

C. Davini
Istituto di Meccanica Teorica ed Applicata, Universita' di Udine, Viale Ungheria 43, 33100 Udine, Italy

R.J. DiPerna
Department of Mathematics, University of California, Berkeley, CA 94720, U.S.A.

J.L. Ericksen
Department of Aerospace Engineering and Mechanics and School of Mathematics, University of Minnesota, Minneapolis, MN 55455, U.S.A.

L.E. Fraenkel
School of Mathematical and Physical Sciences, Mathematics Division, The University of Sussex, Falmer, Brighton, BN1 9QH, U.K.

N. Fusco
Dipartimento di Matematica e Applicazioni "R. Caccioppoli", Università degli Studi di Napoli, V. Mezzocannone, 8, 80134 Napoli, Italy

J.D. Hunt
Department of Metallurgy and Materials Science, University of Oxford, Oxford, U.K.

R. Illner
Department of Mathematics, University of Victoria, Victoria B.C. V8W 2Y2, Canada

R.D. James
Department of Aerospace Engineering and Mechanics, University of Minnesota, Minneapolis, MN 55455, U.S.A.

C.K.R.T. Jones
Department of Mathematics, University of Maryland, College Park, MD 20742, U.S.A.

D. Kinderlehrer
School of Mathematics, University of Minnesota, 206 Church St. S.E., Minneapolis, MN. 55455, U.S.A.

J.L. Lions
Collège de France and CNES, 11 Place Marcelin Berthelot, F-75231 Paris, Cedex 05, France

K.A. Lurie
A.F. Ioffe Physical Technical Institute, Leningrad, USSR

P. Marcellini
Istituto Matematico, Università degli Studi, Viale Morgagni, 67A, 50134 Firenze, Italy

E. Mascolo
Istituto di Matematica, Università di Salerno, 85100 Salerno, Italy

D.G. McCartney
Department of Metallurgy and Materials Science, University of Liverpool, P.O. Box 147, Liverpool L69 3BX, U.K.

A. Novick-Cohen
Department of Mathematics, Michiga State University, East Lansing, Michigan 48824, U.S.A.

O.A. Oleinik
Department of Mathematics, Moscow University, Moscow, B 234, USSR

R.L. Pego
Department of Mathematics, Heriot-Watt University, Riccarton, Edinburgh, EH14 4AS, Scotland, U.K.

O. Penrose
Department of Mathematics, Heriot-Watt University, Riccarton, Edinburgh, EH14 4AS, Scotland, U.K.

M. Pitteri
Istituto di Analisi e Meccanica, Università di Padova, Via Belzoni 7, 35131 Padova, Italy

V. Roytburd
Department of Mathematical Sciences, Rensselaer Polytechnic Institute, Troy, New York, 12180-3590, U.S.A.

J. Saint Jean Paulin
CNRS, Lemta, 2 Rue de la Citadelle, B.P. 850, 54011 Nancy Cedex, France

R. Schianchi
Departimento di Matematica E Applicazion, Università degli Studi di Napoli, Via Mezzocannone, 8-Cap. 80134 Napoli, Italy

M.E. Schonbek
Department of Mathematics, University of California, Santa Cruz, California 95064, U.S.A.

M. Šilhavý
Mathematical Institute, Czechoslovak Academy of Sciences, Žitná 25, 115 67 Praha 1, Czechoslovakia

M. Slemrod
Department of Mathematical Sciences, Rensselaer Polytechnic Institute, Troy, New York, 12180-3590, U.S.A.

F. Tomarelli
Dipartimento di Matematica, Università di Pavia, Corso Carlo Alberto 5, 27100 Pavia, Italy

A. Visintin
Istituto di Analisi Numerica del C.N.R., Corso Carlo, Alberto 5, 27100 Pavia, Italy

1

AN APPROXIMATION LEMMA FOR $W^{1,p}$ FUNCTIONS

EMILIO ACERBI *and* NICOLA FUSCO

In this paper we give a direct proof of the following approximation result:

THEOREM. Let $\Omega \subset \mathbb{R}^n$ be a regular open set, and $p \geqslant 1$. There exists a constant c such that, for every $u \in W^{1,p}(\Omega)$ and every $K > 0$ there exists $v \in W^{1,\infty}(\Omega)$ satisfying

$$\|v\|_{1,\infty} \leqslant K$$

$$\text{meas}\,\{x : u(x) \neq v(x)\} \leqslant c\,\frac{\|u\|^p_{1,p}}{K^p}\,.$$

Unlike several results of Luzin type available ([3],[4]), in this theorem we do not look for an approximating function v which is close to u in the $W^{1,p}$ norm, but we want a precise control on the gradient of v. Thus, when the result is applied to a bounded sequence in $W^{1,p}$ one obtains a sequence of approximating functions which is bounded in $W^{1,\infty}$. This theorem was successfully applied by the authors in two different contexts: in a weaker form ([1]) when dealing with the semicontinuity of quasiconvex integrals, and in the general form ([2]) to obtain a regularity theorem for the same functionals.

The proof was not written explicitly in [2], where the reader was referred to the weaker result of [1], so we decided to present the details here, also thinking that this Lemma might be of interest by itself.

For every $f \in L^1_{\mathrm{loc}}(\mathbb{R}^n)$ we set

$$Mf(x) = \sup_{r>0} \fint_{B_r(0)} |f(x+y)|\, dy \,,$$

and if $u \in W^{1,1}_{\mathrm{loc}}(\mathbb{R}^n)$ we set

$$M'u(x) = M(|Du|)(x)\,.$$

We now show that through the maximal function M' we may control the difference quotient of u outside a small set. We set, for any $u \in W^{1,1}_{\mathrm{loc}}(\mathbb{R}^n)$ and for any $\lambda > 0$

$$H_\lambda = \{x : M'u(x) < \lambda\}\,.$$

LEMMA 1. There exists a constant $c_1 = c_1(n)$ such that, for every $u \in C^\infty_0(\mathbb{R}^n)$, we have

$$\frac{|u(x) - u(y)|}{|x-y|} \leqslant c_1 \lambda$$

for all $x, y \in H_\lambda$.

PROOF. Fix $\lambda > 0$ and $x \in H_\lambda$, and set

$$S_{k,r}(x) = \left\{ y \in B_r(x) : \frac{|u(x) - u(y)|}{|x-y|} \geqslant k\lambda \right\}\,.$$

We prove that, as k increases, $S_{k,r}(x)$ occupies a smaller and smaller portion of $B_r(x)$, independently of the particular function u: indeed,

$$\begin{aligned} k\lambda \frac{\operatorname{meas} S_{k,r}(x)}{\operatorname{meas} B_r(x)} &\leqslant \fint_{B_r(0)} \frac{|u(x+y) - u(x)|}{|y|}\, dy \\ &= \fint_{B_r} \left| \int_0^1 \frac{1}{|y|} \langle Du(x+ty), y \rangle dt \right| dy \\ &\leqslant \fint_{B_r} \int_0^1 |Du(x+ty)|\, dt\, dy \\ &= \int_0^1 \fint_{B_{tr}} |Du(x+y)|\, dy\ dt \leqslant \int_0^1 M'(x)\, dt < \lambda\,, \end{aligned}$$

so that

$$\operatorname{meas} S_{k,r}(x) \leqslant \frac{1}{k} \operatorname{meas} B_r\,. \tag{1}$$

Now fix $x, y \in H_\lambda$, and set $r = |x-y|$. Denote by γ_n the measure of the intersection of two balls of radius 1 in $\mathbb{R}^n$, whose centres are also at distance 1 from each other. If $\omega_n = \text{meas}\, B_1$, it is clear from (1) that for $k = 3\omega_n/\gamma_n$ the measure of $S_{k,r}(x) \cup S_{k,r}(y)$ is less than the measure of $B_r(x) \cap B_r(y)$, so that we may choose

$$z \in [B_r(x) \cap B_r(y)] \setminus [S_{k,r}(x) \cup S_{k,r}(y)] .$$

Then

$$|u(x) - u(z)| < k\lambda r$$

$$|u(y) - u(z)| < k\lambda r ,$$

whence

$$|u(x) - u(y)| < 2k\lambda r = \frac{6\omega_n}{\gamma_n} |x-y| ,$$

and the lemma is proved.

The following result may be found in [5].

LEMMA 2. For every $p \geq 1$ there exists $c_2 = c_2(n,p)$ such that

$$\text{meas}\,\{x : Mf(x) \geq \lambda\} \leq c_2 \frac{\|f\|_p^p}{\lambda^p}$$

for every $f \in L^p(\mathbb{R}^n)$ and $\lambda > 0$.

We may now extend Lemma 1 to the case of $W^{1,p}$ functions.

LEMMA 3. For every $u \in W^{1,p}(\mathbb{R}^n)$ there exists $E \subset \mathbb{R}^n$ with $\text{meas}\, E = 0$ such that

$$\frac{|u(x) - u(y)|}{|x-y|} \leq c_1 \lambda$$

for every $\lambda > 0$ and every $x, y \in H_\lambda \setminus E$.

PROOF. Let $(u_h) \subset C_0^\infty(\mathbb{R}^n)$ be a sequence converging to u strongly in $W^{1,p}(\mathbb{R}^n)$ and a.e. in $\mathbb{R}^n$. We remark that

$$|Mf(x) - Mg(x)| \leq M(f-g)(x) ,$$

so that, using Lemma 2, we have for all $\varepsilon > 0$,

$$\lim_h \text{meas}\,\{x : |M'u(x) - M'u_h(x)| \geq \varepsilon\}$$

$$\leq \lim_h \text{meas}\,\{x : M'(u-u_h)(x) \geq \varepsilon\}$$

$$\leq c_2 \varepsilon^{-p} \lim_h \|Du - Du_h\|_p^p = 0 ,$$

i.e., $M'u_h \to M'u$ in measure. We may then suppose that, at least for a subsequence,

$$M'u_h \to M'u \text{ a.e. in } \mathbb{R}^n .$$

Set

$$\mathbb{R}^n \setminus E = \{x : u_h(x) \to u(x)\ ,\ M'u_h(x) \to M'u(x)\}\ ;$$

then $\operatorname{meas} E = 0$, and for every $x, y \in H_\lambda \setminus E$, and h sufficiently large, we have

$$M'u_h(x) < \lambda,\quad M'u_h(y) < \lambda\ ,$$

so that by Lemma 1

$$\frac{|u_h(x) - u_h(y)|}{|x-y|} \leqslant c_1 \lambda\ ,$$

and the assertion follows by taking the limit as $h \to \infty$.

The following extension result is well known.

LEMMA 4. Let X be a metric space, $E \subseteq X$, and let $f: E \to \mathbb{R}$ be a K-Lipschitz function. There exists a function $g: X \to \mathbb{R}$ which is K-Lipschitz and satisfies

$$g \equiv f \text{ in } E\ ,\quad \sup_X |g| = \sup_E |f|\ .$$

Now we have all the necessary ingredients needed to prove the main result.

PROOF OF THE THEOREM. Fix $u \in W^{1,p}(\Omega)$; since Ω is regular, we may assume that $u \in W^{1,p}(\mathbb{R}^n)$, with

$$\|u\|_{1,p} \leqslant c_3 \|u\|_{1,p,\Omega}$$

and c_3 independent of u. For $K>0$ let $\lambda = K/2c_1$, then consider the set

$$H'_\lambda = \{x : Mu(x) + M'u(x) < \lambda\}\ .$$

By Lemma 2, we have

$$\operatorname{meas}(\mathbb{R}^n \setminus H'_\lambda) \leqslant \operatorname{meas}\left\{x : Mu(x) \geqslant \frac{\lambda}{2}\right\} + \operatorname{meas}\left\{x : M'u(x) \geqslant \frac{\lambda}{2}\right\} \leqslant c_4 \frac{\|u\|^p_{1,p,\Omega}}{K^p}\ .$$

On the other hand, by Lemma 3, we have

$$|u(x)| \leqslant Mu(x) < \lambda \leqslant \frac{K}{2}$$

$$\frac{|u(x) - u(y)|}{|x-y|} < c_1 \lambda = \frac{K}{2}$$

a.e. on H'_λ. Applying Lemma 4 we obtain a function v with Lipschitz constant $K/2$, and L^∞ norm not greater than $K/2$, so that

$$\|v\|_{1,\infty} \leqslant K.$$

Since $v=u$ a.e. in H'_λ the theorem is proved.

As an example, an easy consequence of the above theorem is the following:

COROLLARY 5. Let Ω be a regular open subset of $\mathbb{R}^n$ and $p \geqslant 1$, and let $(u_h) \subset W^{1,p}(\Omega)$ converge strongly to u in $W^{1,p}$. Then for every $\varepsilon > 0$ there exist a subset A_ε of Ω, with meas $A_\varepsilon < \varepsilon$, a subsequence (h_k) and a sequence $(w_k) \subset W^{1,\infty}(\Omega)$ such that

$$w_k \equiv u_{h_k} \quad \text{in} \quad \Omega \setminus A_\varepsilon$$

$$w_k \to w \quad \text{strongly in } W^{1,\infty}(\Omega)$$

and

$$w \equiv u \quad \text{in} \quad \Omega \setminus A_\varepsilon .$$

REFERENCES

1. E. Acerbi and N. Fusco, "Semicontinuity problems in the calculus of variations", *Arch. Rational Mech. Anal.* **86** (1984), pp.125-145.
2. E. Acerbi and N. Fusco, "A regularity theorem for minimizers of quasiconvex integrals", *Arch. Rational Mech. Anal.*, (1987) to appear.
3. F.-C. Liu, "A Luzin type property of Sobolev functions", *Indiana Univ. Math. J.* **26** (1977), pp.645-651.
4. J.H. Michael and W.P. Ziemer, "A Luzin type approximation of Sobolev functions by smooth functions", *Contemporary Mathematics* **42** (1985), pp.
5. E.M. Stein, *Singular Integrals and Differentiability Properties of Functions*, Princeton University Press, Princeton, 1970.

2

HOMOGENIZATION AND PERIODIC STRUCTURES WITH HOLES

EMILIO ACERBI

In the last decade, the concept of Γ-convergence of functionals has been widely investigated; to clarify its link with mechanics, consider the following simple consequence of Γ-convergence.

Assume (F_h) is a sequence of functionals defined, say, on H_0^1, and all satisfying

$$F_h(u) \geqslant \int |Du|^2 \, dx \ , \tag{1}$$

and let F_∞ be another functional. If the sequence (F_h) is $\Gamma^-(L^2)$-convergent to F_∞, then for every $g \in H^{-1}$ and every minimizing sequence (u_h) of

$$F_h(u) + \langle g, u \rangle$$

we may select a subsequence (u_{h_k}) converging in L^2 to a minimum point of

$$F_\infty(u) + \langle g, u \rangle \ .$$

If the functionals F_h are the free energies of some materials, then we may well say that the material represented by F_h tends to behave like the one associated with F_∞.

The usefulness of Γ-convergence lies in the various compactness results available, although the identification of the limit F_∞ is sometimes not straightforward. Most frequently in the literature, Γ-convergence appears in the context of homogenization: suppose you are given a structure, in the unit cube Q, whose free energy is given by

$$F_1(u) = \int_Q f(x, Du)\, dx \ ,$$

and repeat it periodically in space. Rescaling everything by a factor $1/h$, you obtain in the unit cube a finer structure whose energy is

$$F_h(u) = \int_Q f(hx, Du)\, dx \ ,$$

where f is now 1-periodic in x.

It is likely that, looking at this structure from "very far", it will seem to be a homogeneous material.

Every homogenization result thus consists in proving that (F_h) converges, in the Γ-sense, to some functional F_∞, and identifying F_∞ as an energy functional

$$F_\infty(u) = \int_Q \phi(Du)\, dx \ .$$

In this field, many results have been obtained, especially in the scalar case, i.e. u takes its values in $\mathbb{R}$, or when condition (1) is attained through the coerciveness of f:

$$f(x, \xi) \geqslant |\xi|^2 \qquad \forall\, x \ .$$

This condition, unfortunately, rules out many interesting cases: in order to represent an inhomogeneous elastic structure with holes, the integrand

$$f\colon Q \times \mathbb{R}^9 \to \mathbb{R}$$

must satisfy only

$$f(x, \xi) \geqslant |\xi|^2 \ \text{ only if } x \notin H \ , \tag{2}$$

where the hole H is a subset of Q with nice boundary and properly contained in Q.

A homogenization result under these assumptions is contained in [1], where in addition the dependence of f on u is not through the gradient, but through the strain tensor $e(u)$: if f is convex in ξ and satisfies (2) and

$$0 \leqslant f(x,\xi) \leqslant c(1+|\xi|^2) \tag{3}$$

then the functionals

$$F_h(u) = \int_Q f(hx, e(u))\,dx$$

$\Gamma^-(L^p)$-converge to the homogenized functional

$$F_\infty(u) = \int_Q \phi(e(u))\,dx\,,$$

where $\phi(\xi)$ is given by

$$\inf\left\{\int_Q f(x,e(u))\,dx:\ u \in H^1_{loc}(\mathbb{R}^3)\,,\ u-\xi x \text{ is } Q\text{-periodic}\right\}.$$

As a matter of fact, the result is more general: for example, in conditions (2) and (3) we may have a generic growth $p > 1$, instead of 2.

The main tools employed in the proof are an abstract Γ-compactness theorem, various forms of Korn's inequality and the following interesting extension lemma:

Let $p > 1$ and let Ω, ω be bounded open subsets of $\mathbb{R}^n$ with Lipshitz boundary, such that $\omega \subseteq \Omega$. Then there exists a constant $c(\Omega,\omega)$ such that for every $u \in W^{1,p}(\omega;\mathbb{R}^n)$ there exists $\tilde{u} \in W^{1,p}(\Omega;\mathbb{R}^n)$ such that

$$\tilde{u} = u \quad \text{in } \omega$$

$$\int_\Omega |e(\tilde{u})|^p\,dx \leqslant c(\Omega,\omega)\int_\omega |e(u)|^p\,dx\,.$$

Moreover $c(t\Omega, t\omega) = c(\Omega,\omega)$ for every $t > 0$.

REFERENCES

1. E. Acerbi and D. Percivale, "Homogenization of non-coercive functionals: periodic materials with soft inclusions", *Applied Mathematics and Optimization*, to appear.

3

THIN INSULATING LAYERS: THE OPTIMIZATION POINT OF VIEW

GIUSEPPE BUTTAZZO

1. INTRODUCTION

The problem of thin insulating layers has been widely studied by several authors in recent years (see References); we deal here with the related optimization problem, that is to find the "best" way to put a given amount of insulating material around a given conducting body. We shall consider the framework of the stationary heat equation, but the same model applies to electrostatics, and to elastic membranes.

The mathematical model for a thin insulating layer is the following. Let Ω be a regular bounded open subset of $\mathbb{R}^n$, let $d : \partial\Omega \to \mathbb{R}$ be a positive smooth function, and let $f \in L^2(\Omega)$. If Ω is regarded as a conducting body whose boundary $\partial\Omega$ is varnished by an insulating varnish, $d(\sigma)$ represents the density of the varnish at the point $\sigma \in \partial\Omega$, and f is the heat sources density, then the temperature u is the solution of the elliptic problem ($\nu(\sigma)$ is the outward normal vector to Ω at the point $\sigma \in \partial\Omega$)

$$\left.\begin{aligned} -\Delta u &= f & \text{in } \Omega \\ d\,\frac{\partial u}{\partial \nu} + u &= 0 & \text{on } \partial\Omega \end{aligned}\right\}, \tag{1.1}$$

or equivalently, it is the solution of the minimum problem

$$\min\left\{ \int_\Omega |Du|^2\,dx - 2\int_\Omega f(x)\,u\,dx + \int_{\partial\Omega} \frac{u^2}{d}\,d\sigma : \ u \in H^1(\Omega) \right\}. \tag{1.2}$$

Problem (1.2) can be seen to be the "limit problem" of the family of minimum problems

$$\min\left\{\int_\Omega |Du|^2\,dx+\varepsilon\int_{\Sigma_\varepsilon}|Du|^2\,dx-2\int_\Omega f(x)\,u\,dx\ :\ u\in H^1_0(\Omega\cup\Sigma_\varepsilon)\right\} \tag{1.3}$$

where the "thin layer" Σ_ε is given by

$$\Sigma_\varepsilon=\{\sigma+t\nu(\sigma):\sigma\in\partial\Omega\,,\,0\leqslant t<\varepsilon d(\sigma)\}.$$

The correct meaning of "limit-problem" should be given in terms of Γ-convergence (we refer the interested reader to [1], [3],[7],[9],[10]), but the most important feature is that the solutions u_ε of (1.3) converge in $L^2(\Omega)$ to the solution u of (1.2). Taking (1.2) as the limit problem for insulating layers of density $d(\sigma)$, the following question naturally arises:

> given Ω and f, what is the "best" distribution of insulator around $\partial\Omega$?

In order to set mathematically this problem, we have to define our optimality criterion; our choice is the following. For every function d and every $u\in H^1(\Omega)$ denote by $E(u,d)$ the energy associated to the elliptic problem (1.1)

$$E(u,d)=\int_\Omega|Du|^2\,dx-2\int_\Omega f(x)\,u\,dx+\int_{\partial\Omega}\frac{u^2}{d}\,d\sigma\,,$$

let u_d be the solution of problem (1.1), and let $E(d)$ be defined by

$$E(d)=\min\{E(u,d)\ :\ u\in H^1(\Omega)\}\,.$$

By (1.1), (1.2) we have

$$E(d)=E(u_d,d)=-\int_\Omega f(x)\,u_d\,dx. \tag{1.4}$$

Our optimality criterion consists in minimizing $E(d)$ among all functions

$$d:\partial\Omega\to\mathbb{R}^+\quad\text{with}\quad\int_{\partial\Omega}d\,d\sigma=k\,.$$

REMARK 1.1. When the heat sources are uniformly distributed in Ω (i.e. $f\equiv 1$), by (1.4) our optimization problem is equivalent to determining the function d for which the averaged temperature

$\int_\Omega u_d dx$ is maximum. Nevertheless, other criteria, different from the minimization of $E(d)$, should be investigated, such as

$$\min \int_\Omega |u_d - a(x)|^2 \, dx \qquad (a(x) \text{ is the desired temperature}),$$

or more generally

$$\min \int_\Omega g(x, u_d)\, dx + \int_{\partial\Omega} h(\sigma, d, u_d)\, d\sigma$$

for suitable choices of the functions g and h.

2. THE OPTIMIZATION RESULT

In this section we consider the problem

$$\min \{E(d) : d \in \Gamma_k\} \tag{2.1}$$

where Γ_k is the set of functions

$$\Gamma_k = \left\{ d \in L^1(\partial\Omega) : \ d \geq 0 \, , \int_{\partial\Omega} d \, d\sigma = k \right\}.$$

Taking into account the definition of $E(d)$, problem (2.1) is equivalent to

$$\min \left\{ E(u, d) : u \in H^1(\Omega), \ d \in L^1(\Omega), \ d \geq 0, \int_{\partial\Omega} d \, d\sigma = k \right\}. \tag{2.2}$$

PROPOSITION 2.1. For every $u \in L^2(\partial\Omega)$ there exists a solution $d_u \in \Gamma_k$ (which is unique if u is nonzero) of the minimum problem

$$\min \left\{ \int_{\partial\Omega} \frac{u^2}{d} \, d\sigma : d \in \Gamma_k \right\} , \tag{2.3}$$

Moreover, we have

$$d_u = k \, \frac{|u|}{\int_{\partial\Omega} |u| \, d\sigma} \, .$$

PROOF. If $u = 0$, then any function d solves (2.3); on the other hand, if u is nonzero, we claim that the function

$$d_u = k \, \frac{|u|}{\int_{\partial\Omega} |u| \, d\sigma}$$

is the unique solution of (2.3). In fact, by Hölder's inequality, we get for every $d \in \Gamma_k$

$$\left\{\int_{\partial\Omega} |u| d\sigma\right\}^2 \leq \int_{\partial\Omega} \frac{u^2}{d} d\sigma \int_{\partial\Omega} d\, d\sigma = k \int_{\partial\Omega} \frac{u^2}{d} d\sigma\,, \tag{2.4}$$

so that

$$\int_{\partial\Omega} \frac{u^2}{d_u} d\sigma = \frac{1}{k}\left\{\int_{\partial\Omega} |u|\, d\sigma\right\}^2 \leq \int_{\partial\Omega} \frac{u^2}{d} d\sigma$$

which proves that d_u solves (2.3). The uniqueness follows from the strict convexity of the function $d \to 1/d$ and from the fact that any solution of (2.3) must be zero on the set

$$\{x \in \partial\Omega : u(x) = 0\}. \quad \square$$

Coming back to our problem (2.2), by Proposition (2.1) we get that it reduces to the following minimization problem:

$$\min\left\{\int_\Omega |Du|^2 dx - 2\int_\Omega f(x)\, u\, dx + \frac{1}{k}\left(\int_{\partial\Omega} |u|\, d\sigma\right)^2 : u \in H^1(\Omega)\right\} \tag{2.5}$$

which is equivalent (via the Euler-Lagrange equation) to the elliptic equation

$$\begin{cases} -\Delta u = f(x) & \text{in } \Omega \\ 0 \in k \dfrac{\partial u}{\partial \nu} + H(u) \int_{\partial\Omega} |u|\, d\sigma & \text{in } \partial\Omega \end{cases}$$

where $H(t)$ is the multivalued mapping

$$H(t) = \begin{cases} \operatorname{sgn} t & \text{if } t \neq 0 \\ [-1, 1] & \text{if } t = 0\,. \end{cases}$$

PROPOSITION 2.2. The following Poincaré-type inequality holds

$$\int_\Omega u^2\, dx \leq C\left\{\int_\Omega |Du|^2\, dx + \left(\int_{\partial\Omega} |u|\, d\sigma\right)^2\right\} \quad \text{for every } u \in H^1(\Omega) \tag{2.6}$$

for a suitable constant C.

PROOF. By contradiction, if (2.6) is false, we may find a sequence (u_h) in $H^1(\Omega)$ with

$$\int_\Omega u_h^2\, dx = 1 \quad \text{and} \quad \int_\Omega |Du|^2\, dx + \left\{\int_{\partial\Omega} |u_h|\, d\sigma\right\}^2 \to 0\,.$$

In this case, we obtain that $u_h \to 1/|\Omega|$ in $H^1(\Omega)$; hence $u_h \to 1/|\Omega|$ in $L^2(\partial\Omega)$, which is in contradiction with the fact that

$$\int_{\partial\Omega} |u_h|\, d\sigma \to 0 . \square$$

PROPOSITION 2.3. For every $f \in L^2(\Omega)$ there exists a unique solution of problem (2.5).

PROOF. Let (u_h) be a minimizing sequence in $H^1(\Omega)$; then, for every $\varepsilon > 0$ there exists a constant C_ε such that

$$\int_\Omega |Du_h|^2\, dx + \frac{1}{k}\left\{\int_{\partial\Omega} |u_h| d\sigma\right\}^2 \leq C_\varepsilon + \varepsilon \int_\Omega |u_h|^2\, dx \quad \text{for every } h \in \mathbb{N}.$$

By the Poincaré inequality (2.6) we have

$$\int_\Omega |Du_h|^2 dx + \frac{1}{k}\left\{\int_{\partial\Omega} |u_h|\, d\sigma\right\}^2 \leq C_\varepsilon + \varepsilon C \left(\int_\Omega |Du_h|^2\, dx + \left\{\int_{\partial\Omega} |u_h|\, d\sigma\right\}^2 \right),$$

so that (u_h) is bounded in $H^1(\Omega)$. Therefore, possibly passing to a subsequence, we may assume that $u_h \to v$ weakly in $H^1(\Omega)$. By the weak $H^1(\Omega)$-lower semicontinuity of the functional

$$G(u) = \int_\Omega |Du|^2\, dx - 2\int_\Omega f(x)\, u\, dx + \frac{1}{k}\left\{\int_{\partial\Omega} |u|\, d\sigma\right\}^2$$

the function v is a solution of (2.5).

In order to prove the uniqueness, assume v and w are two different solutions of (2.5). It is easy to see that

$$G\left(\frac{v+w}{2}\right) - \frac{G(v) + G(w)}{2} = -\frac{1}{4} \int_\Omega |Dv - Dw|^2\, dx +$$

$$+ \frac{1}{4k}\left\{\int_{\partial\Omega} |v+w| d\sigma\right\}^2 - \frac{1}{2k}\left\{\int_{\partial\Omega} |v| d\sigma\right\}^2 - \frac{1}{2k}\left\{\int_{\partial\Omega} |w| d\sigma\right\}^2,$$

and that the right-hand side is strictly negative if $w - v$ is not constant. Therefore, let us assume $w = v + c$ with c constant; we have only to prove that

$$\left\{\int_{\partial\Omega} |v+w| d\sigma\right\}^2 < 2\left\{\int_{\partial\Omega} |v| d\sigma\right\}^2 + 2\left\{\int_{\partial\Omega} |w| d\sigma\right\}^2 . \tag{2.7}$$

If v and w have a different sign on a nonnegligible set B, we have $|v+w| < |v| + |w|$, a.e. on B, so that

$$\int_{\partial\Omega} |v+w| d\sigma < \int_{\partial\Omega} |v| + |w| \, d\sigma ,$$

and so

$$\left\{ \int_{\partial\Omega} |v+w| d\sigma \right\}^2 < \left\{ \int_{\partial\Omega} |v| + |w| \, d\sigma \right\}^2 \leqslant 2 \left\{ \int_{\partial\Omega} |v| d\sigma \right\}^2 + 2 \left\{ \int_{\partial\Omega} |w| d\sigma \right\}^2 .$$

Assume on the contrary that v and w have the same sign on $\partial\Omega$, say for instance $v \geqslant 0$ and $w \geqslant 0$. In this case we have

$$\left\{ \int_{\partial\Omega} v+w \, d\sigma \right\}^2 - 2 \left\{ \int_{\partial\Omega} v \, d\sigma \right\}^2 - 2 \left\{ \int_{\partial\Omega} w \, d\sigma \right\}^2$$

$$= - \left\{ \int_{\partial\Omega} w-v \, d\sigma \right\}^2 = - c^2 |\partial\Omega|^2$$

and (2.7) is proved.□

We conclude this section by the following result (whose proof follows easily from Propositions 2.1 and 2.3).

THEOREM 2.4. Let $f \in L^2(\Omega)$ be fixed. Then, Problem (2.2) has at least one solution d_{opt}. Moreover, denoting by u_{opt} the corresponding temperature (solution of (2.5)), if $u_{\text{opt}} \not\equiv 0$ on $\partial\Omega$ we have that d_{opt} is unique and

$$d_{\text{opt}}(\sigma) = \frac{k}{\int_{\partial\Omega} |u_{\text{opt}}| \, d\sigma} |u_{\text{opt}}(\sigma)| \qquad \text{for a.e. } \sigma \in \partial\Omega.$$

REMARK 2.5. Note that if f is identically zero, u_{opt} is zero everywhere, so that any function $d \in \Gamma_k$ is a solution of (2.2). Moreover, problem (2.2) does not change if we replace the constraint $\int_{\partial\Omega} d \, d\sigma = k$ by the constraint $\int_{\partial\Omega} d \, d\sigma \leqslant k$. Finally, it is easy to see that Theorem 2.4 holds also for $f \in [H^1(\Omega)]'$.

3. SOME EXAMPLES

In the one-dimensional case let $\Omega =]0,1[$ and $f \geqslant 0$, so that Problem (2.5) becomes

$$\min \left\{ \int_0^1 |u'|^2 dx - 2 \int_0^1 f(x) \, u \, dx + \frac{1}{k} [|u(0)| + |u(1)|]^2 : \; u \in H^1(0,1) \right\}$$

It is easy to see that the solution u_{opt} is nonnegative, and, setting

$$F(x) = \int_0^x f(s)\,ds\,,$$

after some calculations we obtain that for

$$k \geqslant \left|\int_0^1 (f-2F)\,dx\right| \left\{\int_0^1 f\,dx\right\}^{-1}$$

there holds

$$d_{opt}(0) = \frac{k+1}{2} + \left\{\int_0^1 f\,dx\right\}^{-1} \int_0^1 F\,dx\,,$$

$$d_{opt}(1) = \frac{k+1}{2} - \left\{\int_0^1 f\,dx\right\}^{-1} \int_0^1 F\,dx.$$

For instance, if $f(x)$ is the Dirac distribution $\delta_a(x)$ with $a \in [0\,,1]$, our optimization criterion (2.2) coincides with the maximization of the temperature $u(a)$, and if $k \geqslant |1-2a|$, we get (Fig. 1)

$$d_{opt}(0) = \frac{k}{2} + \frac{1-2a}{2}\,, \quad d_{opt}(1) = \frac{k}{2} - \frac{1-2a}{2}\,, \quad u_{opt}(a) = \frac{k+1}{4}\,.$$

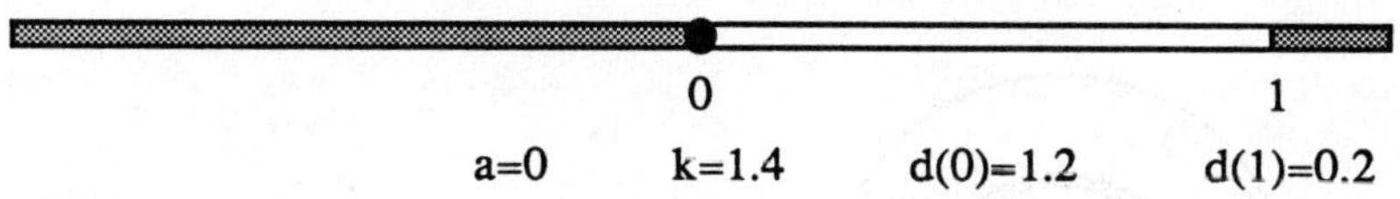

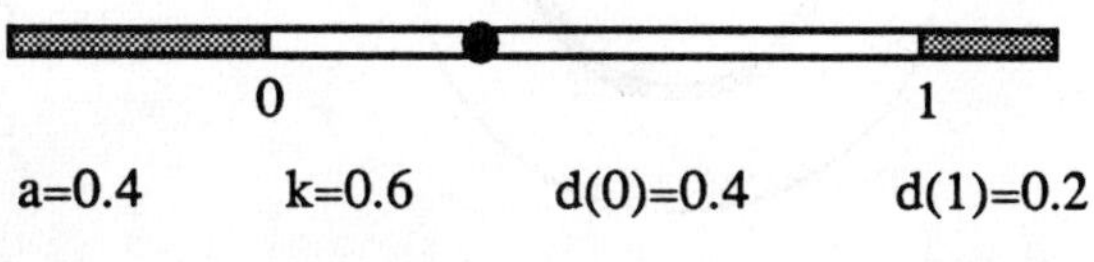

Figure 1

In the two-dimensional case, let Ω be the annulus

$$\Omega = \{x \in R^2 : 1 < |x| < 2\}$$

and let $f \equiv 1$ (uniformly distributed heat sources). In this case, our optimization criterion (2.2) coincides with the maximization of the averaged temperature $\int_{\Omega} u\,dx$, and, if $k \geqslant \pi(3-4\ln 2) \simeq 0.714$ we get (Fig. 2)

$$d_{\text{opt}} = \frac{k}{6\pi} + \frac{3-4\ln 2}{3} \text{ on } \{|x|=1\},\ d_{\text{opt}} = \frac{k}{6\pi} - \frac{3-4\ln 2}{6} \text{ on } \{|x|=2\}.$$

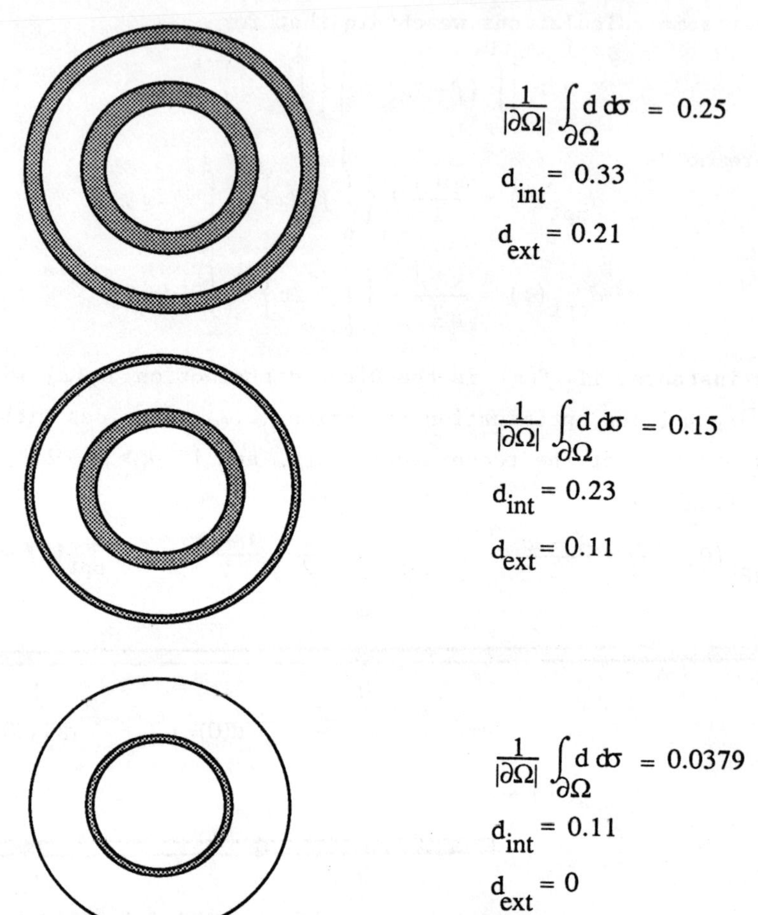

Figure 2

ACKNOWLEDGEMENTS

I am indebted to F. Murat for the many helpful discussions we had while we were visiting Heriot-Watt University.

REFERENCES

1. E. Acerbi and G. Buttazzo, "Reinforcement problems in the calculus of variations", *Ann. Inst. H. Poincaré Anal. Non Linéaire* **3** (1986), pp.273-284.
2. E. Acerbi and G. Buttazzo, "Limit problems for plates surrounded by soft material", *Arch. Rational Mech. Anal.* **92** (1986), pp.355-370.
3. H. Attouch, *Variational Convergence for Functions and Operators*, Pitman, Appl. Math. Ser., Boston (1984).
4. H. Brezis, L. Caffarelli and A. Friedman: "Reinforcement problems for elliptic equations and variational inequalities", *Ann. Mat. Pura Appl.* **123** (1980), pp.219-246.
5. G. Buttazzo, "Elliptic problems with a thin insulating layer around the boundary", *Proceedings of "Integral Functionals in Calculus of Variations"*, Trieste 9-12 September 1985, Springer-Verlag, Lect. Notes in Math., Berlin (to appear).
6. G. Buttazzo, G. Dal Maso and U. Mosco, (Paper in preparation).
7. G. Buttazzo and R.V. Kohn, "Reinforcement by a thin layer with oscillating thickness", *Appl. Math. Optim.* (to appear).
8. L. Caffarelli and A. Friedman, "Reinforcement problems in elasto-plasticity", *Rocky Mountain J. Math.* **10** (1980), pp.155-184.
9. E. de Giorgi, "G-operators and Γ-convergence", *Proceedings of "International Congress of Mathematicians"*, Warsaw 1983, pp.1175-1191.
10. E. de Giorgi and G. dal Maso, "Γ-convergence and calculus of variations", *Proceedings of "Mathematical Theories of Optimization"*, S. Margherita Ligure 1981, Springer-Verlag, *Lect. Notes in Math.* **979**, Berlin (1983), pp.121-143.
11. E. Sanchez-Palencia, "Non-homogeneous media and vibration theory", Springer-Verlag, *Lect. Notes in Phys.* **127**, Berlin (1980).

4

LOWER SEMICONTINUITY AND RELAXATION FOR SOME PROBLEMS IN OPTIMAL DESIGN

ELIO CABIB

INTRODUCTION

The paper deals with some new results about lower semicontinuity and relaxation for functionals arising in the class of problems in structural optimization which have been treated in [3]. Reference to that paper, in which a more extensive bibliography can be found, will be frequent, but sometimes it will be necessary to recall extensively some remarks and results in order to be as clear as possible and independent of it.

Given a bounded open set $\Omega \subset \mathbb{R}^n$, it is required to minimize a functional defined on an admissible set of pairs (A,u), where $A(x) = \{a_{ij}(x)\}$ denotes an $n \times n$, symmetric, real and positive definite matrix on Ω and u denotes the unique solution of the corresponding linear second order Dirichlet problem on Ω with a prescribed datum p and homogeneous boundary conditions. Concerning the physical meaning, the matrix A describes some pointwise material property of the continuous body Ω, such as the electric or thermal conductivity, and u is the corresponding electric potential or the corresponding temperature field, respectively. The functional to be minimized can be interpreted as a cost and it is split into a sum of two terms. One of them, depending only on A, has the meaning of the cost of the material employed, chosen among a set of available materials, while the other one, depending only on the corresponding solution u of the

Dirichlet problem, represents an additional cost due to possible constraints or optimality conditions imposed on the stationary behaviour of the physical system. Such problems are sometimes called control problems, (see [9] and [15]). Their common feature with optimal design problems lies in looking for optimal values of parameters which traditionally arise as data, such as the shape of the domain itself (see [2] and [7]).

The mathematical tool introduced in [3] consists in applying the direct method of the calculus of variations. The admissible set is supposed to be endowed with the G-convergence topology (see [1],[4], [12] and [13]), so that the discussion is focussed on the lower semicontinuity of functionals with respect to that particular topology. In this context, when an optimization problem of this kind does not admit any solution, as in some counter-examples analyzed in detail by Murat (see [9]), a generalized minimum will be a minimum of the relaxed functional with respect to this topology. In this paper, a characterization of lower semicontinuity in the one-dimensional case and a new sufficient condition in general dimension are obtained, but a characterization in general dimension still remains an open problem. Moreover, the relaxed functional is explicitly computed in a new special case in two dimensions where a lack of lower semicontinuity occurs and a general algorithm is given in one dimension. Other methods of relaxation for functionals in similar problems have been studied in [6] and [14].

1. FORMULATION OF THE PROBLEM

Let λ_1 and λ_2 be two real numbers with $\lambda_2 \geq \lambda_1 > 0$. Let $M = M(\lambda_1, \lambda_2)$ denote the set of all $n \times n$ symmetric real matrices whose eigenvalues belong to the interval $[\lambda_1, \lambda_2]$ and $M = M(\lambda_1, \lambda_2)$ denote the set of all measurable functions $A : \mathbb{R}^n \to M$. They are the $n \times n$ symmetric measurable matrices $A(x) = \{a_{ij}(x)\}$ on $\mathbb{R}^n$ satisfying the ellipticity condition

$$\lambda_1|\xi|^2 \leq \sum_{ij=1}^{n} a_{ij}(x)\,\xi_i\,\xi_j = A(x)\,\xi\cdot\xi \leq \lambda_2|\xi|^2 ,$$

for a.e. $x \in \mathbb{R}^n$ and for every $\xi \in \mathbb{R}^n$. Given a bounded open set $\Omega \subset \mathbb{R}^n$ and $p \in H^{-1}(\Omega)$, the dual space of $H_0^1(\Omega)$, we are concerned with optimization problems of the form

$$\min_{A \in M} \left[\int_\Omega f(x, A(x))\,dx + \int_\Omega \phi(x, u_A(x))\,dx \right], \qquad (1.1)$$

where u_A denotes the unique solution of the Dirichlet problem

$$\begin{cases} -\sum_{ij=1}^{n} D_i(a_{ij}(x)\,D_j u) = -D\cdot A(x)\,Du = p, \text{ in } \Omega, \\ u \in H_0^1(\Omega) . \end{cases} \qquad (1.2)$$

Concerning the integrands f and ϕ, they are assumed to fulfil the following hypotheses (see [3]):

(a) $f : \mathbb{R}^n \times M \to \overline{\mathbb{R}}$ is a non-negative Borel function;

(b) $\phi : \mathbb{R}^n \times \mathbb{R} \to \mathbb{R}$ satisfies the classical Carathéodory conditions:

(b_1) $\phi(\cdot, z)$ is measurable on $\mathbb{R}^n$ for every $z \in \mathbb{R}$;

(b_2) $\phi(x, \cdot)$ is continuous on $\mathbb{R}$ for a.e. $x \in \mathbb{R}^n$;

(b_3) $|\phi(x,z)| \leq c_1(x) + c_2(x)|z|^2$, with $c_1 \in L^1_{\text{loc}}(\mathbb{R}^n)$ and $c_2 \in L^\infty_{\text{loc}}(\mathbb{R}^n)$.

It is convenient to localize the functional in problem (1.1) on the Borel sets as follows. If $\mathcal{B}$ denotes the Borel σ-field of $\mathbb{R}^n$, consider, for every $B \in \mathcal{B}$, the functional $F_B : M \to \overline{\mathbb{R}}$ defined by

$$F_B(A) = \int_B f(x, A(x))\,dx , \qquad (1.3)$$

and, for B bounded, the functional $\Phi_B : L^2_{\text{loc}}(\mathbb{R}^n) \to \mathbb{R}$ defined by

$$\Phi_B(u) = \int_B \Phi(x, u(x))\,dx , \qquad (1.4)$$

which is strongly continuous in $L^2(B)$ by condition (b).

The optimization problem can be formulated as follows. Given a bounded open set $\Omega \subset \mathbb{R}^n$ and $p \in H^{-1}(\Omega)$, problem (1.1) can be written in the form

$$\min_{A \in M} \left[F_\Omega(A) + \Phi_\Omega(u_A)\right],$$

where u_A denotes the unique solution of problem (1.2).

In the examples extensively studied in [9], essentially in order to show that nonexistence may arise in such problems, the integrand Φ in (1.4) is

$$\Phi(x, z) = (z - w(x))^2,$$

where $w \in L^2(\Omega)$, so that $\Phi_\Omega(u)$ has the meaning of the distance between u and a prescribed target w. In a case treated in [3]

$$\phi(x, z) = p(x)\, z,$$

with $p \in L^2(\Omega)$, and $\Phi_\Omega(u)$ has the meaning of stored energy.

The space M is assumed to be endowed with G-convergence topology ([1], [4], [12] and [13]), according to the following definition ([3]).

DEFINITION 1.1. A sequence A_k in M is said to G-converge to $A \in M$, and written $A_k \xrightarrow{G} A$, if, for every bounded open set $\Omega \subset \mathbb{R}^n$ and for every $p \in H^{-1}(\Omega)$, the sequence u_k of solutions of the problems

$$\begin{cases} -D \cdot A_k(x)\, Du_k = p, & \text{in } \Omega, \\ u_k \in H_0^1(\Omega), \end{cases}$$

converges weakly in $H_0^1(\Omega)$ to the solution u of the problem

$$\begin{cases} -D \cdot A(x)\, Du = p, & \text{in } \Omega, \\ u \in H_0^1(\Omega). \end{cases}$$

□

It is well known that M, equipped with this topology, is metrizable and compact (see [13], Remark 4). Since Φ_Ω is continuous on $L^2(\Omega)$, the functional $A \to \Phi_\Omega(u_A)$ is continuous on M with respect to G-convergence. The following theorem is then a consequence of the classical direct method in the calculus of variations.

THEOREM 1.1. If $\Omega \subset \mathbb{R}^n$ is a bounded open set and F_Ω is lower semicontinuous with respect to G-convergence (G-l.s.c.), then problem 1.5 admits a solution.

□

The problem is now reduced to the study of G-lower semicontinuity for functionals as in (1.3).

2. LOWER SEMICONTINUOUS FUNCTIONALS

In the one-dimensional case it is possible to characterize all the G-lower semicontinuous functionals. The open set Ω reduces to an open interval and M to the interval $[\lambda_1, \lambda_2]$ itself. The explicit solution of (1.2) shows the well known result according to which the G-convergence of a sequence a_k in M to $a \in M$ is equivalent to the convergence of $1/a_k$ to $1/a$ in $L^\infty(\mathbb{R})$-weak*. By recalling the classical results (see [11]) about the characterization of weakly lower semicontinuous functionals, the following theorem trivially holds.

THEOREM 2.1. Let $f : \mathbb{R}^n \times M \to \overline{\mathbb{R}}$ be a non-negative Borel function and, for every Borel set B, let $F_B : M \to \overline{\mathbb{R}}$ be the functional

$$F_B(a) = \int_B f(x, a(x))\, dx .$$

Then, F_B is G-l.s.c. if and only if

$$f(x, a) = g(x, 1/a)$$

provided $g : \mathbb{R}^n \times [\lambda_2^{-1}, \lambda_1^{-1}] \to \overline{\mathbb{R}}$ is a Borel function, convex and lower semicontinuous on $[\lambda_2^{-1}, \lambda_1^{-1}]$, for a.e. $x \in \mathbb{R}$. □

Concerning the n-dimensional case, it is proved in [3], Theorem 1.2, that, if F_B is defined as

$$F_B(A) = \int_B g(x, \alpha^n(x))\, dx ,$$

where $g(x, \cdot)$ is convex, non-decreasing and lower semicontinuous on $[\lambda_1, \lambda_2]$ for a.e. $x \in \mathbb{R}^n$ and $\alpha^n(x)$ denotes the largest eigenvalue of $A(x)$, then F_B is G-l.s.c. on M. The result is essentially due to an estimate proved in [13], Remark 5, according to which if A and $\tilde{A}$ are respectively a G-limit and an $L^\infty(\mathbb{R}^n)$-weak* limit of a sequence A_k in M, then

$$A(x)\xi\cdot\xi \leqslant \tilde{A}(x)\xi\cdot\xi,$$

for a.e. $x \in \mathbb{R}^n$ and for every $\xi \in \mathbb{R}^n$. A similar lower estimate obtained in [16] can be used to prove the following semicontinuity result.

THEOREM 2.2. Let $g : \mathbb{R}^n \times [\lambda_2^{-1}, \lambda_1^{-1}] \to \overline{\mathbb{R}}$ be a non-negative Borel function and, for every Borel set B, let $F_B : M \to \overline{\mathbb{R}}$ be the functional

$$F_B(A) = \int_B g(x, 1/\alpha^1(x))\,dx\,, \tag{2.1}$$

where $\alpha^1(x)$ denotes the smallest eigenvalue of $A(x)$. If $g(x,\cdot)$ is convex, non-decreasing and lower semicontinuous on $[\lambda_2^{-1}, \lambda_i^{-1}]$ for a.e. $x \in \mathbb{R}^n$, then F_B is G-l.s.c. on M.

PROOF. Assume the Borel set B to be bounded. Let A_k be any sequence in M which G-converges to $A \in M$. If M^{-1} denotes the set of all matrices whose inverse lies in M, it is not restrictive to assume $A_k^{-1} \rightharpoonup \breve{A}^{-1}$ in $L^\infty(\mathbb{R}^n, M^{-1})$-weak*, for a suitable $\tilde{A} \in M$. Let $\alpha_k^1(x)$ and $\tilde{\alpha}^1(x)$ be the smallest eigenvalue of $A_k(x)$ and $\tilde{A}(x)$ respectively. Then, $1/\alpha_k^1(x)$ and $1/\tilde{\alpha}^1(x)$ are the largest eigenvalues of $A_k(x)^{-1}$ and $\tilde{A}(x)^{-1}$. Therefore

$$\int_B g(x, 1/\tilde{\alpha}^1(x))\,dx \leqslant \liminf_{k\to\infty} \int_B g(x, 1/\alpha_k^1(x))\,dx \tag{2.2}$$

by Theorem 1.2 in [3]. Now, the result proved in [16], Theorem 5, yields

$$A(x)^{-1}\xi\cdot\xi \leqslant \tilde{A}(x)^{-1}\xi\cdot\xi\,,$$

for a.e. $x \in \mathbb{R}^n$ and for every $\xi \in \mathbb{R}^n$. Hence

$$1/\alpha^1(x) \leqslant 1/\tilde{\alpha}^1(x)\,. \tag{2.3}$$

From (2.2), (2.3) and monotonicity of g, we obtain

$$F_B(A) = \int_B g(x, 1/\alpha^1(x))\, dx \leq F_B(\tilde{A}) \leq \liminf_{k\to\infty} F_B(A_k) ,$$

which proves the lower semicontinuity of F_B with respect to G-convergence. □

The following result easily follows from Theorem 1.2 proved in [3] and Theorem 2.2.

THEOREM 2.3. Let $g : \mathbb{R}^n \times [\lambda_1, \lambda_2] \times [\lambda_2^{-1}, \lambda_1^{-1}] \to \overline{\mathbb{R}}$ be a non-negative Borel function such that $(\lambda, \mu) \to g(x, \lambda, \mu)$ is convex, non-decreasing as a function of λ or μ alone and lower semi-continuous for a.e. $x \in \mathbb{R}^n$. Then the functional

$$F_B(A) = \int_B g(x, \alpha^n(x) , 1/\alpha^1(x))\, dx \tag{2.4}$$

is G–l.s.c. on M. □

Let us now consider the functional (2.4) in Theorem 2.3 on a bounded open set $\Omega \subset \mathbb{R}^n$, with $p \in L^2(\Omega)$, and the energy functional

$$E(A) = \int_\Omega p u_A\, dx = \int_\Omega A(x)\, Du_A \cdot Du_A\, dx .$$

A generalization of Theorem 3.3 in [3] is the following.

THEOREM 2.4. Given a constant $k \geq 0$, the problem

$$\min_{A \in M} [F_\Omega(A) + kE(A)] \tag{2.5}$$

admits an isotropic solution.

PROOF. Let $A \in M$ be a solution to problem (2.5). If $A(x)$ is not a.e. isotropic, let $\alpha^1(x) \leq \alpha^2(x) \leq \cdots \leq \alpha^n(x)$ be its eigenvalues. If $A(x)$ is replaced by the isotropic matrix $A'(x) = \alpha^n(x) 1$, by the monotonicity assumption in Theorem 2.3 we obtain

$$F_\Omega(A') \leq F_\Omega(A) ,$$

and, by Lemma 3.2 of [3] about a monotonicity property of E, we have

$$E(A') \leq E(A).$$

Therefore, the functional in (2.5) attains its minimum value on the isotropic matrix $A'(x) = \alpha^n(x)1$ too.

□

3. RELAXATION

Let us assume, now, that problem 1.5, in which the functional is bounded from below by assumptions (a) and (b) of Section 1, does not admit any solution. In this case, by a generalized solution we mean a solution of the relaxed problem

$$\min_{A \in M} [\hat{F}_\Omega(A) + \Phi_\Omega(u_A)], \tag{3.1}$$

where $\hat{F}_B$ denotes the largest G-l.s.c. functional on M which does not exceed F_B, namely,

$$\hat{F}_B(A) = \inf \left\{ \liminf_{k\to\infty} F_B(A_k) \,\middle|\, A_k \xrightarrow{G} A, \quad A_k \in M \right\},$$

for every $A \in M$, $B \in \mathcal{B}$. Here the infimum is actually a minimum as can be easily seen by a diagonal argument. The connections between problem 3.1 and problem 1.5 are summarized in [3], Theorem 2.1. According to it problem 3.1 admits a solution, the minimum value in problem 3.1 equals the infimum in problem 1.5 and the minimizing sequences are essentially the same to within subsequences. It has been proved in [3], Theorem 2.2, that $\hat{F}_B$ is an integral functional with a normal integrand $\hat{f} : \mathbb{R}^n \times M \to \overline{\mathbb{R}}$. Moreover, if f does not explicitly depend on x, then the same is true for $\hat{f}$. A consequence of it is the existence of the G-closure $\hat{E} \subset M$ of any $E \subset M$, so that, if M_B^E denotes the set

$$M_B^E = \left\{ A \in M \mid A(x) \in E \text{ a.e. in } B \right\},$$

then $\overline{M_B^E} = M_B^{\hat{E}}$. The problem is now reduced to the identification of the functional $\hat{F}_B$, which is called the G-relaxation of F_B.

In the one-dimensional case, in order to compute the G-relaxation of any functional, it is enough to recall again that G-convergence of a sequence a_k in M to $a \in M$ is equivalent to the convergence of $1/a_k$ to $1/a$ in $L^\infty(\mathbb{R})$-weak*. Indeed, for every $f : \mathbb{R} \times M \to \overline{\mathbb{R}}$ satisfying assumption (a) of Section 1, let $g : \mathbb{R} \times [\lambda_2^{-1}, \lambda_1^{-1}] \to \overline{\mathbb{R}}$ be the function

$$g(x,b) = f(x,1/b)$$

for a.e. $x \in \mathbb{R}$ and $b \in [\lambda_2^{-1}, \lambda_1^{-1}]$. Let, further,

$$g^{**} : \mathbb{R} \times [\lambda_2^{-1}, \lambda_1^{-1}] \to \overline{\mathbb{R}}$$

be its lower convex envelope with respect to the variable b, i.e., the largest convex function such that $g^{**}(x,\cdot) \leq g(x,\cdot)$ for every $x \in \mathbb{R}$. Then, if $F_B : M \to \overline{\mathbb{R}}$ is defined by

$$F_B(a) = \int_B f(x, a(x))\, dx \qquad (3.2)$$

for every $B \in \mathcal{B}$, according to classical results (see [5]), the following representation theorem immediately holds for its G-relaxation.

THEOREM 3.1. The integrand $\hat{f} : \mathbb{R} \times M \to \overline{\mathbb{R}}$ in the G-relaxation $\hat{F}_B : M \to \overline{\mathbb{R}}$ of the functional (3.2) is given by

$$f(x,a) = g^{**}(x, 1/a),$$

for a.e. $x \in \mathbb{R}$ and $a \in M$. □

Concerning explicit integral representations of relaxed functionals in higher dimension, a general algorithm is not available presently. In [3] it has been observed (Proposition 1.1) that, if $I \in M$ denotes the set of isotropic matrices, functionals attaining finite values only on a set M_B^I cannot be G-l.s.c.. Among this kind of functionals, it has been possible to compute the explicit relaxation of a functional F_B which is linear and positive on M_B^I and $+\infty$ otherwise, by using homogenization techniques (see [3]). The basic idea lies in using simultaneously the linear structure itself of the functional and the possibility of finding, for every A in the G-closure $\hat{I} \subset M$ of I (see [8], [10] and [17]), a sequence $a_k(x) \in [\lambda_1, \lambda_2]$ such that $a_k 1$, where 1 denotes the identity matrix, G-converges to A and a_k converges in $L^\infty(\mathbb{R}^n)$-weak* to the largest eigenvalue, α^n, of A. Consequently the scalar a in the integrand of F_B is replaced by α^n in the integrand of $\hat{F}_B$.

A similar method, which holds in two dimensions only, can

be applied to compute explicitly the G-relaxation of the functional

$$F_B(A) = \begin{cases} \displaystyle\int_B 1/a(x)\,dx\,, & \text{if } A(x) = a(x)1 \text{ and } a(x) \in [\lambda_1, \lambda_2] \\ & \text{a.e. on } B\,, \\ +\infty, & \text{otherwise} \end{cases} \tag{3.3}$$

The result, proved in the next theorem, is essentially based upon the property ([18]) of G-convergence in two dimensions, according to which $A_k \xrightarrow{G} A$ if, and only if, $A_k/\det A_k \xrightarrow{G} A/\det A$. On the other hand, this property will not be explicitly mentioned during the proof, because it will be clear by itself for the particular G-converging sequence we need to introduce.

THEOREM 3.2. If $n=2$, the G-relaxation $\hat{F}_B$ of the functional (3.3) is given by

$$\hat{F}_B(A) = \begin{cases} \displaystyle\int_B 1/\alpha^1(x)\,dx\,, & \text{if } A(x) \in \hat{I} \text{ a.e. on } B \\ +\infty\,, & \text{otherwise}\,, \end{cases}$$

for every $A \in M$, where $\alpha^1(x)$ denotes the smallest eigenvalue of the matrix $A(x)$.

PROOF. By comparison with the indicator function of the set M_B^I, whose G-relaxation vanishes on $\overline{M_B^I} = M_B^{\hat{I}}$ and takes the value $+\infty$ otherwise, the G-relaxation, $\hat{F}_B$, of F_B is finite on $M_B^{\hat{I}}$ and equals $+\infty$ otherwise. Let $B \in \mathcal{B}$ and $A \in M_B^{\hat{I}}$. The inequality

$$\hat{F}_B(A) \geq \int_B 1/\alpha^1(x)\,dx$$

follows easily from Theorem 2.2. For the opposite inequality, since $\hat{f}$, the integrand of $\hat{F}_B$, does not explicitly depend on x, it is enough to prove $\hat{f}(A) \leq 1/\alpha^1$ for every $A \in \hat{I}$, which is equivalent to proving that

$$\hat{F}_B(A) \leq \int_B 1/\alpha^1\,dx\,,$$

for every constant matrix $A \in \hat{I}$. It has been proved in [8] that a matrix A, with eigenvalues $\alpha^1 \leq \alpha^2$ in the interval $[\lambda_1, \lambda_2]$, belongs to $\hat{I}$ if, and only if, the estimate

$$\alpha^1 \geq \frac{\lambda_1 \lambda_2}{\lambda_1 + \lambda_2 - \alpha^2} \tag{3.4}$$

holds.

If equality holds in (3.4), let $t \in [0,1]$ be such that

$$\alpha^2 = t\lambda_1 + (1-t)\lambda_2$$

and, consequently,

$$\alpha^1 = (t/\lambda_1 + (1-t)/\lambda_2)^{-1}$$

so that α^1 and α^2 have the meaning, respectively, of weighted harmonic mean and arithmetic mean of λ_1 and λ_2. The matrix A can be viewed as the effective conductivity tensor of an homogenized material obtained as the limit of a sequence of layered two-phase isotropic materials, where λ_1 is taken in proportion t and, consequently, λ_2 in proportion $1-t$. The sequence can be computed as follows. Consider the characteristic functions χ_k and χ'_k, respectively, of the sets in $\mathbb{R}$

$$S_k = \bigcup_{h \in \mathbb{Z}} [h/k, (h+t)/k] \quad \text{and} \quad S'_k = \bigcup_{h \in \mathbb{Z}} [(h+t)/k, (h+1)/k]$$

for every $k \in \mathbb{N}$. If e denotes the unit eigenvector corresponding to the smallest eigenvalue, α^1, of A, let $a_k: \mathbb{R}^2 \to \{\lambda_1, \lambda_2\}$ be the sequence

$$a_k(x) = \lambda_1 \chi_k(x \cdot e) + \lambda_2 \chi'_k(x \cdot e)$$

for every $k \in \mathbb{N}$. Since a_k converges in $L^\infty(\mathbb{R})$-weak* to the arithmetic mean α^2 and the sequence of isotropic matrices $A_k = a_k 1$ G-converges to A, then, the sequence $1/a_k$ converges in $L^\infty(\mathbb{R}^2)$-weak* to the new arithmetic mean $t/\lambda_1 + (1-t)/\lambda_2 = 1/a^1$ and $a_k^{-1} 1$ G-converges to A^{-1}. Therefore,

$$\hat{F}_B(A) \leq \lim_{k \to \infty} \int_B 1/a_k(x)\, dx = \int_B 1/\alpha^1\, dx,$$

by semicontinuity of $\hat{F}_B$.

For the general case it is possible to replace λ_2 by a suitable $\lambda'_2 \in [\lambda_1, \lambda_2]$ such that equality holds again in (3.4) and

$$\alpha^2 = t'\lambda_1 + (1-t')\lambda_2' \tag{3.5}$$

for a suitable $t' \in [0,1]$. In fact, let λ_2' be given by

$$\lambda_2' = \alpha^2 + \lambda_1 \frac{\alpha^2 - \alpha^1}{\alpha^1 - \lambda_1}, \tag{3.6}$$

so that

$$\alpha^1 = \frac{\lambda_1 \lambda_2'}{\lambda_1 + \lambda_2' - \alpha^2}. \tag{3.7}$$

By (3.6), $\lambda_2' \geqslant \alpha^2$, hence there exists $t' \in [0,1]$ such that (3.5) holds and, by (3.5) and (3.7), we obtain

$$\alpha^1 = (t'/\lambda_1 + (1-t')/\lambda_2')^{-1}.$$

The inequality $\lambda_2' \leqslant \lambda_2$ is a trivial consequence of

$$\frac{\lambda_1 \lambda_2}{\lambda_1 + \lambda_2' - \alpha^2} \leqslant \frac{\lambda_1 \lambda_2'}{\lambda_1 + \lambda_2' - \alpha^2}$$

obtained by comparing (3.4) and (3.7). Hence, the same argument as before applies with λ_2 and t replaced by λ_2' and t'.

□

The following is a generalization of Theorem 3.2.

Let $\psi \in L^1_{\text{loc}}(\mathbb{R}^2)$ with $\psi \geqslant 0$ a.e. on $\mathbb{R}^2$. We consider the functional

$$F_B^\psi(A) = \begin{cases} \int_B \psi(x)/a(x)\,dx, & \text{if } A(x) = a(x)1,\ a(x) \in [\lambda_1, \lambda_2] \text{ a.e. on } B \\ +\infty, & \text{otherwise.} \end{cases} \tag{3.8}$$

THEOREM 3.3. If $n=2$, the G-relaxation of the functional (3.8) is given by

$$\hat{F}_B^\psi(A) = \begin{cases} \int_B \psi(x)/\alpha^1(x)\,dx, & \text{if } A(x) \in \hat{I} \text{ a.e. on } B, \\ +\infty, & \text{otherwise,} \end{cases}$$

for every $A \in M$, where $\alpha^1(x)$ denotes the smallest eigenvalue of the matrix $A(x)$.

□

The proof of this theorem can be omitted, since it is formally the same as Theorem 3.2 in [3] and it is based upon the following lemma which is an analogue of Lemma 3.1 in [3]. Here we may let the functional (3.8) be defined in dimension n, with B a Borel set in $\mathbb{R}^n$ and $\psi \in L^1_{loc}(\mathbb{R}^n)$.

LEMMA 3.1. If $\phi, \psi \in L^1_{loc}(\mathbb{R}^n)$ with $\phi, \psi \geqslant 0$ a.e. on $\mathbb{R}^n$, then

$$|\hat{F}^{\phi}_B(A) - \hat{F}^{\psi}_B(A)| \leqslant 1/\lambda_1 \,\|\phi - \psi\|_{L^1(B)}, \tag{3.9}$$

for every $B \in \mathcal{B}$, $A \in M^{\hat{I}}_B$.

PROOF. Let $a_k 1 \xrightarrow{G} A$, with $a_k 1 \in M^I_B$ and $A \in M^{\hat{I}}_B$ be such that $F^{\psi}_B(a_k 1) \to \hat{F}^{\psi}_B(A)$. By semicontinuity (see Theorem 2.2) we have

$$\begin{aligned} \hat{F}^{\phi}_B(A) &\leqslant \liminf_{k\to\infty} \int_B \phi(x)/a_k(x)\,dx \leqslant \\ &\leqslant \limsup_{k\to\infty} \int_B \psi(x)/a_k(x)\,dx + \limsup_{k\to\infty} \int_B (\phi(x)-\psi(x))/a_k(x)dx \leqslant \\ &\leqslant \hat{F}^{\psi}_B(A) + 1/\lambda_1 \,\|\phi-\psi\|_{L^1(B)}, \end{aligned}$$

which yields

$$\hat{F}^{\phi}_B(A) - \hat{F}^{\psi}_B(A) \leqslant 1/\lambda_1 \,\|\phi-\psi\|_{L^1(B)}.$$

By interchanging $\hat{F}^{\phi}_B$ with $\hat{F}^{\psi}_B$ we obtain (3.9). □

ACKNOWLEDGEMENT

I am grateful to the Gruppo Nazionale per la Fisica Matematica of CNR for supporting my attending the "Symposium year on material instabilities". I wish to thank G. Buttazzo and G. Dal Maso for helpful discussions during the preparation of this paper.

REFERENCES

1. A. Bensoussan, J.L. Lions and G. Papanicolaou, *Asymptotic Analysis for Periodic Structures*, North Holland, Amsterdam, (1978).
2. G. Buttazzo,"Thin insulating layers: the optimization point of view", *Proc. of Symp. on Material Instabilities in Continuum Mechanics*, (Ed. J.M. Ball), this volume, (1987), pp. 11-19.
3. E. Cabib and G. Dal. Maso, "On a class of optimum problems in structural design", *J. Optimization Theory Appl.* to appear.

4. E. De Giorgi and S. Spagnolo, "Sulla convergence degli integrali dell'energia per operatori ellitici del secondo ordine", *Boll. Un. Mat. Ital.* **8** (4), (1973), pp.391-411.
5. I. Ekeland and R. Temam, *Convex Analysis and Variational Problems*, North Holland, Amsterdam, (1976).
6. R. Kohn and G. Strang, "Optimal design and relaxation of variational problems," *Communications on Pure and Applied Mathematics* **34**, (1986), Part I, pp.113-137, Part II, pp.139-182, Part III, pp.353-377.
7. R. Kohn and M. Vogelius, "Thin plates with rapidly varying thickness, and their relation to structural optimization," IMA Preprint, (1985).
8. K.A. Lurie and A.V. Cherkaev, "G-closure of a set of anisotropically conducting media in the two-dimensional case". *J. Optimization Theory Appl.* **42**, (1984), pp.305-316.
9. F. Murat, "Contre-exemples pour divers problèmes où le contrôle intervient dans les coefficients," *Ann. Mat. Pura Appl.* **112** (4), (1977), pp.49-68.
10. F. Murat and L. Tartar, "Calcul des variations et homogénéisation", in *Cours de l'Ecole d'Eté d'Analyse Numérique CEA-EDF-INRIA sur l'homogénéisation* (Bréau sans Nappe, juillet 1983). Collection de la Direction des Etudes et Recherches d'Electricité de France, Eyrolles, Paris, (1984).
11. C. Olech, "A characterization of L^1-weak lower semicontinuity of integral functionals," *Bull. Acad. Polon. Sci., Sér. Sci. Math. Astronom. Phys.* **25**, (1977), pp.135-142.
12. E. Sanchez Palencia, "Non homogeneous media and vibration theory," *Lecture Notes in Physics* **127**, Springer, Berlin, (1980).
13. S. Spagnolo, "Convergence in energy for elliptic operators," *Proc. 3rd Symp. Numer. Solut. Partial Diff. Equat. College Park 1975* (ed. Hubbard, R.), Academic Press, New York, (1976) pp.469-498.
14. G. Strang, "Optimal design for a two-way conductor", Conference presented during the Symp. on Material Instabilities in Continuum Mechanics", Edinburgh, 1986.
15. L. Tartar, "Problèmes de contrôle des coefficients dans des équations aux derivées partielles," in *Control Theory, Numerical Methods and Computer Systems Modelling*, (eds. Bensoussan, A. and Lions, J.L.) Lecture Notes Econom. and Math. Systems **107**, Springer, Berlin, (1975), pp.420-426.
16. L. Tartar, "Estimation de coefficients homogénéisés", in *Computing Methods in Applied Sciences and Engineering, 1977, I*, Proceedings, IRIA, Paris, (eds. Glowinski, R. and Lions, J.L.), Lecture Notes in Math. **704**, Springer, Berlin, (1977), pp.364-373.
17. L. Tartar, "Estimations fines des coefficients homogénéisés", in *Ennio De Giorgi Colloquium*, (ed. Krée, P.) Research Notes in Math. **125**, Pitman, London, (1985), pp.168-187.
18. L. Tartar, Personal communication.

5

MATHEMATICAL MODELS OF PHASE BOUNDARIES

G. CAGINALP

1. INTRODUCTION

We consider free boundary problems which arise from phase transitions. Of particular interest are models designed to explore the behaviour of the interface in greater detail. This is possible by using statistical mechanics ideas in order to extend a purely continuum theory one step closer to a molecular theory. The purpose of this article is to discuss the origin of these models, summarize the established rigorous results, motivate some formal conclusions, and compare (formally) these models with those continuum models which may be derived as limiting cases (e.g. the Stefan model defined below).

The physical problem of interest consists of a material which may be in either of two phases, e.g. liquid or solid, separated by an interface. The problem of describing the physics of heat diffusion in the two phases and the latent heat balance across the interface is generally known as the Stefan model [1]. More precisely, suppose the material occupies a region, Ω, with Ω_1, Ω_2 and Γ comprising the set of points comprising the liquid, solid and interface, respectively. Then the Stefan problem is formulated as the set of equations

$$u_t = K\Delta u \quad \text{in} \quad \Omega_1,\ \Omega_2 \tag{1.1}$$

$$l v_n = K(\nabla u_s - \nabla u_L) \cdot n \quad \text{on} \quad \Gamma \tag{1.2}$$

$$u = 0 , \tag{1.3}$$

where u is the temperature, K is the diffusivity (heat capacity per unit volume has been set to unity), l is the latent heat and v_n is the (normal) velocity of the interface.

Equations (1.1) and (1.2) are equivalent [2] in a weak sense to the equation

$$H_t = K\Delta u , \quad H \equiv u + \frac{l}{2}\phi, \quad \phi \equiv \begin{cases} +1 & \text{liquid} \\ -1 & \text{solid} \end{cases} \tag{1.4}$$

in which ϕ is a step function that keeps track of the phase.

In the Stefan model, the sign of the temperature determines the phase so that $u=0$ is the dividing line between the two phases. One feature of the Stefan model is the absence of an intrinsic length scale. From a microscopic prospective, the physical situation has a number of length scales such as the intermolecular spacing and the correlation length [3]. The latter is defined as the length, ξ_0, such that two particles (or spins) a distance ξ_0 apart will have probability $\frac{1}{2}$ of being in the same phase or state (e.g. both spins up or both down for $\pm$ spins).

As a consequence of a non-zero correlation length, the physics is modified in several ways. In particular, the interfacial (or surface) tension and the temperature at the interface are both non-zero. Thus, one may have supercooling, i.e., the presence of liquid of negative temperature and the analogous phenomenon of superheating. One of the most interesting implications of this level of physics is the effect of the stability of the interface. Heuristically, it is easy to see that at the tip of a small protrusion into a supercooled melt, there will be a larger differential in the gradients in (1.2). Hence, the tip will grow faster than nearby flat regions. The instability will therefore grow unless it is checked by other forces. In particular, surface tension acts to stabilize the interface and thereby competes with the effects of supercooling.

An approach which has been implemented in order to understand these phenomena is based on modifying the function ϕ in (1.4) by introducing a second equation for u and ϕ (see [4–7] and references contained therein). This second equation arises from a free energy which follows from Landau-Ginzburg theory [8–9]. The resulting system of second order equations is

$$u_t + \frac{l}{2}\phi_t = K\Delta u \tag{1.5}$$

$$\tau\phi_t = \xi^2\Delta\phi + \frac{1}{a}f(\phi) + 2u\ . \tag{1.6}$$

Here $a^{-1}f(\phi)$ is the derivative of a double-well potential where a is a measure of the depth of the well and is obtained from microscopic considerations. The parameter ξ is a length scale such that $a^{\frac{1}{2}}\xi$ is the correlation length. We will refer to ϕ as the phase field and (1.5), (1.6) as the phase field equations. A sketch of a derivation of (1.7) will be presented in a more general setting in the next section.

With suitable initial and boundary conditions, the system of equations (1.5), (1.6) for u and ϕ may be studied analytically and numerically. The interface now consists of the points

$$\Gamma(t) = \{x \in \Omega:\ \phi(t,x) = 0\}\ . \tag{1.7}$$

2. DERIVATION OF BASIC EQUATIONS IN GENERAL CASE

The derivations of equation (1.6) and analogous higher order equations are based on a Hamiltonian of interactions of spins or occupation functions $\phi(x)$ on a finite lattice of N spins, L_N. The basic Hamiltonian may be expressed as

$$\overline{H} = \tfrac{1}{2}\beta \sum_{x,x'} J(x-x')\,\phi(x)\,\phi(x') - \sum_{x} w(\phi(x)) \tag{2.1}$$

where J is an interaction or coupling function, β is $1/k_BT$ with k_B as Boltzmann's constant, T is absolute temperature, and w is an even fourth order polynomial with extrema at ± 1. One may define the discrete Fourier transforms

$$\hat{\phi}(q) = \sum_{x\in L_N} e^{iq\cdot x}\,\phi(x) \tag{2.2}$$

$$\hat{J}(q) = \sum_{x \in L_N} e^{-iq\cdot x} J(x) \tag{2.3}$$

on the dual lattice L_N^*. Using these Fourier transforms, the first term in (2.1) can be rewritten by means of the identity (see [10] for details)

$$\sum_{x, x' \in L_N} J(x-x')\, \phi(x)\, \phi(x') = N^{-1} \sum_{q \in L_N^*} \hat{J}(q)\, \hat{\phi}(q)\, \hat{\phi}(-q). \tag{2.4}$$

With the interaction part of the Hamiltonian written as a sum over wave vectors, q, the Landau-Ginzburg approximation entails an expression of $\hat{J}(q)$ about $q = 0$, i.e.,

$$\hat{J}(q) = \sum_x J(x)\, e^{-iq\cdot x} = \sum_x J(x) - i \sum_x q\cdot x\, J(x)$$
$$- \tfrac{1}{2} \sum_x (q\cdot x)^2 J(x) + \cdots \tag{2.5}$$

and retention of only finitely many terms. For simplicity, we assume lattice symmetry of reflection about any axis. In this case, all odd terms vanish. The series then consists of the constant term, $\sum J(x)$, a second moment term which forms the coefficient of q^2, etc. Introducing discrete derivatives and using an integration by parts, one may transform q^n terms in the sum (2.4) into n-th derivatives. The constant term in (2.5) results in a ϕ^2 term which may be combined with the double-well potential $w(\phi(x))$.

Truncating this series at q^2 and taking the continuum limit in an appropriate way (in particular so that the energy of the system remains constant [10]) results in (1.6). Retaining higher order wave numbers leads to higher order differential equations [11]. There are two important reasons for considering higher order equations. One is that the higher order equations are a test of the accuracy of the second order equation. The other is that anisotropy can be studied more thoroughly, as the second order equation averages all anisotropy except that which arises from the relative strengths of the axes.

The details of the calculation involved in transforming the Hamiltonian have been presented in [10–11]. The free energy F, is obtained from (2.1) by utilizing (2.4), taking the continuum limit and adding an entropy term $-2u\phi$, so that

$$F\{\phi\} = \int F\, dx_1, \cdots, dx_d \tag{2.6}$$

$$F = \sum_{n=1}^{\infty} \sum_{p_1+\cdots+p_d=2n}{}' (-1)^{n+1} \xi^{2n} b(2n;p_1,\cdots,p_d) \left[D_1^{p_1/2}, \cdots, D_d^{p_d/2} \phi\right]^2 + \frac{1}{a} G(\phi) - 2u\phi$$

where the primed sum is over all sets of positive, even numbers $\{p_1, \cdots, p_d\}$ whose sum is $2n$. The D_i are derivatives in the i-th direction. The coefficients are combinatorial factors based on moments of J:

$$b(2n;p_1,\cdots,p_d) \equiv \frac{1}{p_1!\cdots p_d!} \int J(x)\, x_1^{p_1}, \cdots, x_d^{p_d}\, dx_1, \cdots, dx_d. \tag{2.7}$$

Thus, any anisotropy which is present in the intermolecular interactions will be manifested in some of the coefficients $b(2n;p_1,\cdots,p_d)$. For example, if the interactions are of different strengths along each of the axes x_i, then this anisotropy will be evident in the first coefficient $(n=1)$ corresponding to ξ^2 in (2.6). For a triangular lattice with equal interactions among nearest neighbours the anisotropy would also be manifested in the leading coefficient. However, a set of interactions which are symmetric in the x_i (but are stronger or weaker along the diagonals) would be annihilated in the first order term and would appear for $n=2$ (i.e. ξ^4) and higher terms.

We proceed now to derive the set of differential equations from the free energy (2.6). We truncate the infinite series in (2.6) at some value $n=M$, thereby neglecting higher order wave numbers in the Fourier expansion. This is the Landau-Ginzburg approximation. A basic ansatz in statistical mechanics is that the free energy must be minimized in equilibrium. Symbolically, we write this as $\delta F/\delta\phi = 0$. If the system is not in equilibrium, then standard physical theories (Model A equation in [8]) imply that ϕ must satisfy the evolution equation $\tau\phi_t = \delta F/\delta\phi$, where τ is a relaxation time. Hence, an application of the Euler-Lagrange equations to (2.6) results in

$$\tau\phi_t = \sum_{n=1}^{M} \sum_{p_1+\cdots+p_d=2n}{}' \xi^{2n} b(2n;p_1,\cdots,p_d) D_1^{p_1},\cdots,D_d^{p_d}\phi - G'(\phi) + 2u\,. \tag{2.8}$$

This equation is now coupled with the heat diffusion equation

$$u_t + \frac{l}{2}\phi_t = K\Delta u \tag{2.9}$$

which has the same form as (1.4) except that ϕ is no longer a Heaviside function of u.

In the simplest case of isotropy and truncation after the first term ($M=1$), equation (2.8) reduces to (1.6). The second order system has been studied [4-10, 12-16], subject to initial and boundary conditions, e.g.,

$$u(0,x) = u_0(x)\,, \qquad \phi(0,x) = \phi_0(x), \qquad (x\in\Omega)\,, \tag{2.10}$$

$$u(x,t) = u_\partial(t,x)\,, \qquad \phi(t,x) = \phi_\partial(t,x)\,, \qquad (x\in\partial\Omega)\,. \tag{2.11}$$

Some of the basic results will be summarized in the next section.

3. BASIC RESULTS FOR SECOND ORDER SYSTEM

An existence and uniqueness theorem has been proven [4] for the system of parabolic equations (1.5), (1.6) subject to (2.10), (2.11). The proof is based on classical results for small time combined with the theory of invariant regions [17-21]. We consider domains, Ω, with C^∞ boundaries and define the Banach space $B \equiv BC = \{$bounded, uniformly continuous functions on $\Omega\}$. If the stability inequality

$$\xi^2/\tau < K \tag{3.1}$$

is satisfied, then sufficiently large parallelipipeds (in (u,ϕ) space) form invariant regions. That is, once a solution is within the parallelipiped, it cannot move outside of it. This provides the *a priori* bound needed to prove

THEOREM 3.1: (Existence and uniqueness). Suppose l, K, ξ and τ are any set of positive constants subject to the stability inequality (3.1). If u and ϕ are in $\mathcal{B}$ and $T \in (0,\infty)$, then there exists a unique solution (u,ϕ) to the system (1.5), (1.6), (2.10), (2.11) for all $t \in [0,T]$ such that $u(\cdot,x)$ and $\phi(\cdot,x)$ are in $\mathcal{B}$.

Regularity of solutions may also be proven for this system. We make the assumption

$$C_1 \leqslant \xi^2/\tau \tag{3.2}$$

in addition to (3.1), and use the standard metric

$$d(P,Q) \equiv [|x_1 - x_2|^2 + |t_1 - t_2|]^{\frac{1}{2}} \tag{3.3}$$

where $P = (t_1, x_1)$ and $Q = (t_2, x_2)$ in $\Lambda \equiv \bar{\Omega} \times [0,\bar{1}]$. Using the usual norm defined via (3.3) on the Banach space $C_{2+\alpha}(\Lambda)$, the basic result [4] may be stated as

THEOREM 3.2: (Gradient bounds). Suppose the initial data and boundary conditions (2.10), (2.11) are in $C_{2+\alpha}$. Then the solution (u,ϕ) to (1.5), (1.6), (2.10), (2.11) is also in $C_{2+\alpha}$. Furthermore, one has the bounds

$$\left|\frac{\partial\phi}{\partial x_s}\right| \leqslant C \qquad \left|\frac{\partial^2\phi}{\partial x_s^2}\right| \leqslant C \tag{3.4}$$

where $x_s \equiv x/\xi$ is a stretched variable and C is a constant which depends on $l, K, \Omega, T, u_0, u_\partial, \phi_0, \phi_\partial$ and C_1 but not on ξ or τ,

Physically, the importance of the result (3.4) lies in the fact that the interfacial region (i.e. transition between $\phi \approx -1$ and $\phi \approx +1$) does not become significantly sharper in time.

Within this model, the interface and conditions on the interface have been incorporated directly into the system of equations. In particular, since the interface is simply described by (1.7), one may enquire about the value of the temperature at these points. For a system in equilibrium the physical expectation is that at any point on the interface, the temperature should be proportional to the local sum of principal curvatures of the interface. The result, which is known from equilibrium statistical mechanics as the Gibbs-Thomson relation, is usually expressed in the form

$$\Delta s\, u(x) = -\sigma\, \kappa(x) \tag{3.5}$$

where σ is the interfacial (or surface) tension, Δs is the entropy difference between the solid and liquid, which in this model is 4, and $\kappa(x)$ is the sum of principal curvatures at the point x.

It is of interest to know whether this relation is implied by the equilibrium analogs of (1.5), (1.6), which are obtained by setting the time-derivatives equal to zero. In this case, equation (1.5) is just Laplace's equation, and u is determined uniquely by the boundary condition. In equilibrium, equation (1.6) is

$$0 = \xi^2 \Delta\phi + \frac{1}{a} f(\phi) + 2\xi \bar{u}(x) \tag{3.6}$$

where u is now a known function. The domain, Ω, is taken to be an annular region which is constrained (by means of boundary conditions) to be solid in the interior and liquid on the exterior boundary. We assume $a^{-1} f(\phi) + 2u$ has three roots and let ϕ_+, ϕ_- be the largest and smallest respectively. A case of interest is the prototype $a^{-1} f(\phi) = \frac{1}{2}(\phi - \phi^3)$.

The mathematical problem then consists of examining the transition layer behaviour for small ξ. Three different analytical methods have been applied to this problem.

(A) Asymptotic expansion for fixed Γ. If the set of points for which $\phi = 0$ is fixed, then we may consider the expansions on each side separately. One may define normal and tangential coordinates, r and s, respectively. In terms of the scaled normal coordinate $\rho \equiv r/\xi$, the inner expansion has as its $O(1)$ term, ψ_0, the solution to the equation

$$\frac{d^2\psi}{d\rho^2} + \tfrac{1}{2}(\psi - \psi^3) = 0 . \tag{3.7}$$

Hence, $\psi_0(\rho) = \tanh \rho/2$. For the outer expansion, the $O(1)$ term is $\phi_0 = \pm 1$, with the sign depending on the side of the boundary. The higher order terms are formally calculated by subtracting from the original equation. In this way, one constructs a sequence of functions

$$\phi_M(x,\xi) = \sum_{j=0}^{M} \xi^j \phi_j(x,\xi) \tag{3.8}$$

which approximate the true solution $\phi(x)$. One may then prove (p.235 of [4])

THEOREM 3.3: (Asymptotic expansion for positive ϕ). Given a positive integer $M \in \{0, \cdots, M_*\}$ and $\xi \in (0, \xi_*(M_*))$ for some $\xi_*(M_*)$, the function

$$\phi \equiv \Phi_M + \tilde{\phi}_M \tag{3.9}$$

is a solution of (3.6) such that

$$\sup |\tilde{\phi}_M| = O(\xi^{M+1}) . \tag{3.10}$$

The surface tension, σ, may be calculated from the free energy (2.6) as

$$\sigma = \tfrac{2}{3}\xi + O(\xi^2) \tag{3.11}$$

and, hence, one may prove that the Gibbs-Thompson condition, (3.5), is necessary for a solution ϕ (p.238 of [4]).

(B) Construction of transition layers in arbitrary dimension and spherical symmetric solutions to (3.6) leads to the equation (with $\tfrac{1}{2}k = 2\xi\bar{u}$)

$$\xi^2\phi'' + \xi^2 \frac{(N-1)}{r}\phi^1 + \tfrac{1}{2}(\phi - \phi^3) + \tfrac{1}{2}k = 0 \tag{3.12}$$

in an annular domain $\Omega = \{x \in \mathbb{R}^d \,;\, a \leqslant r \leqslant b\}$ subject to the boundary conditions

$$\phi(a) = \phi_- , \qquad \phi(b) = \phi_+ . \tag{3.13}$$

Existence and monotonicity of solutions to (3.12), (3.13) have been proven in [12]. By means of shooting methods and the maximum principle, it has been shown [13] that the point of crossover, R_0, (i.e. $\phi(R_0)=0$) must occur at the value which corresponds to a curvature that agrees with the Gibbs-Thompson relation, i.e.

$$R_0(\xi) = -\frac{(N-1)}{k}\frac{\xi}{4}\int_{-\infty}^{\infty} \operatorname{sech}^4 \frac{x}{2}\, dx + O(\xi) . \qquad (3.14)$$

(C) Interior transition layer in two-dimensions without assumption of spherical symmetry. This problem has been considered [14,15] in a formulation similar to (B). A prerequisite to resolving this problem is to prove the existence of a curve, Γ, such that

$$4\,u(x) = -\sigma\,\kappa(x) . \qquad (3.15)$$

Using polar coordinates, (r,θ), and defining $s = 1/r$, this problem may be written as the following. Find a 2π-periodic function $s(\theta)$ such that

$$N[s] \equiv s'' + s - F(s,\theta)\left[1 + \frac{s'^2}{s^2}\right]^{3/2} = 0 \qquad (3.16)$$

$$a < s(\theta) < b \qquad (0 \leqslant \theta \leqslant 2\pi) .$$

Existence of a solution to (3.16) has been proven using sub- and super-solutions. The existence question for (3.16) can then be resolved using a generalization of Theorem 1.5.1 [22] of Bernfeld and Lakshmikantham.

For dimensions higher than two, one expects similar results, but the construction of solutions is an open problem.

There is no assertion of uniqueness in the equilibrium results discussed in (B) and (C). In the case of spherical symmetry, i.e. (B), however, the solution is restricted to be in a narrow band of width ξ. Hence, if two distinct solutions exist, they would not be significantly different physically.

A detailed study of the internal layer in the dynamic problem has thus far been formal [23]. In analyzing the system (1.5), (1.6) for small ξ, one must consider inner and outer expansions for both u and ϕ, using dynamical matching conditions. One may enquire again about the value of the temperature of the interface. The result (with anisotropy incorporated into the model as $\Delta\phi + \xi_1^2\,\phi_{xx}$) is

$$\Delta s\, u(r,\theta) = -[\sigma(\theta) + \sigma''(\theta)]\kappa - \frac{\tau v \sigma(\theta)}{\xi_A^2(\theta)} + O(\xi^2) \qquad (3.17)$$

$$\xi_A^2(\theta) \equiv \xi^2 + (\xi_1^2 - \xi^2)\cos\theta \,.$$

A first step in making this rigorous is to consider a moving plane [23]. The system of partial differential equations then becomes a system of ordinary differential equations. It appears that travelling wave solutions (i.e. constant velocity) do not usually exist, and the solutions are more likely to have a variable velocity with constant $O(1)$ term.

4. RELATIONSHIPS BETWEEN PHASE FIELD MODELS AND STEFAN MODELS

It is of considerable mathematical interest to know the relationship between the phase field equations (1.5), (1.6) and the Stefan model. In addition to the classical formulation of the Stefan problem [(1.1)−(1.3)], various models have been considered to approximate the effects of surface tension and supercooling. In particular, one may replace the condition (1.3) for the temperature at the interface with one of the following:

$$u = -\sigma\kappa/\Delta s \qquad (4.1)$$

$$u = -\sigma\kappa/\Delta s - cv \qquad (4.2)$$

where c is a constant and v is the signed magnitude of the normal velocity of the interface (plus sign if motion is toward the liquid). The additional term in (4.2) is due to dynamical supercooling (see [24-26]). The relative importance of the dynamical undercooling term, which is often neglected, depends on microscopic parameters. This dependence is most easily seen via the phase field equations as we will see below.

The system of equations (1.1), (1.2), (4.1) or (1.1), (1.2),

(4.2) then provide modified Stefan problems which incorporate, to some extent, the effects of surface tension and supercooling. Although these models are often useful in practical problems, their fundamental limitation is the absence of a length scale in (1.1), (1.2). The heat released by freezing is thus concentrated on a set of measure zero.

The models [(1.1)-(1.3)], [(1.1), (1.2), (4.1)] and [(1.1), (1.2), (4.2)] can all be formally obtained as particular limits of the phase field model [(1.5), (1.6)]. In the phase field model, the scales ξ and τ are measures of length and time in diffusion, while a^{-1} is a measure of well depth in the double-well potential and is an indication of the preference of the material for either of the phases as opposed to the interfacial region. Thus, the scaling of ξ, τ and a is crucial in the limiting procedure. If a also approaches zero, then the asymptotic analysis then implies a first order solution

$$\phi_0 = \tanh \frac{r-R_0}{2\sqrt{a}\xi} \tag{4.3}$$

with interfacial thickness of order $\sqrt{a}\xi$ and surface tension

$$\sigma = \frac{2}{3}\frac{\xi}{\sqrt{a}} + o(\xi a^{-\frac{1}{2}}) . \tag{4.4}$$

Finally, the temperature of the interface is then

$$u(x) \cong -\frac{2}{3}\frac{\xi}{\sqrt{a}}(\kappa + \alpha v) \tag{4.5}$$

where α is defined by $\tau = \alpha\xi^2$.

One can then obtain the following limits.

(A) The Stefan limit [(1.1)-(1.3)]. This limit can be attained if $\xi, a \to 0$ while α remains fixed provided

$$\xi a^{-\frac{1}{2}} \to 0 . \tag{4.6}$$

By (4.3)-(4.5), it is clear (formally) that the interfacial tension, thickness and temperature all approach zero. Equation (1.6) then becomes a triviality, $0=0$, while equation (1.5) is just the heat equation with a source at the interface. In taking this limit, however, the sign of u near the interface serves to determine the sign of ϕ. Hence, ϕ_0 approaches ± 1 with the value +1 attained at those points (t, x) for which $u > 0$ while -1 is attained at points for which $u < 0$. Equation (1.5) then has the limiting form (1.4). That is, the heat equation applies to all points which are not on the interface. The latent heat condition (1.2) must apply across the interface due to the equivalence (in a weak sense) between (1.1), (1.2) and (1.4).

(B) Modified Stefan limit (with velocity term) [(1.1), (1.2), (4.2)]. Suppose we keep α fixed and let ξ, a approach zero as in (A) but replace (4.6) with

$$\xi a^{-\frac{1}{2}} = c_1 = \text{fixed}. \qquad (4.7)$$

Then we obtain a limit which differs from (A) in that the temperature at the interface is not zero, but satisfies

$$u(x) \cong -\tfrac{2}{3}\, c_1(\kappa + \alpha v)\,. \qquad (4.8)$$

The surface tension is given by $\sigma \cong \frac{2}{3}\, c_1$ and the interfacial thickness approaches zero. The other aspects (i.e. heat diffusion and latent heat) of the limit are identical to those of (A). Hence, we obtain a formal limit in which the interface becomes sharp while the surface tension and interfacial temperature remain finite.

(C) Alternative modified Stefan limit (without velocity term) [(1.1), (1.2), (4.1)]. We let ξ and a approach zero as in (B) while (4.7) remains in effect. If, in addition, we take the limit as α approaches 0 appropriately, e.g. $\alpha = \xi^{\frac{1}{4}}$, then we obtain a temperature at the interface given by

$$u(x) \cong -\tfrac{2}{3} c_1\, \kappa\,. \qquad (4.9)$$

Once again, this results in a limit in which the interface is sharp while the surface tension and temperature at the interface are finite.

The limits (A), (B), (C) are interesting to consider for a number of reasons. Physically, they are extreme cases in that the role of one or more quantities is suppressed. In (B) the role of interfacial tension is neglected. In (C) the role of thickness and relaxation time are neglected. Finally, in (A), thickness, relaxation time and interfacial tension are all neglected. The justification for neglecting any of these physical quantities depends on the microscopic parameters for the particular substance. There is a consensus that interfacial tension is important for the suppression of instabilities. Furthermore, it is clear from either the phase field equations (1.5), (1.6) or from (4.4), (4.5) that setting a small surface tension equal to zero is a very singular perturbation which results in significantly different behaviour of the equations.

It has been suggested that for an anisotropic material, the effects of interfacial tension and a finite interfacial thickness play a role in determining the growth direction and in selectively suppressing instabilities in all but the preferred directions [10]. The presence of a velocity term in the interfacial temperature [e.g. (4.5)] has been suggested for many years [24-26]. It is often neglected in computations, and its role in stability does not appear to be one-sided.

The system (1.5), (1.6), which provides a unified way of considering a broad spectrum of the physical quantities discussed above, allows an easy interpretation of limiting cases such as (A)-(C). Furthermore, an existence, uniqueness and regularity theory is also not difficult for (1.5), (1.6) [see Section 3], and may facilitate the understanding of the limit models.

The limits (A)-(C) have not yet been obtained rigorously, except in the equilibrium case [14]. A careful formal analysis has justified some of the limiting cases [23]. A rigorous analysis of these limits remains as an important mathematical problem in this area.

5. HIGHER ORDER EQUATIONS

We now discuss the higher order phase field equations (2.8),

(2.9). A general existence theory for these equations has not yet been presented. However, some formal analysis [11] indicates a number of interesting features. In particular, one may enquire about the temperature at the interface and determine whether it is different from that of the second order equation (3.17). The result is that it is identical for isotropic systems, although the value of the interfacial tension differs. For an anisotropic system the relation (3.5) is modified as is the interfacial tension.

In the M-th order equation, the leading order behaviour is given by $\psi(\rho)$ where ψ is a solution of

$$\sum_{n=1}^{M} \frac{J_{2n}}{(2n)!} \frac{\partial^{2n}\psi}{\partial\rho^{2n}} - G'(\phi) = 0 \tag{5.1}$$

$$J_{2n} \equiv \int J(x)\,(x_1^2 + \cdots + x_d^2)^n\,dx, \quad \rho \equiv r/\xi$$

where ψ must approach distinct limits as in the second order case.

By means of various identities, one can compute the interfacial tension as

$$\sigma = \sum_{n=1}^{M} (-1)^{n+1} \frac{J_{2n}}{(2n)!}\, n\,\|\psi_n\|^2 \tag{5.2}$$

where ψ_n is the n-th derivative with respect to ρ and

$$\|f\|^2 \equiv \int_{-\infty}^{\infty} f(\rho)\,d\rho.$$

Using some relations between various norms, one may then show (for an isotropic system) that the Gibbs-Thomson relation (3.5) is identical in form.

For an anisotropic system, the situation is more complicated. The analogue of (3.5) is replaced by an identity which involves several L_2 norms.

As a first step in placing the theory of the higher order equations on a firm mathematical basis would be to address the following problems:

(1) Existence, uniqueness and monotonicity of (5.1) subject to appropriate boundary conditions;

(2) Existence, uniqueness and regularity for the system (2.8), (2.9) subject to suitable initial and boundary conditions.

(3) In the equilibrium case $[u_t = \phi_t = 0]$, a rigorous asymptotic analysis for the transition layer similar to [14].

6. NUMERICAL ANALYSIS

A primary interest in any of the equations discussed in preceding sections is the nature of the growth of the interface. In particular, one would like to know the conditions under which one has unstable growth or anisotropic growth. Equations (1.5), (1.6) [with $a=1$ and $f(\phi)=\frac{1}{2}(\phi-\phi^3)$] have been studied using finite difference schemes [27]. Unstable growth is observed when the surface tension (i.e. ξ) is small and the undercooling large. The effects of anisotropy have also been studied [28]. For xy anisotropy, i.e. equation (1.6) is modified by changing the Laplacian term $\xi^2\Delta\phi$ to $\xi^2\Delta\phi+\xi_1^2\phi_{xx}$, and coupling with (1.5). In this case we find preferential growth for all values of ξ. For large ξ, and initial "seed" in the form of a circle evolves into an ellipse. For smaller values of ξ, one has almost all growth in the preferred direction, i.e. an instability in the form of a spike.

Other types of anisotropy may be modelled using second order equations using more phenomenological ideas than those suggested in the preceding section. These have been reported in [28] along with the numerical results.

Note: Supported by NSF Grant DMS-8601746.

REFERENCES

1. L.I. Rubinstein, "The Stefan problem", *Am. Math. Soc. Transl.* **27**, AMS Providence, R.I. (1971). See also references in [4].
2. O.A. Oleinik, "A method of solution of the general Stefan problem", *Sov. Math. Dokl.* **1**, (1960), pp.1350-1354.
3. J. Glimm and A. Jaffe, *Quantum Physics*, Springer, Berlin, (1981).
4. G. Caginalp, "An analysis of a phase field model of a free boundary", *Arch. Rat. Mech. Analysis* **92**, (1986), pp.205-242.
5. G. Caginalp, "Surface tension and supercooling in solidification theory", *Lecture Notes in Physics* **216**, (1984), pp.216-226; *Proc. Appl. of Field Theory to Statistical Mechanics*, Sitges, Spain, June (1984), Springer, N.Y.
6. G. Caginalp, "Phase field models of solidification: free boundary problems as systems of nonlinear parabolic differential equations", in *Free Boundary Problems: Applications and Theory*, (eds. Bossavit, A. *et al.*), (Proc. Col. International Maubuisson-Carcans, France, June (1984), vol.3, pp.107-121, Pitman, Boston.
7. G. Caginalp, "Solidification problems as systems of nonlinear differential equations", *Lectures in Applied Mathematics* **23**, Proceedings of Santa Fe, N.M. AMS-SIAM Conference, July (1984), (ed. Nickolaenko, B.), pp.347-369, AMS Providence, R.I.
8. P.C. Hohenberg and B.I. Halpern, "Theory of dynamic critical phenomena", *Reviews of Modern Physics* **49**, (1977), pp.435-480.
9. D. Jasnow, "Critical phenomena of interfaces", *Reports on Progress in Physics* **47**, (1984), pp.1059-1132.
10. G. Caginalp, "The role of microscopic anisotropy in the macroscopic behaviour of a phase boundary", *Annals of Physics*, (to appear).
11. G. Caginalp and P.C. Fife, "Higher order phase field models and detailed anisotropy", *Physical Review B* **34**, (1986), pp.4940-4943.
12. G. Caginalp and S. Hastings, "Properties of some ordinary differential equations related to free boundary problems", *Proc. Royal Society of Edinburgh*, (to appear).
13. G. Caginalp and B. McLeod, "The interior transition layer for an ordinary differential equation arising from solidification theory", *Quarterly of Applied Mathematics* **44**, (1986), pp.155-168.
14. G. Caginalp and P.C. Fife, "Elliptic problems involving phase boundaries satisfying a curvature condition", University of Arizona Preprint, (1985).
15. G. Caginalp and P.C. Fife, "Elliptic problems with layers representing phase interfaces", in *Nonlinear Parabolic Equations: Qualitative Properties of Solutions*, (Proc. Rome Conf., June (1985)), Pitman, Boston, (in press).
16. G. Caginalp and P.C. Fife, "Phase field methods for interfacial boundaries", *Phys. Rev. B* **33**, (1986), pp.7792-7794.

17. J. Smoller, *Shock Waves and Reaction-Diffusion Equations*, Springer-Verlag, N.Y. (1983).
18. H. Weinberger, "Invariant sets for weakly coupled parabolic and elliptic systems", *Rend. Mat.* **8**, (1975), pp.295-310.
19. K. Chueh, C. Conley and J. Smoller, "Positively invariant regions for systems of nonlinear diffusion equations", *Indiana Univ. Math. J.* **26**, (1977), pp.373-392.
20. J. Bebernes, K. Chueh and W. Fulks, "Some applications of invariance for parabolic systems", *Indiana Univ. Math. J.* 28, (1979), pp.269-277.
21. H. Amann, "Invariant sets and existence theorems for semi-linear parabolic and elliptic systems", *J. Math. Annal. Appl.* **65**, (1978), pp.432-467.
22. S. Bernfeld and V. Lakshmikantham, *An Introduction to Non-linear Boundary Value Problems*", Academic Press, N.Y. (1974).
23. G. Caginalp and P.F. Fife, (in preparation).
24. G. Horvay and J.W. Cahn, "Dendritic and Spheroidal Growth", *Acta. Met.* **9**, (1961), pp.695-705.
25. G.F. Bolling and W.A. Tiller, "Growth of the Melt III. Dendritic Growth", *Journal of Applied Physics* **32**, (1961), pp.2587-2605.
26. B. Chalmers, *Principles of Solidification*, p.106, Krieger, Huntington, N.Y. (1977).
27. Lin and Fix, (paper in preparation).
28. G. Caginalp and J.T. Lin, "A numerical analysis of an aniso-tropic phase field model", *University of Pittsburgh Preprint*, (1986).

6

CONTINUA WITH CONSTRAINED OR LATENT MICROSTRUCTURE

G. CAPRIZ

1. INTRODUCTION

The proposal has been repeatedly advanced to consider certain non-simple materials as continua endowed with some form of microstructure, the evolution of which is regulated by the gross motion through the action of appropriate internal constraints (see, for instance, Toupin's remarks in [1], on the possibility of viewing hyperelastic materials of second grade as Cosserat's continua with constrained microstructure). Then one mechanical balance equation suffices to determine the motion: it takes the form of Cauchy's equation, where, however, some traditional tenets are abandoned (e.g., the stress tensor need not be symmetric and may depend on higher derivatives of displacement such as acceleration gradients).

In two recent papers I have examined some cases of such "latent microstructure", principally to show how certain apparent inconsistencies with rational thermodynamics can be resolved [2], [3].

Here, after an introductory paragraph (where the general balance equations for continua with general lagrangian microstructure are recalled), I return to the matter, concentrating on mechanical issues. I consider also a class of continua with latent microstructure, which comprises some important material types (e.g., continua with voids, dilatant granular materials, etc.)

2. CONTINUA WITH LAGRANGIAN MICROSTRUCTURE

In a continuum $\mathcal{B}$ with microstructure one must assign 'order parameters' v^α ($\alpha = 1, 2, \cdots, m$) to characterize the mechanical state of each material element p; the real numbers v^a are best interpreted as the coordinates in a local chart of a member $\underset{\sim}{v}$ of a differentiable manifold $\mathfrak{m}$ of dimension m.

The choice of $\underset{\sim}{v}$ appropriate for a particular p is sometimes affected by the observer: the number which measures locally the void fraction in continua with voids is obviously observer-independent, but the unit vector which specifies the present preferred direction in a droplet of liquid crystal is not. In other words, the value of $\underset{\sim}{v}$ read on an element at the same instant in two motions of $\mathcal{B}$ differing one from the other by a rigid motion may be different. Of course, in principle, there are many alternative ways to measure the local state and one might be devised which is observer-independent: for instance, in liquid crystals one could specify the preferred direction with respect to a local material reference. But we are interested here only in the class of choices which lead to an expression of the power of microstresses linear and homogeneous in $\underset{\sim}{v}$ and grad $\dot{\underset{\sim}{v}}$ (the dot indicating total time derivative); so the only alternatives we admit from the start are those consistent with that requirement.

Let us indicate by $\underset{\sim}{v} \to \underset{\sim}{v}_{(Q)}$ the group action which maps the value $\underset{\sim}{v}$ of the order parameters in a certain placement into its value $\underset{\sim}{v}_{(Q)}$ subsequent to a rigid rotation specified by the proper orthogonal tensor Q and by $\mathfrak{a}$ the infinitesimal generator of the group. If q is the vector associated with Q (the rotation is around an axis parallel to q and of an angle $|q|$ or $-|q|$ depending on the orientation of q), then $\mathfrak{a}$ is defined by the following property:

$$\underset{\sim}{v}_{(Q)} = \underset{\sim}{v} + \boldsymbol{a}q + o(q) \,. \tag{2.1}$$

As a consequence, the most general rigid velocity distribution for $\boldsymbol{B}$ is expressed by the formulae

$$\begin{aligned} \dot{x} &= c + w \times (x - \bar{x}) \,, \\ \dot{\underset{\sim}{v}}(x) &= \boldsymbol{a}(\underset{\sim}{v}(x))\, w \,. \end{aligned} \tag{2.2}$$

where c and w are the translational and angular velocities respectively. Component notation will often be used below: Cartesian components of vectors in Euclidean space will be marked by latin indices, whereas components of vectors in the tangent space $\tau_{\underset{\sim}{v}}\boldsymbol{m}$ of $\boldsymbol{m}$ at $\underset{\sim}{v}$ will be marked by Greek exponents, so that $(2.2)_2$ will be also written as follows:

$$\dot{v}^{\alpha} = \boldsymbol{a}^{\alpha}_{\ i} w_i \,;$$

the indices reflect the fact that, by definition (2.1), $\boldsymbol{a}$ is a linear operator from the space V of vectors into $\tau_{\underset{\sim}{v}}\boldsymbol{m}$.

The mechanical balance equations for $\boldsymbol{B}$ are: the usual equation of balance mass

$$\dot{\rho} + \rho \operatorname{div} \dot{x} = 0 \,, \tag{2.3}$$

the usual equation of Cauchy

$$\rho \ddot{x} = \rho f + \operatorname{div} T \tag{2.4}$$

(ρ, mass density; f, density per unit mass of body forces; T, Cauchy's stress tensor) and a balance equation for micromomentum. The latter equation involves vectors of the cotangent space $\tau^{*}_{\underset{\sim}{v}}\boldsymbol{m}$:

(i) a Lagrangian acceleration

$$(\partial\kappa/\partial\dot{\underset{\sim}{v}})^{\bullet} - (\partial\kappa/\partial\underset{\sim}{v}) \,,$$

deduced as a Lagrangian derivative from the extra kinetic energy density per unit mass due to the micromotion

$$\kappa = \tfrac{1}{2}\mu_{\alpha\beta}(v)\, \dot{v}^{\alpha} \dot{v}^{\beta} \,; \tag{2.5}$$

(ii) resultant densities per unit volume of internal $(-\underset{\sim}{\zeta})$ and external $(\rho\underset{\sim}{\beta})$ actions;

(iii) the divergence of a microstress $\boldsymbol{S}$ (with the property that on any surface element of unit normal vector n, $\boldsymbol{S}n$ is the action per unit area on the microstructure).

The balance equation is written in the form

$$\rho((\partial\kappa/\partial\dot{\underset{\sim}{v}})^{\bullet} - \partial\kappa/\partial\underset{\sim}{v}) = \rho\underset{\sim}{\beta} - \underset{\sim}{\zeta} + \operatorname{div}\boldsymbol{S}\,; \tag{2.6}$$

the general character of the microstructure justifies the formal differences between equations (2.4), (2.5).

From (2.3) — (2.6) a theorem of kinetic energy can be deduced, valid during smooth motions

$$\left(\int_{B}\rho\kappa\right)^{\bullet} = \int_{B}\rho(b\cdot\dot{x} + \underset{\sim}{\beta}\cdot\dot{\underset{\sim}{v}}) + \int_{\partial B}((Tn)\cdot\dot{x} + (\boldsymbol{S}n)\cdot\dot{\underset{\sim}{v}}) -$$

$$-\int_{B}(T\cdot\operatorname{grad}\dot{x} + \underset{\sim}{\zeta}\cdot\dot{\underset{\sim}{v}} + \boldsymbol{S}\cdot\operatorname{grad}\dot{\underset{\sim}{v}})\,;$$

here B is the region occupied by the body $\boldsymbol{B}$, ∂B its boundary and the notation for scalar product must be appropriately interpreted. The quantity,

$$T\cdot\operatorname{grad}\dot{x} + \underset{\sim}{\zeta}\cdot\dot{\underset{\sim}{v}} + \boldsymbol{S}\cdot\operatorname{grad}\dot{\underset{\sim}{v}}$$

or, in components,

$$T_{ij}\dot{x}_{ij} + \zeta_{\alpha}\dot{v}^{\alpha} + \boldsymbol{S}_{\alpha i}\dot{v}_{\alpha,i} \tag{2.7}$$

is seen to be the opposite of the total power density of the internal actions. The condition that this density vanished in all rigid motions (2.2) leads to the final balance equation of moment of momentum

$$\boldsymbol{e}_{ijk}T_{jk} = \boldsymbol{a}^{\alpha}_{\ i}\zeta_{\alpha} + \boldsymbol{a}^{\alpha}_{\ ij}\boldsymbol{S}_{\alpha j}\,, \tag{2.8}$$

where $\boldsymbol{e}$ is Ricci's permutation tensor.

REMARK. The operator $\boldsymbol{a}$ has an important rôle also in the statement of conditions of invariance; for instance, the density of kinetic energy κ must be Galilean invariant

$$\kappa(\underset{\sim}{v}_{(Q)}, (\underset{\sim}{v}_{(Q)})^{\bullet}) = K(\underset{\sim}{v}, \dot{\underset{\sim}{v}}),$$

for all choices of the constant proper orthogonal tensor Q and this relation implies and is implied by the condition

$$(\partial\kappa/\partial v^{\alpha})\, a^{\alpha}_{\ i} + (\partial\kappa/\partial \dot{v}^{\alpha})(\partial a^{\alpha}_{\ i}/\partial v^{\beta})\, \dot{v}^{\beta} = 0 . \tag{2.9}$$

3. CONSTRAINED MICROSTRUCTURE

One consequence of the balance equation (2.8) is an objective version of (2.7)

$$(\mathrm{sym}\, T)_{ij} D_{ij} + \zeta_{\alpha}(\dot{v}^{\alpha} - a^{\alpha}_{\ i} r_i) + \mathbf{S}_{\alpha i}(\dot{v}^{\alpha}_{\ ,i} - a^{\alpha}_{\ j,i} r_j) ; \tag{3.1}$$

here

$$D = \mathrm{sym\,grad}\, \dot{x} , \qquad 2r = -\mathbf{e}\,(\mathrm{grad}\, \dot{x}) . \tag{3.2}$$

On the basis of (3.1) it is possible to give a convenient definition of a continuum subject to internal constraints: $\mathcal{B}$ is said to be internally constrained if the allowed macro- and micro-velocity distributions are such that not all values of the factors

$$D_{ij},\ \dot{v}^{\alpha} - a^{\alpha}_{\ i} r_i,\ \dot{v}^{\alpha}_{\ ,i} - a^{\alpha}_{\ j,i} r_j$$

in (3.1) are accessible.

When $\mathcal{B}$ is constrained, the stress and the microstresses are each the sum of an active and a reactive component

$$T = \overset{a}{T} + \overset{r}{T} ,\ \zeta = \overset{a}{\zeta} + \overset{r}{\zeta} ,\ \mathbf{S} = \overset{a}{\mathbf{S}} + \overset{r}{\mathbf{S}} . \tag{3.3}$$

Only the active components can be specified through appropriate constitutive laws; the reactive components will depend on the specific process occurring in $\mathcal{B}$.

Actually, here, only perfect constraints are considered; they are characterized by the property that the density of power of reactive actions vanishes for all velocity distributions allowed by the constraints

$$(\mathrm{sym}\, \overset{r}{T})_{ij} D_{ij} + \overset{r}{\zeta}_{\alpha} (\dot{v}^{\alpha} - a^{\alpha}_{\ i} r_i) + \overset{r}{\mathbf{S}}_{\alpha i} (\dot{v}^{\alpha}_{\ ,i} - a^{\alpha}_{\ j,i} r_j) = 0 . \tag{3.4}$$

Then it is possible to use some of the equations (2.4), (2.6), (2.8), (3.4) for the elimination of $\overset{r}{T}, \overset{r}{\zeta}, \overset{r}{\mathbf{S}}$ and thus arrive at a subset of *pure* equations, which are alone sufficient to study the evolution of $\mathcal{B}$.

Suppose, for instance, that the microstructure is so

constrained that at each instant τ only *microvelocities of the rigid type* are allowed, i.e.

$$\dot{\underset{\sim}{v}} = \boldsymbol{a}(v)\, h(x,\tau)\,, \tag{3.5}$$

where only the choice of the vector function h is left open.

Then the macromotion is not constrained at all and hence D remains arbitrary in (3.4), whereas the factors between brackets are free only to the extent that the vector $r-h$ and the tensor $\operatorname{grad} h$ locally are. The following conditions ensue

$$\begin{aligned} &\operatorname{sym} \overset{r}{T} = 0\,, \\ &\boldsymbol{a}^{\alpha}_{\;\;i} \overset{r}{\zeta}_{\alpha} + \boldsymbol{a}^{\alpha}_{\;\;i,j} \overset{r}{\mathbf{S}}_{\alpha j} = 0\,, \\ &\overset{r}{\mathbf{S}}_{\alpha i}\, \boldsymbol{a}^{\alpha}_{\;\;j} = 0\,. \end{aligned} \tag{3.6}$$

The first two of these conditions, together with (2.8), give

$$\overset{r}{T} = -\operatorname{skw} \overset{a}{T} + \tfrac{1}{2}\, \boldsymbol{e}\left(\boldsymbol{a}^{T} \overset{a}{\underset{\sim}{\zeta}} + (\operatorname{grad} \boldsymbol{a}^{T})\, \overset{a}{\mathbf{S}}\right) \tag{3.7}$$

(where the superscript T indicates transposition). This result leads to an expression of Cauchy's stress in terms of constitutive components only

$$T = \operatorname{sym} \overset{a}{T} + \tfrac{1}{2}\, \boldsymbol{e}\left(\boldsymbol{a}^{T} \overset{a}{\underset{\sim}{\zeta}} + (\operatorname{grad} \boldsymbol{a}^{T})\, \overset{a}{\mathbf{S}}\right), \tag{3.8}$$

and hence renders 'pure' Cauchy's equation (2.4).

The evolution of the microstructure depends still on the field $h(x,\tau)$; a pure equation in h is easily obtained by transforming both sides of (2.6) with $\boldsymbol{a}^{T}$ and using the second and third equation (3.6).

$$\rho\, \boldsymbol{a}^{T} \left((\partial\kappa/\partial\dot{\underset{\sim}{v}})^{\cdot} - (\partial\kappa/\partial\underset{\sim}{v})\right) = \rho\, \boldsymbol{a}^{T} \underset{\sim}{\beta} - \boldsymbol{a}^{T} \overset{a}{\underset{\sim}{\zeta}} + \boldsymbol{a}^{T} (\operatorname{div} \overset{a}{\mathbf{S}})\,.$$

Actually use can also be made of (2.9) to achieve the more compact form

$$\rho\left(\boldsymbol{a}^{T} (\partial\kappa/\partial\dot{\underset{\sim}{v}})\right)^{\cdot} = \boldsymbol{a}^{T} (\rho\beta - \overset{a}{\zeta} + \operatorname{div} \overset{a}{\mathbf{S}})\,. \tag{3.9}$$

For instance, in the case of liquid crystals, $\underset{\sim}{v}$ is a unit vector d and $\boldsymbol{a}$ is the tensor $\boldsymbol{e}d$, so that (3.5) expresses the constancy of the length of the vector d

$$\dot{d} = (\boldsymbol{e} d)\, h\,.$$

Also $\underset{\sim}{\beta}, \overset{a}{\underset{\sim}{\zeta}}$ are vectors (b, z, say), $\overset{a}{\boldsymbol{S}}$ is a second order tensor S. Equation (3.8) puts in evidence an extra stress

$$T = \operatorname{sym} \overset{a}{T} + \operatorname{skw}(d \otimes z + (\operatorname{grad} d)\, S^T)$$

and eqn (3.9) says that the difference

$$\rho(\partial\kappa/\partial\dot{d})^{\bullet} - (\rho b - z + \operatorname{div} S)$$

is a vector parallel to d.

4. LATENT MICROSTRUCTURE

If, in (3.5), one requires of h to coincide with the spin r (see $(3.2)_2$), the evolution of the microstructure becomes totally dictated by the macromotion. The vector $r-h$ vanishes, rather than being arbitrary, so that the second equation is crossed out from the set (3.6) and there is no condition on $\overset{r}{\underset{\sim}{\zeta}}$; instead eqns $(3.6)_1$, $(3.6)_3$ still apply.

The first equation (3.6) and the balance equation (2.8) lead to a less severe restriction than (3.7)

$$\overset{r}{T} = -\operatorname{skw} \overset{a}{T} + \tfrac{1}{2}\, \boldsymbol{e}(\boldsymbol{a}^T \underset{\sim}{\zeta} + (\operatorname{grad} \boldsymbol{a}^T)\boldsymbol{S}). \tag{4.1}$$

On the other hand, once $\overset{r}{\underset{\sim}{\zeta}}$ and $\overset{r}{\boldsymbol{S}}$ were also eliminated from this expression of $\overset{r}{T}$, Cauchy's equation would suffice to study the evolution of $\boldsymbol{B}$. Thus (2.6) can be used to eliminate $\underset{\sim}{\zeta}$ altogether from the right-hand side of (4.1)

$$\overset{r}{T} = -\operatorname{skw} \overset{a}{T} + \tfrac{1}{2}\, \boldsymbol{e}\Big(\boldsymbol{a}^T (\rho\underset{\sim}{\beta} + \operatorname{div} \boldsymbol{S} - \rho((\partial\kappa/\partial\dot{\underset{\sim}{\nu}})^{\bullet} - (\partial\kappa/\partial\underset{\sim}{\nu}))) + (\mathit{grad}\, \boldsymbol{a}^T)\, \boldsymbol{S}\Big).$$

Using $(3.6)_3$ and again (2.9) one obtains finally

$$T = \operatorname{sym} \overset{a}{T} + \tfrac{1}{2}\, \boldsymbol{e}\Big(\rho\, \boldsymbol{a}^T \underset{\sim}{\beta} - \rho(\boldsymbol{a}^T (\partial\kappa/\partial\dot{\underset{\sim}{\nu}}))^{\bullet} + \operatorname{div}(\boldsymbol{a}^T \overset{a}{\boldsymbol{S}})\Big). \tag{4.2}$$

where $\underset{\sim}{\nu}$ must be expressed in terms of the macromotion through the evolution equation

$$\dot{\underset{\sim}{\nu}} = \boldsymbol{a}(\underset{\sim}{\nu})\, r\,. \tag{4.3}$$

In conclusion only an indirect trace of the existence of the microstructure remains: *the microstructure is latent.* However, one dramatic fact emerges: T may depend on acceleration gradients, need not be symmetric and not even objective.

There is an alternative way of looking at this matter; terms can be arranged in Cauchy's equation as follows

$$\rho\ddot{x} + \tfrac{1}{2}\operatorname{rot}\left(\rho\,(\boldsymbol{a}^T\,(\partial\kappa/\partial\dot{\underset{\sim}{v}}))^{\bullet}\right) = \rho f + \tfrac{1}{2}\operatorname{rot}(\rho\boldsymbol{a}^T\,\underset{\sim}{\beta}) + \operatorname{div}\hat{T}, \tag{4.4}$$

where

$$\hat{T} = \operatorname{sym}\overset{a}{T} + \tfrac{1}{2}\,\boldsymbol{e}\left(\operatorname{div}\boldsymbol{a}^T\overset{a}{\boldsymbol{S}}\right) \tag{4.5}$$

can be interpreted as the stress to be specified through a constitutive relation, so as to be objective. But then the density of the inertia force departs from the classical value; one must presume the existence also of a surface flux of inertia.

The constraint (4.3) is, of course, special and rather peculiar. We consider below a more general case, which comprises a number of materials already studied in the literature; it is the case when the microstate is completely determined by the microstrain

$$\underset{\sim}{v} = \underset{\sim}{\omega}(F)\,, \quad F = \nabla x\,. \tag{4.6}$$

To represent an internal constraint this relation must be objective, i.e. the following property must hold

$$\underset{\sim}{v}_{(Q)} = \underset{\sim}{\omega}(QF)\,, \tag{4.7}$$

for all proper orthogonal tensors Q. A necessary and sufficient condition for (4.7) to apply is as follows:

$$\boldsymbol{a}^{\alpha}{}_{i} = -\,\boldsymbol{e}_{ipq}\,B^{\alpha}_{pq}\,, \tag{4.8}$$

where

$$B^{\alpha}_{pq} = (d\omega^{\alpha}/dF_{pA})\,F_{qA}\,. \tag{4.9}$$

Condition (4.8) has an interesting consequence; in fact, (4.6) implies that

$$\dot{v}^{\alpha} = B^{\alpha}_{pq}\,\dot{x}_{p,q} = B^{\alpha}_{pq}\,(D_{pq} - \boldsymbol{e}_{pqc}\,r_c)$$

and, by (4.8),

$$\dot{v}^{\alpha} = B^{\alpha}_{pq} D_{pq} + \boldsymbol{a}^{\alpha}_{p} r_{p} . \tag{4.10}$$

Thus, the present constraint is much less restrictive than (4.3); it reduces to the latter if and only if B^{α}_{pq} is skew in the indices p,q.

Introduction of (4.10) into (3.4) and perusal of (4.8) leads to the condition

$$\left(\overset{r}{T}_{(ij)} + \overset{r}{\zeta}_{\alpha} B^{\alpha}_{(ij)} + \overset{r}{\boldsymbol{S}}_{\alpha k} B^{\alpha}_{(ij),k}\right) D_{ij} + \overset{r}{\boldsymbol{S}}_{\alpha k} B^{\alpha}_{ij} (D_{ij,k} - \boldsymbol{e}_{ijm} r_{m,k}) = 0 , \tag{4.11}$$

where the round brackets comprising indices indicate symmetrization. As the last factor between brackets in the left-hand side equals $\dot{x}_{i,jk}$, (4.11) implies

$$\begin{aligned} \overset{r}{T}_{(ij)} + \overset{r}{\zeta}_{\alpha} B^{\alpha}_{(ij)} + \overset{r}{\boldsymbol{S}}_{\alpha k} B^{\alpha}_{(ij),k} &= 0 \\ \overset{r}{\boldsymbol{S}}_{\alpha(k} B^{\alpha}_{ij)} &= 0 . \end{aligned} \tag{4.12}$$

On the other hand, eqns (2.6), (2.8), (2.9) give

$$T_{[ij]} = \tfrac{1}{2} \boldsymbol{e}_{ijk} \left(\rho \boldsymbol{a}^{\alpha}_{k} \beta_{\alpha} - \rho (\boldsymbol{a}^{\alpha}_{k} (\partial \kappa / \partial \dot{v}^{\alpha}))^{\boldsymbol{\cdot}} + (\boldsymbol{a}^{\alpha}_{k} \boldsymbol{S}_{\alpha m})_{,m} \right),$$

where square brackets comprising indices indicate anti-symmetrization. Separation of reactive and active components and use of (4.9) leads to

$$\overset{r}{T}_{[ij]} = -\overset{a}{T}_{[ij]} - \rho B^{\alpha}_{[ij]} \beta_{\alpha} + \rho (B^{\alpha}_{[ij]} (\partial \kappa / \partial \dot{v}^{\alpha}))^{\boldsymbol{\cdot}} - \left(B^{\alpha}_{[ij]} \boldsymbol{S}_{\alpha m} \right)_{,m} .$$

operating similarly on (4.12) one obtains

$$\overset{r}{T}_{(ij)} = - B^{\alpha}_{(ij)} (\rho \beta_{\alpha} - \overset{a}{\zeta}_{\alpha}) + \rho (B^{\alpha}_{(ij)} (\partial \kappa / \partial \dot{v}^{\alpha}))^{\boldsymbol{\cdot}} - (\boldsymbol{S}_{\alpha m} B^{\alpha}_{(ij)})_{,m} + B^{\alpha}_{(ij),m} \overset{a}{\boldsymbol{S}}_{\alpha m}$$

with the conclusion that

$$\overset{r}{T}_{ij} = -\overset{a}{T}_{[ij]} - \rho B^{\alpha}_{ij} \beta_{\alpha} + \rho (B^{\alpha}_{ij} (\partial \kappa / \partial \dot{v}^{\alpha}))^{\boldsymbol{\cdot}} + B^{\alpha}_{(ij)} \overset{a}{\zeta}_{\alpha} + B^{\alpha}_{(ij),m} \overset{a}{\boldsymbol{S}}_{\alpha m} - (B^{\alpha}_{ij} \boldsymbol{S}_{\alpha m})_{,m} .$$

Because of $(4.12)_2$ the last term of this expression differs from

$$-(B^{\alpha}_{ij}\overset{a}{\mathbf{S}}_{\alpha m})_{,m}.$$

only by a quantity, the divergence of which vanishes.

Hence the pure equation is Cauchy's equation

$$\rho\ddot{x} = \rho b + \operatorname{div}\hat{T},$$

where

$$\hat{T}_{ij} = \overset{a}{T}_{(ij)} - \rho B^{\alpha}_{ij}\beta_{\alpha} + \rho(B^{\alpha}_{ij}(\partial\kappa/\partial\dot{v}^{\alpha}))^{\bullet} - B^{\alpha}_{(ij)}(\overset{a}{\mathbf{S}}_{\alpha m,m} - \overset{a}{\zeta}_{\alpha}) - (B^{\alpha}_{[ij]}\overset{a}{\mathbf{S}}_{\alpha m})_{,m}.$$

5. SOME EXAMPLES

An important subcase of the constraint (4.6) is where $\underset{\sim}{v}$ depends on F only through $\iota = \det F$

$$\underset{\sim}{v} = \underset{\sim}{\gamma}(\iota). \qquad (5.1)$$

Then

$$B^{\alpha}_{ij} = \iota(d\gamma^{\alpha}/d\iota)\,\delta_{ij}, \qquad (5.2)$$

$$\boldsymbol{a}^{\alpha}_{i} = 0, \qquad (5.3)$$

and

$$\hat{T}_{ij} = \overset{a}{T}_{(ij)} - \iota(d\gamma^{\alpha}/dt)(\rho\beta_{\alpha} + \overset{a}{\mathbf{S}}_{\alpha m,m} - \overset{a}{\zeta}_{\alpha})\,\delta_{ij} + \rho\,\delta_{ij}(\mu_{\alpha\beta}(d\gamma^{\alpha}/d\iota)(d\gamma^{\beta}/d\iota)\,\iota\,\dot{\iota})^{\bullet}. \qquad (5.4)$$

In instances quoted below as examples of the constraint class (5.1) the microstructure is scalar; $\underset{\sim}{v}$ is a real number in an appropriate interval, $\underset{\sim}{\zeta}$ and $\underset{\sim}{\beta}$ are also scalar (and correspondingly the same letters but without tilde will be used) whereas $\mathbf{S}$ is a vector s, so that (5.4) takes the form

$$\hat{T} = \operatorname{sym}\overset{a}{T} - \iota(d\gamma/d\iota)(\rho\beta + \operatorname{div}\overset{a}{\mathbf{S}} - \overset{a}{\zeta})I + \rho(\mu(d\gamma/d\iota)^2\,\iota\dot{\iota})^{\bullet}\,I \qquad (5.5)$$

A continuum with finely distributed voids (in particular a liquid with bubbles) is often modelled as a mixture of an incompressible matrix of constant density ρ_m and a compressible dispersed phase. If there is no diffusion, then the void fraction

$\vartheta(0 \leq \vartheta < 1)$ can be interpreted as an internal parameter characterizing the state of the element. The compressibility of the element is due to the compressibility of the dispersed phase only; if the density of that phase can be disregarded with respect to ρ_m, then conservation of mass requires that

$$(1-v)\iota = 1-v_*,$$

where v_* is the value of v in the reference state.

Thus we have a case of a continuum with scalar latent microstructure, where (5.1) applies in the form

$$v = 1-(1-v_*)/\iota \,. \tag{5.6}$$

Reference to a formula of Rayleigh suggests, for the case when v is small, the following expression of the extra kinetic energy density

$$\kappa = \tfrac{1}{2}(\nu_*/48\,\pi^2)^{1/3}\, v^{-1/3}\, \dot{v}^2 \,,$$

where ν_* is the number density of the bubbles in the reference state. Appropriate expressions are also advanced for the active components of stress and microstress (see, for instance [4],[5]).

A dual case is represented by dilatant granular materials; i.e. suspensions of incompressible spherical heavy particles in a thin fluid, when, within the usual approximation, the constraint is expressed by

$$v\iota = v_* \tag{5.7}$$

and κ is given by

$$\tfrac{1}{2}\alpha(v_*/\nu_*)^{2/3}\, v^{-8/3}\, \dot{v}^2$$

(see, for instance, [6] for other details).

The last case is the perfect Korteweg fluid, for which the model of a continuum with latent scalar microstructure (and $v = \iota$) was proposed implicitly in [7] and explicitly in [2] to escape a thermodynamic paradox. The perfect Korteweg fluid is characterized by: (i) the existence of a potential of internal actions ϕ such that

$$\rho\dot{\phi} = \overset{a}{T}\cdot D + \overset{a}{\zeta}\dot{\iota} + \overset{a}{s}\cdot \operatorname{grad}\dot{\iota} \tag{5.8}$$

for all choices of $\mathcal{D}$ and $\dot{\iota}$, and (ii) the assumption that ϕ depend on the process through ι and $\operatorname{grad}\iota$ only.

The use of the identities

$$\dot{\iota} = \iota \operatorname{tr} D$$

and

$$(\operatorname{grad}\iota)^{\bullet} = \operatorname{grad}\dot{\iota} - (\operatorname{grad}\dot{x})^T \operatorname{grad}\iota$$

in (5.8), and the remark that (5.8) must then hold for any choice of D and $(\operatorname{grad}\iota)^{\bullet}$ leads to the formulae

$$\overset{a}{T} = (-\iota\overset{a}{\zeta} + \rho\iota(\partial\phi/\partial\iota)I - \operatorname{grad}\iota \otimes \overset{a}{s},$$

$$\overset{a}{s} = \rho(\partial\phi/\partial(\operatorname{grad}\iota)).$$

Finally, the nominal Cauchy stress $\hat{T}$ has the expression

$$\hat{T} = \Big(\rho(\mu\iota\dot{\iota})^{\bullet} - \rho\iota\beta + \rho(\partial\phi/\partial\iota) - \iota \operatorname{div}\Big(\rho(\partial\phi/\partial(\operatorname{grad}\iota))\Big)\Big)I$$
$$- \rho(\operatorname{grad}\iota) \otimes (\partial\phi/\partial(\operatorname{grad}\iota)).$$

ACKNOWLEDGEMENT

This work is part of a research programme under the auspices of the Italian Ministry of Education. The Author acknowledges the support of the Department of Mathematics of Heriot-Watt University for a visit during which this paper was written.

REFERENCES

1. R.A. Toupin, "Theory of elasticity with couple-stress", *Arch. Rat. Mech. An.* **17** (1964), pp.85-112.
2. G. Capriz, "Continua with latent microstructure", *Arch. Rat. Mech. An.* **90** (1985), pp.43-66.
3. G. Capriz, "Latent microstructure", (to appear in *Proc. of Symposium on Continuum Models of Discrete Structures*).
4. G. Capriz and H. Cohen, "The bubbly fluid as a continuum with microstructure", *Mech. Res. Comm.* **10** (1983), pp.359-367.
5. J.W. Nunziato and S. Cowin, "A nonlinear theory of elastic materials with voids", *Arch. Rat. Mech. An.* **72** (1979), pp.175-201.
6. M.A. Goodman and S.C. Cowin, "A continuum theory for granular materials", *Arch. Rat. Mech. An.* **44** (1972), pp.249-265.
7. J.E. Dunn and J. Serrin, "On the thermodynamics of interstitial working", *Arch. Rat. Mech. An.* **88** (1985), pp.267-292.

7

ELASTIC BEHAVIOUR OF VERY THIN CELLULAR STRUCTURES

DOINA CIORANESCU *and* JEANNINE SAINT JEAN PAULIN

Very thin cellular structures are encountered in many fields, for example, in civil engineering, aeronautics and electrotechnics. These two or three dimensional structures consist of identical cells periodically distributed in all directions (reticulated structures – see Fig. 1, 2), in two directions (networks – Fig. 4) or in one direction (tall structures – Fig. 3). The material is concentrated along layers (honeycomb structures Fig. 1) or along bars (reinforced structures Fig. 2, 3).

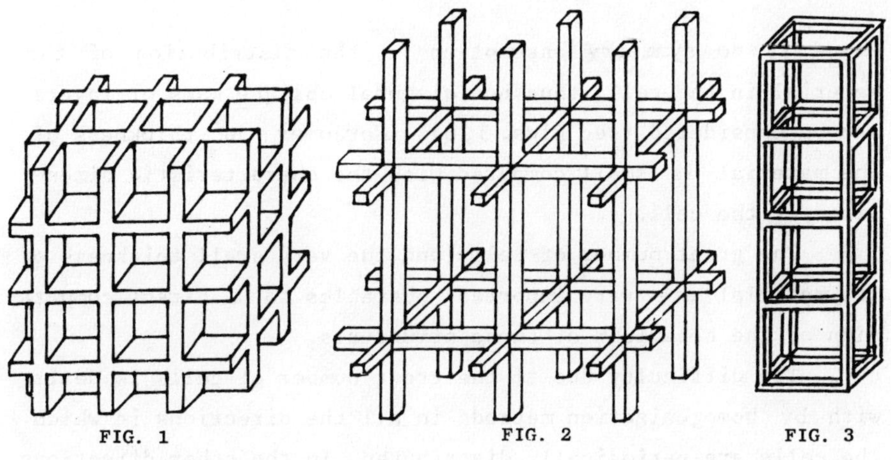

FIG. 1 FIG. 2 FIG. 3

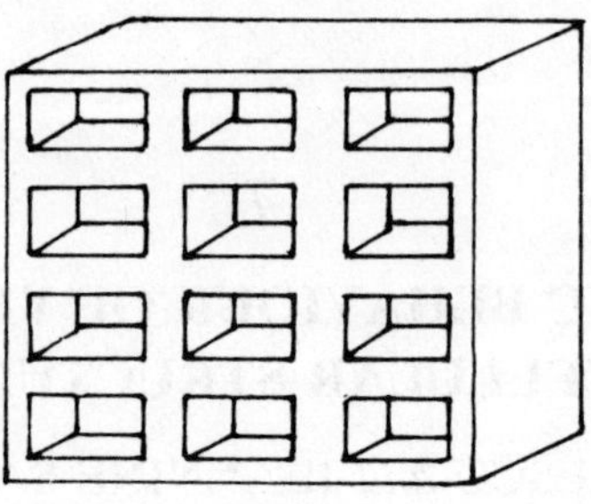

FIG. 4

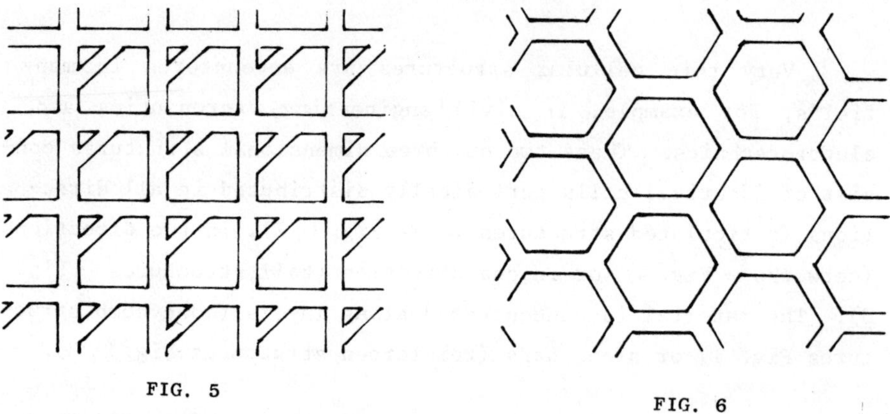

FIG. 5

FIG. 6

There is no symmetry assumption on the distribution of the material in the cell; thus non diagonal oblique bars or layers can be considered (see Figs. 5, 6). Moreover the thickness of the material is small compared with the characteristic dimensions of the cell.

The great number of cells and the very small thickness of the material are very important obstacles to a direct computation of the behaviour of these structures.

The difficulty due to the great number of cells is dealt with by homogenization methods in all the directions in which the cells are periodically distributed. In the other directions, if any, the width of the material is equal to the width of the cell and we use dilatations in these directions.

Whatever the kind of structure we study, this homogenization step gives an approximation of the real cellular material by an abstract homogeneous one. The effective or homogenized coefficients of this global homogeneous material are calculated by means of solutions of a partial differential system defined on a representative cell. Here arises the second difficulty — this system is defined on a domain whose thickness is very small (compared with the global dimensions of the representative cell). Thus a direct numerical computation of its solution is still very expensive.

We look for a more easily computable approximation of the homogenized coefficients. In this step the fact that the material is distributed along layers or bars is quite essential. We prove that we can work on each direction of layers or bars independently of the others and just add the results thus obtained. In this second step we give a mathematical method to obtain the global coefficients of the structures — the small thickness of the material being the essential parameter. We make explicit the effective global coefficients by means of the characteristic constants of the material.

We treat for these structures thermal or elasticity problems with Neumann and Dirichlet boundary conditions. We also treat eigenvalue problems. As an example we give here some results concerning elasticity problems.

We assume that the structure is contained in a domain Ω, is clamped on Γ, the exterior boundary of Ω, is subjected to applied body forces f and that there is no applied surface force on the boundary of the holes. Let us denote by $\Omega^*_{\varepsilon\eta}$ the part of Ω occupied by the material, ε being the period and $\varepsilon\eta$ the thickness of the material.

Each cell in the real material $\Omega^*_{\varepsilon\eta}$ is homothetic of ratio $\varepsilon : 1$ to a fixed representative cell Y^*_η.

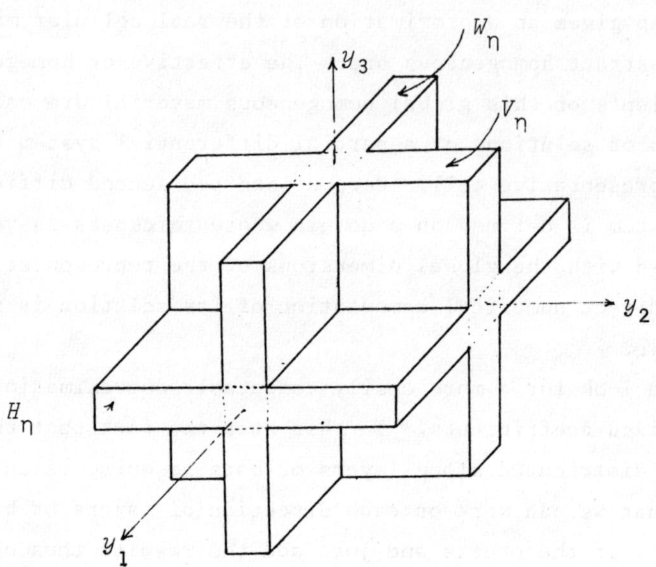

FIG. 7 Representative cell for a honeycomb structure

The displacement $u^{\varepsilon\eta}$ is given by solving the system

$$\left.\begin{aligned} &-\frac{\partial}{\partial x_j}\left(a_{ijpr}\frac{\partial u_p^{\varepsilon\eta}}{\partial x_r}\right) = f_i \quad \text{in} \quad \Omega^*_{\varepsilon\eta} \\ &u^{\varepsilon\eta} = 0 \quad \text{on} \quad \Gamma \\ &a_{ijpr}\frac{\partial u_p^{\varepsilon\eta}}{\partial x_r} n_j = 0 \quad \text{on the boundary of the holes.} \end{aligned}\right\} \qquad (1)$$

The elasticity constants a_{ijpr} satisfy the usual hypotheses of symmetry and coercivity.

In a first step we study the dependence of $u^{\varepsilon\eta}$ on the period ε. This is done by a homogenization method which consists in replacing problem (1) by two simpler problems: the first one is defined on the cell Y_η and the second concerns a theoretical "homogenized" material which would occupy the whole domain Ω.

We have

THEOREM 1. For $\varepsilon \to 0$, there exists an extension operator

$$P^{\varepsilon\eta} \in L\,(H^1(\Omega^*_{\varepsilon\eta}), H^1_0(\Omega))$$

such that $P^{\varepsilon\eta} u^{\varepsilon\eta} \to u^\eta$ in $\boldsymbol{H}^1_0$ weakly where u^η is the solution of the system

$$\left.\begin{aligned} -q^\eta_{ijlm} \frac{\partial^2 u^\eta_m}{\partial x_l \partial x_j} &= \frac{\text{meas } Y^*_\eta}{\text{meas } Y} f_i \quad \text{in } \Omega \\ u^\eta &= 0 \quad \text{on } \Gamma . \end{aligned}\right] \tag{2}$$

The homogenized coefficients q^η_{ijlm} are defined by

$$q^\eta_{ijlm} = \frac{1}{\text{meas } Y} \int_{Y^*_\eta} \left(a_{ijlm} - a_{ijpr} \frac{\partial \chi^{lm}_{\eta,p}}{\partial y_r} \right) dy \tag{3}$$

The correcting terms χ^{lm}_η are solutions of the system

$$\left.\begin{aligned} &-\frac{\partial}{\partial y_j}\left(a_{ijpr} \frac{\partial \chi^l_{\eta,p}}{\partial y_r} \right) = 0 \quad \text{in } Y^*_\eta \\ &\chi^{lm}_\eta \ \text{are } Y \text{ periodic} \\ &a_{ijpr} \frac{\partial \left(\chi^{lm}_{\eta,p} - \Pi^{lm}_p \right)}{\partial y_r} \eta_j = 0 \end{aligned}\right\} \tag{4}$$

where $\Pi^{lm}_p = y_l\, \delta_{mp}$. □

The proof of the theorem is an application of the energy method of Tartar [5] — for details see also [2].

We can also apply the multiple scale method to problem (1). We obtain:

$$u^{\varepsilon\eta} = u^\eta - \varepsilon \chi^{lm}\left(\frac{x}{\varepsilon}\right) \frac{\partial u^\eta_l}{\partial x_m} + \varepsilon^2 g^\eta$$

(see Bensoussan-Lions-Papanicolaou [1] and Sanchez-Palencia [4] for details).

In a second step we study the dependence of u^η on the thickness η. The result given by Theorem 1 is general in the sense that there is no geometric assumption on the shape of the holes. We shall now suppose that the material is distributed along bars or layers — this hypothesis enables us to give explicit formulas for the coefficients as $\eta \to 0$.

To simplify formulas we consider the case of a honeycomb structure with an isotropic material with a_{ijpr} given in terms of Lamé constants:

$$a_{ijpr} = \lambda \delta_{ij} \delta_{pr} + \mu(\delta_{ip} \delta_{jr} + \delta_{ir} \delta_{jp}) \tag{5}$$

We have:

THEOREM 2. For a honeycomb structure with the coefficients given by (5) the following convergence holds:

$$u^\eta \to u^* \quad \text{in } H^1_0(\Omega) \text{ weakly}$$

where u^η is the solution of the problem (4) and u^* is the solution of the system

$$\left.\begin{aligned} -q^*_{ijpr} \frac{\partial^2 u^*_p}{\partial x_j \, \partial x_r} &= 3 f_i \quad \text{in } \Omega \\ u^* &= 0 \quad \text{on } \Gamma . \end{aligned}\right\} \tag{6}$$

The coefficients q^*_{ijpr} satisfy the symmetries of elasticity and are defined by:

$$\left.\begin{aligned} q^*_{1111} &= q^*_{2222} = q^*_{3333} = \frac{8\mu(\lambda+\mu)}{\lambda+2\mu} \\ q^*_{1122} &= q^*_{1133} = q^*_{2233} = \frac{2\lambda\mu}{\lambda+2\mu} \\ q^*_{1212} & \quad q^*_{1313} = q^*_{2323} = \mu \\ q^*_{ijpr} &= 0 \quad \text{in the other cases.} \end{aligned}\right\} \tag{7}$$

Moreover:

$$\frac{1}{(\text{meas } \Omega^*_{\varepsilon\eta})^{1/2}} \|u^{\varepsilon\eta} - u^*\|_{H^1(\Omega^*_{\varepsilon\eta})} \leqslant C(\varepsilon^{1/2} + \eta^{1/2})$$

where C is a constant independent of ε and η. □

Sketch of the proof (for more details see [3]).

From system (4) one obtains easily the *a priori* estimate

$$|\nabla \chi^{lm}_{\eta,p}|_{[L^2(Y^*_\eta)]^3} \leqslant C\eta^{1/2} \tag{9}$$

which implies

$$\eta^{-1} q^{\eta}_{ijlm} \to q^{*}_{ijlm}.$$

We denote by H^η (resp. W^η, V^η) the layer of thickness η orthogonal to the Oy_3-axis (resp. Oy_2-axis, Oy_1-axis). Thus (see Fig. 7)

$$Y^*_\eta = H^\eta \cup W^\eta \cup V^\eta \setminus R^\eta$$

where meas $R^\eta = \eta^2(3-\eta)$ and

$$\eta^{-1} q^{\eta}_{ijlm} = (3-3\eta+\eta^2)\, a_{ijlm} - \\ -\eta^{-1}\left(\int_{H^\eta} + \int_{W^\eta} + \int_{V^\eta}\right)\left(a_{ijpr}\frac{\partial \chi^{lm}_{\eta,p}}{\partial y_r}\right) + \eta O(1) \tag{10}$$

Now we transform the domains H^η, W^η and V^η into the fixed domain $\bar{Y}$. For instance, the affine transformation with axis Oy_3 and ratio $1:\eta$ transforms H^η into $\bar{Y}$ and any function ϕ defined in $\mathbb{R}^3$ is changed in ϕ_H by

$$\phi_H\left(y_1, y_2, \frac{y_3}{\eta}\right) = \phi(y_1, y_2, y_3).$$

The *a priori* estimate (9) implies then that

$$\left.\begin{aligned} &\frac{\partial(\chi^{lm}_{\eta,p})_H}{\partial z_1} \rightharpoonup h^{lm}_{p1} \\ &\frac{\partial(\chi^{lm}_{\eta,p})_H}{\partial z_2} \rightharpoonup h^{lm}_{p2} \\ &\eta^{-1}\frac{\partial(\chi^{lm}_{\eta,p})_H}{\partial z_3} \rightharpoonup h^{lm}_{p3} \end{aligned}\right\} \text{ in } L^2(Y) \text{ weakly} \tag{11}$$

$$\left(z_1 = y_1,\ z_2 = y_2,\ z_3 = \frac{y_3}{\eta}\right).$$

The notations w_{pi}^{lm}, v_{pi}^{lm} $(i=1,2,3)$ have a similar meaning. Moreover, the periodicity of χ^{lm} in Y implies

$$\int_Y h_{p1}^{lm}\,dy = \int_Y h_{p2}^{lm}\,dy = \int_Y w_{p1}^{lm}\,dy = \int_Y w_{p3}^{lm}\,dy =$$

$$= \int_Y v_{p2}^{lm}\,dy = \int_Y v_{p3}^{lm}\,dy = 0 . \tag{12}$$

Using (11) and (12) in formula (10), by passing to the limit with $\eta\to 0$ one obtains:

$$q_{ijlm}^* = 3\,a_{ijlm} - a_{ijp3}\int_Y h_{p3}^{lm}\,dy - a_{ijp2}\int_Y w_{p2}^{lm}\,dy -$$

$$- a_{ijp1}\int_Y v_{p1}^{lm}\,dy . \tag{13}$$

To calculate

$$\int_Y h_{p3}^{lm}\,dy,\ \int_Y w_{p2}^{lm}\,dy \quad\text{and}\quad \int_Y v_{p1}^{lm}\,dy$$

we multiply the system (4) defining χ_η^{lm} by test functions depending only on one variable. For instance if we take $\varphi(y)=\phi(y_3)$, ϕ smooth and periodic, it follows that:

$$\eta^{-1}\int_{Y_\eta^*} a_{i3pr}\frac{\partial\chi_{\eta,p}^{l}}{\partial y_r}\frac{\partial\phi}{\partial y_3}\,dy = \eta^{-1}\int_{H^\eta} a_{i3lm}\frac{\partial\phi}{\partial y_3}\,dy =$$

$$= a_{i3lm}\int_{-\frac12}^{\frac12}\int_{-\frac12}^{\frac12} dy_1\,dy_2\left\{\frac{1}{\eta}\left[\int_{-\eta/2}^{\eta/2}\frac{\partial\phi}{\partial y_3}\,dy_3\right]\right\} .$$

As $\eta\to 0$, the right hand side converges to

$$a_{i3lm}\frac{\partial\phi}{\partial y_3}(0) .$$

To pass to the limit in the left hand side we use the affine transformations defined above and get

$$\lim_{\eta\to 0}\eta^{-1}\int_{Y_\eta^*} a_{i3pr}\frac{\partial\chi_{\eta,p}^{lm}}{\partial y_r}\frac{\partial\phi}{\partial y_3}\,dy =$$

$$= \int_Y \left[a_{i3p1}\, v_{p1}^{lm} + a_{i3p2}\, w_{p2}^{lm} + a_{i3p3}\left(v_{p3}^{lm} + w_{p3}^{lm}\right)\right] \frac{\partial\phi}{\partial y_3}\, dz +$$

$$+ \left[a_{i3p3} \int_Y h_{p3}^{lm}\, dy\right] \frac{\partial\phi}{\partial y_3}(0)\,.$$

We have finally

$$\int_Y \psi(z)\, \frac{\partial\phi}{\partial y_3}\, dz + \left[a_{i3p3} \int_Y h_{p3}^{lm}\, dy - a_{i3lm}\right] \frac{\partial\phi}{\partial y_3}\,(0) = 0\,,$$

where $\psi(z)$ contains the limits v_{pk}^{lm} and w_{pk}^{lm} $(k=1,2,3)$. This identity implies by making use of Lemma 2.1 [3] that:

$$a_{i3p3} \int_Y h_{p3}^{lm}\, dy = a_{i3lm}\,. \tag{14}$$

Similar arguments lead to:

$$a_{i2p2} \int_Y w_{p2}^{lm} = a_{i2lm}\,, \quad a_{i1p1} \int_Y v_{p1}^{lm} = a_{i1lm}. \tag{15}$$

The explicit formulas (7) are now obtained by using (14) and (15) in (13). A simple calculation shows that the matrix (q^*_{ijlm}) is coercive, so

$$u^\eta \rightharpoonup u^* \quad \text{in } H_0^1(\Omega) \text{ weakly}$$

and the limit u^* is a solution of system (6).

As for the error estimate, we refer the reader to [1], the technique being the same.

REMARK 3. Formula (13) and equations (14)-(15) hold for general elasticity coefficients; it is only for simplicity that we took Lamé constants. In the general case, the coefficients q^*_{ijlm} are much more complex.

REMARK 4. If we consider a reinforced structure (Fig. 1) where the material is distributed along beams we still obtain convergence results, namely

$$\eta^{-2}\, q^\eta_{ijlm} \rightharpoonup q^*_{ijlm}$$

with

$$q^*_{ijlm} = 3a_{ijlm} - a_{ijp1}\int_Y \left(h^{lm}_{p1} + v^{lm}_{p1}\right) dy -$$

$$-a_{ijp2}\int_Y \left(w^{lm}_{p2} + v^{lm}_{p2}\right) dy - a_{ijp3}\int_Y \left(w^{lm}_{p3} + h^{lm}_{p3}\right) dy\,. \quad (16)$$

We also obtain the systems

$$\begin{cases} a_{i2p2}\displaystyle\int_Y w^{lm}_{p2}\, dy + a_{i2p3}\int_Y w^{lm}_{p3}\, dy = a_{i2lm} \\ a_{i3p2}\displaystyle\int_Y w^{lm}_{p2}\, dy + a_{i3p3}\int_Y w^{lm}_{p3}\, dy = a_{i3lm} \end{cases}$$

$$\begin{cases} a_{i1p1}\displaystyle\int_Y h^{lm}_{p1}\, dy + a_{i1p3}\int_Y h^{lm}_{p3}\, dy = a_{i1lm} \\ a_{i3p1}\displaystyle\int_Y h^{lm}_{p1}\, dy + a_{i3p3}\int_Y h^{lm}_{p3}\, dy = a_{i3lm} \end{cases}$$

$$\begin{cases} a_{i1p1}\displaystyle\int_Y v^{lm}_{p1}\, dy + a_{i1p2}\int_Y v^{lm}_{p2}\, dy = a_{i1lm} \\ a_{i2p1}\displaystyle\int_Y v^{lm}_{p1}\, dy + a_{i2p2}\int_Y v^{lm}_{p2}\, dy = a_{i2lm}\,. \end{cases}$$

In the particular case of Lamé constants, (16) and the above systems lead to

$$\begin{cases} q^*_{1111} = q^*_{2222} = q^*_{3333} = \dfrac{\mu(3\lambda + 2\mu)}{\lambda + \mu} \\ q^*_{ijkh} = 0 \quad \text{for all the other cases}\,. \end{cases}$$

The matrix (q^*_{ijkh}) is not coercive anymore so we do not have results analogous to (6) and (8). (It is known by engineers that spatial reinforced structures are not stable — they add oblique bars to stabilize them.)

REFERENCES

1. A. Bensoussan, J.L. Lions and G. Papanicolaou, *Asymptotic Analysis for Periodic Structures*, North Holland, Amsterdam, 1978.
2. D. Cioranescu and J. Saint Jean Paulin, "Homogenization in open sets with holes", *J. Math. Anal. Appl.* **71** (1979), pp.590-607.
3. D. Cioranescu and J. Saint Jean Paulin, "Reinforced and honeycomb structures", *J. Math. Pures Appl.* **65** (1986), pp.403-422.
4. E. Sanchez-Palencia, *Nonhomogeneous media and vibrations*, Lecture Notes in Physics, No. 127, Springer-Verlag, Berlin, 1980.
5. L. Tartar, "Problèmes d'homogénéisation dans les équations aux dérivées partielles", *Cours Peccot*. Collège de France, 1977.

8

A COUNTEREXAMPLE IN THE VECTORIAL CALCULUS OF VARIATIONS

BERNARD DACOROGNA *and* PAOLO MARCELLINI

Let us consider the integral

$$I(u) = \int_{\Omega} f(\nabla u(x))\, dx , \tag{1}$$

where Ω is a bounded open set of $\mathbb{R}^n$, $u : \Omega \subset \mathbb{R}^n \to \mathbb{R}^m$, ∇u is the $n \times m$ matrix of the gradient of u, and $f : \mathbb{R}^{nm} \to \mathbb{R}$ is a continuous function.

When one studies the weak lower semicontinuity of I in a Sobolev space $W^{1,p}$, one is led to consider a necessary condition for f, called *rank one convexity*, and a sufficient condition known as *polyconvexity* (*cf.* below for the definitions). Both conditions reduce to the ordinary convexity of f if either $m = 1$ or $n = 1$. The two conditions are known not to be equivalent if $m \geqslant 3$ and $n \geqslant 3$ (see, Terpstra [4], Serre [3]). Recently Aubert [1] gave an example showing that these conditions are not equivalent if $m = n = 2$; the example is expressed in terms of isotropic functions, i.e.:

$$f(\nabla u) = g(\lambda, \mu), \tag{2}$$

where λ and μ are the eigenvalues of $(\nabla u^T \nabla u)^{\frac{1}{2}}$, the function g being defined (and finite) only if $\lambda, \mu > 0$.

We show here that the example remains valid even if one suppresses the condition of positivity of λ and μ. Moreover, we do not need to consider the representation of f in terms of

g like in (2); therefore we have a direct proof.

First let us define precisely the notions of rank one convexity and polyconvexity; we limit ourselves to the case $m=n=2$.

We denote $\mathbb{R}^4$ the set of 2×2 matrices ξ. We will use for the matrix ξ also the vectorial notation in terms of components:

$$\xi \equiv (\xi_1, \xi_2, \xi_3, \xi_4). \tag{3}$$

The scalar product and the norm are defined as usual:

$$(\xi,\lambda) = \sum_{i=1}^{4} \xi_i \lambda_i, \quad |\xi| = \left(\sum_{i=1}^{4} \xi_i^2\right)^{\frac{1}{2}}. \tag{4}$$

Finally, the determinant of the matrix ξ is

$$\det \xi = \xi_1 \xi_4 - \xi_2 \xi_3. \tag{5}$$

DEFINITION 1: <u>A function</u> $f:\mathbb{R}^4 \to \mathbb{R}$ <u>is said to be rank one convex if</u>

$$f(t\xi + (1-t)\lambda) \leqslant tf(\xi) + (1-t)f(\lambda) \tag{6}$$

<u>for every</u> $t\in[0,1]$ <u>and for every</u> $\xi,\lambda \in \mathbb{R}^4$ <u>with</u> rank $\{\xi-\lambda\} \leqslant 1$, <u>i.e. with</u> $\det(\xi-\lambda)=0$.

If $f\in C^2$ then (6) is equivalent to the well known *Legendre-Hadamard* (or *ellipticity*) *condition*:

$$\sum_{i,j,\alpha,\beta=1}^{2} f_{\xi^i_\alpha \xi^j_\beta}(\xi)\mu_\alpha \mu_\beta \eta^i \eta^j \geqslant 0 \tag{7}$$

for every $\xi = (\xi^i_\alpha) \in \mathbb{R}^4$ (here we use the matricial notation) and for every $\mu\ \eta \in \mathbb{R}^2$.

Since a matrix $\lambda \in \mathbb{R}^4$ can be represented in the form $(\mu_1\eta_1, \mu_1\eta_2, \mu_2\eta_1, \mu_2\eta_2)$ if and only if $\det\lambda = 0$, the Legendre-Hadamard condition (7) is equivalent to:

$$\sum_{i,j=1}^{4} f_{\xi_i \xi_j}(\xi)\lambda_i \lambda_j \geqslant 0 \tag{8}$$

for every $\xi \in \mathbb{R}^4$ and for every $\lambda \in \mathbb{R}^4$ with $\det\lambda = 0$.

Ball in [2] proposed the following:

DEFINITION 2: A function $f: \mathbb{R}^4 \to \mathbb{R}$ is said to be polyconvex if there exists a convex function $g: \mathbb{R}^5 \to \mathbb{R}$ so that, for every $\xi \in \mathbb{R}^4$:

$$f(\xi) = g(\xi, \det \xi) . \tag{9}$$

We note that in general we have:

$$f \text{ polyconvex} \Longrightarrow f \text{ rank one convex} . \tag{10}$$

THEOREM: Let f be defined, for $\xi \in \mathbb{R}^4$, by

$$f(\xi) = |\xi|^4 - \frac{4}{\sqrt{3}} |\xi|^2 \det \xi . \tag{11}$$

Then f is rank one convex but not polyconvex.

REMARKS

(i) The function f defined in (11) gives essentially Aubert's example if we restrict to diagonal matrices with positive eigenvalues.

(ii) It is not known if the integral $I(u)$ defined in (1), with f as in (11), is weakly lower semicontinuous in $W^{1,p}$ for some $p \geqslant 1$.

PROOF: — Part 1: The fact that f is not polyconvex can be easily seen. To this end, let us assume, for contradiction, that f is polyconvex and that the representation formula in (9) holds. The function g, like every proper convex function, must be bounded from below by an affine function; that is

$$g(\xi_1, \xi_2, \xi_3, \xi_4, \delta) \geqslant a_0 + \sum_{i=1}^{4} a_i \xi_i + a_5 \delta \tag{12}$$

for some $a_i \in \mathbb{R}$ $(i=0, \cdots, 5)$ and for every $\xi \in \mathbb{R}^4$, $\delta \in \mathbb{R}$.

Since $\delta = \det \xi$ is quadratic with respect to ξ, by (12) and (9) there exists a constant c such that

$$\frac{f(\xi)}{1+|\xi|^2} \geqslant c , \qquad \forall\, \xi \in \mathbb{R}^4 . \tag{13}$$

This is absurd; in fact, for $\xi \equiv (t,0,0,t)$, with t a real parameter, we have $|\xi|^2 = 2t^2$, $\det \xi = t^2$, and thus

$$\frac{f(\xi)}{1+|\xi|^2} = \frac{4t^4 - (8/\sqrt{3})t^4}{1+2t^2} = \frac{4(\sqrt{3}-2)}{\sqrt{3}} \frac{t^4}{1+2t^2}, \tag{14}$$

which goes to $-\infty$ as $t \to \infty$.

Part 2: To prove that f is rank one convex is more involved and we decompose the proof into three steps. We will show that f satisfies the Legendre-Hadamard condition as in (8).

<u>Step 1</u>: With the aim of computing the quadratic form in (8) we introduce some notation. To every $\xi \in \mathbb{R}^4$ we associate $\hat{\xi} \in \mathbb{R}^4$ defined by:

$$\text{if } \xi \equiv (\xi_1, \xi_2, \xi_3, \xi_4) \quad \text{then } \hat{\xi} \equiv (\xi_4, -\xi_3, -\xi_2, \xi_1). \tag{15}$$

The components of $\hat{\xi}$ will be denoted by $\hat{\xi}_i$, $i = 1, \cdots, 4$. We observe that:

$$\begin{cases} |\xi| = |\hat{\xi}| \\ \det \xi = \det \hat{\xi} \\ (\xi, \hat{\xi}) = 2 \det \xi \\ (\xi, \hat{\lambda}) = (\hat{\xi}, \lambda). \end{cases} \tag{16}$$

With this notation we can easily express the gradient of the determinant of ξ:

$$(\det \xi)_{\xi_i} = \hat{\xi}_i \text{ for } i = 1, \cdots, 4; \quad \text{i.e. } \nabla(\det \xi) = \hat{\xi}. \tag{17}$$

For the second derivatives of the determinant we use the notation

$$\frac{\partial^2}{\partial \xi_i \partial \xi_j}(\det \xi) = \frac{\partial}{\partial \xi_j}(\hat{\xi}_i) = \hat{\delta}_{ij}. \tag{18}$$

Since the quadratic form associated to the matrix of the second derivatives is equal to twice the original quadratic form, we have

$$\sum_{i,j=1}^{4} \hat{\delta}_{ij} \lambda_i \lambda_j = \sum_{i,j=1}^{4} (\det \xi)_{\xi_i \xi_j} \lambda_i \lambda_j = 2 \det \lambda. \tag{19}$$

Finally, as usual, we denote $\delta_{ij}=0$ if $i\neq j$ and $\delta_{ii}=1$.

Step 2: We compute the quadratic form in (8). Let us begin with the derivatives of f:

$$f_{\xi_i} = \frac{\partial}{\partial\xi_i}\left(|\xi|^4-\frac{4}{\sqrt{3}}|\xi|^2\det\xi\right)$$

$$= 4|\xi|^2\xi_i - \frac{8}{\sqrt{3}}\xi_i\det\xi - \frac{4}{\sqrt{3}}|\xi|^2\hat{\xi}_i . \tag{20}$$

$$f_{\xi_i\xi_j} = 8\,\xi_i\xi_j + 4|\xi|^2\delta_{ij} - \frac{8}{\sqrt{3}}(\delta_{ij}\det\xi+\xi_i\hat{\xi}_j) - \frac{4}{\sqrt{3}}(2\,\hat{\xi}_i\xi_j+|\xi|^2\hat{\delta}_{ij}) .$$

Then the quadratic form in (8) is equal to

$$\sum_{i,j=1}^{4} f_{\xi_i\xi_j}\lambda_i\lambda_j = 8(\xi,\lambda)^2+4|\xi|^2|\lambda|^2-\frac{8}{\sqrt{3}}\det\xi\,|\lambda|^2 - \frac{16}{\sqrt{3}}(\xi,\lambda)(\hat{\xi},\lambda)-\frac{8}{\sqrt{3}}|\xi|^2\det\lambda . \tag{21}$$

In the case of interest ($\det\lambda=0$) we have

$$\psi(\xi,\lambda) \overset{\text{def}}{=} \frac{1}{4}\sum_{i,j=1}^{4} f_{\xi_i\xi_j}\lambda_i\lambda_j = |\xi|^2|\lambda|^2+2(\xi,\lambda)^2-\frac{2}{\sqrt{3}}\det\xi|\lambda|^2-\frac{4}{\sqrt{3}}(\xi,\lambda)(\hat{\xi},\lambda). \tag{22}$$

Step 3: The rank one convexity of $f(\xi)$ is reduced to showing the nonnegativity of $\psi(\xi,\lambda)$ for every ξ,λ, with $\det\lambda=0$.

Observe first that $\psi(\xi,\lambda)$ is homogeneous of degree 2 in ξ (and in λ). Therefore, in order to show that ψ is nonnegative, it is sufficient to prove that

$$\min_{\xi}\left\{\psi(\xi,\lambda):\ |\xi|=1\right\} \geqslant 0, \qquad \forall\ \lambda:\ \det\lambda=0 . \tag{23}$$

It is clear that the minimum of $\psi(\cdot,\lambda)$ exists on the manifold $|\xi|=1$. Let α be a Lagrange multiplier; we will prove that $\psi(\xi_0,\lambda)\geqslant 0$ for every critical point ξ_0 of the function of ξ:

$$\psi(\xi,\lambda)-\alpha(|\xi|^2-1) . \tag{24}$$

The gradient (with respect to ξ) of the above function is equal to zero when:

$$|\lambda|^2 \xi_0 + 2(\xi_0,\lambda)\lambda - \frac{1}{\sqrt{3}}|\lambda|^2 \hat{\xi}_0 - \frac{2}{\sqrt{3}}[(\hat{\xi}_0,\lambda)\lambda + (\xi_0,\lambda)\hat{\lambda}] = \alpha\xi_0 . \tag{25}$$

Upon multiplication by ξ_0, bearing in mind that $|\xi_0|^2 = 1$ and that $(\hat{\xi}_0,\xi_0) = 2\det\xi_0$, we obtain

$$\psi(\xi_0,\lambda) = \alpha . \tag{26}$$

Upon multiplication of (25) first by λ, then by $\hat{\lambda}$, bearing in mind that $(\hat{\lambda},\lambda) = 2\det\lambda = 0$ and that $(\hat{\xi},\hat{\lambda}) = (\xi,\lambda)$, we have

$$\begin{cases} (3|\lambda|^2 - \alpha)(\xi_0,\lambda) - \sqrt{3}|\lambda|^2(\hat{\xi}_0,\lambda) = 0 \\ -\sqrt{3}|\lambda|^2(\xi_0,\lambda) + (|\lambda|^2 - \alpha)(\hat{\xi}_0,\lambda) = 0 , \end{cases} \tag{27}$$

which leads to either $(\xi_0\lambda) = (\hat{\xi}_0,\lambda) = 0$, or

$$(3|\lambda|^2 - \alpha)(|\lambda|^2 - \alpha) - 3|\lambda|^4 = \alpha^2 - 4|\lambda|^2\alpha = 0 . \tag{28}$$

In this second case we would have either $\alpha = 0$ or $\alpha = 4|\lambda|^2$, and thus $\psi(\xi_0,\lambda) = \alpha \geqslant 0$. In the first case $(\xi_0,\lambda) = (\hat{\xi}_0,\lambda) = 0$ we would have

$$\begin{aligned} \psi(\xi_0,\lambda) &= |\lambda|^2\left(|\xi_0|^2 - \frac{2}{\sqrt{3}}\det\xi_0\right) \\ &\geqslant |\lambda|^2|\xi_0|^2\left(1 - \frac{1}{\sqrt{3}}\right) \geqslant 0 , \end{aligned} \tag{29}$$

since $2|\det\xi| \leqslant |\xi|^2$ for every $\xi \in \mathbb{R}^4$. This completes the proof.

REFERENCES

1. G. Aubert, *Contributions aux Problèmes du Calcul des Variations et Application à l'Élasticité Nonlinéaire*, Thèse de Doctorat, Paris VI, 1986.
2. J.M. Ball, "Convexity conditions and existence theorems in nonlinear elasticity", *Arch. Rat. Mech. Anal.* **67**, pp.337-403, (1977).
3. D. Serre, "Formes quadratiques et calcul des variations", *J. Math. Pures et Appl.* **62**, pp.177-196, (1983).
4. F.J. Terpstra, "Die Darstellung biquadratischer Formen als Summen von Quadraten mit Anwendung auf die Variations - *rechnung*, *Math. Ann.* **116**, pp.116-180,(1938).

9

ELASTIC INVARIANTS IN CRYSTAL THEORY

CESARE DAVINI

1. INTRODUCTION

Continuous theories of crystals rely upon the assumption that for most applications it is sufficient to describe the average state of the material over regions large enough for the discrete nature of the crystal to be ignorable. Accordingly, we picture crystals as continuous bodies and we use elasticity theory to describe them. With defects also we sometimes take the continuum viewpoint. This presumes that we consider averages over regions that contain a great number of defects. In some materials, e.g. crystalline solids, defects occur at a much finer length scale than that at which their effects become of practical relevance, so that the continuum approach can be a reasonable way to tackle the problem. The passage from the discrete to the continuous level of description is a crucial one in theories to be formulated directly. In this paper I deal with aspects of a continuum theory of defects in crystalline solids focused on this point.

The difference in length scales, and the need to envisage evolution of defects in the lattice for such a theory to be of some interest, in fact render the continuous model conceptually independent of the molecular one and leave some freedom in making the choices that lead to its full definition. As I see it, the main problems arise in connection with the constitutive aspects and with the kinematical meaning of the notion of defect. My purpose is to address these points. After revisiting the molecular theories of elasticity, I start from a suitable notion of crystallinity and adapt ideas from the topological theory of defects [1] in order to provide some rationale for the definition of the notion of defect. This approach to defects seems to me better grounded than others that have exploited ideas from the pioneering works of Kondo and Bilby on dislocations, see, e.g. [2].

Basically, the ideas expounded in the following have been discussed in [3]. In this paper I reconsider them in a more general way. In particular, in [3] the analysis was based on a constitutive choice that to some extent anticipated prejudices about the defects to be described. The present approach shows that we need not anticipate views on defects and that the choice of the constitutive variables can be founded on a more systematic basis.

2. CAUCHY'S MOLECULAR THEORY REVISITED

In order to motivate views discussed later it is useful to review concepts from the molecular theories of elasticity. Let us take Cauchy's approach that has the advantage of presenting those concepts within the simplest context.

Cauchy considered monatomic crystals with a simple Bravais lattice. Ignoring boundaries and thermal motion, he envisaged the case of an unbounded assemblage of identical atoms at rest, interacting with each other by means of central forces depending upon their mutual distances. For this model Cauchy proposed his molecular interpretation of the stress of continuum mechanics — having chosen an ideal surface through the lattice, the stress acting on it is regarded as the total, per unit area, of forces exchanged between atoms sited on the two sides of the surface. The stress is then calculated by adding those forces up, assumed that they die off rapidly enough with distance for the sum to be finite.

Details of Cauchy's analysis are discussed for instance in

Love [4, note B]. Denoting by $\phi(r)$ the mutual attraction between atoms at distance r, the analysis yields that the Cartesian components of the stress tensor $\boldsymbol{T}$ with respect to some chosen reference frame are given by

$$T_{ij} = \frac{n}{2} \sum \lambda_i \lambda_j r\phi(r) \tag{1}$$

with the usual convention that T_{ij} denote components of the stress on the jth coordinate plane. In (1) the summation $\sum$ extends over all pairs of atoms whose joining lines cross that plane within a portion of surface of unit area and large with respect to the atomic spacing; the λ_i are the direction cosines of the joining lines and n is the number of atoms per unit volume in the lattice. From the theory of simple lattices, having fixed one lattice point as the origin, the position vector of any other point in the lattice is given by

$$\boldsymbol{r} = n^a \boldsymbol{d}_a , \tag{2}$$

with $n^a \in \mathbb{Z}$ integers and $\boldsymbol{d}_a$ $a = 1,2,3$ given linearly independent vectors called lattice vectors. It follows that the summation in (1) extends to terms for which the values of r and λ_i are given by

$$r = (n^a n^b \boldsymbol{d}_a \cdot \boldsymbol{d}_b)^{\frac{1}{2}} . \tag{3}$$

with the dot denoting the inner product and

$$\lambda_i = n^a d_{ai}/r . \tag{4}$$

If we introduce the pair potential $\psi = \psi(r)$, $\phi = d\psi/dr$, the expression (2) becomes

$$T_{ij} = \frac{n}{2} \sum_{n^a, n^b \in \mathbb{Z}} \frac{1}{r} (d\psi/dr)\, n^a n^b d_{ai} d_{bj} . \tag{5}$$

Hence, by putting

$$\psi = \hat{\psi}(r^2) . \tag{6}$$

one obtains from (5)

$$T_{ij} = n \sum (d\hat{\psi}/dr^2)\, n^a n^b d_{ai} d_{bj} = n\,(\partial\Psi/\partial d_{bi})\, d_{bj}\,, \tag{7}$$

with

$$\Psi \equiv \tfrac{1}{2} \sum \hat{\psi}\,(n^a n^b \boldsymbol{d}_a \cdot \boldsymbol{d}_b).$$

It follows that $\boldsymbol{T}$ depends upon the lattice vectors only, as is natural for an infinite lattice.

Notice that the stress is a macroscopic notion because it represents an average of interatomic forces over a large portion of a surface. The whole approach aims at a macroscopic theory and, in order to obtain it, Cauchy tacitly introduced two assumptions. First, he identified the macroscopic density with quantity

$$\rho \equiv mn\,, \tag{8}$$

with m the atomic mass. Second, introducing a reference lattice with lattice vectors $\boldsymbol{d}_a^0$, he assumed that the present lattice vectors were given by

$$\boldsymbol{d}_a = \boldsymbol{F}\boldsymbol{d}_a^0\,, \tag{9}$$

with $\boldsymbol{F}$ regarded as the gradient of a macroscopic (homogeneous) deformation taking the reference lattice into the present one.

Cauchy used (8) and (9) in the framework of the linear theory but the restriction is irrelevant. When (8) and (9) are plugged into (7), (7) becomes the general constitutive equation for the stress tensor in classical elasticity provided we interpret the quantity Ψ/m as the Helmholtz free energy per unit mass of the crystal. Then, via suitable identification, we recover the formal scheme of continuum theory, so getting from a different perspective motivation for several theoretical issues.

Besides providing support to theory, the molecular approach proves helpful in accounting for a class of gross deformations, usually small in most materials, that we call macroscopically elastic. On the other hand, it fails to provide good estimates for a number of material properties associated with deformations that go beyond elasticity. It is known, for instance, that classical molecular calculations of the yield limit give values that are $10^3 - 10^4$ times higher that those measured in the experiments; cf. Frenkel [5]. This discrepancy is now attributed to the behaviour of lattice defects in the crystal. Interpreting somewhat extensively ideas of Taylor on dislocations [6], let us call *Taylor's conjecture* the assumption that the changes of defects produce macroscopically anelastic deformations of materials.

Allowing for defects leads to a revision of the kinematical basis of Cauchy's theory and to a questioning of assumptions (8) and (9) that draw their motivation from the notion of an underlying perfect lattice. It seems then sensible to concentrate on them in elaborating a continuous model for defective crystals. I concentrate on this point in the following.

REMARK 1. Since $\boldsymbol{d}_1 \cdot \boldsymbol{d}_2 \times \boldsymbol{d}_3$ is the cell volume, the number of atoms per unit volume is given by

$$n = (\boldsymbol{d}_1 \cdot \boldsymbol{d}_2 \times \boldsymbol{d}_3)^{-1}. \tag{10}$$

When (10) is used in (8), assumption (9) implies the familiar rule for the conservation of mass

$$\rho \det \boldsymbol{F} = \rho_R, \tag{11}$$

with $\rho_R \equiv m/(\boldsymbol{d}_1^0 \cdot \boldsymbol{d}_2^0 \times \boldsymbol{d}_3^0)$ the density in the reference configuration. Conversely, (11) and (9) together imply (8). It is perhaps because the validity of (11) is never questioned under the ordinary conditions of continuum mechanics that the name of *Cauchy's hypothesis* is usually reserved for assumption (9), although the two appear conceptually independent and equally important when the kinematical scheme of a perfect lattice is abandoned.

3. THE CONTINUOUS CRYSTALS

Hereafter I discuss a continuous model for crystals with defects. As in Cauchy's theory, the approach is addressed to monatomic crystals with a simple lattice and consideration is restricted to defects that do not generate singularities in the fields of physical quantities describing the model. In fact the reference to defects is vague because, as is made clear later, there is as yet no natural notion of defect. By this statement I rather anticipate that the model is not intended to cover physical situations where singularities are likely to occur, as in the case of grain boundaries or mechanical twinning.

Crystals are made of indistinguishable constituents. More strongly than for other physical systems, probably, this suggests that the most natural way to represent a crystal is to avoid labelling its constituent parts and to picture it as it is seen at the present time by suitable means of observation. Following [3], we assume that the state of a crystalline body is given by the assignment of the list $\Sigma = \{\mathcal{B}, \boldsymbol{d}_a, \rho\}$ with: $\mathcal{B}$ an open region in a three-dimensional Euclidean space E; $\boldsymbol{d}_a$, $a = 1,2,3$, three smooth vector fields over $\mathcal{B}$ obeying the condition

$$\boldsymbol{d}_1 \cdot \boldsymbol{d}_2 \times \boldsymbol{d}_3 > 0 \quad \text{for all} \quad \boldsymbol{x} \in \mathcal{B}\,; \tag{12}$$

and ρ a smooth scalar field over $\mathcal{B}$ with positive values. $\mathcal{B}$ is the macroscopic placement of the body in physical space E, ρ (*mass density*) describes the gross distribution of mass throughout $\mathcal{B}$; and the vectors $\boldsymbol{d}_a$ (*lattice vectors*) are quantities carrying information on the crystalline structure at points $\boldsymbol{x}$ in $\mathcal{B}$. According to the theory of lattices, any triad of lattice vectors $\boldsymbol{d}'_a$ given by $\boldsymbol{d}'_a = n_a^b \boldsymbol{d}_b$, n_a^b an integer matrix with $\det n_a^b = \pm 1$, defines the same lattice. Then, the smoothness of the $\boldsymbol{d}_a$ implies that one and the same family of lattice vectors is selected throughout $\mathcal{B}$. We assume that we can recognize that family and refer to it when different states of the body are compared in the following.

Although we use the term lattice vectors its meaning is not the same as in Cauchy's approach because here lattice vectors are intended to represent averages of some kind rather than geometrical points in a lattice. The idea underlies any continuum approach to crystals. A clearcut discussion of this aspect in the context of crystal plasticity is given by Havner [7],[8]. He points out how continuum theories reflect observations at two distinct levels of accuracy (called *macroscopic* and *microscopic*, respectively) and notices that the microscopic level can never be too accurate if one wishes to avoid that a highly inhomogeneous deformation of the lattice shows up: a point that "(...) has been discussed by Hill (1956) and recently by Lin (1973), but [that] has been often overlooked by other theorists."; cf. [7, p.386].

The macroscopic level corresponds to the typical viewpoint of continuum mechanics. At this level the region occupied by the body appears continuous and its points x are recognizable during continuous changes of the state. This allows us to use for them the conventional term of "*material points*" and define familiar notion such as deformation, velocity, material domains, etc. The quantities ρ and $\mathbf{d}_a$ are assumed to be functions of x, implying that we regard the macroscopic scale as the characteristic scale of the continuum description. Infinitesimal elements of lines or volumes and, in general, differential calculus reflect notions that refer to the macroscopic length scale. In particular, ρ represents the mass per unit (macroscopic) volume. The microscopic level, on the other hand, must reflect observations by an apparatus that can distinguish features of the lattice but not single atoms or defects. An X-ray microscopy that can resolve distances of 10^{-4} cm could be appropriate. Lattice vectors are averaged over a spot of these dimensions and account in a rough way, only, for the behaviour of the matter at a submicroscopic level. In representing the d_a's as functions of $\boldsymbol{x}$ one tacitly assumes that they do not vary too wildly on the macroscopic length scale, which seems a reasonable restriction on the situations a continuum theory can account for.

The previous considerations support the view that material points and lattice vectors may behave independently of each other and that the fields ρ and d_a can account for the occurrence of defects only in a way that is not easily predictable from the outset. They also indicate that lattice vectors are rather conventional notions in the model so that we can hardly derive constitutive equations or the very notion of defect from molecular reasonings.

In [3] I have proposed to call a body crystalline if: (i) a lattice is distinguishable; (ii) the constitutive functions are the same at all material points $\boldsymbol{x}$; (iii) for each $\boldsymbol{x}$, they depend upon the microscopic arrangement of the atoms at $\boldsymbol{x}$ rather than upon the macroscopic behaviour of the matter. In a sense, the last requirement expresses a view recommended by (7). We can interpret it as the assumption that the constitutive functions at $\boldsymbol{x}$ depend upon the fields $\boldsymbol{d}_a$ and ρ around $\boldsymbol{x}$, reference to these fields being motivated by the view that, in some way, they are determined by the microscopic arrangement of atoms and defects.

A local theory can be obtained systematically by expanding these fields at $\boldsymbol{x}$ and by assuming that the response functions depend upon their values at $\boldsymbol{x}$ and that of their partial gradients up to a certain order. Here we consider a model described by the gradients up to the first order. Introducing, for later convenience, the dual lattice vectors $\boldsymbol{d}^a$ defined by the conditions

$$\boldsymbol{d}^a \cdot \boldsymbol{d}_b = \delta^a_b \qquad a, b = 1, 2, 3 , \tag{13}$$

we assume that the local state σ is described by the list

$$\sigma = \{\boldsymbol{d}^a , \partial \boldsymbol{d}^a / \partial \boldsymbol{x} , \rho , \partial \rho / \partial \boldsymbol{x}\} . \tag{14}$$

Material properties in crystals are determined by the interactions between atoms. Since these are believed to decay rapidly with distance, including gradients up to the first order could be a reasonable compromise between a theory that involves the values of $\boldsymbol{d}^a$ and/or ρ only, like in Cauchy's approach, and others that could encompass more extended nonlocality effects.

With the above notion of crystallinity, in particular, we need not mention the macroscopic deformation in the constitutive theory nor the notion of a reference configuration. This viewpoint is familiar to people working in the micromechanics of plasticity, but it is not the common way macroscopic plasticity is treated.

Another consequence of the consideration developed so far concerns deformation and its connections with the notion of defect.

A general deformation from the present state Σ is described by the assignment of a new placement $\mathcal{B}^+$ and new fields $\boldsymbol{d}^+_a = \boldsymbol{d}^+_a(\boldsymbol{x}^+)$ and $\rho = \rho^+(\boldsymbol{x}^+)$ on $\mathcal{B}$, together with the macroscopic correspondence

$$\boldsymbol{x}^+ = \boldsymbol{x}^+(\boldsymbol{x}) \tag{15}$$

between points of $\mathcal{B}$ and $\mathcal{B}^+$. We require that the field $\boldsymbol{x}^+(\boldsymbol{x})$ is smooth and invertible and that the usual rule for the conservation of mass holds

$$\rho^+ \det \boldsymbol{F} = \rho \ , \tag{16}$$

with $\boldsymbol{F}$ the macroscopic deformation gradient and quantities on the two sides of (16) evaluated at the same material point.

Since the macroscopic and microscopic behaviour come from averaging at different scales, $\boldsymbol{x}^+(\boldsymbol{x})$ and $\boldsymbol{d}^+_a(\boldsymbol{x}^+)$ are not expected to be simply related. It follows from (16), on the contrary, that the same view does not apply to the mass density that is always determined by the assignment of $\boldsymbol{x}^+(\boldsymbol{x})$. The nature of the model, with a submicroscopic structure that does not reflect directly upon the observable fields, would probably justify us to postulate independence of the mass density too on the macroscopic deformation. However, I am not aware of cases where the validity of (16) has been questioned, which might indicate that circumstances requiring such greater generality do not occur under the most common conditions a continuum theory is addressed to. By this reason (16) is assumed to hold for all deformations of the body.

The model is asked to encompass deformations corresponding to both macroscopically elastic and macroscopically anelastic behaviour. Accordingly, we expect that some of the deformations above keep the defects frozen and others change them, although it is not clear which are which. Observing that the molecular theory proves helpful in describing the macroscopically elastic deformations, Cauchy's hypotheses must hold for them. On the other hand, in macroscopically anelastic deformations something must go wrong with Cauchy's approach and observations say that hypothesis (9) is violated in many cases. This leads us to define as *elastic* a deformation from Σ such that

$$\boldsymbol{d}_a^+ = \boldsymbol{F}\boldsymbol{d}_a\,, \text{ with } \boldsymbol{F} = \partial\boldsymbol{x}^+/\partial\boldsymbol{x}\,. \tag{17}$$

Then, from Taylor's conjecture, we are also led to assume that defects do not change under the elastic deformations.

In the next section, we study quantities involving the fields $\boldsymbol{d}_a$ and ρ and invariant under the elastic deformations. This yields necessary conditions for two states of the body to be elastically connectable and provides a characterization that naturally leads to a definition of the notion of defect. In a sense, this is the course taken in the topological theory of defects [1]. I borrow the idea, as I grasp it, but without getting into technicalities of homotopy theory. I develop an analysis where the elastic deformations are the topological transformations and where attention is confined to quantities that depend upon the constitutive variables only.

4. ELASTIC INVARIANTS

Let us turn our attention to the integral quantities

$$\oint_c \boldsymbol{g}(\sigma)\cdot d\boldsymbol{x}\,, \quad \int_S \boldsymbol{g}(\sigma)\cdot\boldsymbol{n}\,ds \quad \text{and} \quad \int_v g(\sigma)\,dv \tag{18}$$

defined over the circuits c, the closed surfaces S and the volumes v in $\mathcal{B}$. We look for the most general functions g and

g that make the integrals invariant under elastic deformations when they are evaluated over material domains in $\mathcal{B}$, i.e. such that

$$\oint_{c} \boldsymbol{g}(\sigma)\cdot d\boldsymbol{x} = \oint_{c^+} \boldsymbol{g}(\sigma^+)\cdot d\boldsymbol{x}^+ ,$$

$$\int_{S} \boldsymbol{g}(\sigma)\cdot \boldsymbol{n}\, ds = \int_{S^+} \boldsymbol{g}(\sigma^+)\cdot \boldsymbol{n}^+ ds^+ , \tag{19}$$

$$\int_{v} g(\sigma)\, dv = \int_{v^+} g(\sigma)\, dv^+$$

for all c, S and v and for all elastic deformations driven by some $\boldsymbol{x}^+ = \boldsymbol{x}^+(\boldsymbol{x})$ defined over $\mathcal{B}$. Following a common view in the topological theory, these quantities (*elastic invariants*) are believed to reflect the strength of the defects (of some kind) contained in or embraced by the respective material domains. They are also expected to lead either directly, c.f. $(8)_3$, or via the Stokes and divergence theorems, cf. $(18)_1$ and $(18)_2$, to suitable densities providing local measures of the defectiveness at the points of $\mathcal{B}$.

In characterizing the invariants let us restrict attention to sufficiently smooth functions and notice that the integrands in $(18)_1$ and $(18)_2$ may not depend upon the gradients $\partial \boldsymbol{d}^a/\partial \boldsymbol{x}$ and $\partial \rho/\partial \boldsymbol{x}$ if the associated densities are asked not to contain higher order derivatives than is allowed for in the constitutive list (14). Then, use of mappings $\boldsymbol{x}^+ = \boldsymbol{x}^+(\boldsymbol{x})$ that reduce to the identity map outside an arbitrarily small neighbourhood of any given $\boldsymbol{x}$, implies that the invariance conditions (19) are equivalent to

$$\boldsymbol{g}(\boldsymbol{d}^a, \rho)\cdot d\boldsymbol{x} = \boldsymbol{g}(\boldsymbol{d}^{a+}, \rho^+)\cdot d\boldsymbol{x}^+ ,$$

$$\boldsymbol{g}(\boldsymbol{d}^a, \rho)\cdot \boldsymbol{n}\, ds = \boldsymbol{g}(\boldsymbol{d}^{a+}, \rho^+)\cdot \boldsymbol{n}^+ ds^+ \tag{20}$$

$$g(\boldsymbol{d}^a, \partial\boldsymbol{d}^a/\partial\boldsymbol{x}, \rho, \partial\rho/\partial\boldsymbol{x})\, dv = g(\boldsymbol{d}^{a+}, \partial\boldsymbol{d}^{a+}/\partial\boldsymbol{x}^+, \rho^+, \partial\rho^+/\partial\boldsymbol{x}^+)\, dv^+ ,$$

where the equalities must hold for all material elements $d\boldsymbol{x}$, $\boldsymbol{n}\, ds$ and dv and all the elastically related values of the arguments.

Since under the elastic deformations d^a and ρ transform according to, cf. (16), (13) and (17)

$$\rho^+ = \rho \det \boldsymbol{F}^{-1} \quad \text{and} \quad \boldsymbol{d}^{a+} = \boldsymbol{F}^{-1T}\boldsymbol{d}^a \tag{21}$$

with $\boldsymbol{F}^{-1} = \partial\boldsymbol{x}/\partial\boldsymbol{x}^+$, from the usual transformation rules for the material elements it follows that $(20)_1$ and $(20)_2$ respectively yield

$$\begin{aligned} \boldsymbol{g}(\boldsymbol{d}^a,\rho) &= \boldsymbol{F}^T\boldsymbol{g}(\boldsymbol{F}^{-1T}\boldsymbol{d}^a, \rho\det\boldsymbol{F}^{-1}) \\ \boldsymbol{g}(\boldsymbol{d}^a,\rho) &= (\det \boldsymbol{F})\boldsymbol{F}^{-1}\boldsymbol{g}(\boldsymbol{F}^{-1T}\boldsymbol{d}^a, \rho\det\boldsymbol{F}^{-1}), \end{aligned} \tag{22}$$

with $\boldsymbol{F} \in \mathrm{Inv}^+$ any invertible tensor with positive determinant. Then, by choosing $\boldsymbol{F} = \boldsymbol{d}_b^0 \otimes \boldsymbol{d}^b$ with the $\boldsymbol{d}_b^0$'s any fixed positively oriented triad of vectors, one obtains after some manipulations

$$\boldsymbol{g}(\boldsymbol{d}^a,\rho) = h_b(\rho\boldsymbol{d}_1\cdot\boldsymbol{d}_2\times\boldsymbol{d}_3)\boldsymbol{d}^b \tag{23$_1$}$$

and

$$\begin{aligned} \boldsymbol{g}(\boldsymbol{d}^a,\rho) &= h^b(\rho\boldsymbol{d}_1\cdot\boldsymbol{d}_2\times\boldsymbol{d}_3)(\boldsymbol{d}^1\cdot\boldsymbol{d}^2\times\boldsymbol{d}^3)\boldsymbol{d}_b = \\ &= \tfrac{1}{2}\varepsilon_{bcd}h^b(\rho\boldsymbol{d}_1\cdot\boldsymbol{d}_2\times\boldsymbol{d}_3)\boldsymbol{d}^c\times\boldsymbol{d}^d, \end{aligned} \tag{23$_2$}$$

with ε_{bcd} Ricci's symbol and h_b and h^b any two functions of the argument

$$m \equiv \rho(\boldsymbol{d}_1\cdot\boldsymbol{d}_2\times\boldsymbol{d}_3)\,. \tag{24}$$

Physically m can be interpreted as the mass of the average cell. Formula (24) recalls the first assumption of Cauchy (8), but here the cell mass m can take on different values in different states for the same crystalline body whereas in Cauchy's model it cannot because it represents the mass of a single atom in the lattice.

It follows that line and surface invariants must take the form

$$\begin{aligned} \oint_c \boldsymbol{g}(\sigma)\cdot d\boldsymbol{x} &= \oint_c h_b(m)\,\boldsymbol{d}^b\cdot d\boldsymbol{x}\,, \\ \int_S \boldsymbol{g}(\sigma)\cdot\boldsymbol{n}\,ds &= \tfrac{1}{2}\varepsilon_{bcd}\int_S h^b(m)\,\boldsymbol{d}^c\times\boldsymbol{d}^d\cdot\boldsymbol{n}\,ds\,. \end{aligned} \tag{25}$$

In order to get an analogous representation formula for the volume invariants, let us describe the local state of the material by means of an equivalent set of variables. It is easily seen from (24) and from

$$\partial m/\partial \boldsymbol{x} = (\boldsymbol{d}_1 \cdot \boldsymbol{d}_2 \times \boldsymbol{d}_3)\,\partial\rho/\partial\boldsymbol{x} + \rho\,\partial(\boldsymbol{d}_1 \cdot \boldsymbol{d}_2 \times \boldsymbol{d}_3)/\partial\boldsymbol{x} \tag{26}$$

that there is a one to one correspondence between the pairs $(\rho, \partial\rho/\partial\boldsymbol{x})$ and $(m, \partial m/\partial\boldsymbol{x})$ when $\boldsymbol{d}^a$ and $\partial\boldsymbol{d}^a/\partial\boldsymbol{x}$ are given. Furthermore, we can also replace the gradients $\partial\boldsymbol{d}^a/\partial\boldsymbol{x}$ with their symmetric and skewsymmetric parts defined by

$$\begin{aligned} \boldsymbol{D}^a &\equiv \tfrac{1}{2}\,(\partial\boldsymbol{d}^a/\partial\boldsymbol{x} + (\partial\boldsymbol{d}^a/\partial\boldsymbol{x})^T)\,,\\ \boldsymbol{W}^a &\equiv \tfrac{1}{2}\,(\partial\boldsymbol{d}^a/\partial\boldsymbol{x} - (\partial\boldsymbol{d}^a/\partial\boldsymbol{x})^T)\,. \end{aligned}$$

Therefore, we can think of g as a function of the $\boldsymbol{d}^a$'s and of the new variables and write $(20)_3$ in the form

$$g(\boldsymbol{d}^a, \boldsymbol{D}^a, \boldsymbol{W}^a, m, \partial m/\partial\boldsymbol{x})\,dv = g(\boldsymbol{d}^{a+}, \boldsymbol{D}^{a+}, \boldsymbol{W}^{a+}, m^+, \partial m^+/\partial\boldsymbol{x}^+)\,dv^+ \tag{28}$$

that yields a representation formula once the transformation rules for the new variables are determined.

From (24), (16) and (17) the cell mass is an invariant scalar

$$m^+ = m\,. \tag{29}$$

It follows that $\partial m/\partial\boldsymbol{x}$ transforms as a covariant vector

$$\partial m^+/\partial\boldsymbol{x}^+ = (\partial m/\partial\boldsymbol{x})(\partial\boldsymbol{x}/\partial\boldsymbol{x}^+)\,. \tag{30}$$

Analogously, from $(21)_2$ we find

$$\begin{aligned} \boldsymbol{D}^{a+} &= (\partial\boldsymbol{x}/\partial\boldsymbol{x}^+)^T\,\boldsymbol{D}^a\,(\partial\boldsymbol{x}/\partial\boldsymbol{x}^+) + \boldsymbol{d}^a(\partial^2\boldsymbol{x}/\partial\boldsymbol{x}^{+2})\,,\\ \boldsymbol{W}^a &= (\partial\boldsymbol{x}/\partial\boldsymbol{x}^+)^T\,\boldsymbol{W}^a(\partial\boldsymbol{x}/\partial\boldsymbol{x}^+)\,, \end{aligned} \tag{31}$$

i.e. $\boldsymbol{W}^a$ transforms as a covariant tensor whereas $\boldsymbol{D}^a$ does not, the transformation rule involving the second gradient of the inverse mapping $\boldsymbol{x} = \boldsymbol{x}(\boldsymbol{x}^+)$.

In (28) the transformation rules of all the variables but the D^a's involve the first gradient of the deformation at most, or of its inverse. Since this can be fixed independently of the second gradient, locally, it is not difficult to see that in (28) the D^{a+}'s can be changed at will, the other variables remaining unchanged. It follows that g cannot depend upon D^a if (28) is to hold.

Introducing the axial vector curl $\boldsymbol{d}^a$ for $\boldsymbol{W}^a$ and using the ensuing transformation rule (see also [3, formula (6.3)])

$$(\operatorname{curl} \boldsymbol{d}^a)^+ = (\det \boldsymbol{F})^{-1} \boldsymbol{F} \operatorname{curl} \boldsymbol{d}^a , \tag{32}$$

condition (28) becomes

$$\begin{aligned} g(\boldsymbol{d}^a, \operatorname{curl} \boldsymbol{d}^a, m, \partial m/\partial \boldsymbol{x}) &= \\ = g(\boldsymbol{F}^{-1T} \boldsymbol{d}^a, (\det \boldsymbol{F}^{-1}) \boldsymbol{F} \operatorname{curl} \boldsymbol{d}^a, m, \boldsymbol{F}^{-1T} \partial m/\partial \boldsymbol{x}) \det \boldsymbol{F} \end{aligned} \tag{33}$$

for all $\boldsymbol{F} \in \mathrm{Inv}^+$. By choosing $\boldsymbol{F} = \boldsymbol{d}_b^0 \times \boldsymbol{d}^b$ again, (33) yields

$$g = h(\boldsymbol{d}^b \cdot \operatorname{curl} \boldsymbol{d}^a / (\boldsymbol{d}^1 \cdot \boldsymbol{d}^2 \times \boldsymbol{d}^3), m, (\partial m/\partial \boldsymbol{x}) \cdot \boldsymbol{d}_a) \, \boldsymbol{d}^1 \cdot \boldsymbol{d}^2 \times \boldsymbol{d}^3 \tag{34}$$

with h an arbitrary function. It follows that the volume invariants must be of the form

$$\int_v g(\sigma) \, dv = \int_v h(\zeta^{ab}/n, m, g_a) \, dv \tag{35}$$

with

$$n \equiv \boldsymbol{d}^1 \cdot \boldsymbol{d}^2 \times \boldsymbol{d}^3 , \quad g_a \equiv (\partial m/\partial \boldsymbol{x}) \cdot \boldsymbol{d}_a , \quad \zeta^{ab} \equiv \operatorname{curl} \boldsymbol{d}^a \cdot \boldsymbol{d}^b . \tag{36}$$

5. INTERPRETATION OF THE INVARIANTS

All possible choices of the integrands in (25) and (35) generate an infinite set of elastic invariants that characterize the content and distribution of defects throughout the region $\mathcal{B}$ occupied by the body in the state Σ. This statement amounts to a definition of the notion of defect — we identify the distribution of defectiveness over $\mathcal{B}$ with the assignment of the elastic invariants over all circuits, closed surfaces and volumes in $\mathcal{B}$. In this section we discuss a number of invariants that fully characterize the defect distribution, cf. Section 6, and interpret them in terms of topological properties of defects as they are pictured at the submicroscopic level. While this is not needed in principle, the interpretation of the invariants helps making the notion of defect less abstract, suggesting what physical situations the present model can cover.

Taking the constant and linear functions for h_b, h^b and h in $(25)_{1,2}$ and (35), respectively, yields

$$B^a[c] = \oint_c \boldsymbol{d}^a \cdot d\boldsymbol{x}, \qquad D^a[c] = \oint_c m\, \boldsymbol{d}^a \cdot d\boldsymbol{x}; \qquad a = 1,2,3 \tag{37}$$

$$J^{ab}[S] = \int_S \boldsymbol{d}^a \times \boldsymbol{d}^b \cdot \boldsymbol{n}\, ds, \qquad K^{ab}[S] = \int_S m\, \boldsymbol{d}^a \times \boldsymbol{d}^b \cdot \boldsymbol{n}\, ds; \qquad a,b = 1,2,3\,(a \neq b) \tag{38}$$

$$N[v] = \int_v n\, dv, \quad Z^{ab}[v] = \int_v (\zeta^{ab}/n)\, n\, dv \quad a,b = 1,2,3, \qquad M[v] = \int_v m n\, dv, \quad G_a[v] = \int_v g_a\, n\, dv \quad a = 1,2,3. \tag{39}$$

The invariants D^a, K^{ab}, G_a are related to local measures of defectiveness involving the gradient of the mass density as is better seen in the next section (see $(36)_1$ and (51)). I do not see any clear interpretation for them, but their role seems to be important in order to get a reasonably complete description of defectiveness, as is suggested by a study of the so-called neutral deformations, cf. [3], that still is in progress. The remaining invariants have been already discussed in [3], and can be interpreted in terms of continuous distribution of dislocations and of vacancies or interstitials. Let us recall that discussion.

The B^{a}'s are the Burgers integrals of the continuum theory of dislocations. According to an interpretation introduced in [9], for a given circuit c terms $\boldsymbol{d}^{a}\cdot d\boldsymbol{x}$ represent the number of lattice steps in the a-lattice direction corresponding to the line elements $d\boldsymbol{x}$ and the $B^{a}[c]$'s (Burgers numbers of c) are the total number of these when going round c. Integration of the differential forms $dy^{a} = \boldsymbol{d}^{a}\cdot d\boldsymbol{x}$ along c gives the parametric representation of a curve that is pictorially interpreted as the image of c in a perfect lattice and whose failure from closure is measured by its Burgers numbers. The argument recalls the notion of Burgers circuit and the way crystallographers describe isolated dislocations. That is why Bilby, Bullough and Smith propose the use of integrals $(37)_1$ to describe continuously dislocated crystals.

From Stokes' theorem,

$$B^{a}[c] = \int_{S[c]} \operatorname{curl}\boldsymbol{d}^{a}\cdot\boldsymbol{n}\, ds\,, \tag{40}$$

with $S[c]$ any surface with boundary c. Therefore, distributed dislocations reflect the anholonomy of the dual lattice vectors. Their strength is locally described by

$$\boldsymbol{b}^{a} \equiv \operatorname{curl}\boldsymbol{d}^{a}\,, \tag{41}$$

that measure the closure failure (per unit area) of infinitesimal circuits on a surface with normal $\boldsymbol{n}$ through the numbers

$$b^{a} = \boldsymbol{b}^{a}\cdot\boldsymbol{n}\,. \tag{42}$$

In the theory of Bilby, Bullough and Smith (41) and (42) are often replaced by the dislocation density tensor $\boldsymbol{S}\equiv\boldsymbol{d}_{a}\otimes\operatorname{curl}\boldsymbol{d}^{a}$ and the local Burgers vector $\boldsymbol{b} = b^{a}\boldsymbol{d}_{a}$, two quantities that seem to me less meaningful in characterizing dislocation defects when the lattice vectors are not known.

The invariants J^{ab} are also related to dislocations as follows from applying the divergence theorem

$$J^{ab}[S] = 2 \int_{v[S]} \mathrm{curl}\,\boldsymbol{d}^{[a}\boldsymbol{d}^{b]}\,dv, \tag{43}$$

where $v[S]$ is the volume inside S and brackets denote alternation with respect to indices a and b. They, however, put in evidence a different topological consequence of anholonomy that apparently has not been noticed in the continuous theory of dislocations.

Let us decompose the vector surface element $\boldsymbol{n}\,ds$ in $(38)_1$ along the local lattice planes

$$\boldsymbol{n}\,ds = \frac{\boldsymbol{d}^c}{|\boldsymbol{d}^c|}\,ds_c \tag{44}$$

where ds_c is the area with sign of the surface element in the (d,e)-plane with $d \neq e \neq c$. Observing that

$$\frac{\boldsymbol{d}^c}{|\boldsymbol{d}^c|} = \tfrac{1}{2}\,\varepsilon^{cde}\,\frac{\boldsymbol{d}_d \times \boldsymbol{d}_e}{|\boldsymbol{d}_d \times \boldsymbol{d}_e|}, \tag{45}$$

(44) becomes

$$\boldsymbol{n}\,ds = \frac{\boldsymbol{d}_d \times \boldsymbol{d}_e}{|\boldsymbol{d}_d \times \boldsymbol{d}_e|}\,ds^{de} \quad \text{with} \quad ds^{de} = \tfrac{1}{2}\,\varepsilon^{cde}\,ds_c. \tag{46}$$

Hence

$$\boldsymbol{d}^a \times \boldsymbol{d}^b \cdot \boldsymbol{n}\,ds = \frac{ds^{ab}}{|\boldsymbol{d}_a \times \boldsymbol{d}_b|} \qquad (a \neq b). \tag{47}$$

Since $|\boldsymbol{d}_a \times \boldsymbol{d}_b|$ is the area of the (a,b)-lattice face, the left hand side of (47) is the number (with sign) of lattice faces of this type corresponding to $\boldsymbol{n}\,ds$. It follows that J^{ab} measures their overall number if we assume, as is tacitly done for the B^a's, that we can add them up when going round S. It is not difficult to convince ourselves that this number is different from zero when considering surfaces crossed by a single edge dislocation line.

Finally the quantities ζ^{ab} are the lattice components of Bilby's dislocation density tensor. From $(39)_2$, ζ^{ab} is the density of the dislocations of type (a,b) per unit volume and

Z^{ab} their total in v.

Of course, the kind of defect described so far need not be related to dislocations as we picture them at the submicroscopic scale, but the interpretation is suggestive. So is the analogy that leads one to look at ζ^{aa} and ζ^{ab}, $a \neq b$, as density measures for the screw and edge dislocation defect, respectively. The fact that only the latter gives rise to surface invariants provides a topological criterion for distinguishing between the two in the continuum theory.

The invariants N and M do not depend upon curl $\boldsymbol{d}^a$, so we expect that they detail properties of the state Σ not related to dislocations. From (24), and observing that $n = (\boldsymbol{d}_1 \cdot \boldsymbol{d}_2 \times \boldsymbol{d}_3)^{-1}$, we get

$$M[v] = \int_v \rho \, dv.$$

Thus, $M[v]$ is the mass within v. In molecular terms it estimates the number of atoms contained in v, ρ describing their number per unit volume at points of $\mathcal{B}$. Since n is the inverse of the volume of the average cell, $N[v]$ can be interpreted as the overall number of cells in v, roughly. Then, if we compare $M[v]$ with the mass of $N[v]$ perfect cells, we get an estimate of the matter not organized in lattice positions, which seems a natural expression for the content of vacancies and interstitials (counted together) in v.

From (16) $M[v]$ is an invariant under general deformations. Then, the evolution of vacancies is described by the changes of $N[v]$ and it is possible under nonelastic deformations only. If we accept this view, vacancies turn out to be topological defects, in contrast with a conclusion drawn in [1]. As I see it, this simply points out that the classes of topological transformations in the two approaches are different.

REMARK 2. With the above interpretation, assumption (16) implies that the same number of atoms remains for ever in any material volume under deformations from Σ. It then excludes that hidden flux of matter may occur and confines the mechanism of vacancy growth to the local exchange of atoms between lattice and non-lattice positions. As noticed in Section 3, the requirement (16) is more restrictive than the nature of the model would allow for. It might be worthwhile reconsidering (16) to account for situations where massive atomic migration over macroscopic distances becomes important. If so, (16) should be added as a requirement for elastic deformations, the only ones that are pretty well understood in the model, and the analysis of the invariants would remain valid.

6. LOCAL MEASURES OF DEFECTIVENESS

Assigning invariants (37) and (39) over all circuits and volumes in $\mathcal{B}$ is equivalent to give the following fields:

$$n,\ \zeta^{ab},\ m,\ g_a \tag{48$_1$}$$

and, from Stokes' theorem in (37):

$$b^a \quad \text{and} \quad \mathcal{d}^a \equiv \operatorname{curl}(m\,\mathcal{d}^a) = (\partial m/\partial x)\times \mathcal{d}^a + m b^a\ . \tag{48$_2$}$$

Knowledge of these fields determines the values of all the line and volume invariants obtained from the representation formulae (25) and (35). Moreover, it also fixes the values of the surface invariants, that always reduce to volume invariants via the divergence theorem†. Therefore, the invariants (37) and (39) fully characterize the distribution of defects throughout B and the fields (48) describe the local defectiveness at each point.

When the dual lattice vectors at one point are known, the defect measures (48) are determined, for instance, by the quantities

$$\zeta^{ab}/n\ ,\ m\ ,\ g_a \tag{49}$$

† **This aspect and the role of surface invariants in the analysis of the defect notion has been discussed in Section 5 for the J^{ab}'s.**

evaluated at that point. From $(39)_{2,3,4}$ these quantities are defect measures for the average cell. Since there is a one to one correspondence between the n-tuples $\{\zeta^{ab}/n, m, g_a\}$ and $\{\mathrm{curl}\, d^a, m, \partial m/\partial x\}$, the local state is equivalently described by

$$\sigma = \{d^a, D^a, \zeta^{ab}/n, m, g_a\}, \tag{50}$$

with the advantage that the new variables are invariant under elastic deformations and two local states can be elastically deformed into each other only if they share the same values of defect measures (49). Conversely, the argument given in [3] could be adjusted in order to show that two states can be elastically deformed into each other if their measures (49) are equal. Thus , (49) provides a classification of the states that can be elastically connected.

Developing views discussed by Ericksen in [10], I have noticed in [3] that such a classification has a constitutive relevance because the thermodynamic potentials such as the specific entropy can only be defined within subsets of local states that can be obtained from one another by an elastic deformation. Pursuing Ericksen's argument we can conclude that, for a model insensitive to defects that depend upon higher order gradients of d^a and ρ, the specific entropy function for the crystalline material is to be defined to within an arbitrary function of variables (49). While the question on whether accounting for invariants that depend upon higher order gradients would yield additional measures of defect remains open, it seems that (49) are a minimal set of defect measures to consider in the constitutive theory in order to account for Ericksen's views in some generality. It is worth noticing, on the contrary, that the D^a's have an elastic meaning. They simply describe second grade elasticity effects in the model and one may exclude them from the constitutive list without implications for thermodynamics.

ACKNOWLEDGEMENTS

This material is based on work sponsored by the Italian M.P.I. I am also grateful to G.N.F.M. of C.N.R. for supporting my participation in the Symposium Year on Material Instabilities and Continuum Mechanics.

REFERENCES

1. N.D. Mermin, "The topological theory of defects in ordered media", *Reviews of Modern Physics* **51** (1979), pp.591-648.
2. E. Kröner, "Differential geometry of defects in condensed systems of particles with only translationaly mobility", *Int. J. Eng. Sci.* **19** (1981), pp.1507-1515.
3. C. Davini, "A proposal for a continuum theory of defective crystals", *Arch. Rational Mech. Anal.* **96** (1986), pp.295-317.
4. A.E.H. Love, *A treatise on the mathematical theory of elasticity*, 4th edition, Dover Publications, New York (1944).
5. J. Frenkel "Zur Theorie der Elastizitätsgrenze und der Testigkeit kristallinischera körpera", *Zeit. Phys.* **37** (1926), pp.572-609.
6. G.I. Taylor, "The mechanism of plastic deformation of crystals", Part I and II, *Proc. Roy. Soc. A* **145**, (1934) pp.362-387 and pp.388-404.
7. K.S. Havner, "On the mechanics of crystalline solids", *J. Mech. Phys. Soilds* **21** (1973), pp.383-394.
8. K.S. Havner, "The theory of finite plastic deformation of crystalline solids", in *Mechanics of Solids*, (eds. Hopkins, H.G. and Sewell, M.J.), Pergamon Press, Oxford and New York, (1982).
9. B.A. Bilby, R. Bullough and E. Smith, "Continuous distributions of dislocations: a new application of the methods of non-Riemannian geometry", *Proc. Roy. Soc. London* **A231** (1955), pp.263-273.
10. J.L. Ericksen, "Thermoelastic considerations for continuously dislocated crystals", *Proc. Int. Symp. on Mechanics of Dislocations*, Houghton, Mi, 28-31 August 1983.

10
CONCENTRATIONS IN SOLUTIONS TO CONSERVATIVE SYSTEMS

RONALD J. DiPERNA†

We shall discuss several aspects of a general program with A. Majda, dealing with concentration phenomena in conservative systems. One of the basic examples is provided by the Euler equations in two space dimensions:

$$\partial_t u + \operatorname{div} u \otimes u + \nabla p = 0$$
$$\operatorname{div} u = 0 \ , \qquad x \in R^2 . \tag{1}$$

We shall first recall two classical problems and then discuss how they fit into the general program.

The first deals with existence of solutions to the Cauchy problem for (1) in the case of data with finite total kinetic energy and total vorticity:

$$\int_{R^2} |u_0|^2 \, dx < \infty$$

$$\int_{R^2} |\operatorname{curl} u_0| \, dx < \infty .$$

Initial configurations of this type arise in the study of vortex sheets.

The second problem deals with the zero diffusion limit for the 2-D Navier-Stokes equations:

---

†Partially supported by N.S.F. Grant 83-01135

$$\partial_t u_\varepsilon + \operatorname{div} u_\varepsilon \otimes u_\varepsilon + \nabla p_\varepsilon = \varepsilon \Delta u_\varepsilon$$
$$\operatorname{div} u_\varepsilon = 0$$

with initial data u_0 satisfying (2) and (3). Do the Navier-Stokes solutions u_ε converge strongly in L^2 in the infinite Reynolds number limit, i.e. as the diffuson coefficient ε tends to zero? If convergence is strong, then the limit solves the Euler equations.

Both of these topics motivate the study of sequences of solutions (exact or approximate) to the 2-D Euler equations. In the context of the existence problem, a natural strategy is to regularize the data and attempt to pass to the limit along a sequence.

In the setting of two space dimensions the Hamiltonian structure leads to two basic *a priori* estimates which assert that the total energy and total vorticity are uniformly bounded:

$$\int |u_\varepsilon(x,t)|^2 \, dx \leqslant C \tag{2}$$

$$\int |\omega_\varepsilon(x,t)| \, dx \leqslant C; \quad \omega_\varepsilon \equiv \operatorname{curl} u_\varepsilon \tag{3}$$

These estimates hold at time $t>0$ for both Euler and Navier-Stokes sequences, u_ε, provided that they hold at time $t=0$.

As a consequence of the uniform energy bound it is possible to extract a subsequence, which we shall not bother to relabel, that converges weakly in L^2:

$$u = w - \lim_{\varepsilon \to 0} u_\varepsilon .$$

In view of this fact, it is natural to ask whether or not u_ε converges strongly to u. If the answer is negative, the problem arises of determining the size of the set in physical space $R^2 \times R^+$ on which L^2 compactness can be lost, i.e. the size of the *exceptional sets* E for which

$$\lim \int_E |u_\varepsilon - u|^2 \, dx \neq 0 .$$

In general, losses of compactness may be associated with the persistence of oscillations and/or the development of concentrations. One of the main problems is to represent and analyse oscillations and concentrations in L^2 weakly convergent Euler and Navier-Stokes sequences u_ε for which the total vorticity is uniformly bounded in the sense of (3). For the present purpose, we shall begin by discussing Euler sequences.

For background we shall recall two examples of oscillating and concentrating function sequences. Let σ be a scalar function of one variable. If σ is periodic, i.e. $\sigma(x) = \sigma(x+p)$, then the normalized function $u_\varepsilon = \sigma(x/\varepsilon)$ executes the same structure as σ with progressively smaller and smaller period as ε tends to zero. In the limit, u_ε fluctuates around a constant c given by the Lebesgue area of the graph of σ:

$$w - \lim_{\varepsilon \to 0} u_\varepsilon = c \equiv \frac{1}{p} \int_0^p \sigma(y)\, dy .$$

Here the weak limit of u_ε is a constant function. This situation provides an elementary example of the persistence of oscillations in a limiting process.

On the other hand, if σ has compact support then the normalized function $u_\varepsilon = \varepsilon^{-1}\sigma(x/\varepsilon)$ converges weakly to a δ-function. As ε tends to zero, the mass is preserved while the support shrinks. In the limit, all of the "energy" is concentrated at a point. This process provides an example of the development of a concentration. A general problem in the context of conservative systems is to determine the extent to which similar structures can arise in solution sequences.

In the context of the Euler equations (1) in two space dimensions, oscillations and concentrations can be ruled out by imposing additional regularity on the data. For example, if the initial vorticity ω_0^ε lies in L^p with $p > 1$, i.e. if

$$\int_{R^2} |\omega_0^\varepsilon|^p \, dx \leqslant c$$

where the constant c is independent of ε, then the corresponding solutions u_ε to the Cauchy problem for the 2-D Euler equations are precompact in L^2 : there exists a subsequence which converges strongly in L^2, i.e. in the energy norm. In short, no loss of L^2 compactness can occur in 2-D Euler sequences if the vorticity lies in a bounded set of L^p, $p > 1$, [5].

A proof of this fact can be given with the aid of the Calderon-Zygmund inequality. It is motivated by the classical vorticity stream formulation of the 2-D Euler equations which recasts system (1) into a coupled system of two equations for two scalar quantities, namely the scalar vorticity ω and the stream function ψ:

$$\partial_t \omega + \operatorname{div}(\omega u) = 0, \qquad \Delta\psi = \omega. \tag{4}$$

The velocity field u is recovered from the vorticity ω by inverting the Laplacian in space and taking an (orthogonal) gradient:

$$u = (\nabla\psi)^{\perp} \qquad \psi = \Delta^{-1}\omega$$
$$\omega = (\operatorname{curl} u)\cdot\hat{k}.$$

The velocity field in 2-D flow lies in a plane since

$$u = \{u_1(x_1, x_2), u_2(x_1, x_2), 0\}.$$

The vorticity vector, $\operatorname{curl} u$, is perpendicular to this plane and consequently is described by one scalar quantity, the third component associated with the unit vector $\hat{k} = (0,0,1)$.

Equation (4) expresses the fact that the vorticity is constant along particle paths of a divergence-free field. Thus, the total amount of vorticity, as measured in any L^p space for example, is formally conserved with time for 2-D flow:

$$\int_{R^2} |\omega(x,t)|^p \, dx = \int_{R^2} |\omega(x,0)|^p \, dx.$$

It follows that, for each fixed value t, the associated stream function $\psi(\cdot, t)$ lies in a bounded set of $W^{2,p}$: uniformly elliptic operators of second order gain two derivatives in the scale of Lebesgue spaces if $1<p<\infty$. Thus, the velocity field u lies in a bounded set of $W^{1,p}$, since it is obtained from the stream function by taking one derivative. Finally, the Sobolev embedding theorem, which asserts that $W^{1,p}(R^n)$ is compactly contained in L^q if

$$1 \leqslant q < np/(n-p),$$

implies that u lies in a compact subset of L^2 since $n=2$ and $p>1$.

In particular, we conclude that if u_ε is a sequence of 2-D Euler solutions whose initial vorticity lies in a bounded set of L^p with $p>1$ then u_ε lies in a compact set of L^2 for each fixed t. A standard argument involving time derivatives in a negative Sobolev space leads to the conclusion that u_ε lies in a compact subset of $L^2_{\mathrm{loc}}(R^2 \times R^+)$.

The proof indicates a deterioration in compactness as the index p approaches one. Both the Calderon-Zygmund inequality and the Sobolev inequality deteriorate as p approaches one. It turns out that the anticipated loss of compactness in the case $p=1$ can be demonstrated through explicit examples of 2-D Euler sequences u_ε with uniformly bounded energy and vorticity in the sense of (2) and (3). In [4,5] special vortex sequences u^*_ε are constructed and analyzed which exhibit a loss of compactness in L^2 due to the development of concentrations of energy. The sequences u^*_ε consist of collapsing vortices of steady, rotating and phantom type which deposit a finite amount of kinetic energy on a small set in the limit.

With motivation from these examples, the general problem arises of estimating the size of the exceptional sets in space and in space-time on which energy can concentrate in the weak L^2 limit of 2-D Euler sequences with initial vorticity uniformly bounded in L^1, or more generally with initial vorticity given by a sequence of Radon measures with uniformly bounded total mass. The latter case is of particular interest in the study of the geometrically unstable flow of vortex sheets.

In [6] we prove that the exceptional sets for energy concentration in 2-D flow have small Hausdorff dimension. Before discussing this result we recall a characterization of the loss of compactness in the Sobolev embedding at the critical exponent which has been derived by P.L. Lions [7,8]. This characterization is part of a general theory of concentration-compactness [7,8] which analyzes losses of compactness in minimizing sequences of elliptic variational problems due to the action of noncompact groups.

Let v_ε be an arbitrary weakly convergent function sequence in $W^{1,p}(R^n)$. As a consequence of the Sobolev inequality, namely

$$\int_{R^n} |v|^q \, dx \leqslant c \int_{R^n} |\nabla v|^p \, dx ,$$

with $q = np/(n-p)$, we may extract a subsequence with the following two properties: v_ε converges weakly to v in $W^{1,p}(R^n)$ and $|v_\varepsilon|^q$ converges in the weak-star topology of measure to some bounded measure which we denote by μ:

$$\mu \equiv w^* - \lim |v_\varepsilon|^q .$$

We recall that weak-star convergence means convergence of averages of continuous (test) functions ϕ:

$$\lim_{\varepsilon \to 0} \int \phi \, |v_\varepsilon|^q \, dx = \int \phi \, d\mu ,$$

for all ϕ in $C_0(R^n)$. Here we have identified the function $|v_\varepsilon|^8$ with the measure which associates the value

$$\int_E |v_\varepsilon|^q \, dx$$

with each Borel set E.

As a consequence of the lower semicontinuity of convex functions it follows that $\mu \geqslant |v|^q$ in the sense of measures. The difference σ is referred to as the weak-star defect measure.

$$w^* - \lim_{\varepsilon \to 0} |v_\varepsilon|^q = |v|^q + \sigma .$$

Thus, $\sigma \geqslant 0$ in general and $\sigma \equiv 0$ if and only if v_ε converges strongly to v in L^q. A theorem of P.L. Lions asserts that σ is purely atomic, i.e.

$$\sigma = \sum_{j=1}^{\infty} a_j \, \delta(x_j)$$

where $\delta(x_j)$ denotes the Dirac mass at x_j. In short, the loss of compactness in the context of the Sobolev embedding at the critical exponent is confined to a countable set, from the viewpoint of the L^q weak-star defect measure σ.

A special case of the general situation above asserts that if u_ε is a sequence of vector fields which converges weakly in $L^2(R^2)$ to a vector field u and if the Jacobian matrix ∇u_ε is uniformly bounded in $L^1(R^2)$ then the associated L^2 defect measure in the weak-star topology, namely

$$\sigma = w^* - \lim |u_\varepsilon - u|^2$$

is concentrated at a countable number of points.

In [5] we show that this result is sharp in the context of the 2-D Euler equations by constructing a sequence of smooth steady solutions u^*_ε such that

$$\int_{R^2} |\nabla u^*_\varepsilon| \, dx \leqslant c$$

and the associated weak-star defect measure is concentrated at a countable number of points.

In the study of incompressible flows, the natural uniform estimate deals with the vorticity, i.e. the antisymmetric part of the Jacobian matrix, rather than with the entire Jacobian ∇u_ε. This diminished level of control permits a more substantial loss of compactness. Indeed one can construct examples where the weak-star defect measure is spread over the whole domain of definition.

In order to investigate losses of compactness in general circumstances, a notion of reduced defect has been formulated which records the loss of compactness in a different topology, cf. [5,6]. The reduced defect measure θ associated with a weakly convergent sequence of functions u_ε in L^p is defined by the limiting weight that the difference $u_\varepsilon - u$ gives to a general set:

$$\theta(E) = \limsup_{\varepsilon\to 0} \int_E |u_\varepsilon - u|^p \, dy$$

if E is a Borel subset of R^m. If F is an arbitrary closed set then $\theta(F) \leqslant \sigma(F)$. In general the gap between the two can be rather large.

It follows from the definition that θ is an outer measure, i.e. a subadditive set function, which vanishes precisely on those sets of E for which u_ε converges strongly to u in the L^p topology.

In the context of the Poisson equation on R^2,

$$\Delta\psi_\varepsilon = \omega_\varepsilon ,$$

where ψ_ε and ω_ε are arbitrary distributions, not necessarily connected with a Euler flow, the reduced defect measure θ is concentrated on a set with zero Hausdorff dimension. In [6] we prove that if ψ_ε converges weakly to ψ in H^1, i.e. if the gradient of ψ_ε converges weakly to $\nabla\psi$ in L^2 and if the total mass of the right side ω_ε is bounded uniformly with respect to ε then the L^2 reduced defect measure associated with the sequence $\nabla\psi_\varepsilon$ is concentrated on a set with zero Hausdorff dimension.

In order to analyze time dependent solutions of 2-D Euler equations this result on the size of spatial defects is combined with estimates on the uniform temporal regularity of solutions to prove the following. The reduced L^2 defect measure of an arbitrary Euler sequence satisfying the uniform bounds (2) and (3) on the total kinetic energy and vorticity is concentrated on a set with Hausdorff dimension $\leqslant 1$. Thus, for time dependent flows, kinetic energy may concentrate at most on a one-dimensional set.

From the viewpoint of Hausdorff dimension, the cut-off at dimension one is sharp. In [5] we construct explicit examples of steady 2-D Euler solutions with the desired uniform bounds for which the reduced defect measure is concentrated at a point. Regarding these solutions as functions on space-time leads to an L^2 exceptional set in the form of a line. We also construct time dependent examples consisting of rotating Kirchhoff elliptic vortices for which the dimensional upper bound on the concentration set of the reduced defect measure is achieved.

A second aspect of the program deals with the sensitivity of various nonlinear maps to defects in the solution. Suppose g is a continuous real-valued map on the state space R^2 and suppose that u_ε is a L^2 weakly convergent sequence of Euler solutions. The problem is to determine whether or not g is continuous in the weak topology, i.e. to determine whether or not

$$w-\lim g(u_\varepsilon) = g(w-\lim u_\varepsilon).$$

Of course, in general, nonlinear maps are not continuous in the weak topology. For example the sequence $\sin(x/\varepsilon)$ fluctuates around zero while its square fluctuates around half: the limit of the square is not the square of the limit. The operators of composition and local averaging do not commute.

The theory of compensated compactness is [9,16] is concerned in part with the identification of special nonlinear maps which are insensitive to oscillations. Our current program is concerned in part with maps which are insensitive with respect to concentrations. In this connection we prove that the inertial terms for the 2-D Euler equations are insensitive to defects in the energy field. In order to formulate a more precise statement we first recall the conventional notion of solution to the Euler equations.

A divergence-free vector field u in $L^2(\Omega)$ is a weak solution of (1) if

$$\iint (\partial_t \phi, u) + \nabla\phi : u \otimes u \, dx\, dt = 0 \tag{5}$$

for all compactly supported divergence-free test fields ϕ. Here $\nabla\phi$ denotes the Jacobian matrix of ϕ and colon denotes tensor contraction. Thus the nonlinear terms are represented by the form

$$T(\phi, u) = \iint \nabla\phi : u \otimes u \, dx\, dt = \iint \frac{\partial \phi_i}{\partial x_j} u_i u_j \, dx\, dt .$$

The weak formulation (5) is obtained by multiplying (1) by a suitable test function and using the L^2 orthogonality of divergence-free and curl-free fields.

A surprising process of concentration cancellation occurs in the aforementioned special vortex sequences, namely u^*_ε converges weakly to u^* and

$$\lim T(\phi, u^*_\varepsilon) = T(\phi, u^*) ,$$

if $\operatorname{div}\phi = 0$. In this situation u^*_ε does not converge strongly to u due to a limiting concentration of energy at the origin but the residuals, which appear in the limit, cancel to produce the continuity of the inertial terms as represented by T.

One of the consequences of the continuity of T on the special sequences u^*_ε is that the limiting field u^* is also a solution. In general the weak limit of a sequence of solutions to a nonlinear p.d.e. is not a solution. However certain special circumstances may arise for which a solution set is closed in the weak topology.

In the setting of the Euler equations the continuity of T on general sequences u_ε with uniformly bounded vorticity would imply that the solution set is closed when restricted to solutions with uniformly bounded vorticity. It remains an open problem to determine whether or not the form T is continuous on general sequences.

Despite the fact that the continuity of T is still undecided, we have shown in [6] that the solution sets for the steady and the quasisteady 2-D Euler equations are closed in the L^2

weak topology when restricted to solutions with uniformly bounded vorticity. The proof relies on the construction of a shadow sequence ϕ_ε which shields the inertial terms from defects in the energy field. Given an L^2 weakly convergent solution sequence u_ε and a test field ϕ with zero divergence there exists a sequence of test fields ϕ_ε with zero divergence such that

$$\lim T(\phi_\varepsilon, u_\varepsilon) = T(\phi, u)$$

$$w-\lim \phi_\varepsilon = \phi .$$

The construction of ϕ_ε makes use of the fact that the L^2 defects in u_ε are associated with a small set and that the inertial terms are rotationally invariant.

A similar construction is possible for general time dependent Euler sequences u_ε provided that the associated reduced defect measure is concentrated on a set in space-time with dimension less than one. However, as we mentioned above it is possible for the concentration set to have dimension equal to one.

REFERENCES

1. R.J. DiPerna, "Convergence of approximate solutions to conservation laws", *Arch. Rat. Mech. Anal.* **82** (1983), pp.27-70.
2. R.J. DiPerna, "Convergence of the viscosity method for isentropic gas dynamics", *Comm. Math. Phys.* **91** (1983), pp.1-30.
3. R.J. DiPerna, "Compensated compactness and general systems of conservation laws", *Trans. Amer. Math. Soc.* **292** (1985), pp.383-420.
4. R.J. Diperna and A. Majda, "Oscillations and concentrations in weak solutions of the incompressible fluid equations", *Comm. Math. Physics* (1987) (to appear).
5. R.J. DiPerna and A. Majda. "Concentrations in regularizations for 2-D incompressible flow", *Comm. Pure Appl. Math.* (1987) (to appear).
6. R.J. DiPerna and A. Majda, "Reduced Hausdorff dimension and concentration-cancellation for 2-D incompressible flow", *Journal of the AMS* (1988) (to appear).

7. P.L. Lions, "The concentration-compactness principle in the calculus of variations, the locally compact case, Parts I an II", *Ann. Inst. H. Poincaré*, (1984), pp.109-145 and pp.223-283.
8. P.L. Lions, "The concentration-compactness principle in the calculus of variations, the limit case, Parts I and II", *Riv. Mat. Iberoamericana*, (1984), pp.145-201 and (1985) pp.45-121.
9. C. Morawetz, "On a weak solution of a transonic flow problem", *Comm. Pure Applied Math.* **38** (1986)
10. F. Murat, "Compacité par compensation", *Ann. Scuola Norm. Sup. Pisa* **5** (1978), pp.489-507.
11. M. Rascle and D. Serre, "Compacité par compensation et systèmes hyperboliques de lois de conservation", *Compte Rendus Acad. Sci.*, **299** (1984), pp.673-676.
12. V. Roytburd and M. Slemrod, "Dynamic phase transitions and compensated compactness", *Proc. IMA Workshop on Dynamic Problems in Continuum Mechanics*, (ed. Bona, J.) to appear in *Springer Lecture Notes in Math.*
13. V. Roytburd and M. Slemrod, "An application of the method of compensated compactness to a problem in phase transitions", to appear in *Indiana Math. J.*
14. D. Serre, "La compacité par compensation pour les systèmes hyperboliques nonlinéaires de deux équations à une dimension d'espace", Preprint *Equip d'Analyse Numerique*, Université de St. Etienne, France, (1985).
15. L. Tartar, "Compensated compactness and applications to partial differential equations", in *Nonlinear Analysis and Mechanics*, Heriot-Watt Symposium, IV, pp.136-192 Research Notes in Math., Pitman, (1976).
16. L. Tartar, "The compensated compactness method applied to systems of conservation laws", in *Systems of Nonlinear Partial Differential Equations*, (ed. Ball, J.), NATO ASI Series, Reidel Pub. (1983).

11

SOME CONSTRAINED ELASTIC CRYSTALS

J. L. ERICKSEN

1. INTRODUCTION

Often, crystals undergo phase transformations involving some change of symmetry. When such transformations are of second-order, and often when they are of first-order but weakly so, it is observed that some linear elastic moduli become quite small compared to others, near transition. It then seems rather reasonable to try using simpler, idealized theory, roughly considering the larger moduli as infinite or, more properly regarding the crystals as constrained materials. Motivated by observations of cubic-tetragonal transformations in A-15 superconductors, Ericksen [1] formulated a thermoelasticity theory of this kind, which also shows some promise as a theory for somewhat similar transformations occuring in Indium-Thallium systems. A variety of different constraints would be in reasonable agreement with the observations. Ericksen [1] gives one theoretical argument leading to a unique choice. Here, we give a rather different argument leading to the same choice.

In the Indium-Thallium systems, but not in the A-15 superconductors, cubic and tetragonal phases have been observed to coexist, by Burkart and Read [2], and they form rather complicated configurations. Ball and James [3] found a way to deduce algorithms which enable one to describe coarser features of such configurations. We will present and discuss the results of such calculations, for the theory of constrained crystals. The reasoning of Ball and James does not suggest why such configurations should be stable enough to be observed. Burkart and Read [2] attempt to rationalize this, but a satisfactory theory for this remains to be developed. It should explain why such coexistence is not observed in A-15 superconductors. It would be easy to explain this, if such transformations are of second-order, as was suggested by the first workers to observe them, Batterman and Barrett [4]. As is discussed by Ericksen [5], reasoning accepted by many physicists leads to the conclusion that they should not be, but there is some room for argument about this. We have not seen clear experimental proof that the implied discontinuities in deformation occur. In any event, such stability questions remain open.

2. CONSTRAINTS

For the aforementioned crystals, we have cubic crystals at higher temperatures, transforming to crystals of tetragonal form, as the temperature T is lowered through a critical value T_c and, commonly, the tetragonal phase contains twins. The transformation involves a discontinuity in deformation which is quite small. In the A-15 superconductors, it might be zero or just small enough to be obscured by experimental errors. In the Indium-Thallium alloys, data presented by Burkart and Read [2] indicate strains of the order 10^{-2}, definitely non-zero, but not terribly large. Briefly, we seek a theory capable of describing both phases, twinning in the tetragonal phase, as well as the effects of small loadings and temperature changes. Roughly, the deformations of interest should be considered as finite, although they are not very large; conversion of one twin to another involves finite rotations, for example.

In the cubic phase, the linear elastic strain energy W is of the form

$$2W = (\hat{C}_{11} - \hat{C}_{12})(\eta_{11}^2 + \eta_{22}^2 + \eta_{33}^2) + [(\hat{C}_{11} + 2\hat{C}_{12})/3](\varepsilon_{11} + \varepsilon_{22} + \varepsilon_{33})^2 + 4\hat{C}_{44}(\varepsilon_{12}^2 + \varepsilon_{23}^2 + \varepsilon_{31}^2) , \tag{2.1}$$

where

$$\eta_{ij} = \varepsilon_{ij} - \tfrac{1}{3}(\varepsilon_{11} + \varepsilon_{22} + \varepsilon_{33})\,\delta_{ij} , \tag{2.2}$$

and ε is the infinitesimal strain tensor. The $\hat{C}_{ij}$ are elastic moduli, labelled as most experimentalists do, except that we have added carets to distinguish these from components of a tensor C to be used later. Ericksen [1], noting observations indicating that, for T near T_c,

$$\left.\begin{aligned} (\hat{C}_{11} - \hat{C}_{12})/(\hat{C}_{11} + 2\hat{C}_{12}) &\ll 1 , \\ (\hat{C}_{11} - \hat{C}_{12}) / \hat{C}_{44} &\ll 1 , \end{aligned}\right\} \tag{2.3}$$

thought of the denominators as effectively infinite, to get

$$\varepsilon_{11} + \varepsilon_{22} + \varepsilon_{33} = 0 , \tag{2.4}$$

and

$$\varepsilon_{12} = \varepsilon_{23} = \varepsilon_{31} = 0 , \tag{2.5}$$

as a first estimate of likely constraints. In some way, we need to extrapolate these, to apply to finite measures of strain.

From the viewpoint of nonlinear thermoelasticity theory, it is convenient to take as a reference configuration the (unstressed) cubic phase at the transition temperature T_c. Refer this to rectangular Cartesian coordinates $\boldsymbol{x} = (x_1, x_2, x_3)$, with the orthonormal base vectors $\boldsymbol{e}_i$ parallel to the usual cubic lattice vectors, as is presumed in (2.1). A deformation maps $\boldsymbol{x}$ to

$$\boldsymbol{u} = \boldsymbol{u}(\boldsymbol{x}) , \tag{2.6}$$

with

$$F = \nabla \boldsymbol{u} , \quad \det F > 0 \tag{2.7}$$

the usual deformation gradient, and the symmetric tensor

$$C = F^T F \tag{2.8}$$

is a commonly used measure of finite deformation. The only

reasonable extrapolation of (2.5) seems to be

$$C_{12} = C_{23} = C_{31} = 0 . \qquad (2.9)$$

As will be made clear, if it is not already so, a variety of reasonable extrapolations of (2.4) exist. Ericksen [1] gives an argument leading to the choice

$$C_{11} + C_{22} + C_{33} = 3 , \qquad (2.10)$$

and, shortly, we will argue this in a different way. Given the experimental errors involved in measuring small strains, it is hard to say which extrapolation might best fit the deformations experienced by the real crystals. From what little experience we have, if one picked the constraint to make the mathematical theory simplest, one might well pick (2.10).

3. JUMP DISCONTINUITIES

Consider a material surface, with unit normal $\boldsymbol{N}$ in the reference configuration, across which the deformation gradient F suffers a finite discontinuity, with $\boldsymbol{u}$ remaining continuous. If the two limiting values are denoted by $\bar{F}$ and F, the usual kinematical conditions of compatibility read

$$\bar{F} = F(1 + A \otimes N) , \qquad (3.1)$$

where A is the so-called amplitude vector. From this, we get

$$\begin{aligned} \bar{C} = \bar{F}^T \bar{F} &= (1 + N \otimes A)\, C(1 + A \otimes N) \\ &= C + N \otimes M + M \otimes N , \end{aligned} \qquad (3.2)$$

where

$$2M = 2CA + A \cdot CAN \qquad (3.3)$$

We assume that (2.9) applies, so that

$$\bar{C} - C = \alpha \boldsymbol{e}_1 \otimes \boldsymbol{e}_1 + \beta \boldsymbol{e}_2 \otimes \boldsymbol{e}_2 + \gamma \boldsymbol{e}_3 \otimes \boldsymbol{e}_3 . \qquad (3.4)$$

For the present, we ignore the remaining, ambiguous constraint. From (3.2), if V is a non-zero vector satisfying

$$V \cdot M = V \cdot N = 0, \qquad (3.5)$$

then

$$(\bar{C}-C)\,\boldsymbol{V} = 0\,. \tag{3.6}$$

Using (3.4), we then get

$$\alpha V_1 = \beta V_2 = \gamma V_3 = 0\,, \tag{3.7}$$

with $V_i = \boldsymbol{V}\cdot\boldsymbol{e}_i$, the components of $\boldsymbol{V}$.

If $\alpha=\beta=\gamma=0$, $\bar{C}=C$, implying the existence of a rotation matrix R such that

$$\bar{F} = RF\,. \tag{3.8}$$

With (3.1), this gives

$$\begin{aligned} R &= F(1+\boldsymbol{A}\otimes\boldsymbol{N})\,F^{-1} \\ &= 1 + F\boldsymbol{A}\otimes F^{-T}\boldsymbol{N}\,. \end{aligned} \tag{3.9}$$

This implies that any vector perpendicular to $F^{-T}\boldsymbol{N}$ is an eigenvector of R, with eigenvalue one. Since a nontrivial rotation has only one axis, we have

$$\alpha = \beta = \gamma = 0 \Rightarrow R = 1 \Rightarrow \boldsymbol{A} = 0\,, \tag{3.10}$$

the trivial case. This result is well-known to those familiar with twinning theory.

If two of the three quantities vanish, say

$$\alpha = \beta = 0 \neq \gamma\,, \tag{3.11}$$

it is easy to show that (3.2) – (3.4) imply that $\boldsymbol{N}$, $\boldsymbol{M}$ and $\boldsymbol{A}$ must all be parallel to $\boldsymbol{e}_3$, so we can take

$$\boldsymbol{N} = \boldsymbol{e}_3\,, \quad \boldsymbol{A} = \delta\boldsymbol{e}_3\,. \tag{3.12}$$

With the determinants of F and $\bar{F}$ positive, (3.1) implies that

$$\det(1+\boldsymbol{A}\otimes\boldsymbol{N}) = 1+\boldsymbol{A}\cdot\boldsymbol{N} > 0\,, \tag{3.13}$$

For the particular case at hand, this gives $1+\delta>0$, a mild restriction.

If just one of the three quantities vanish, there is no important loss of generality in assuming that

$$\gamma = 0\,, \quad \alpha\beta \neq 0\,, \quad \boldsymbol{V} = \boldsymbol{e}_3\,. \tag{3.14}$$

Then, with (3.5), we can represent the unit vector $\boldsymbol{N}$ in the form

$$\boldsymbol{N} = \cos\phi\, \boldsymbol{e}_1 + \sin\phi\, \boldsymbol{e}_2 . \tag{3.15}$$

Now $\boldsymbol{M}$, also perpendicular to $\boldsymbol{V}$, must satisfy

$$\boldsymbol{M} \otimes \boldsymbol{N} + \boldsymbol{N} \otimes \boldsymbol{M} = \alpha \boldsymbol{e}_1 \otimes \boldsymbol{e}_1 + \beta \boldsymbol{e}_2 \otimes \boldsymbol{e}_2 . \tag{3.16}$$

An elementary analysis then shows that $\boldsymbol{M}$ has the form

$$\boldsymbol{M} = m\boldsymbol{P} , \quad \boldsymbol{P} = \cos\phi\, \boldsymbol{e}_1 - \sin\phi\, \boldsymbol{e}_2 \tag{3.17}$$

where m is some non-zero scalar. This gives

$$\begin{aligned} \bar{C}_{11} - C_{11} &= \alpha = 2m \cos^2\phi , \\ \bar{C}_{22} - C_{22} &= \beta = -2m \sin^2\phi , \\ \bar{C}_{33} &= C_{33} . \end{aligned} \tag{3.18}$$

The diagonal matrices C and $\bar{C}$ must be positive definite, implying that

$$C_{11} + 2m\cos^2\phi > 0 , \quad C_{22} - 2m\sin^2\phi > 0 . \tag{3.19}$$

Suppose that we are given (diagonal) C, m and ϕ satisfying these conditions. Then (3.15) determines $\boldsymbol{N}$, and (3.17) determines $\boldsymbol{P}$. From (3.3), we must have

$$\boldsymbol{A} = C^{-1} (m\boldsymbol{P} - a\boldsymbol{N}) , \tag{3.20}$$

with

$$2a = \boldsymbol{A} \cdot C\boldsymbol{A} = C^{-1} (m\boldsymbol{P} - a\boldsymbol{N}) \cdot (m\boldsymbol{P} - a\boldsymbol{N}) , \tag{3.21}$$

giving a quadratic equation for a, viz

$$a^2 \boldsymbol{N} \cdot C^{-1} \boldsymbol{N} - 2(m\boldsymbol{P} \cdot C^{-1} \boldsymbol{N} + 1) a + m^2 \boldsymbol{P} \cdot C^{-1} \boldsymbol{P} = 0 . \tag{3.22}$$

It is elementary to verify that our assumptions imply that the two roots are real. Condition (3.13) implies that

$$1 + m\boldsymbol{P} \cdot C^{-1} \boldsymbol{N} > a\boldsymbol{N} \cdot C^{-1} \boldsymbol{N} , \tag{3.23}$$

which selects one of the two roots, that given by

$$\boldsymbol{N} \cdot C^{-1} \boldsymbol{N} a = 1 + m\boldsymbol{P} \cdot C^{-1} \boldsymbol{N} - \sqrt{\Delta} \tag{3.24}$$

$$\Delta = (m\boldsymbol{P} \cdot C^{-1} \boldsymbol{N} + 1)^2 - m^2 (C^{-1} \boldsymbol{P} \cdot \boldsymbol{P})(C^{-1} \boldsymbol{N} \cdot \boldsymbol{N}) . \tag{3.25}$$

With (3.20), this determines $\boldsymbol{A}$. One can take $F = \sqrt{C}$, the usual positive definite square root of the matrix C, use (3.1) to determine $\bar{F}$ and verify that the corresponding $\bar{C}$ is diagonal, etc. So, this summarizes the kind of solutions of (3.1) which are permitted by the constraint (2.9).

4. THE REMAINING CONSTRAINT

In the A-15 superconductors and Indium-Thallium alloys, discontinuities of the kind considered above commonly occur in the tetragonal phase, associated with twinning. For such twins, observations indicate that $\boldsymbol{N}$ takes on one of the six directions given by

$$\sqrt{2}\boldsymbol{N} = \boldsymbol{e}_i \pm \boldsymbol{e}_j \,, \qquad i \neq j \,. \tag{4.1}$$

Clearly, these do not conform to (3.12), but fit (3.15), or equivalents obtained by replacing $\boldsymbol{e}_1$ and $\boldsymbol{e}_2$ by two other base vectors, or their negatives. Also, one has

$$\cos^2\phi = \sin^2\phi = \tfrac{1}{2} \,. \tag{4.2}$$

Briefly, observations indicate that, in the unstressed cubic phase

$$C = c1 \,, \tag{4.3}$$

c being a function of temperature, describing thermal expansion, with $c(T_c) = 1$ a consequence of our choice of reference configuration. In the unstressed tetragonal phase, C can take on any one of three values, of the form

$$\left.\begin{aligned} C_{(1)} &= \operatorname{diag}\,(\nu^2\,, \mu^2\,, \mu^2)\,, \\ C_{(2)} &= \operatorname{diag}\,(\mu^2\,, \nu^2\,, \mu^2)\,, \\ C_{(3)} &= \operatorname{diag}\,(\mu^2\,, \mu^2\,, \nu^2)\,, \end{aligned}\right\} \tag{4.4}$$

with μ and $\nu \neq \mu$ positive functions of temperature. Clearly, these conform to (2.9). If experimentalists had doubts about this, they would not be likely to call these tetragonal phases.

As is discussed in some detail by Ericksen [1], the

possibility of twinning is linked to the assumption that governing constitutive equations are invariant under the cubic point group. It would seem unreasonable not to require this of the constraints. Assuming (2.9) holds, we thus look for a constraint extrapolating (2.8), of the form

$$f(C_{11}, C_{22}, C_{33}) = 0, \tag{4.5}$$

with f a smooth symmetric function of its arguments. Our assumption, that the reference configuration is attainable, gives the condition

$$f(1,1,1) = 0. \tag{4.6}$$

Any such function will satisfy

$$\frac{\partial f}{\partial C_{11}} = \frac{\partial f}{\partial C_{22}} = \frac{\partial f}{\partial C_{33}} = b \quad \text{when} \quad C = 1, \tag{4.7}$$

for some value of b. For infinitesimal deformations, we can linearize f, writing

$$C_{ii} = 1 + 2\varepsilon_{ii}, \quad \text{(no sum)} \tag{4.8}$$

to introduce infinitesimal strains, and use

$$f \approx 2b(\varepsilon_{11} + \varepsilon_{22} + \varepsilon_{33}) = 0. \tag{4.9}$$

Assuming only that $b \neq 0$, we thus get (2.4) in this approximation, so the linear estimate is of little help, in selecting particular choices of f.

With f a symmetric function, if $f=0$ holds for $C = C_{(1)}$, it will hold for $C_{(2)}$ and $C_{(3)}$. Briefly, this is enough to ensure that the standard analysis of twinning in the unstressed tetragonal phase will go through, for any reasonable choice of f. Naturally, $f=0$ should hold for a value of $C_{(1)}$ well approximating observed values, but (4.9) comes close to guaranteeing this, for the crystals considered.

Application of small loads to a twinned crystal may or may not cause the discontinuities to move through the material, perhaps to be created or destroyed. What they seem not to want to do is rotate through the material. That is, they prefer the material planes listed in (4.1), although the loadings cause C to change. In the A-15 superconductors, I have not seen evidence of other kinds of jump discontinuities in F. In Indium-Thallium alloys, cubic and tetragonal phases coexist, in a complicated way. Possibly, some differently oriented discontinuity is in this picture, but I have not seen clear evidence for it, so will ignore this.

So, let us find forms of f which force $\boldsymbol{N}$ to take on just the values given by (4.1). First, suppose that, for all values of C of interest

$$\frac{\partial f}{\partial C_{11}} = \frac{\partial f}{\partial C_{22}} = \frac{\partial f}{\partial C_{33}} \,. \tag{4.10}$$

A general integral of these equations is an arbitrary function of

$$y = C_{11} + C_{22} + C_{33} - 3 \,, \tag{4.11}$$

so, with (4.6), we have

$$f = \phi(y) \,, \quad \phi(0) = 0 \,. \tag{4.12}$$

Since we are only interested in the set where $f=0$ we can, with very little loss of generality, take $f = y$ giving the constraint as (2.10). Bearing in mind (3.4), we see that this excludes possibilities like (3.11), as would various reasonable alternatives. The rest are like (3.18). If $\bar{C}$ and C satisfy (2.10), we see from (3.18) that either $m=0$, the trivial case, or (4.2) holds, implying that $\boldsymbol{N}$ is included in the set given by (4.1). So, this form of f has the desired property.

If f is a symmetric function not covered by the above argument, there will be some value of C such that

$$f(C_{11}\,, C_{22}\,, C_{33}) = 0 \tag{4.13}$$

and

$$\frac{\partial f}{\partial C_{11}} \neq \frac{\partial f}{\partial C_{22}} \Rightarrow C_{11} \neq C_{22} \,. \tag{4.14}$$

Now fix such a C and consider

$$f(\tilde{C}_{11}\,,\tilde{C}_{22}\,,C_{33}) = 0\,, \tag{4.15}$$

with

$$\begin{aligned}\tilde{C}_{11} &= C_{11} + 2m\cos^2\phi\,,\\ \tilde{C}_{22} &= C_{22} + 2m\cos^2\phi\,,\end{aligned} \tag{4.16}$$

as an equation to be solved for m, in terms of ϕ. Since f is a symmetric function, (4.15) will be satisfied at $(m,\phi) = (\hat{m},\hat{\phi})$, where

$$\begin{aligned}\hat{m} &= C_{22} - C_{11} \neq 0\,,\\ \hat{\phi} &\ni \cos^2\hat{\phi} = \sin^2\hat{\phi} = \tfrac{1}{2}\,,\end{aligned} \tag{4.17}$$

since this makes $\tilde{C}_{11} = C_{22}$ etc. Calculating the partial derivative of f with respect to m at this place, we get

$$\frac{\partial f}{\partial m} = \frac{\partial f}{\partial C_{22}} - \frac{\partial f}{\partial C_{11}} \neq 0\,. \tag{4.18}$$

Thus, by the implicit function theorem, we can solve (4.15)–(4.16) for m, locally. Thus, we can pick ϕ near $\hat{\phi}$, not satisfying (4.2), and find a corresponding m. With the analysis given in Section 3, it is then clear that we can construct jump continuities, with N not included in the set given by (4.1).

Essentially, then, (2.10) is the only extrapolation of (2.8) which forces all jump discontinuities to be on the twin planes. As is discussed by Ericksen [1], relevant equilibrium equations have a hyperbolic character and, with the constraints selected, and only with these, the characteristic surfaces also become the twin planes. Clearly, this indicates that these become the possible bearers of various other kinds of weaker discontinuities. As he indicates, one can construct relatively simple Helmholtz free energy functions which, near $T = T_c$, have absolute or relative minima at the cubic phase, described by

$$C = 1\,, \tag{4.19}$$

the form of (4.3) satisfying (2.10), and at the tetragonal phase, described by (4.4), with

$$\nu = \sqrt{3 - 2\mu^2}\;,\quad 0 < \mu < \sqrt{3/2}\;, \tag{4.20}$$

to satisfy (2.10). At least qualitatively, such forms are capable of describing the cubic-tetragonal transformation, twinning, effects of small shear stresses, etc.

There is a rather interesting interpretation of the constraints, not mentioned by Ericksen [1]. In the cubic reference configuration, a diagonal of the unit cell is parallel to the unit vector

$$\boldsymbol{E} = (\pm \boldsymbol{e}_1 \pm \boldsymbol{e}_2 \pm \boldsymbol{e}_3)/\sqrt{3}\,, \tag{4.21}$$

where the algebraic signs can be chosen arbitrarily. If the constraints are satisfied,

$$||\boldsymbol{FE}||^2 = \boldsymbol{E}\cdot\boldsymbol{CE} = (C_{11}+C_{22}+C_{33})/3 = 1 \tag{4.22}$$

so the material is inextensible in all of these directions. Conversely, if it is extensible in all, the constraints (2.9) and (2.10) must hold. After I mentioned this in a lecture, David Parker noted that, for the plane deformations considered by Ericksen [1], the material behaves as if it were inextensible in two in-plane directions, making these deformations analogous to the plane deformations encountered in textiles. In three dimensions, one has something more like a block of foam rubber, with four appropriately oriented systems of inextensible fibres running through it. Possibly, this will help some to picture possible deformations.

5. PHASE INTERFACES

An example of twinning can be obtained as follows. Set

$$F = \sqrt{C_{(3)}} = \mu(\boldsymbol{e}_1 \otimes \boldsymbol{e}_1 + \boldsymbol{e}_2 \otimes \boldsymbol{e}_2) + \nu \boldsymbol{e}_3 \otimes \boldsymbol{e}_3\,, \tag{5.1}$$

with μ and ν satisfying (4.20). Introduce the orthonormal basis

$$\boldsymbol{f}_1 = \boldsymbol{e}_1,\ \boldsymbol{f}_2 = (\boldsymbol{e}_2+\boldsymbol{e}_3)/\sqrt{2}\,,\ \boldsymbol{f}_3 = (\boldsymbol{e}_2-\boldsymbol{e}_3)/\sqrt{2}\,, \tag{5.2}$$

$\boldsymbol{f}_3$ being one of directions listed in (4.1). Set

$$\underline{A} = \sigma \boldsymbol{f}_2,\ N = \boldsymbol{f}_3\,. \tag{5.3}$$

With a little calculation, one finds that

$$\bar{F} = F(1 + A \otimes N) \tag{5.4}$$

satisfies

$$\bar{F}^T\bar{F} = C_{(2)} \tag{5.5}$$

provided

$$\sigma = 6(1-\mu^2)/(3-\mu^2)\,. \tag{5.6}$$

Often, one sees sets of these layers separated by parallel planes of discontinuity, with the deformation gradient alternating between the two values. The thickness of these layers can vary in a random or regular manner.

In Indium-Thallium alloys, observations of coexistence of tetragonal and cubic phases, by Burkart and Read [2], involve such a system of twins in the tetragonal phase. Those layers with F seem to be of about the same thickness, as do those with value $\bar{F}$. However, the two thicknesses are not the same, one being about twice the other. This configuration meets the cubic phase at an interface which is roughly planar, close enough to a plane that one can estimate its crystallographic orientation. As is discussed by Ericksen [1], one cannot fit the above F to a value consistent with another fitting (4.19). This presumed that the constraints are satisfied, but this is not vital. So, indications are that the interface or transition region is more complicated, probably involving inhomogeneous deformation near the interface. Burkart and Read [2] give a "schematic representation" of the interface, perhaps to be regarded as a guess as to how atoms could adjust their positions and avoid great distortions of the latter. Roughly similar configurations are observed in various crystals which undergo martensitic transformations, and we have not yet found any theory to describe any of them, in detail.

Ball and James [3] noticed that one can construct minimizing sequences for the total energy which resemble such configurations, involving regular spaced twins, with the thickness of the layers approaching zero. Such sequences do not converge to energy minimizers. It is not so clear why configurations resembling some of these should be stable enough to be observed.

However, by analyzing the limit, they derive formulae, which seem to describe coarser features of such interfaces rather well. With our constraints, there is doubt as to whether the kinds of deformations needed in the derivation can be constructed. However, it seems interesting to see what the end result predicts.

Involved is a scalar $\lambda \in (0,1)$, interpretable as the fractional thickness of the layers in which $\bar{F}$ occurs, F occurring in the proportion $1-\lambda$. Then, in the obvious sense, the average deformation gradient in the tetragonal phase is

$$\begin{aligned}\langle F \rangle &= \lambda \bar{F} + (1-\lambda) F \\ &= F(1 + \lambda \boldsymbol{A} \otimes \boldsymbol{N}). \end{aligned} \tag{5.7}$$

In the cubic phase, (4.19) should hold, implying that the deformation gradient should be R, some rotation. According to the Ball-James analysis, we should have

$$\langle F \rangle = R(1 + \boldsymbol{B} \otimes \boldsymbol{D}) \tag{5.8}$$

where $\boldsymbol{D}$ is the normal to the nominal plane phase interface, and $\boldsymbol{B}$ is some vector. If $\langle F \rangle$ were the actual deformation gradient in the tetragonal phase, this would, of course, be the usual kinematical conduction of compatibility. Burkart and Read [2] deduce a kind of approximation to this, using a notion of small average strain, neglecting some small rotations, etc.

Here, F, $\boldsymbol{A}$ and $\boldsymbol{N}$ can be regarded as given, but λ, R, $\boldsymbol{B}$ and $\boldsymbol{D}$ are not, so one uses (5.7) and (5.8), i.e.

$$R(1 + \boldsymbol{B} \otimes \boldsymbol{D}) = F(1 + \lambda \boldsymbol{A} \otimes \boldsymbol{N}), \tag{5.9}$$

to try to determine them, it being understood that F, $\boldsymbol{A}$ and $\boldsymbol{N}$ are given by (5.1) and (5.3). We have found no way to avoid some tedious calculations in doing this, and will not record the grim details. It is helpful to begin by noting that, if $\boldsymbol{E}$ is a unit vector perpendicular to $\boldsymbol{D}$ and $\boldsymbol{N}$, (5.9) implies that

$$F\boldsymbol{E} = R\boldsymbol{E} \Rightarrow \| F\boldsymbol{E} \| = 1 . \tag{5.10}$$

It is easy to check that, as the notation suggests, $\boldsymbol{E}$ must be one of the inextensible directions, given by (4.21), one in the subset perpendicular to $\boldsymbol{N}$. This gives two possible directions,

$$\boldsymbol{E} = (\boldsymbol{e}_1 + \boldsymbol{e}_2 + \boldsymbol{e}_3)/\sqrt{3}\,, \tag{5.11}$$

or

$$\boldsymbol{E} = (\boldsymbol{e}_1 - \boldsymbol{e}_2 - \boldsymbol{e}_3)/\sqrt{3}\,. \tag{5.12}$$

One can select either and proceed, with the knowledge that $\boldsymbol{D}$ must be perpendicular to $\boldsymbol{E}$. By taking determinants, one gets

$$1 + \boldsymbol{B}\cdot\boldsymbol{D} = \det F = \mu^2\nu\,. \tag{5.13}$$

Then, one requires other conditions guaranteeing that R, obtained by solving (5.9), is a rotation, the tedious part. With (5.12) as the choice, one gets two solutions. One is of the form:

$$\lambda = (3 + \sqrt{4\mu^2 - 3})/6\,, \tag{5.14}$$

$$\sqrt{2}\,\mu\boldsymbol{D} = \boldsymbol{f}_1 + (\boldsymbol{f}_2 + \sqrt{4\mu^2 - 3}\,\boldsymbol{f}_3)/\sqrt{2}\,, \tag{5.15}$$

$$\boldsymbol{B} = \frac{\sqrt{2}\,\mu(\mu^2 - 1)}{1+\nu}\,[\nu\boldsymbol{f}_1 - (\boldsymbol{f}_2 + \sqrt{4\mu^2 - 3}\,\boldsymbol{f}_3)/\sqrt{2}\,]\,, \tag{5.16}$$

R being calculated from (5.9). Clearly, these exist only if

$$4\mu^2 \geqq 3\,, \tag{5.17}$$

a condition not guaranteed by (4.20). To get the second solution, simply replace $\sqrt{4\mu^2-3}$ by $-\sqrt{4\mu^2-3}$ in these prescriptions. As long as (5.17) holds, the two values of λ lie in $(0,1)$, as they should, and any value in the interval can be obtained, by suitable choice of μ. When equality holds in (5.17), the two solutions coalesce, with $\lambda = \frac{1}{2}$. For a fixed value of μ, the two values of λ sum to one. One gets two similar solutions using (5.11). The two values of $\boldsymbol{E}$ are related by an element of the cubic point group, reversing $\boldsymbol{e}_2$ and $\boldsymbol{e}_3$. Clearly, this leaves F and $\boldsymbol{A}\otimes\boldsymbol{N}$ invariant; one can get this by transforming the entries in (5.9) by the indicated group element. Actually, one of the previous solutions can be converted to the other, by a group element interchanging $\boldsymbol{e}_2$ and $\boldsymbol{e}_3$ but this is a bit more subtle. Altogether, this gives four solutions, the number we should have, according to the work of Ball and James [3], although the breakdown of solutions as $\lambda \to \frac{1}{2}$ was not something we anticipated from the general considerations.

In the observations of Burkart and Read [2], μ is close to one. From Eqn. (5.14), we have

$$\lim_{\mu \to 1} \lambda = \frac{2}{3}, \tag{5.18}$$

in good agreement with their observations. Similarly,

$$\begin{aligned} \lim_{\mu \to 1} D &= [f_1 + (f_2 + f_3)/\sqrt{2}]/\sqrt{2} \\ &= (e_1 + e_2)/\sqrt{2} \end{aligned} \tag{5.19}$$

also in good agreement with their estimate of the crystallographic orientation of the normal to the nominal interface. Interestingly, in this limit, D becomes normal to a twin plane, different from that associated with the twin planes in the tetragonal phase, a possible bearer of discontinuities. If the result were precise, one might have such a plane of discontinuity as the real interface, separating the two phases, with some kind of inhomogeneous deformation near the interface. Conceivably, re-examination of the derivation of (5.9), with the constrained deformation, would modify the result to be in agreement with this. In any event, this is what seems to be suggested, by the "schematic representation" of Burkart and Read [2]. Another possibility might have the twin planes with the normals indicated by (5.19) as surfaces of discontinuity, arranged in stair-step fashion. In places, the twin planes involved in the tetragonal phase might extend, to provide risers, separating one set of twins from the cubic phase. In this picture, the real interface becomes crinkled, but need not stray far from the nominal plane. It is not so obvious that one must have such walls separating the phases; F could vary smoothly, at least in places, to effect the transition. Also, some different kind of crinkling might occur, as suggested below. With electron microscopy, one might decide what kind of picture of the interface fits best, but I have not seen such observations. With our highly constrained deformations, it is not very clear whether one can construct deformations fitting any of these pictures. If one can, it is likely that one can find Lagrange multipliers enabling one to satisfy equations of equilibrium in a weak sense, from Ericksen's [1] discussion of such matters. One also needs some analysis of their stability, which is likely to be tricky. So, we are far from having satisfactory analysis of such phase mixtures.

It seems worthwhile to note that the twin planes in the tetragonal phase intersect those with normal indicated by (5.19) in a line parallel to $\boldsymbol{E}$, given by (5.12), also the direction of the line of intersection of either set of twin planes with the plane with normal $\boldsymbol{D}$. Another set of twin planes containing such lines exist, those with normal parallel to $\boldsymbol{e}_1 + \boldsymbol{e}_3$. With three such systems of planes, one can form an interesting variety of crinkled planes, triangular cylinders, etc., geometrically. Of course, arranging patterns of discontinuities also involves considerations of compatibility conditions, complicating matters.

ACKNOWLEDGEMENT

Much of the research covered in Sections 3 — 5 was done while I was visiting Heriot-Watt University, where I greatly appreciated the warm hospitality, lively discussions, and having some time to think. Especial thanks go to John Ball, for much help in arranging this. Also, discussions with Richard James helped me to make a little progress in understanding the phase mixtures in Indium-Thallium alloys.

REFERENCES

1. J.L. Ericksen, "Constitutive theory for some constrained elastic crystals", to appear in *Int. J. Solids and Structures*.
2. M.W. Burkart and T.A. Read, "Diffusionless phase change in Indium-Thallium systems", *Trans. AIME J. Metals* **197**, (1953), pp.1516-1524.
3. J.M. Ball and R.D. James, "Fine phase mixtures as minimizers of energy", to appear in *Arch. Ratl. Mech. Anal.*
4. B.W. Batterman and C.S. Barrett, "Crystal structure of superconducting V_3Si", *Phys. Rev. Lett.* **13**, (1964), pp.390-392.
5. J.L. Ericksen, "Some phase transitions in crystals", *Arch. Ratl. Mech. Anal.* **73**, (1980), pp.99-124.

12

SOME RESULTS FOR A LINEAR, PARTLY HYPERBOLIC MODEL OF VISCOELASTIC FLOW PAST A PLATE

L. E. FRAENKEL

1. THE PROBLEM

The model in question is due to Joseph [2]. Let $M > 1$ be the viscoelastic Mach number (the fluid velocity at infinity relative to the propagation speed of certain shear waves). The equation governing the vorticity is that preceding (12.5) on p. 285 of [2]. We write x_J, y_J for the coordinates in that equation, introduce a reference length $l := \eta/\rho U$ (in the notation of [2]), and define $b := (M^2-1)^{\frac{1}{2}}$, $x := x_J/b^2 l$, $y := y_J/b^2 l$. Our total stream function is $cy + \psi(x,y)$, where $(c, 0)$ is a dimensionless velocity at infinity; thus the total velocity is $(c+\psi_y, -\psi_x)$ and the vorticity is $-\psi_{xx} - \psi_{yy}$. Denoting the plate and (linearized) shock wave of vorticity respectively by

$$P: = \{(x,y) \in \mathbb{R}^2 \mid x > 0 \ , \quad y = 0\} \, ,$$

and

$$S: = \{(x,y) \in \mathbb{R}^2 \mid x \geqslant 0 \ , \ |y| = x/b\} \, ,$$

as in Fig. 1, we seek $\psi \in C^4(\mathbb{R}^2 \backslash P \backslash S) \cap C^1(\mathbb{R}^2)$ such that

$$\left(\frac{\partial^2}{\partial x^2} + \frac{\partial}{\partial x} - \frac{1}{b^2}\frac{\partial^2}{\partial y^2}\right)\left(\frac{\partial^2}{\partial x^2} + \frac{\partial^2}{\partial y^2}\right)\psi = 0 \ \text{ in } \ \mathbb{R}^2 \backslash P \backslash S \, ; \qquad \text{(1a)}$$

$$\psi_{xx} + \psi_{yy} = 0 \ \text{ in } \ \{(x,y) \mid x < b|y|\} \, ; \qquad \text{(1b)}$$

$$\left(\frac{\partial}{\partial x} + \frac{1}{b}\frac{\partial}{\partial |y|} + \tfrac{1}{2}\right)\left(\psi_{xx} + \psi_{yy}\right) \longrightarrow 0 \ \text{ as } \ x \rightarrow b|y|+ \, ; \qquad \text{(1c)}$$

$$\text{on } \bar{P}, \quad \psi(x,0) = 0 \quad \text{and} \quad \psi_y(x,0) = -c \quad (x \geqslant 0); \tag{1d}$$

$$|\nabla\psi(x,y)| \longrightarrow 0 \quad \text{as} \quad |(x,y)| \longrightarrow \infty \quad \text{with} \quad x < b|y| . \tag{1e}$$

Here (1c) plays the role of a conservation statement for the shock; it ensures that, on the downstream side of S, the tangential derivative of vorticity is not more singular than the vorticity itself.

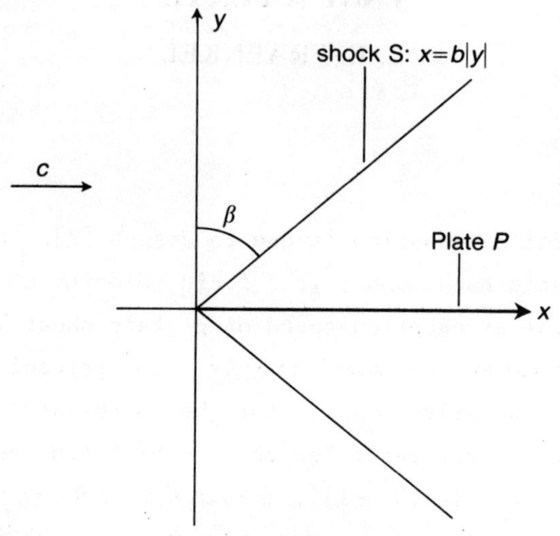

FIG. 1 Notation

We expect ψ to be an odd function of y, and that $|\nabla\psi| \to 0$ at infinity not merely upstream of the shock but outside some boundary layer.

2. REMARKS

Following the lead of Lewis and Carrier [3] for the elliptic problem of Oseen flow past the plate P, we introduce a damping constant $\varepsilon > 0$, solve a perturbed form $(1)_\varepsilon$ of the problem (1) by the Wiener-Hopf technique [4],[7], and then let $\varepsilon \to 0$. The details are in [1]; of course, they are more intricate than for the Oseen problem. It is proved in [1] (for the wider problem in which a general shear kernel G replaces the exponential one of the present Maxwell model) that the function ψ obtained for $\varepsilon = 0$ by this procedure is a solution of (1).

The boundary layer turns out to be parabolic once again:

$$|\nabla\psi(x,y)| \not\to 0 \quad \text{as} \quad x \to \infty \text{ with } \eta := \frac{by}{2\sqrt{x}} \text{ bounded,} \tag{2a}$$

$$\nabla\psi(x,y) = -\frac{2}{\sqrt{\pi}}\frac{c}{b}\nabla\left(r^{\frac{1}{2}}\cos\frac{\theta}{2}\right) + O(r^{-3/2}) \quad \text{as} \quad r \to \infty$$

$$\text{with } \frac{|y|}{x} \geqslant \text{const.} > 0 \text{ when } x > 0\,; \tag{2b}$$

here $x + iy = re^{i\theta}$, $0 < \theta < 2\pi$. The dominant term in (2b) is the disturbance velocity of irrotational flow, with velocity $(c, 0)$ at infinity, past the parabola

$$y^2 = \frac{4}{\pi b^2}x + \frac{4}{\pi^2 b^4},$$

which represents (for $x \to \infty$) the displacement thickness of the retarded fluid near the plate P. Indeed, the far velocity field, both inside and outside the boundary layer, is to the lowest order (as $r \to \infty$) precisely that of the Oseen solution in [3] with an appropriate viscosity.

Uniqueness presents difficulties because at infinity $|\nabla\psi| \not\to 0$ in the boundary layer, and tends to zero only slowly outside it. We add to (1) the following condition on the vorticity along the plate. As

$$x \longrightarrow \infty, \quad \psi_{yy}(x, 0\pm) = a^{\pm}x^{-\frac{1}{2}} + O(x^{-1-\delta}), \quad \delta > 0\,, \tag{3}$$

where the constants a^+ and a^- are not prescribed but may depend on ψ. The result is: *there exists exactly one function* ψ *in* $C^4(\mathbb{R}^2\backslash P\backslash S) \cap C^1(\mathbb{R}^2)$ *that satisfies* (1) *and* (3).

The additional condition (3) may seem contrived. Therefore we point out that, if (0) denotes the counterpart of (1) for the Oseen problem, there exist infinitely many functions ψ in $C^4(\mathbb{R}^2\backslash\bar{P}) \cap C^1(\mathbb{R}^2)$ that satisfy (0) and (3), even if a^+ and a^- are assigned the values of the solution in [3]; see [5] and [6].

3. NOTATION AND SPECIAL FUNCTIONS

(i) Recalling that $b = (M^2 - 1)^{\frac{1}{2}}$, we let

$$\beta : = \tan^{-1} b \in (0, \tfrac{\pi}{2}), \quad \gamma := \tfrac{1}{2} + \tfrac{\beta}{\pi} \in (\tfrac{1}{2}, 1).$$

Then $\frac{\pi}{2} - \beta$ is the Mach angle, and we shall see that the velocity gradients $\psi_{yy}(x, 0+)$, $x > 0$, and $\psi_{yx}(x, 0)$, $x < 0$, are singular like $|x|^{-\gamma}$ as $x \to 0$. (For any function f, the symbol $f(x, 0+)$ denotes the limiting value as y tends to zero from above: $y \to 0+$.)

(ii) We use the Fourier transform

$$\tilde{f}(s, y) : = \int_{-\infty}^{\infty} e^{isx} f(x, y)\, dx, \quad s = \sigma + i\tau,$$

of appropriate functions f, and are led to consider functions holomorphic only outside certain cuts in the complex plane $\mathbb{C}$. Therefore we define (see Fig. 2)

$$\Omega_1 : = \mathbb{C} \setminus \{ i\tau \mid \tau \leqslant -1 \}, \quad -\frac{\pi}{2} < \arg(s + i) < \frac{3\pi}{2} \text{ when } s \in \Omega_1,$$

$$\Omega_+ : = \mathbb{C} \setminus \{ i\tau \mid \tau \leqslant 0 \}, \quad -\frac{\pi}{2} < \arg s < \frac{3\pi}{2} \text{ when } s \in \Omega_+,$$

$$\Omega_- : = \mathbb{C} \setminus \{ i\tau \mid \tau \geqslant 0 \}, \quad -\frac{3\pi}{2} < \arg s < \frac{\pi}{2} \text{ when } s \in \Omega_-.$$

For fractional powers and logarithms of s, we write $s = s_+$ when $s \in \Omega_+$, and $s = s_-$ when $s \in \Omega_-$.

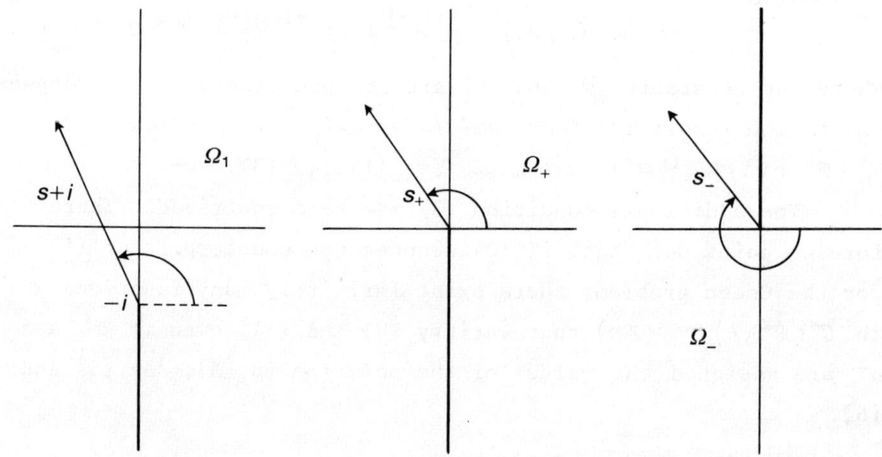

FIG. 2 Cuts and branches of arg in the complex plane

(iii) If we define

$$\left.\begin{aligned} p = p(s) &:= (s_+)^{\frac{1}{2}}(s+i)^{\frac{1}{2}}, \quad s \in \Omega_+, \\ q = q(s) &:= (s_+)^{\frac{1}{2}}(s_-)^{\frac{1}{2}}, \quad s \in \Omega_+ \cap \Omega_-, \end{aligned}\right\} \tag{4}$$

then the Fourier transform on the real s-axis of equation (1a) has solutions $\exp(ibp|y|)$ and $\exp(-q|y|)$ which tend to zero as $|y| \to \infty$ when $s \neq 0$. Note that $\operatorname{Re} ip < 0$ for $\tau = \operatorname{Im} s \geqslant 0$ and $s \neq 0$, and that $\operatorname{Re} q > 0$ for $\sigma = \operatorname{Re} s \neq 0$. The key step in the present application of the Wiener-Hopf technique is to construct a function E, holomorphic and non-zero in Ω_1 and tending to 1 at infinity there, and a function F, holomorphic and non-zero in Ω_- and tending to 1 at infinity there, such that

$$\frac{(s_-)^{\frac{1}{2}} - ib(s+i)^{\frac{1}{2}}}{1-ib} = (s+i)^{\beta/\pi}(s-i)^{\frac{1}{2}-\beta/\pi} E(s) F(s), \qquad s \in \Omega_1 \cap \Omega_-, \tag{5}$$

where $-3\pi/2 < \arg(s-i) < \pi/2$. In the present paper we use only E, but we remark that the Fourier transform of the upstream velocity $\psi_y(x,0)$, $x < 0$, is expressed more naturally in terms of F. We define

$$E(s) := \exp\left\{\frac{1}{\pi}\int_1^\infty \left[\beta - \tan^{-1}\left(b\sqrt{\frac{\rho-1}{\rho}}\right)\right]\frac{d\rho}{\rho - is}\right\}, \qquad s \in \Omega_1; \tag{6}$$

the function $\tan^{-1}$ takes values in $[0, \pi/2]$ throughout the paper. The function E has distinct (but decent) limiting values as $\sigma \to 0+$ or $\sigma \to 0-$ with $\tau < -1$.

As $s \to 0$, $\quad E(s) = E(0)\{1 + \Gamma_1 \text{ is } + O(s^2)\}$, $\qquad$ (7a)

where

$$\Gamma_1 := \frac{1}{\pi}\int_1^\infty \left[\beta - \tan^{-1}\left(b\sqrt{\frac{\rho-1}{\rho}}\right)\right]\frac{1}{\rho^2}\,d\rho. \tag{7b}$$

As $s \to \infty$ in Ω_1,

$$E(s) = 1 + a_1\frac{i}{s}\left(\log s_+ - \frac{i\pi}{2}\right) + A_1\frac{i}{s} + O\left(\left[\frac{\log s}{s}\right]^2\right), \tag{8a}$$

where

$$a_1 := \frac{b}{2\pi(1+b^2)}, \tag{8b}$$

$$A_1 := \frac{1}{\pi}\int_1^\infty \left[\beta - \tan^{-1}\left(b\sqrt{\frac{\rho-1}{\rho}}\right) - \frac{\pi a_1}{\rho}\right] d\rho\,. \tag{8c}$$

(iv) We introduce the function B in the merely formal statement

$$\tilde{\psi}(s,y) \overset{\text{i.s.r.}}{=} \pm(s_+)^{-3/2}B(s)\left\{e^{-q|y|} - e^{ibp|y|}\right\}, \quad \left.\begin{array}{l} + \text{ for } y \geqslant 0, \\ - \text{ for } y < 0, \end{array}\right\} \tag{9}$$

where i.s.r. means 'in some respects'. (The inversion operator

$$\frac{1}{2\pi}\int_{-\infty}^{\infty} e^{-ixs}(\cdot)\, ds$$

cannot be applied directly to (9) if convergent integrals are desired.) Let

$$K := \frac{cb}{E(0)}\exp\left(\frac{i\pi}{4} - \frac{i\beta}{2}\right), \tag{10a}$$

$$B(s) := \frac{K(s+i)^{\beta/\pi}E(s)}{s + b^2(s+i)}, \quad s \in \Omega_1\,; \tag{10b}$$

then

$$B(0) = \frac{c}{b}\exp\left(-\frac{i\pi}{4}\right). \tag{10c}$$

The function B is meromorphic in Ω_1 with a pole only at $-ib^2/(b^2+1)$. Its behaviour as $s \to 0$ and $s \to \infty$ is implied by that of E; in particular, $B(s) = O(s^{-1+\beta/\pi})$ as $s \to \infty$ in Ω_1.

4. RESULTS

(i) *The Stream Function before Differentiation.* Let T and L denote, respectively, the telegraph and Laplace operators in (1a) (which then becomes $TL\psi = 0$ in $\mathbb{R}^2 \setminus P \setminus S$). We present the solution in the form

$$\psi = \psi_L + \psi_T, \quad \text{where} \quad L\psi_L = 0 \quad \text{in} \quad \mathbb{R}^2 \setminus \bar{P},$$
$$T\psi_T = 0 \quad \text{in} \quad \mathbb{R}^2 \setminus P \setminus S\,.$$

The functions ψ_L and ψ_T combine nicely on the x-axis, but for $y \neq 0$ they play rather different rôles. Both are odd functions of y; *we restrict attention to values* $y \geqslant 0$. First,

$$\psi_L(x,y) := \frac{1}{2\pi}\int_{-\infty}^{\infty} e^{-ixs-yq}\,\frac{B(s)-B(0)}{(s_+)^{3/2}}\,ds + \phi_L(x,y)\,, \quad (11)$$

where

$$\phi_L(x,y) := -\frac{2}{\sqrt{\pi}}\,\frac{c}{b}\,r^{\frac{1}{2}}\cos\frac{\theta}{2}\,, \quad (12)$$

with $x+iy = re^{i\theta}$, $0 \leqslant \theta \leqslant \pi$. Next, for any constant $\mu \geqslant 0$,

$$\psi_T(x,y) := -\frac{1}{2\pi}\int_{i\mu-\infty}^{i\mu+\infty} e^{-ixs+ibyp}\,\frac{B(s)-B(0)}{(s_+)^{3/2}}\,ds + \phi_T(x,y)\,, \quad (13)$$

where

$$\phi_T(x,y) := \frac{2}{\sqrt{\pi}}\,\frac{c}{b}\left\{(x-by)^{\frac{1}{2}}e^{-\frac{1}{2}by} + \tfrac{1}{4}by\int_0^{x-by} \xi^{\frac{1}{2}}k(x-\xi,y)\,d\xi\right\}, \quad (14a)$$

$$k(t,y) := e^{-\frac{1}{2}t}\,\frac{I_1\left(\frac{1}{2}\sqrt{t^2-b^2y^2}\right)}{\frac{1}{2}\sqrt{t^2-b^2y^2}}\,, \quad (14b)$$

with I_1 denoting the modified Bessel function of the first kind and of order 1. It is to be understood that $\phi_T(x,y) = 0$ for $x < by$, and this is true also for ψ_T.

The functions ϕ_L and ϕ_T are *far-field* stream functions: the dominant parts of ψ_L and ψ_T for $r\to\infty$. Note that

$$\phi_L(x,0) = -\phi_T(x,0) = \begin{cases} 0, & x \leqslant 0\,, \\ -\dfrac{2}{\sqrt{\pi}}\,\dfrac{c}{b}\,x^{\frac{1}{2}}\,, & x \geqslant 0\,, \end{cases}$$

and that, for $y=0$, the integrals in (11) and (13) also sum to zero. In (11), the dominance of ϕ_L as $r\to\infty$ is uniform in θ; but in (13), ϕ_T dominates the integral only in the boundary layer (that is, as $x\to\infty$ with $\eta := by/2\sqrt{x}$ bounded). For example,

$$\phi_{T,y}(x,y) = -c\{1-\operatorname{erf}\eta\} + O(x^{-1})\,,$$

$$\psi_{T,y}(x,y) - \phi_{T,y}(x,y) = O(x^{-1})\,,$$

as $x\to\infty$ with η bounded. If $x\to\infty$ with $y/x \geqslant$ const. > 0, then $\psi_T(x,y)$ and $|\nabla\psi_T(x,y)|$ are $o(x^{-n})$ for every n. The behaviour of $\nabla\phi_T$ near the shock S is *not* the behaviour of $\nabla\psi_T$ (however large x may be, if $x-by$ is sufficiently small).

(ii) *Derivatives of the Stream Function.* Let $m=(m_1, m_2)$ be a double index, so that $D^m := (\partial/\partial x)^{m_1}(\partial/\partial y)^{m_2}$ is a differential operator of order $|m| := m_1+m_2$. Let $X := \mathbb{R}\times[0,\infty)$ and $X_0 := X\setminus\{(0,0)\}$. For $|m|\geqslant 1$ we have

$$D^m\psi_L(x,y) = \frac{1}{2\pi}\int_{P_L} D^m(e^{-ixs-yq})\,\frac{B(s)}{(s_+)^{3/2}}\,ds, \quad (x,y)\in X_0 \tag{15}$$

$$D^m\psi_T(x,y) = -\frac{1}{2\pi}\int_{P_T} D^m(e^{-ixs+ibyp})\,\frac{B(s)}{(s_+)^{3/2}}\,ds, \quad (x,y)\in X\setminus S, \tag{16}$$

provided that we choose a suitable path of integration P_L or P_T. Referring to Fig. 3, we note that possible paths (ensuring exponential decay of the integrand as $|s|\to\infty$) are $P_L = a$ if $x<0$ and $y\geqslant 0$, or $P_L = R$ (the real axis) if $y>0$, or $P_L = B$ if $x>0$ and $y\geqslant 0$; and P_T as in the figure if $x>by$. For $|m|=1$, the formulae are valid also if $(x,y)=(0,0)$ in the case of (15), or if $(x,y)\in S$ in the case of (16). In fact, ψ_L, ψ_T and their *first* derivatives are continuous on the closed set X; for higher derivatives we have $\psi_L\in C^\infty(X_0)$ and $\psi_T\in C^\infty(X\setminus S)$.

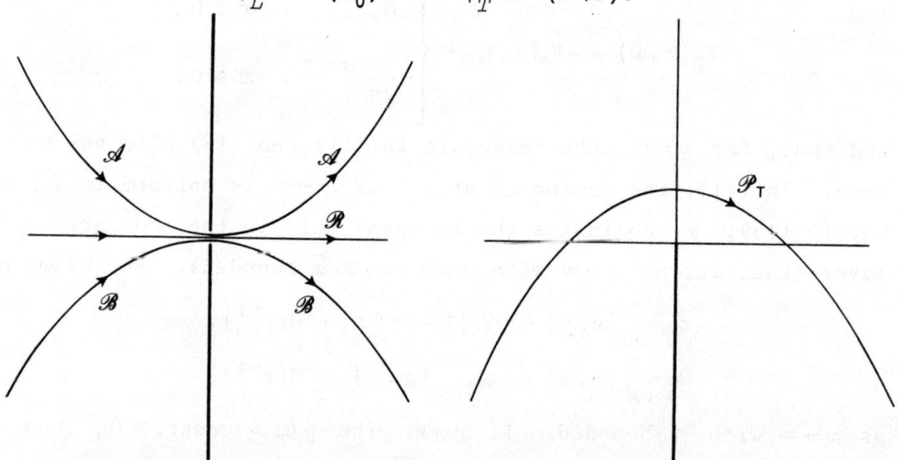

FIG. 3 Possible paths of integration in various contexts

(iii) *The Velocity Gradient* $\psi_{yy}(x,0+)$, $x>0$, *and the Velocity* $\psi_y(x,0)$, $x<0$, *for Small or Large Values of* $|x|$. In the traditional way, we have reversed the natural order of things: the

Wiener-Hopf technique enabled us to find $\tilde{\psi}_{yy}(s,0+)$ and $\tilde{\psi}_y(s,0)$, and these transforms then led to the formulae (11) to (14). Be that as it may, the results (15) and (16) imply the following asymptotic approximations. The factorial and digamma functions

$$\alpha! := \int_0^\infty e^{-t}t^\alpha\,dt\,, \qquad \operatorname{Re}\alpha > -1\,,$$

$$D(\alpha) := \frac{d}{d\alpha}\log(\alpha!)\,,$$

are more convenient than the gamma and psi functions in this context. In the dreadful coefficients that arise, the constants $c, b, \cdots, \Gamma_1$ are as in (1), 3(i) and 3(iii).

As $x \to 0+$,

$$\psi_{yy}(x,0+) = \frac{cb}{E(0)(-\gamma)!}\,x^{-\gamma}\left\{1 + \frac{a_1}{1-\gamma}\,x\log\frac{1}{x} + \frac{a_1 D(1-\gamma) + A_1 + \beta/\pi}{1-\gamma}\,x + O\left(\left[x\log\frac{1}{x}\right]^2\right)\right\}. \tag{17}$$

As $x \to \infty$,

$$\psi_{yy}(x,0+) = \frac{cb}{\sqrt{\pi}}\,x^{-\frac{1}{2}}\left\{1 + \tfrac{1}{2}\left(\Gamma_1 - \frac{\beta}{\pi}\right)x^{-1} + O(x^{-2})\right\}. \tag{18}$$

As $x \to 0-$,

$$\psi_y(x,0) = -c + \frac{cb}{(1+b^2)^{\frac{1}{2}}E(0)(1-\gamma)!}\,|x|^{1-\gamma}\left\{1 + \frac{a_1}{2-\gamma}\,x\log\frac{1}{|x|} + \frac{a_1 D(2-\gamma) - \pi a_1 b + A_1 + \beta/\pi}{2-\gamma}\,x + O\left(\left[x\log\frac{1}{|x|}\right]^2\right)\right\}. \tag{19}$$

As $x \to -\infty$,

$$\psi_y(x,0) = -\frac{c}{\sqrt{\pi}\,b}\,|x|^{-\frac{1}{2}}\left\{1 + \tfrac{1}{2}\left(\Gamma_1 - \frac{\beta}{\pi} + \frac{1+b^2}{b^2}\right)x^{-1} + O(x^{-2})\right\}. \tag{20}$$

Acknowledgements

This paper summarizes calculations and proofs of considerable length, begun at my own university (Sussex), continued at the Centre for Mathematical Analysis of the Australian National University, Canberra, and concluded at Heriot-Watt University. I am most grateful for the friendly hospitality of these last two institutions.

References

1. L.E. Fraenkel, "On a linear, partly hyperbolic model of visco-elastic flow past a plate" (to appear).
2. D.D. Joseph, "Hyperbolic phenomena in the flow of viscoelastic fluids", in *Viscoelasticity and Rheology* (Eds. Lodge, A.S., Renardy, M. and Nohel, J.A.), Academic Press, 1985, pp.235-321.
3. J.A. Lewis and G.F. Carrier, "Some remarks on the flat plate boundary layer", *Quart. Appl. Math.* **7** (1949), pp.228-234.
4. B. Noble, *The Wiener-Hopf Technique*, Pergamon, 1958.
5. W.E. Olmstead, "A homogeneous solution for viscous flow around a half-plane", *Quart. Appl. Math.* **33** (1975), pp.165-169.
6. W.E. Olmstead and D.L. Hector, "On the nonuniqueness of Oseen flow past a half plane", *J. Math. and Phys.* **45** (1966), pp.408-417.
7. R.E.A.C. Paley and N. Wiener, *Fourier Transforms in the Complex Domain*, Amer. Math. Soc., 1934.

13

DERIVATION AND VALIDITY OF THE BOLTZMANN EQUATION: SOME REMARKS ON REVERSIBILITY CONCEPTS, THE H-FUNCTIONAL AND COARSE-GRAINING

REINHARD ILLNER†

Abstract. It is discussed how various concepts of reversibility are lost in the transition from a microscopic to a macroscopic description of a hard sphere particle system, and what parts H-functionals and coarse-graining procedures play in this transition.

INTRODUCTION

The irreversibility of many natural phenomena — like, e.g. heat transfer, particle diffusion or gas flows — remains one of the mysteries of physics, in spite of successful theories like thermodynamics, which contain irreversibility as an ingredient and work well on a macroscopic scale (see, for example, [1]). The difficulty lies at a more fundamental level: the evolution of large particle systems or fluids is *a priori* governed by the laws of mechanics, and these laws are reversible. The emergence of irreversible behaviour must therefore be related to the transition from a microscopic to a macroscopic description of the system, and reversibility is lost in this transition.

Rare gases, whose time evolution is described by the non-linear Boltzmann equation, are the only case I know for which this transition has been analysed rigorously from a mathematical

---

† Research supported in part by grant no. A 7847 from the National Science and Engineering Research Council of Canada.

point of view. The papers [2] and [3] show how solutions of the Boltzmann equation arise from Hamiltonian dynamics, the Liouville equation and the BBGKY-hierarchy in the Boltzmann-Grad limit.

In the present paper, we will review the situation from a more general point of view. More specifically, it will be made precise what we mean by "reversibility" and "transition from microscopic to macroscopic description". The necessary concepts were already listed and applied to the Boltzmann equation in [4] (by H. Neunzert and myself). However, we did not comment on several aspects of the theory, like, e.g., coarse-graining procedures, and I will use this article, among other things, to discuss these aspects.

Since this is a review, I have tried to make it as self-contained as possible (unfortunately, this also means that it is longer than necessary). Section 1 is used to repeat the necessary concepts of reversibility and the notion of an abstract H-functional. The N- particle- hard- sphere gas is introduced as an example. In Section 2, we discuss the transition to a statistical description of the underlying system as a first step in which reversibilities can be lost — a simple discrete example due to Penrose and Goldstein [5] is used to illustrate this.

In Section 3 we review the derivation of the Boltzmann equation for hard spheres from the Liouville equation (through the BBGKY-hierarchy). The reduction to joint distribution densities and the Boltzmann-Grad limit are seen as a second and third step in the transition from microscopic to macroscopic description, and we discuss these steps and the accompanying reversibility losses in connection with the validity result from [3]. The next section is devoted to a philosophical discussion, in which we collect arguments for the statistical description first introduced in Section 2. Section 5 shows how the Boltzmann H-functional fits into the general picture, and we also present a result which shows why the function $h(x) := x \ln x$ arises naturally in this theory. Finally, Section 6 is used to define coarse-graining procedures and to discuss their significance for the Boltzmann equation.

1. FUNDAMENTAL CONCEPTS

Here, we repeat the concepts introduced in [4]. We consider a (physical) system whose state at any time is given as an element of a phase space Γ. It will generally be assumed that Γ carries a Hausdorff topology induced by a metric (it is possible that this assumption can be relaxed, but I know of no relevant example where this is necessary), and that μ is a finite Borel measure on Γ. We shall be concerned with two types of evolution.

DEFINITION 1.1. A family of mappings $(T_t):\Gamma\to\Gamma$, $t\geqslant 0$, is called a semi-flow if all T_t are injective and if $T_t\circ T_s = T_{t+s}$ for all $s,t\geqslant 0$, $T_0 = id$.

A family of mappings $(T_t):\Gamma\to\Gamma$, $t\in\mathbb{R}$, is called a flow if all the T_t are bijective and if $T_t\circ T_s = T_{t+s}$ for all $t,s\in\mathbb{R}$. □

Flows are more important for us than semi-flows, but we will encounter examples for both. All the examples of flows to be considered here will be measurable with respect to μ and even leave μ invariant, i.e. $\mu(T_tA)=\mu(A)$ for all Borel sets A in Γ.

EXAMPLE 1. We introduce Γ and μ associated with a mechanical system of N hard spheres of diameter $d>0$ moving in a bounded domain $\Lambda\subset\mathbb{R}^3$ (with specular reflection at the boundary):

$$\begin{aligned}\Gamma &= (\Lambda\times\mathbb{R}^3)^N\setminus\tilde{\Gamma}, \\ \mu &= \exp\left(-\beta\sum_{i=1}^{N}\xi_1^2\right)dx_1\,d\xi_1\cdots dx_N\,d\xi_N\end{aligned}\tag{1.1}$$

x_i and ξ_i denote the position and velocity of the i-th particle, and β is a positive constant. The evolution (T_t) is the flow defined by free motion of the particles plus elastic collisions at encounters. The set $\tilde{\Gamma}$ is the part of the phase space for which two or more spheres overlap, plus all the phase points whose (forward or backward) evolution leads to a triple or grazing collision. Alexander [6] has shown that the latter set is of measure 0 (with respect to Lebesgue measure and hence with respect to μ) in Γ.

(T_t) is well defined on Γ. The elements of Γ will be called "admissible phase points".

Now we introduce three different reversibility concepts.

(a) *Reversibility with Time Inversion*. Every flow is reversible with time inversion in the sense that $T_{-t} \circ T_t = id$ for all t.

This reversibility concept is the least important for us.

(b) *S- Reversibility*. Suppose that we are given an involution $S:\Gamma \to \Gamma$, $S^2 = id$, which is μ-measurable and actually leaves μ invariant. The flow (T_t) is called S-reversible if $(ST_t)^2 = id$ for all t (or equivalently, $T_t S T_t = S$ for all t).

REMARK. If we drop the condition that S be μ-measurable, then, as shown in [7], every flow admits an involution S such that we have S-reversibility. All the relevant examples, however, satisfy μ-invariance of S.

Continuation of Example 1. For the case of N hard spheres in $\Lambda \subset \mathbb{R}^3$, choose S to be velocity inversion:

$$S(x_1, \xi_1, \cdots, x_N, \xi_N) = (x_1, -\xi_1, \cdots, x_N, -\xi_N).$$

Clearly S leaves μ invariant, is involutive and the flow (T_t) introduced earlier is S-reversible.

(c) *Poincaré-Reversibility*. A (semi-) flow on Γ is called Poincaré-reversible if for each measurable set $A \subset \Gamma$ and almost each $x \in A$ there is a sequence $(t_k)_{k \in N}$ with $t_k \to \infty$, such that $T_{t_k} x \in A$ for all k.

As we have assumed that $\mu(\Gamma) < \infty$ and that all our examples (T_t) preserve μ, the Poincaré recurrence theorem (see [8]) implies that all of them display Poincaré-reversibility. This is in particular true for the hard-sphere system of Example 1.

Irreversible behaviour of a system is usually linked to

the existence of an entropy (or H-) functional which displays strictly monotonic behaviour under the evolution in question. We next give a quite general (and very restrictive) definition of such functionals.

DEFINTIION 1.2. An H-functional belonging to the (semi-) flow (T_t) is a mapping $H: \Gamma \to \mathbb{R}$ such that the functions $t \to H(T_t x)$ are strictly monotonic decreasing as long as $t \to T_t x$ is not constant.

REMARKS

(1) The following construction, which is due to M. Scheutzow [9], shows that every flow for which all the mappings $t \to T_t x$, $x \in \Gamma$, are one to one (in particular, we have neither loops nor equilibria), admits an H-functional. This functional is in general not measurable.

Choose any strictly decreasing function $\beta: \mathbb{R} \to \mathbb{R}_+$. On Γ, we introduce an equivalence relation $\sim$ by $x \sim y$ iff there is a $t \in \mathbb{R}$ such that $x = T_t y$. Γ is the disjoint union of all equivalence classes. Choose exactly one element x from each equivalence class (axiom of choice!), set $H(x) = \beta(0)$ for all x in this set, and $H(y) = \beta(t)$ if y is in the equivalence class $[x]$ and $y = T_t x$.

Clearly, H is an H-functional.

(2) Every strict Ljapunov function associated with an autonomous system of ordinary differential equations is an H-functional; *hence the existence of H-functionals and reversibility with time inversion are compatible.*

(3) Unless μ- almost all elements of Γ are equilibria points with respect to (T_t), the existence of a measurable H-functional and Poincaré-reversibility are <u>not</u> compatible. Hence none of our Poincaré-reversible examples will admit a measureable H-functional.

To prove this, suppose that H is a measurable H-functional on Γ. Then, for a suitably chosen $\alpha \in \mathbb{R}$, we will have

$$\mu\{x; H(x) > \alpha\} > 0 .$$

Let $A' = \{x; H(x) > \alpha\}$ and $A = A' \setminus \{x;\ x$ is an equilibrium point under $(T_t)\}$. It is no restriction of the generality to assume that $\mu(A) > 0$.

As H decreases along all trajectories starting in A, it follows that

$$\mu\Big(\bigcup_{t \geq m} (T_t A \cap A)\Big) < \mu(A)$$

for $m \in N$ sufficiently large. On the other hand, Poincaré recurrence says that

$$\mu\Big(\bigcap_{m \in N}\Big(\bigcup_{t \geq m} (T_t A \cap A)\Big)\Big) = \mu(A) ,$$

and we have a contradiction.

(4) Finally, we note that S-reversibility of a flow and the existence of an S-invariant H-functional for that flow are incompatible. In fact, if H were such a functional, we had the chain of inequalities $H(T_t S T_t x) \leq H(S T_t x) = H(T_t x) < H(x) = H(Sx)$ for all non-equilibria x, and hence $T_t S T_t x \neq Sx$.

2. TRANSITION TO $L^1_+(\Gamma)$

We now take the first of two steps which will transfer us from a microscopic description (given in Section 1) to a macroscopic description of the system under consideration. Motivated by the observation that for realistic systems we can never determine the initial state with ultimate precision, we choose to pass to a statistical description.

(T_t) and S are now generally assumed to leave μ invariant.

Suppose we have an absolutely continuous (with respect to μ) probability measure on Γ, which has density $f \in L^1_+(\Gamma, \mu)$ and gives a statistical distribution over all possible states. The flow (T_t) and the involution S induce a flow and an involution on $L^1_+(\Gamma, \mu)$, which, for convenience, we again denote by (T_t) and S:

$$T_t f(x) = f(T_{-t}x) \quad \text{for all } t \text{ and } \mu\text{- almost all } x \tag{2.1}$$

$$Sf(x) = f(Sx) \quad \text{for } \mu\text{- almost all } x\,. \tag{2.2}$$

Note that (2.1) and (2.2) make sense because of the assumed μ- invariance.

(2.1) defines a flow (T_t) on $L^1_+(\Gamma, \mu)$; this flow is S-reversible if the original flow is, and it is clearly reversible with time inversion. The concept of Poincaré-reversibility, however, does not transfer to this level. We need not expect a transfer of this concept because (a) the Poincaré cycles are in general no real cycles, but bring points only back to small neighbourhoods of the original states, and (b) the "period" of these "cycles" may depend sensitively on the point under consideration, i.e. we may get completely different recurrence times after small perturbations. This loss of Poincaré-reversibility alone can serve as a motivation to make the transition to $L^1_+(\Gamma)$.

What about H-functionals? Straightforward generalization of our earlier definition suggests that an H-functional should now be a functional $H: L^1_+(\Gamma, \mu) \to \mathbb{R}$ such that $t \to H(T_t f)$ decreases unless f is an equilibrium density. It turns out, however, that this definition would exclude some very interesting examples. We shall therefore take a less rigorous approach here and only give a semi-formal definition of an H-functional for the induced flow; we use Example 2 below to explain some of the difficulties.

First, an H-functional need in general only be defined on a dense subset of $L^1_+(\Gamma, \mu)$ (like, e.g., $L^1 \cap L^\infty$). Secondly, we only require that for all f in this subset the functions $t \to H(T_t f)$ are non-increasing, and that $\lim_{t\to\infty} H(T_t f) < H(f)$ "in general".

The semi-formal part lies mainly in the expression "in general"; what we mean here is the generic case, which can have a different meaning in different examples. The popular meaning is that H must eventually decrease in all but a few exceptional situations. Then, just like before, S-reversibility of the flow

(T_t) and the existence of an S-invariant H-functional are not compatible — see Remark 4 at the end of Section 1. On the other hand, if we do not insist on S-invariance, then the following example, due to Goldstein and Penrose [5], displays all three kinds of reversibility on the microscopic level, but admits H-functionals (not S-invariant, of course!) for the induced flow.

EXAMPLE 2. Time is here discretized: $t \in Z$. The flow is defined by means of Baker's transformation T on the phase space

$$\Gamma = \{(p,q);\ 0 \leq p,q \leq 1\}.$$

For $(p,q) \in \Gamma$, we define

$$T(p,q) = \begin{cases} (2p, q/2) \text{ if } p < 1/2 \\ (2p-1, q/2 + 1/2 \text{ if } p \geq 1/2 \end{cases}$$

and the flow (T_t) by $T_t = T^t$, $t \in Z$.

Here, μ will denote the ordinary two-dimensional Lebesgue measure on Γ, and we define S by $S(p,q) = (q,p)$. One easily checks that (T_t) has all three reversibility properties.

If we abbreviate $x := (p,q)$, then the induced flow on L^1_+ is given by $f(t,x) := f_0(T_{-t}x)$. If we define

$$\begin{aligned} \rho(t,p) &= \int_0^1 f(t,p,q)\,dq\,, \\ H[f](t) &= \int_0^1 \rho \ln \rho\ (t,p)\,dp \end{aligned} \tag{2.3}$$

the evolution formula for p (see [5])

$$\rho(t+1,p) = \tfrac{1}{2}[\rho(t, \tfrac{1}{2}p) + \rho(t, \tfrac{1}{2}p + \tfrac{1}{2})] \tag{2.4}$$

and the convexity of the function $h(x) = \begin{cases} 0 & \text{for } x = 0 \\ x \ln x & \text{for } x > 0 \end{cases}$

imply that $H[f](t+1) \leq H[f](t)$, so H is non-increasing along trajectories. By induction, (2.4) implies that

$$\rho(t+k,p) = \frac{1}{2^k} \sum_{i=0}^{2^k-1} p\left(t, \frac{p}{2^k} + \frac{i}{2^k}\right).$$

Suppose for the moment that $H[f](t+k) = H[f](t)$, i.e.

$$\int_0^1 h\left(\frac{1}{2^k}\sum_{i=0}^{2^k-1} \rho\left(t, \frac{p+i}{2^k}\right)\right) dp$$

$$= \int_0^1 h \circ \rho(t,p)\, dp$$

$$= \int_0^1 \frac{1}{2^k}\sum_{i=0}^{2^k-1} (h\circ\rho)\left(t, \frac{p+i}{2^k}\right) dp .$$

Since we have strict convexity of h, it follows that there is for almost all p a value $c(t,p)$ such that

$$\rho\left(t, \frac{p+i}{2^k}\right) = c(t,p) , \quad i = 0, \cdots, 2^k - 1 .$$

In other words, $\rho(t,p)$ must be periodic with period $1/2^k$ (with the exception of a set of measure zero). We conclude that the only densities f for which $H[f](t)$ will never decrease are those for which $\rho(t,\cdot)$ is periodic with period $1/2^k$ for every k, i.e. ρ must be constant a.e. This classifies the exceptional densities for which H will never decrease; the equilibrium $f \equiv 1$ is one such density, but obviously not the only one.

3. THE IDEAL N-PARTICLE GAS, THE BOLTZMANN-GRAD LIMIT AND THE VALIDITY OF THE BOLTZMANN EQUATION

We now return to the setting of Example 1. As before, the induced flow on $L^1_+(\Gamma)$ (see (1.1) for the definition of Γ) is given by

$$T_t^N\, f_0^N(X) = f_0^N\, T_{-t}^N(X) . \tag{3.1}$$

Here, $X \in \Gamma$ denotes an (admissible) phase point. It is, however, convenient to define $f_0^N(X) = 0$ if X is not an admissible phase point; this extends f_0^N to an L^1-function on $(\Lambda \times \mathbb{R}^3)^N$. Similarly, we define $T_t^N f_0^N(X) = 0$ if X in non-admissible.

We have added an index N to the flow operators and densities because we will shortly investigate the limit $N \to \infty$. In reality, we are interested in numbers of the order of magnitude

10^{23}, but it is of course hopeless to use (3.1) in any practical sense then. Therefore, we may as well assume $N\to\infty$ and ask what kind of information can still be extracted from (3.1).

First, let $f_t^N := f_0^N \circ T_{-t}^N$. By (3.1), we find

$$\frac{d}{dt}\left[f_t^N\left(T_t^N X\right)\right] = 0 . \tag{3.2}$$

This is the famous Liouville equation, written in Lagrangian coordinates. Next suppose that f_0^N (and hence f_t^N) is symmetric with respect to all the particles, i.e. invariant under permutations of all the position-velocity coordinate pairs $x_1, \xi_1, \cdots, x_N, \xi_N$; in other words, we assume that we have no way to distinguish the particles from each other. To overcome the difficulty of the astronomically large N, we study the so-called joint distribution densities of f_t^N, defined by

$$f_{k,t}^N(x_1,\xi_1,\cdots,x_k,\xi_k) := \int f_t^N\, dx_{k+1}\, d\xi_{k+1}, \cdots, dx_N\, d\xi_N . \tag{3.3}$$

$f_{k,t}^N$ depends only on k of the N particles and is invariant with respect to permutations of these particles. Note that (3.3) defines $f_{k,t}^N$ only for almost all $(x_1, \xi_1, \cdots, x_k, \xi_k)$, because $f_t^N \in L_+^1(\Gamma)$ and the integration on the right of (3.3) is over a null set in Γ.

Under some mild additional additional assumptions on f_0^N (see [10] for details), the $f_{k,t}^N$ satisfy the so-called BBGKY-hierarchy of equations

$$\frac{d}{dt}\left[f_{k,t}^N\left(T_t^k X^k\right)\right] = \left(C_{k+1} f_{k+1,t}^N\right)\left(T_t^k X^k\right), \tag{3.4}$$

where $f_{k,t}^N := 0$ for $k > N$ and

$$C_{k+1} f_{k+1}^N(X^k) = \sum_{j=1}^{k} (N-k)\, d^2 \int_{S^2}\int_{\mathbb{R}^3} \omega(\xi_{k+1} - \xi_j) \times f_{k+1}^N(x_1,\xi_1,\cdots,x_k,\xi_k,x_j+\omega d,\xi_{k+1})\, d\xi_{k+1}\, d\omega . \tag{3.5}$$

We explain the notation in (3.4 – 5). X^k denotes a k-particle phase point, S^2 is the unit sphere in $\mathbb{R}^3$ and $d\omega$ the surface measure on the sphere. Note that the integration in (3.5) is again over a null set in $(\Lambda \times \mathbb{R}^3)^k$; it is for this reason that additional restrictions on f_0 are needed to derive (3.4) rigorously from (3.2) — see [10].

Let $k=1$ in (3.4). The equation is

$$\frac{d}{dt}\left[f^N_{1,t}\left(x_1+t\xi_1,\xi_1\right)\right] = (N-1)\,d^2 \int_{S^2}\int_{\mathbb{R}^3} \omega(\xi_2-\xi_1)\times$$

$$\times f^N_{2,t}(x_1+t\xi_1,\xi_1,x_1+t\xi_1+\omega d,\xi_2)\,d\xi_2\,d\omega .$$

In the Boltzmann-Grad limit $Nd^2 \to 1/\varepsilon$ (as $N\to\infty$, $d\to 0$), this formally turns into the Boltzmann equation for hard spheres if

$$\lim_{N\to\infty} f^N_{1,t} = f_t$$

(we will not bother to specify the quality of convergence here because at this level we are only concerned with formal manipulations), and, most important,

$$\lim_{N\to\infty} f^N_{2,t}(x_1,\xi_1,x_1+\omega d,\xi_2) = f_t(x_1,\xi_1)\,f_t(x_1,\xi_2)$$

for incoming collisions, i.e. if $\omega(\xi_2-\xi_1) < 0$ (the reverse inequality describes outgoing collisions). The second hypothesis is known as "Stosszahlansatz" or "hypothesis of molecular chaos".

For the limit f_t, we obtain the Boltzmann equation in mild formulation, i.e. $f_t^{\#}(x,\xi) := f_t(x+t\xi,\xi)$ satisfies

$$\frac{d}{dt}\, f_t^{\#} = \frac{1}{\varepsilon}\, C(f_t,f_t)^{\#}\ , \tag{3.6}$$

with

$$C(f,f)(x,\xi) = \int_{\mathbb{R}^3}\int_{S^2_+} (\xi-\eta)[f(x,\xi')\,f(x,\eta') - -f(x,\xi)\,f(x,\eta)]\,d\omega\,d\eta .$$

S^2_+ denotes the hemisphere $\{\omega\in S^2\ ;\ \omega(\xi-\eta)\geqslant 0\}$, and ξ',η' are the post-collisional velocities

$$\xi' = \xi-\omega(\omega(\xi-\eta)),\quad \eta' = \eta+\omega(\omega(\xi-\eta)) .$$

The details of this very sketchy derivation of the

Boltzmann equation can be found in [2] or [11]. For the present discussion, the most important feature of the Boltzmann equation is the existence of a functional

$$H[f](t) := \iint h \circ f_t(x,\xi)\, d\xi\, dx$$

$(h(y) = y \ln y$ for $y > 0$, 0 for $y = 0)$ such that

$$\frac{d}{dt} H[f](t) < 0$$

unless f is an equilibrium solution (for details, see again [2]). This H is the famous Boltzmann H-functional, whose decrease demonstrates the irreversible behaviour of a rarified gas. We will shortly discuss the relation between this H-functional and the H-functionals introduced in Sections 1 and 2. First, however, we will mention results which lift the formal derivation of the Boltzmann equation given above to the rigorous level.

We start with the remark that the $f^N_{k,t}$ given by (3.3) are well defined and satisfy the BBGKY-hierarchy if f_0 is continuous along trajectories (i.e. we require $t \to f^N_t X$ to be continuous for almost all X) and decays fast enough at infinity (see [10]).

Hence the hierarchy is rigorous. As for the limit $N \to \infty$, O. Lanford [2] was the first who gave conditions on the initial data $f^N_{0,k}$ for the hierarchy and f_0 for the equation such that the joint distribution densities $f^N_{k,t}$ would converge to the solution f_t (belonging to initial datum f_0) in a sense sufficiently strong to make the transition from the hierarchy to the equation. His theorem has been formulated and discussed in a number of papers, so I will not repeat it here. The validity of the Boltzmann equation was (for a serious technical reason) only verified for a time interval on the order of magnitude of 1/5 the mean free time between collisions. This does not limit the philosophical significance of the result, because the limiting evolution given by the Boltzmann equation is irreversible even on such a short time interval. Lanford's theorem was reconstructed, with a simplified proof, by Shinbrot [12].

A global result was proven for a two-dimensional gas of hard disks for suitable data and sufficiently large mean free paths in [3]. We use this case here to discuss the necessary conditions and the assertion.

In two space dimensions, the Boltzmann-Grad limit is $N \to \infty$, $d \to 0$, $N \cdot d \to 1/\varepsilon$. We introduce the notation

$$(\mathbb{R}^2 \times \mathbb{R}^2)^{j,d}_{\neq} := \{X^j;\ |x_i - x_k| > d \text{ for } i \neq k\}$$

and make two asumptions on the data:

A1. $f^N_{0,j}$ is continuous on $(\mathbb{R}^2 \times \mathbb{R}^2)^{j,d}_{\neq}$, and

$$\lim_{N\to\infty} f^N_{0,j}(X^j) = \prod_{i=1}^{j} f(x_i, \xi_i) \text{ uniformly on compact}$$

subsets of $(\mathbb{R}^2 \times \mathbb{R}^2)^{j,d}_{\neq}$.

A2. $$\sup_{X^j} |f^N_{0,j}(X^j)| \cdot \exp\left(\beta\left(\sum_1^j (x_i^2 + \xi_i^2)\right)\right) \leqslant \text{const.} \cdot z^j$$

for some $\beta \geqslant 2\exp(1/\varepsilon)$ and some $z > 0$.

We then have the following

THEOREM 3.1, Reference [3]. If z/ε is sufficiently small, then, for all $t > 0$, $j > 0$, we have

$$\lim_{N\to\infty} f^N_{j,t} = \prod_{i=1}^{j} f_t(x_i, \xi_i)$$

a.e., where f_t is the solution of the Boltzmann equation for the initial value f.

For the proof see [3].

Note that the convergence which we assume in A1 is stronger than the convergence given in the assertion of the Theorem. This loss of convergence quality (which is also a feature of Lanford's Theorem) is indispensable in this kind of validity result, as was already pointed out by Lanford. For a thorough discussion, see [4]; we will also discuss this point below.

Let us now return to our reversibility concepts. We took two significant steps in going from the Liouville equation to the Boltzmann equation: First, we studied the joint distribution densities $f^N_{j,t}$, with a particular interest in the evolution of the one-particle distribution density $f^N_{1,t}$. Then we let N go to infinity and arrived at the Boltzmann equation.

We discuss what happens to our different concepts of reversibility in these steps. First, recall that Poincaré-reversibility may already have been lost in the transition to $L^1(\Gamma)$; it will certainly not be recovered by the reduction to joint distribution densities or the limit $N \to \infty$. Second, reversibility with time inversion is obviously preserved in the transition to $L^1(\Gamma)$. Whether it is preserved or lost in the next two steps (consideration of the hierarchy and Boltzmann-Grad limit) is not clear; in particular, we do not know whether the evolution given by the Boltzmann equation is a flow or semi-flow; it is true that if $f(t,\cdot)$ is a solution of the Boltzmann equation with initial value f_0 on $[0,T]$, then

$$\tilde{f}(\tau,\cdot) := f(T-\tau,\cdot)$$

is a solution of the "anti"-Boltzmann equation

$$\partial_t f + \xi\cdot\nabla f = -C(f,f) \tag{3.7}$$

with initial value $f(T,\cdot)$ on $[0,T]$.

The evolution given by (3.7) may (even locally!) be defined on a different (probably smaller) domain than the Boltzmann evolution, but we know too little about the Cauchy problem for the Boltzmann equation to check this. As an example where an evolution fails to be a flow for the reason given above, we mention the evolution defined by the heat equation $\partial_t f = \Delta f$ on $L^1_+(\mathbb{R})$.

Finally, we consider the concept of S-reversibility. For our example, S denotes velocity inversion and has therefore a natural counterpart for the joint distribution densities f^N_j and the (limit) Boltzmann density f:

$$Sf^N_j(x_1,\xi_1,\cdots,x_j,\xi_j) = f^N_j(x_1,-\xi_1,\cdots,x_j,-\xi_j) \tag{3.8}$$

$$Sf(x,\xi) = f(x,-\xi)\ . \tag{3.9}$$

Note that the Boltzmann H-functional is invariant with

respect to S given by (3.9). This, together with the decrease of H along a generic trajectory $t \to f_t$, demonstrates that the Boltzmann evolution is not S-reversible with respect to this S, and it implies that the loss in convergence quality given by Theorem 3.1 is unavoidable. Consider the following diagram (see also [4]):

$$\begin{array}{ccc} f_0^N & \xrightarrow[(L)]{} & f_0^N \circ T_{-t}^N \\ \downarrow s & & \downarrow w \\ f_0 & \xrightarrow[(B)]{} & f_t . \end{array} \tag{3.10}$$

Here s denotes the (strong) convergence given by A1 of Theorem 3.1, and w the (weaker) convergence given in the assertion. The horizontal arrows denote the time evolutions given by the Liouville (L) and Boltzmann (B) equations respectively.

The diagram gives a concise reformulation of Theorem 3.1. The convergence w *cannot* be replaced by s, because we could otherwise set up a diagram

$$\begin{array}{ccc} S(f_0^N \circ T_{-t}^N) & \xrightarrow[(L)]{} & (S(f_0^N \circ T_{-t}^N)) \circ T_{-t}^N \\ \downarrow s & & \downarrow w \\ S(f_t) & \xrightarrow[(B)]{} & (S(f_t))_t \end{array}$$

which would lead to a contradiction: by S-reversibility, we have $(S(f_0^N \circ T_{-t}^N)) \circ T_{-t}^N = Sf_0^N$ and clearly $Sf_0^N \xrightarrow[s]{} Sf_0$, but $H[(S(f_t))_t] < H[Sf_0]$ unless f_0 happens to be an equilibrium solution of the Boltzmann equation. Therefore, the sequence (Sf_0^N) cannot have both Sf_0 and $(S(f_t))_t$ as a limit, and our assumption of s-convergence must have been wrong.

In Section 5 we will see how the loss of convergence quality can actually explain the decrease of H.

4. A PHILOSOPHICAL DIGRESSION

Let us now touch an issue that is quite important for our analysis, but largely inaccessible to the tools of the mathematician. I mean the arguments which can be thought of to justify the "transition to $L^1(\Gamma)$" which was so fundamental for us. Recall what I wrote at the beginning of Section 2: "Motivated by the observation that for realistic systems we can never determine the initial state with ultimate precision, we ... pass to a statistical description." If we reflect on this "justification", we might draw an outrageous conclusion. If the statistical description is the foundation of the emergence of irreversibility (like loss of Poincaré-reversibility, for example), and if our inability to observe microscopic states is the only reason that forces us into statistics, then irreversibility is no natural phenomenon at all, but something that depends on the level of precision which we use to observe nature.

This, of course, is a statement which defies experimental experience. Nobody would believe that the approach to equilibrium (or the direction of heat flow from hot to cold) has anything to do with the perception abilities of the humans who observe it.

I see two better reasons that actually force us to pass to a statistical description. First — even if the evolution of a particle system were only given by the laws of mechanics — there is no ideal gas. Molecules have inner degrees of freedom, they experience friction when they interact with each other, and they interact in many ways with the boundaries of the system. These effects alone make it an impossibility to describe nature by isolated microscopic states and their evolution. Secondly, for those unsatisfied by this argument (strictly speaking, one could still call it an excuse for our inability to observe inner degrees of freedom, friction, etc.), the indeterminacy principle of quantum mechanics gives an ultimate barrier to the precision with which we can describe initial conditions.

The transition $L^1(\Gamma)$ is a possible way to incorporate these observations.

Now let us reflect on the second law of thermodynamics in view of these arguments. Is the second law a fundamental law like, say, energy conservation? The above analysis forces us to make a reinterpretation. Imagine a rare gas of 10^{23} ideal hard spheres (it helps to think of perfect tennis balls), which do not possess inner degrees of freedom or quantum mechanical properties. As mentioned above, there is no such gas, but the thought experiment is still useful. There are clearly some initial states for which this gas will display behaviour incompatible with entropy increase, for example, periodic oscillations. On the other hand, we would still expect that some kind of equilibrium will be approached "for most" initial states; to verify this, we again have no choice but to pass to a statistical description — after all, it is impossible to check the data one by one.

The second law, then, becomes a law which allows exceptions; the exceptions, however, are so rare that we never observe them, and if nature by accident pushed our (non-ideal) gas to an initial state with periodic trajectory, then inner degrees of freedom, quantum mechanical effects, etc. would shortly force it away from this trajectory.

5. SOME REMARKS ON THE BOLTZMANN H-FUNCTIONAL

We have seen how the loss of S-reversibility in the derivation of the Boltzmann equation from the BBGKY-hierarchy, and the decrease of the Boltzmann H-functional, are related to the loss of convergence quality given by Theorem 3.1 (or the diagram (3.10)). More precisely, the loss of convergence quality is a consequence of the decrease of H and can conversely (a) explain the decrease of H and (b) shed some light on the special part of the function

$$h(x) = \begin{cases} x \ln x & \text{for } x > 0 \\ 0 & \text{for } x = 0 \end{cases}$$

in the theory.

For (a) define a functional H^N on a suitable subset of $L^1(\Gamma)$ by

$$H^N(f_0) := \frac{1}{N} \int h \circ f_0^N \, dx_1 d\xi_1, \cdots, dx_N d\xi_N .$$

Suppose that $f_0^N \xrightarrow[s]{} f_0$, that $f_0 \in L_+^1(\Lambda \times \mathbb{R}^3)$ and that $H^N(f_0^N) \longrightarrow H(f_0)$ as $N \longrightarrow \infty$

$$\left(\text{note that } H^N(f_0^N) = H(f_0) \text{ if } f_0(X) = \prod_{i=1}^{N} f_0(x_i, \xi_i)\right),$$

and let $f_t^N (= f_0^N \circ T_{-t}^N)$ and f_t be the solutions of the Liouville equation with initial value f_0^N and the Boltzmann equation with initial value f_0 respectively. We can set up the following diagram

$$\begin{array}{ccc} H^N(f_0^N) & \underset{\text{(Liouville's Theorem)}}{=} & H^N(f_t^N) \\ N\to\infty \big\downarrow & & \\ H(f_0) & > & H(f_t) \end{array} .$$

It follows that

$$\liminf_{N\to\infty} H^N(f_t^N) \geqslant H(f_t), \tag{5.1}$$

with strict inequality in general. On the other hand, if f_0^N and f_0 satisfy the conditions of Theorem 3.1, then $f_t^N \xrightarrow[w]{} f_t$, and (5.1) is also a consequence of the following proposition.

PROPOSITION 5.1. Suppose that $f^N \xrightarrow[w]{} f$ and that there are constants $c, \beta > 0$ such that $f_1^N(x,\xi) \leqslant c\, e^{-\beta(x^2+\xi^2)}$ for all N. Then $H^N(f^N) \geqslant H(f)$, and this inequality is in general strict.

PROOF. From the elementary inequality

$$x - y \geqslant y \,\ln \frac{x}{y} \quad (x \geqslant 0, y \geqslant 0)$$

(we set the right hand side equal to zero for $y = 0$, and equal to $-\infty$ for $y > 0$, $x = 0$) we get

$$0 = \int \cdots \int [f_1^N(x_1,\xi_1) \cdots f_1^N(x_N,\xi_N) - f^N(x_1 \cdots \xi_N)]\, dx_1 \cdots d\xi_N$$

$$\geqslant \int \cdots \int f^N(x_1 \cdots \xi_N)\, \ln \left[\frac{f_1^N(x_1,\xi_1) \cdots f_1^N(x_N,\xi_N)}{f^N(x_1 \cdots \xi_1)}\right] dx_1 \cdots d\xi_N .$$

This implies

$$\int h \circ f_1^N \, dx\, d\xi \leqslant \frac{1}{N} \int \cdots \int h \circ f^N \, dx_1 \cdots d\xi_N .$$

Equality applies if and only if $f^N(X) = \prod_{i=1}^{N} f_1^N(x_i, \xi_i)$ a.e.

Now note that $f^N \underset{w}{\longrightarrow} f$ implies that $f_1^N \longrightarrow f$ a.e. This together with the bounds on the f_1^N, shows

$$\int h \circ f_1^N \, dx \, d\xi \longrightarrow \int h \circ f \, dx \, d\xi ,$$

and the assertion follows.

We see that the decrease of the Boltzmann H-functional is a consequence of the asymptotic loss of factorization of the functions f_t^N in the limit $N \to \infty$. The function h arises quite naturally in this context. Let us ask for which continuous functions $\phi : \mathbb{R}_+ \longrightarrow \mathbb{R}$, differentiable for $x > 0$, we have the inequalities

$$\int \phi \circ f_1^N(x, \xi) \, dx \, d\xi \leq \frac{1}{N} \int \cdots \int \phi \circ f^N(x_1, \xi_1, \cdots, x_N, \xi_N) \, dx_1 \cdots d\xi_N \tag{5.2}$$

for all N and for all normalized symmetric $f^N \in L_+^1(\Gamma)$ for which the integrals on the right exist, with equality if

$$f^N(X) = \prod_{i=1}^{N} f_1^N(x_i, \xi_i) .$$

PROPOSITION 5.2. The only functions satisfying (5.2) are $\phi(x) = c \cdot h(x)$, where c is a nonnegative constant.

PROOF. The proof of Proposition 5.1 shows that $\phi(x) = c \cdot h(x)$ satisfies (5.2). For the converse, suppose that (5.2) holds, choose $f_1^N(x, \xi) = (1/\varepsilon)^6 \cdot \aleph_{W(\varepsilon)}(x, \xi)$, where $\aleph_{W(\varepsilon)}$ denotes the characteristic function of a cube of volume ε^6 in $\Lambda \times \mathbb{R}^3$, and let

$$f^N(x_1, \cdots, \xi_N) = \prod_{i=1}^{N} f^N(x_i, \xi_i) .$$

By assumption, we have equality in (5.2), i.e.

$$\varepsilon^6 \cdot \phi((1/\varepsilon)^6) = \frac{1}{N} \varepsilon^{6N} \phi((1/\varepsilon)^{6N}) \tag{5.3}$$

for all $\varepsilon > 0$ and all N. Let $x = (1/\varepsilon)^6$, then (5.3) becomes

$$\frac{1}{N} \frac{\phi(x^N)}{x^N} = \frac{\phi(x)}{x} \tag{5.4}$$

for all $x > 0$. Writing this as $\phi(x^N) = N \cdot x^{N-1} \cdot \phi(x)$, we read off that $\phi(x) \longrightarrow 0$ as $x \longrightarrow 0$, i.e. $\phi(0) = 0$. On the other hand, let $g(x) := \frac{\phi(x)}{x}$ for $x > 0$, then (5.4) becomes

$$\frac{1}{N} g(x^N) = g(x) . \tag{5.5}$$

Taking derivatives we find

$$g'(x^N) \cdot x^{N-1} = g'(x) , \tag{5.6}$$

or

$$g'(y) = y^{-1+1/N} \cdot g'(y^{1/N}) \tag{5.7}$$

for all $y > 0$. From (5.5) we see that

$$g(1) = 0 . \tag{5.8}$$

Let $g'(1) = c$. Taking the limit $N \longrightarrow \infty$ in (5.7) leads to

$$g'(y) = c/y ,$$

i.e. $g(y) = c \,\ell n\, y + d$. (5.8) implies $d = 0$, and we find

$$\phi(x) = \begin{cases} c \cdot x \cdot \ell n\, x & \text{for } x > 0 \\ 0 & \text{for } x = 0 \end{cases}$$

c cannot be negative because otherwise we would get the wrong inequality in (5.2).

6. COARSE-GRAINING

The usual justification for coarse-graining is again our (subjective or objective — see Section 4) inability to observe microscopic states. Instead, we concentrate on quantities that we hope to be able to measure, like, e.g., the number of particles in a (small) cell. To this end, suppose that we split the *one particle*-phase space $\Lambda \times \mathbb{R}^3$ into a countable set of cells (say, hypercubes) $\Delta_1, \Delta_2, \cdots$, and suppose that the only information we can get about a phase point $X \in \Gamma$ are the numbers $N_i(X)$ of particles in the cell Δ_i, $i = 1, \cdots$.

This is the standard way of setting up a coarse-graining procedure for rarefied gases. The procedure displays invariance with respect to permutations of the particles — an essential feature.

On the statistical level (i.e. after transition to $L^1_+(\Gamma)$) the procedure must also eliminate all distinction between phase points having identical occupation numbers $N_i(X)$; in other words, the feasible L^1_+-densities are those which satisfy $f(X) = f(Y)$ if $N_i(X) = N_i(Y)$ for all i. Such densities are called "coarse-grained".

If we denote the class of admissible (non-negative and symmetric) densities by $L^1_{+,s}$, then the coarse-graining procedure is given by an operator

$$C:\ L^1_{+,s} \longrightarrow L^1_{+,s}$$
$$f \longrightarrow f_C$$

by

$$f_C(x_1,\xi_1,\cdots,x_N,\xi_N)$$
$$= \frac{1}{\lambda^{6N}(\Delta)} \int_\Delta f(x_1,\xi_1,\cdots,x_N,\xi_N)\, dx_1 \cdots d\xi_N, \quad (6.1)$$

where $\Delta = \Delta_1 \times \Delta_2 \times \cdots \times \Delta_N$ and Δ_i is the cell containing (x_i,ξ_i). One easily checks that f_C is symmetric and coarse-grained.

It is convenient to introduce the following notation. Let $\nu_{(y_i,\eta_i)}$ be the probability measure on $\Lambda \times \mathbb{R}^3$ given by

$$\frac{1}{\lambda^6(\Delta_i)} \cdot \aleph_{\Delta_i}(x_i,\xi_i)\, dx_i\, d\xi_i\ ,$$

where Δ_i is the cell containing (y_i,η_i), and, for

$$Y = (y_1,\eta_1,\cdots,y_N,\eta_N)\ ,$$

let

$$\nu_Y = \prod_{i=1}^{N} \nu_{(y_i,\eta_i)}\ . \quad (6.2)$$

This allows us to rewrite (6.1) as

$$f_C(X) = \langle \nu_X, f \rangle\ , \quad (6.3)$$

with ν_X given by (6.2).

Equation (6.3) offers a convenient way to define general coarse-graining procedures for hard sphere particle systems. Suppose that for every point (y,η) in the physical space $\Lambda \times \mathbb{R}^3$, we are given a probability measure $\nu_{(y,\eta)}$ on $\Lambda \times \mathbb{R}^3$. We require that these probability measures are absolutely continuous, that their densities $\rho_{(y,\eta)}$ depend in a measurable way on (y,η) and that they are also normalized such that

$$\int \rho_{(y,\eta)}(x,\xi)\, dy\, d\eta = 1 . \tag{6.4}$$

For $Y=(y_1,\eta_1,\cdots,y_N,\eta_N)$, let ν_Y be the product measure

$$\nu_Y := \prod_{i=1}^{N} \nu_{(y_i,\eta_i)} .$$

Note that ν_Y is absolutely continuous with density

$$R_Y(X) = \prod_{i=1}^{N} \rho_{(y_i,\eta_i)} .$$

DEFINITION 6.1. The family $\{\nu_{(y,\eta)};\ (y,\eta)\in\Lambda\times\mathbb{R}^3\}$ is said to induce a coarse-graining procedure by the formula

$$C_\nu : f \longrightarrow \langle \nu_Y, f\rangle . \tag{6.5}$$

C_ν is called the coarse-graining operator, and the densities $\rho_{(y,\eta)}$ coarse-graining densities.

REMARKS. Condition (6.4) guarantees that C_ν is isometric in $L^1_+(\Gamma)$, because

$$\int \langle \nu_Y, f\rangle\, dY = \iint f(X)\, R_Y(X)\, dX\, dY = \int f(X) \int R_Y(X)\, dY\, dX$$
$$= \int f(X)\, dX .$$

Note that C_ν is in general not a projection; it is a projection if the densities $\rho_{(y,\eta)}$ satisfy

$$\int \rho_{(z,\zeta)}(x,\xi)\cdot\rho_{(x,\xi)}(y,\eta)\, dx\, d\xi = \rho_{(z,\zeta)}(y,\eta), \tag{6.6}$$

because then, as one easily checks, $C_\nu^2 = C_\nu$. Also, we remark that the formula (6.5) defines C_ν for every $N=1,2,\cdots$; this will be useful when we study the limit $N\longrightarrow\infty$.

EXAMPLE 1. For the example which we introduced at the beginning of this section

$$\rho_{(y,\nu)}(x,\xi) = \begin{cases} 1/\lambda^6(\Delta_{y,\eta}) & \text{if } (x,\xi)\in\Delta_{y,\eta} \\ 0 & \text{otherwise} \end{cases}$$

i.e $\rho_{(y,\eta)}(x,\xi) = 1/\lambda^6(\Delta_{y,\eta})\ \ (=1/\lambda^6(\Delta_{x,\xi}))$ if (y,η) and (x,ξ) are in the same cell, and zero otherwise. (6.4) follows from this; an easy calculation shows that (6.6) is also satisfied, so the C_ν induced by these densities is a projection.

EXAMPLE 2. Suppose that μ is a probability density on $\Lambda \times \mathbb{R}^3$, and let $\rho_{(y,\eta)}(x,\xi) = \mu(y-x, \eta-\xi)$. (6.4) is then trivial, but (6.6) is in general violated.

EXAMPLE 3. (A limiting case). We can choose $\rho_{(y,\eta)} = \delta_{(y,\eta)}$. Obviously C_ν becomes the identity operator. The requirement of absolute continuity of the measures $\nu_{(y,\eta)}$ is violated but this case is important as the limiting case where the cells have shrunk to points. (6.4) and (6.5) are satisfied.

Next we mention the action of coarse-graining procedures on joint distribution densities. We assume C_ν is given and abbreviate $f_C := C_\nu f$. The simple relation between f_j and $f_{C,j}$ is

$$f_{C,j}(X^j) = \int f_j(Y^j) \prod_{i=1}^{j} \rho_{(x_i,\xi_i)}(y_i,\eta_i)\, dY^j . \qquad (6.7)$$

(6.7) is an easy consequence of (6.4).

We see that the coarse-graining procedure reduces in a straightforward way to the joint distribution densities.

Formula (6.7) enables us to investigate how a coarse-graining procedure relates to the convergence concepts "s" and "w" introduced in Section 3. For simplicity, we return to a setting similar to the one of Theorem 3.1. Suppose that

$$f^N \xrightarrow[\mathrm{w}]{} f \left(\text{i.e., } \lim_{N\to\infty} f_j^N(X^j) = \prod_{i=1}^{j} f(x_i,\xi_i) \text{ a.e. in } (\Lambda\times\mathbb{R}^3)^j \text{ for all } j=1,2,\cdots\right), \text{ that}$$

$$\sup_{X^j} |f_j^N(X^j)| \leqslant \text{const. } z_0^j \exp\left(-\beta \sum_{i=1}^{j} (x_i^2+\xi_i^2)\right)$$

for some $\beta > 0$ and some $z_0 > 0$, and that we are given a family of coarse-graining densities $\rho_{(y,\eta)}$ on $\Lambda \times \mathbb{R}^3$. Moreover, we assume that there is a constant $K > 0$ such that $\|\rho_{(y,\eta)}\|_{L^\infty} \leqslant K$ for almost all (y,η). We denote by C_ν the coarse-graining operator associated with the densities $\rho_{(y,\eta)}$. Under these conditions, we have

PROPOSITION 6.1. $C_\nu f^N \xrightarrow[s]{} g$ as $N \longrightarrow \infty$, where

$$g(x,\xi) = \int f(y,\eta)\, \rho_{(x,\xi)}(y,n)\, dy\, d\eta .$$

REMARK. We recall here that (6.5) defines C_ν for every N. We will not add an index to the operator.

PROOF. We have to show (see A1 in section 3) that

$$(C_\nu f^N)_j(X^j) \xrightarrow[N\to\infty]{} \prod_{i=1}^{j} g(x_i,\xi_i)$$

uniformly on compact subsets of $(\Lambda \times \mathbb{R}^3)^{j,0}_{\neq}$. Actually, we obtain uniform convergence everywhere. By (6.6), we have

$$(C_\nu f^N)_j(X^j) = \int f_j^N(Y^j) \prod_{i=1}^{j} \rho_{(x_i,\xi_i)}(y_i,\eta_i)\, dy_1 \cdots d\eta_j .$$

Hence

$$\begin{aligned} |(C_\nu f^N)_j(X^j) - \prod_{i=1}^{j} g(x_i,\xi_i)| &= \Big| \int \Big(f_j^N(Y^j) - \prod_{i=1}^{j} f(y_i,\eta_i)\Big) \cdot \\ &\quad \cdot \prod_{i=1}^{j} \rho_{(x_i,\xi_i)}(y_i,\eta_i)\, dy_1 \cdots d\eta_j \Big| \\ &\leq K^j \int \Big| f^N(Y^j) - \prod_{i=1}^{j} f(y_i,\eta_i)\Big|\, dy_1 \cdots d\eta_j , \end{aligned}$$

and the last expression clearly approaches 0 as $N \longrightarrow \infty$ by Lebesgue's Theorem. This completes the proof.

The representation (6.5) makes it easy to investigate how the functionals H^N introduced in Section 5 react to coarse-graining. By Jensen's inequality and (6.4)

$$H^N(C_\nu f^N) \leq H^N(f^N) . \tag{6.8}$$

We get equality in (6.8) if

$$\langle \nu_X, f^N \rangle = f(X^N) \quad \text{for almost all } X . \tag{6.9}$$

Conversely, equality in (6.8) implies

$$h(\langle \nu_X, f^N \rangle) = \langle \nu_X, h \circ f^N \rangle \text{ for almost all } X, \tag{6.10}$$

but it is not clear whether (6.10) entails (6.9). I found neither a proof nor a counterexample.

Formula (6.9) will always be satisfied if C_ν is a projection and if f is already coarse-grained, or if f is simply an eigenvector of C_ν belonging to the eigenvalue 1.

What about the physical meaning of coarse-graining procedures? What we really want is a procedure that "blurs" the microscopic picture, but leaves the macroscopic description of the system unaffected — in other words, the coarse-graining procedure should not be noticeable in the macroscopic description as $N \longrightarrow \infty$. Let us return for the moment to the cell construction at the beginning of this chapter. There, the above requirement means that the cell size (i.e., the maximal diameter of all the cells, also called "mesh") should shrink to zero as $N \longrightarrow \infty$, but such that the number of particles in each cell still goes to ∞. In this way we may hope to get an "unblurred" macroscopic and a "blurred" microscopic description in the limit $N \longrightarrow \infty$.

This target forces us to abandon the idea of using the same coarse-graining procedure for every N. Instead, we shall consider a family of coarse-graining operators $(C^N)_{N=1,2,\cdots}$, given by a family $(\rho^N_{(y,\eta)}(x,\xi))$ of probability densities on the one-particle phase space as in (6.5). To avoid cumbersome language, we will from now on refer to such a "family of coarse-graining procedures" simply as coarse-graining procedures.

The dependence of ρ on N is *a priori* arbitrary; we have to propose criteria which will specify feasible dependencies. A basic requirement which we make is that the assertion of Proposition 6.1 must remain true; more precisely, suppose that the hypotheses of Proposition 6.1 are satisfied for (f^N) and f, then the coarse-graining procedure (C^N) must be such that

$$C^N f^N \xrightarrow[s]{} f \tag{6.11}$$

as $N \longrightarrow \infty$.

This requirement is quite delicate. For example, it excludes that all the C^N are equal to the identity. On the other hand, it implies that the C^N approach the identity operator in the sense that $\rho^N_{(x,\xi)} \longrightarrow \delta_{(x,\xi)}$ in the sense of distributions.

The action of such coarse-graining procedures on the functionals H^N and H from Section 5, is best illustrated by means of the following diagram:

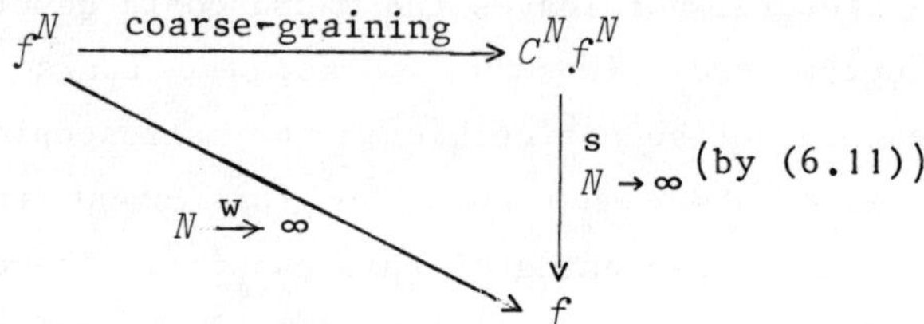

By Proposition 5.1 we have

$$\liminf_{N \to \infty} H^N(f^N) \geqslant H(f) ,$$

and the inequality is in general strict. Also, (6.8) shows that $H^N(f^N) \geqslant H^N(C^N f^N)$. Summarizing, we see that a coarse-graining procedure satisfying (6.11) will supply us with a family of densities $C^N f^N$ which approach f in the strong sense required in A1 of Theorem 3.1. The expense is that the functionals H^N are lowered, i.e. the improvement in convergence quality is paid for by an increase in entropy.

Suppose that

$$\liminf_{N \to \infty} H^N(f^N) > H(f) .$$

Under this assumption, it is an interesting question where the entropy jump actually happens. In the coarse-graining procedure, i.e. do we have

$$\liminf_{N \to \infty} H^N(C^N f^N) < \liminf_{N \to \infty} H^N(f^N) \ ? \ ;$$

or in the limit, i.e. do we have

$$H(f) < \liminf_{N \to \infty} H^N(C^N f^N) \ ? \ ;$$

or do we even have both ?

I have not investigated this question, but it is likely that the answer depends on the particular coarse-graining procedure under consideration. The question is one of many in this theory that seem to deserve more investigation.

Finally, I mention that the concept of coarse-graining developed here displays some differences from the concepts discussed in [5] and [13]. I have confined my attention to hard sphere particle systems, whereas the work in [13] applies to more general flows. Also, I have avoided to allow any time-dependence of the procedures; the point of view taken here is that irreversibility is not *a priori* connected to coarse-graining. Instead, coarse-graining procedures are a natural way to correct the loss of convergence quality, which was here treated as the main feature of irreversibility.

ACKNOWLEDGEMENTS

I am indebted to John Ball for inviting me to the Department of Mathematics at Heriot-Watt University, and to Oliver Penrose for a number of stimulating discussions.

REFERENCES

1. C. Truesdell, *Rational Thermodynamics*, McGraw-Hill, (1969).
2. O. Lanford III, *Time Evolution of Large Classical Systems.* (ed. Moser, E.J.), Lecture Notes in Physics, vol. 38, pp.1-111. Berlin, Heidelberg, New York: Springer (1975).
3. R. Illner and M. Pulvirenti, "Global validity of the Boltzmann equation for a two-dimensional rare gas in vacuum", *Comm. Math. Phys.* **105**, (1986), pp.189-203.
4. R. Illner and H. Neunzert, "The concept of irreversibility in the kinetic theory of gases". *Transport Theory Stat. Phys.* **16**(1), (1987), pp.89-112.
5. S. Goldstein and O. Penrose, "A nonequilibrium entropy for dynamical systems", *Journal Stat. Phys.* **24** (2), (1981), pp.325-343.
6. R.K. Alexander, *The Infinite Hard Sphere System*, Ph.D. Thesis, Department of Mathematics, Berkeley, University of California, (1975).
7. S. Schmid, *S-Reversibilitaet*, Diplomarbeit, Kaiserslautern, (1986).
8. M. Kac, "Probability and related topics in physical sciences", *Interscience* (1959).

9. M. Scheutzow, private communication, Kaiserslautern (1985).
10. R. Illner and M. Pulvirenti, "A derivation of the BBGKY-hierarchy for hard sphere particle systems", *Transport Theory Stat. Phys.* (to appear).
11. C. Cercignani, *Theory and Application of the Boltzmann Equation*, Scottish Academic Press, (1975).
12. M. Shinbrot, "Local validity of the Boltzmann Equation". *Math. Meth. Appl. Sci.* **6** (1984), pp.539-549.
13. O. Penrose, "Entropy and Irreversibility", *Ann. New York Acad. Sci.* **373** (1981), pp.211-219.

14

MICROSTRUCTURE AND WEAK CONVERGENCE

R. D. JAMES

1. INTRODUCTION

Displacive phase transformations are solid-state phase transformations which are accompanied by a change of shape. Often in such transformations the crystal or polycrystal does not change shape uniformly but rather by different amounts on different parts, separated from each other by phase boundaries. The characteristic shapes of these parts are referred to as the microstructure of the material. The goal of the theory discussed in this paper is to predict the detailed microstructure of a material caused by a displacive phase transformation. An equally compelling goal (not discussed here; see [10]) is to predict the behaviour of the crystal under applied loads or fields, which often have a significant and highly specific influence on microstructure.

An extremely diverse set of microstructures is represented by even the simplest class of materials — those undergoing reversible displacive phase transformations without diffusion. In this paper we confine attention to such materials. A large class of mostly metallic alloys have a microstructure known as martensite [7, 17]. The basic feature of this microstructure is a plane interface with fine ($< \sim 20\mu m$) parallel bands known as twins on one side (Fig. 1a). Such interfaces comprises the boundaries of wedge-shaped regions or platelets characteristic

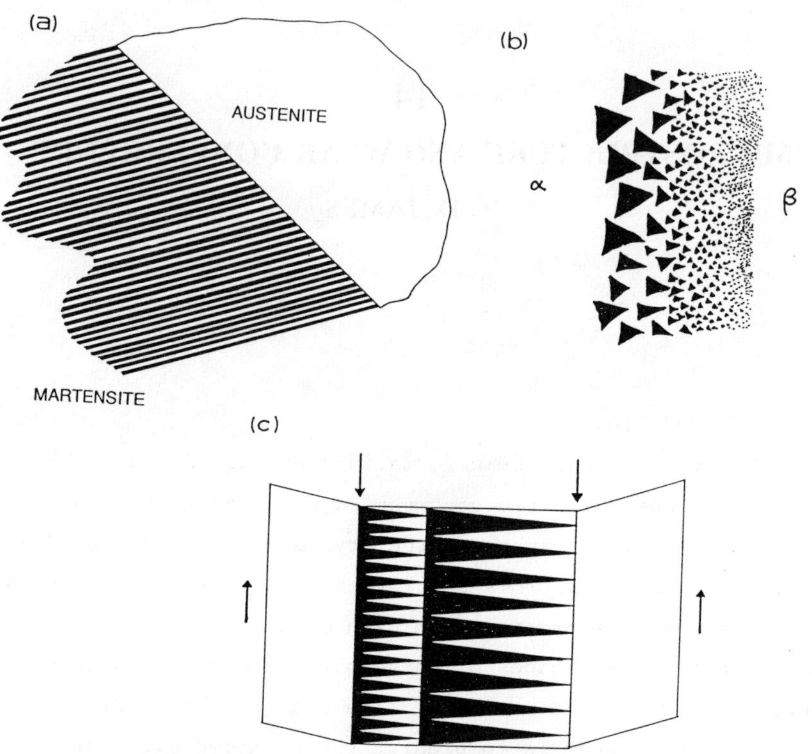

FIG. 1 (a) martensite, (b) Dauphiné twins in quartz, (c) zig-zag domains in neodymium pentaphosphate

of the material. A different kind of microstructure occurs in quartz plates held in a temperature gradient which includes the $\alpha-\beta$ transformation temperature. It consists of variants of one phase known as Dauphiné twins [16]. These variants occupy triangles which get finer and finer in the direction of increasing temperature. A third example is neodymium pentaphosphate [10] which undergoes an orthorhombic to monoclinic transformation at about 146°C. The monoclinic phase can be coaxed with small forces to adopt the zig-zag pattern shown in Fig. 1c and this pattern persists in the unloaded crystal. Many more examples of this general nature could be cited. In fact, it is difficult to find any microstructure produced by a displacive phase transformation which includes both parent and transformed phases but which does not involve some fine geometric features.

A possible reason for the existence of such fine features, explored in the case of internally twinned martensite by Ball and James [2], is the following. By virtue of symmetry, the free energy function of a material of this kind, regarded as a function of the deformation gradient F ($F = Dy$, $y:\Omega \to \mathbb{R}^3$, $\Omega \subset \mathbb{R}^3$), has various potential wells. The free energy is also a function of temperature θ and the wells vary with θ. At the transformation temperature θ_0 the minimizers of all of the wells have equal free energy. For energetic reasons the material likes to be at or near potential well minima. As the crystal is cooled through the transformation temperature, it must pass from one set of wells to the complementary set. Since the wells are disjoint and the functions y that we consider are continuous (no fracture), then in passing from one well to others, surfaces of discontinuity of Dy must appear in the crystal. It is then a matter of calculation whether the minimizers of these potential wells satisfy classical conditions of compatibility which allow such surfaces of discontinuity to form. The location of these minima are known from measurements of a certain transformation strain matrix and from the given symmetry of the crystal. It is found that these classical conditions of compatibility are *not* satisfied between the appropriate potential wells in most cases. However, while this precludes the possibility of classical interfaces, it is possible in some cases to mix deformation gradients from one set of potential wells in finer and finer arrays so as to achieve compatibility with other wells in the limit of infinite fineness. This idea is worked out for internally twinned martensite by Ball and James [2] and it is conjectured that a similar phenomenon is responsible for the microstructures shown in Figs. 1(b) and 1(c).

A useful notion for the study of such problems is the idea of weak convergence. Suppose that $S_p \cup S_t$ represents the set of deformation gradient matrices at potential well minima with p associated with the parent and t with the transformed phase. The crystal would like to have $Dy \in S_p \cup S_t$ on Ω, with each subset represented on a set of positive volume in the crystal. With S_p and S_t arising from measured transformation strains for most crystals, this is not possible with continuous, piecewise differentiable functions as explained above. However, one can consider a sequence of functions $\{y_k\}$, bounded in $W^{1,\infty}(\Omega, \mathbb{R}^3)$, such that for each k, $Dy_k \in S_p \cup S_t$ on $\Omega_k \subset \Omega$ with $\mathrm{vol}(\Omega - \Omega_k)$ tending to zero as $k \to \infty$ and with the volume fraction of Ω in S_p fixed. Such sequences have subsequences which have weak $*$ limits in $W^{1,\infty}$. In reality the crystal does not go to the limit but lives with a very small set on which its deformation y does not satisfy $Dy \in S_p \cup S_t$. Nevertheless, the notion of weak convergence goes far toward explaining many details of fine microstructures including observed orientations of approximate interfaces.

The analysis of these sequences raises some questions which appear to have a fundamental status. The analysis of the microstructures of Figs. 1(b) and 1(c) as well as more complicated martensitic microstructures would be aided greatly by a knowledge of what are the limits of such sequences and what are the appropriate nonclassical conditions of compatibility. The purpose of this paper is to review the evidence for this approach and to highlight these open questions in the context of models for phase transformations.

2. MINIMIZING SEQUENCES ASSOCIATED WITH MARTENSITE

We confine attention to cubic to tetragonal transformations to explain the ideas. See Ball and James [2] for a detailed discussion. Consider a single crystal which is cubic above a certain *transformation temperature* θ_0. Its symmetry is summarized by the group P_c of twenty-four rotation matrices which map a cube into itself. In discussions of such transformations the cubic phase is termed *austenite*. Let $\{e_1, e_2, e_3\}$

be a right-handed orthonormal basis which represents the cubic axes and let the regular region $\Omega \subset \mathbb{R}^3$ represent an undistorted homogeneous body of austenite at a uniform temperature just above θ_0.

The tetragonal phase can be thought of as arising from the cubic phase by a change of shape

$$y(x) = U_3 x, \quad x \in \Omega, \tag{2.1}$$

where U_3 is a positive-definite symmetric matrix having the form

$$U_3 = \eta_1 1 + (\eta_2 - \eta_1)\, e_3 \otimes e_3. \tag{2.2}$$

Typical values for the scalars η_1 and η_2 are $\eta_1 \doteq 0.987$ and $\eta_2 \doteq 1.026$ for the alloy InTl [5-7] at about 20% Tl, one of the most convenient for study because the transformation temperature is near room temperature and the martensite bands are relatively fat and can be seen with an optical microscope. Other alloys, MnCu [4] for example, have the property that $\eta_1 < 1 < \eta_2$ so that the tetragonal phase is obtained by a compression along the e_3 axis rather than an extension as in InTl. Both alloys have qualitatively the same microstructure consisting mainly of interfaces as in Fig. 1(a). For the discussion below we assume that $\eta_1 > 0$, $\eta_2 > 0$, $\eta_1 \neq \eta_2$.

Let y be a deformation of Ω. We assume $y \in W^{1,\infty}(\Omega, \mathbb{R}^3)$ with $\det Dy(x) > 0$ a.e. We also assume that the free energy per unit volume in Ω of a crystal subject to a deformation y is

$$\phi(Dy(x), \theta), \quad x \in \Omega, \tag{2.3}$$

where θ is the temperature. The function $\phi(\cdot, \theta)$ is defined for all matrices with positive determinant.

We assume $\phi(\cdot, \theta)$ is Galilean invariant: for each rotation matrix R and every matrix F with $\det F > 0$,

$$\phi(RF, \theta) = \phi(F, \theta). \tag{2.4}$$

To formalize the requirement that every rotation in the cubic group P_c applied to Ω before deforming it does not affect the energy (recall that Ω is interpreted as an undistorted cubic crystal), we assume that for every $\bar{R} \in P_c$,

$$\phi(F\bar{R}, \theta) = \phi(F, \theta), \tag{2.5}$$

which holds for all F, $\det F > 0$. Let

$$\phi_0(F) = \phi(F, \theta_0) \tag{2.6}$$

denote the free energy per unit volume in Ω at the transformation temperature.

Since both the cubic and tetragonal phases are observed to coexist at θ_0, we assume that ϕ_0 has potential wells at 1 corresponding to austenite and at the matrix U_3 corresponding to the tetragonal phase. However, if 1 and U_3 minimize ϕ_0, so will any matrices of the form $RU_3\bar{R}$ or of the form R, where R is a generic rotation and $\bar{R} \in P_c$. A little calculation shows that this reduces to four distinct potential wells with minima at

$$R1, \; RU_1, \; RU_2, \; RU_3, \tag{2.7}$$

where R is a generic rotation and

$$U_i = \eta_1 1 + (\eta_2 - \eta_1)\, e_i \otimes e_i \quad \text{(no sum)}. \tag{2.8}$$

Specifically, we have assumed

$$\phi_0(1) = \phi_0(U_1) = \phi_0(U_2) = \phi_0(U_3) < \phi_0(U) \tag{2.9}$$

for every positive definite symmetric matrix U which is unequal to $1, U_1, U_2, U_3$. The 2nd and 3rd equalities in (2.9) are redundant by invariance and (2.9) is consistent with the invariance embodied in (2.4) and (2.5).

According to Gibbs [7], *stable* deformations $y: \Omega \to \mathbb{R}^3$ of an unloaded crystal at the transformation temperature minimize

$$\int_\Omega \phi_0(Dy(x))\, dx. \tag{2.10}$$

The minimizers of (2.10) in the space

$$\{y \in W^{1,\infty}(\Omega, \mathbb{R}^3) : \det Dy(x) > 0 \text{ a.e.}\}$$

are all such functions satisfying almost everywhere

$$Dy^T Dy \in \{1, U_1^2, U_2^2, U_3^2\}. \tag{2.11}$$

Subsets of Ω on which $Dy^T Dy \in \{U_1^2, U_2^2, U_3^2\}$ are associated with *variants of martensite* while subsets on which $Dy^T Dy = 1$ are associated with *austenite*.

A nice characterization of the deformations satisfying (2.11) appears to be unavailable; we discuss this further in Section 5. However, Reshetnyak [14, Corollary of Lemma 3] shows that any $y \in W^{1,\infty}(\hat{\Omega}, \mathbb{R}^3)$ satisfying

$$Dy^T Dy = \hat{U} \tag{2.12}$$

where $\hat{U}$ is a constant positive definite matrix and $\hat{\Omega} \subset \Omega$ is open and connected is in fact the affine deformation

$$y = \hat{R}\hat{U}x + c, \quad x \in \hat{\Omega}, \tag{2.13}$$

$\hat{R}$ being a constant rotation and c a constant vector. Hence, it is natural to look for piecewise affine deformations satisfying (2.11). For this purpose, it is useful to note the following.

PROPOSITION 2.1 (Ball and James). <u>Necessary and sufficient conditions for a symmetric</u> 3×3 <u>matrix</u> C <u>with eigenvalues</u> $\lambda_1 \leqslant \lambda_2 \leqslant \lambda_3$ <u>to be expressible in the form</u>

$$C = (1 + m \otimes b)(1 + b \otimes m) \tag{2.14}$$

<u>for nonzero</u> b <u>and</u> m <u>are that</u> $\lambda_1 \geqslant 0$ <u>(i.e.</u> $C \geqslant 0$) <u>and</u> $\lambda_2 = 1$. <u>If</u> $1 + b \cdot m > 0$ <u>then</u> $C \neq 1$. <u>The solutions with</u> $1 + b \cdot m > 0$ <u>are given</u> by

$$\begin{aligned} b &= \rho\left[\sqrt{\frac{\lambda_3(1-\lambda_1)}{\lambda_3-\lambda_1}}\, e_1 + \kappa\sqrt{\frac{\lambda_1(\lambda_3-1)}{\lambda_3-\lambda_1}}\, e_3\right], \\ m &= \rho^{-1}\left[\left(\frac{\sqrt{\lambda_3}-\sqrt{\lambda_1}}{\sqrt{\lambda_3-\lambda_1}}\right)\left(-\sqrt{1-\lambda_1}\, e_1 + \kappa\sqrt{\lambda_3-1}\, e_3\right)\right], \end{aligned} \tag{2.15}$$

where $\rho \neq 0$ is constant and e_1, e_3 are normalized eigenvectors of C corresponding to λ_1, λ_3 respectively, and where κ can take on the values ± 1.

With this in mind we summarize the possible classical interfaces below.

(1) *Austenite/Austenite Interfaces.* Referring to (2.11) these are governed by the equation

$$R^+ - R^- = a \otimes n, \tag{2.16}$$

where R^+ and R^- are rotation matrices, or equivalently

$$R = 1 + a' \otimes n, \tag{2.17}$$

where $R = (R^-)^T R^+$ is another rotation. By the proposition, $R = 1$; that is, $R^+ = R^-$ and there are no austenite/austenite interfaces. No such interfaces are ever observed in austenite.

(2) *Martensite/Martensite Interfaces.* These are governed by an equation of the form

$$R^+ U_i - R^- U_j = a \otimes n, \qquad i, j \in \{1, 2, 3\}, \tag{2.18}$$

where R^+ and R^- are rotations. If $i = j$ then (2.18) can be immediately reduced to an equation of the form (2.17) and no interfaces are possible. If $i \neq j$ we can apply the proposition with $C = U_j^{-1} U_i^2 U_j^{-1}$ which by (2.8) equals

$$\frac{\eta_2^2}{\eta_1^2} e_i \otimes e_i + e_k \otimes e_k + \frac{\eta_1^2}{\eta_2^2} e_j \otimes e_j, \tag{2.19}$$

with $k \in \{1, 2, 3\}$, $k \neq i$, $k \neq j$.

Since we have assumed $\eta_1 \neq \eta_2$ and $\eta_1 > 0$ and $\eta_2 > 0$, the hypotheses of the proposition are satisfied. In summary, given any matrix of the form RU_i, there are exactly two rank-one connections to each of the other two martensitic potential wells and no rank-one connections between one variant and itself. To write down these solutions we take $R^- = 1$ without loss of generality; all remaining solutions are obtained by premultiplying (2.18) by a rotation matrix. By the proposition the solutions of (2.18) with $R^- = 1$ are

$$n = \frac{1}{\sqrt{2}} (\pm e_i + e_j) , \quad a = \frac{\sqrt{2}\,(\eta_2^2 - \eta_1^2)}{\eta_2^2 + \eta_1^2} (\pm \eta_1 e_i - \eta_2 e_j) , \tag{2.20}$$

with $\pm$ signs taken in parallel. R^+ is obtained from (2.18) and (2.20). The solution represented by (2.20) are called twins (cf. Gurtin [9]). They describe the bands to the left in Fig. 1(a). In fact these bands are always found to occur on planes represented by $(2.20)_1$, the so called $\{110\}$ family of planes, and the changes of shape implied by (2.20) agree precisely with experiment. Furthermore, no twins are ever observed between one variant of martensite and itself.

(3) *Austenite/Martensite Interfaces*. These are governed by an equation of the form

$$R^+ U_i - 1 = a \otimes n . \tag{2.21}$$

Again, with $\eta_1 \neq 1$ and $\eta_2 \neq 1$ the equation (2.21) has no solutions by the proposition. However, the region to the right of the bands in Fig. 1(a) is austenite, so some kind of austenite/ martensite is represented in Fig. 1(a).

To understand the austenite/martensite interface, we consider the following line of thought. Because of the existence of martensite/martensite interfaces, the free energy ϕ_0 is not rank-one convex and the associated functional (2.10) is not lower semicontinuous on $W^{1,\infty}(\Omega, \mathbb{R}^3)$. To see this, we can consider a laminate having layers of alternating thickness $\lambda, 1-\lambda, \lambda, 1-\lambda, \cdots, \lambda \in (0,1)$, on which is defined

$$Dy(x) = \begin{cases} F^+ & \text{for} \quad k \leqslant x \cdot n < k+\lambda , \\ F^- & \text{for} \quad k+\lambda \leqslant x \cdot n < k+1 , \end{cases} \tag{2.22}$$

k being an integer. We assume that this deformation is twinned martensite, that is,

$$F^{+} = R^{+}U_i\,, \qquad F^{-} = R^{-}U_j$$
$$F^{+} - F^{-} = a \otimes n\,, \quad i,j \text{ fixed} \tag{2.23}$$

Because of $(2.23)_3$ we may choose $y \in W^{1,\infty}(\mathbb{R}^3, \mathbb{R}^3)$ and $y(0) = 0$. Now let

$$y^{(j)} = j^{-1}\, y(jx)\,. \tag{2.24}$$

Since the transformation (2.24) preserves gradients, $y^{(j)}$ is a bounded sequence in $W^{1,\infty}(\Omega, \mathbb{R}^3)$ and therefore has a weak * limit in $W^{1,\infty}(\Omega, \mathbb{R}^3)$:

$$y^{(j)} \xrightarrow{\;*\;} \bar{y}\,. \tag{2.25}$$

The function $\bar{y}$ can be calculated from the specific form of $y(x)$, either by calculating the constants of integration implied by (2.22) and (2.23) or by using ideas of the weak convergence of periodic functions [15]. The result is

$$\bar{y}(x) = (\lambda F^{+} + (1-\lambda) F^{-})\, x\,, \quad x \in \Omega\,. \tag{2.26}$$

Because of (2.25),

$$\sup_{x \in \Omega} \left| y^{(j)} - \bar{y} \right| \to 0 \quad \text{as} \quad j \to \infty\,. \tag{2.27}$$

Generally, $\bar{y}$ is not a minimizer of the total free energy (2.10). In particular, if $\eta_1 \neq 1$ and $\eta_2 \neq 1$ it follows from Ball and James [2, Theorem 7] that $\lambda F^{+} + (1-\lambda) F^{-}$, $\lambda \in (0,1)$, is not a rotation with F^{+} and F^{-} satisfying the twinning relations (2.23). Also, by looking at the trace of $\lambda F^{+} + (1-\lambda) F^{-}$, we find that if $\eta_1 \neq 1$, $\eta_2 \neq 1$ and $\eta_1 \neq \eta_2$, then $\lambda F^{+} + (1-\lambda) F^{-}$, $\lambda \in (0,1)$, does not assume the values $R_1 U_1$, $R_2 U_2$ or $R_3 U_3$ for any rotation matrices R_1, R_2, R_3. Hence, under these mild conditions on η_1 and η_2, $\bar{y}$ is not a minimizer.

Returning to the austenite/martensite interface, we now consider the possibility that a finely twinned martensite laminate achieves compatibility with austenite. On one side of a plane interface $\{x : x \cdot m = 0\}$ we construct a laminate y as in (2.22) and put the identity $y = x$ on the other side. The resulting deformation $y:\mathbb{R}^3 \rightarrow \mathbb{R}^3$ is not continuous because (2.20) has no solutions. However, we can change the deformation by putting a transition layer in the region $\{x: -1 < x \cdot m < 1\}$. The resulting continuous deformation, still called y, is not a minimizer because the deformation gradients in the interpolated layer are not at potential well minima. However, it is found that this interpolated layer can be constructed with a uniformly bounded gradient if and only if

$$(\lambda F^{+} + (1-\lambda) F^{-}) - 1 = b \otimes m. \tag{2.28}$$

Assuming the layer has been constructed with a uniformly bounded gradient the sequence of deformations

$$y^{(j)}(x) = j^{-1}\, y(jx), \quad x \in \Omega, \tag{2.29}$$

is a minimizing sequence for the total energy. That is, all deformation gradients are at potential well minima except for those in the layer. However, those in the layer are bounded and the volume of the layer goes to zero as $j \to \infty$. Equation (2.28) is just the condition that the limiting deformation of the laminate $\bar{y}$ is compatible with the deformation $y(x) = x$ in a classical sense.

The condition (2.28) is also necessary in a certain sense ([2, Theorem 3]). Roughly, if a sequence of weakly convergent deformations takes on gradients near two matrices F^{+} and F^{-} on one open connected set and near the matrix 1 on another open connected set with $F^{+} - 1$ and $F^{-} - 1$ not rank-one, then the limiting deformation has the essential features of Fig. 1(a) — a plane interface separating an infinitesimal laminate of martensite from austenite — and (2.28) holds.

When the twinning relations (2.23) are substituted into (2.28), it becomes an algebraic problem for the determination of a rotation $\hat{R}$ (which premultiplies (2.28)), the vectors b and m, and the spacing parameter λ. After rearrangement, (2.28) becomes

$$U_i + \lambda a \otimes n = \hat{R}^T(1 + b \otimes m). \tag{2.30}$$

The following governs solutions $\{\hat{R}, \lambda, b, m\}$ of (2.30) and is not restricted to the special form of U_i we have chosen.

THEOREM 2.2 (Ball and James). Let the positive definite symmetric 3×3 matrix U satisfy the twinning relation

$$RU\bar{R} = U + a \otimes n \tag{2.31}$$

for some pair of rotations R and $\bar{R}$ and for vectors $a \neq 0$, n, $|n| = 1$.

I. Assume U does not have an eigenvalue equal to 1. Necessary and sufficient conditions that

$$U + \lambda a \otimes n = \hat{R}^T(1 + b \otimes m) \tag{2.32}$$

has a solution consisting of a rotation $\hat{R}$ a scalar λ and a rank one matrix $b \otimes m$ (hereafter termed a solution) are that

$$1 + \tfrac{1}{2}\delta^* \leqslant 0 \tag{2.33}$$

and that

$$\operatorname{tr} U^2 - \det U^2 - 2 + \frac{1}{2\delta^*}|a|^2 \geqslant 0, \tag{2.34}$$

where

$$\delta^* = a \cdot U(U^2 - 1)^{-1} n. \tag{2.35}$$

If further

$$1 + \tfrac{1}{2}\delta^* < 0, \tag{2.36}$$

then strict inequality holds also in (2.34) and there are exactly four distinct solutions, these having the form

$$\begin{aligned} &(\hat{R}_1, \lambda^*, b_1^+ \otimes m_1^+), \\ &(\hat{R}_2, \lambda^*, b_1^- \otimes m_1^-), \\ &(\hat{R}_3, 1-\lambda^*, b_2^+ \otimes m_2^+), \\ &(\hat{R}_4, 1-\lambda^*, b_2^- \otimes m_2^-), \end{aligned} \tag{2.37}$$

where

$$\lambda^* = \tfrac{1}{2}\left[1 - \sqrt{1 + \frac{2}{\delta^*}}\right], \tag{2.38}$$

so that $0<\lambda^* < \frac{1}{2}$. If

$$1+\frac{1}{2}\delta^* = 0, \tag{2.39}$$

then all solutions have $\lambda = \frac{1}{2}$; if strict inequality holds in (2.34) then there are exactly two distinct solutions, while if equality holds in (2.34) there is just one solution.

II. Assume U has an eigenvalue equal to 1. A necessary and sufficient condition that (2.32) has a solution is that

$$\mu^* \overset{\text{def}}{=} \operatorname{tr} U^2 - \det U^2 - 2 > 0 . \tag{2.40}$$

All solutions are given as follows. If $\mu^* > (|a|^2/4)$ then for each $\lambda \in (0,1)$ there are exactly two distinct solutions

$$(\hat{R}^+_\lambda, \lambda, b^+_\lambda \otimes m^+_\lambda), \quad (\hat{R}^-_\lambda, \lambda, b^-_\lambda \otimes m^-_\lambda). \tag{2.41}$$

If $0<\mu^* \leqslant (|a|^2/4)$ and $\bar{\lambda} \overset{\text{def}}{=} \frac{1}{2}(1-\sqrt{1-(4\mu^*/|a|^2)})$, so that $0<\bar{\lambda} \leqslant \frac{1}{2}$, then for each $\lambda \in (0,\bar{\lambda}) \cup (1-\bar{\lambda},1)$ there are exactly two distinct solutions of the form (2.41), whereas if $\lambda = \bar{\lambda}$ or $\lambda = 1-\bar{\lambda}$ with either $\bar{\lambda} \neq \frac{1}{2}$ (i.e. $\mu^* < (|a|^2/4)$) or $\bar{\lambda} = \frac{1}{2}$ and $\det U \neq 1$ then there is one solution $(\hat{R}_\lambda, \lambda, b_\lambda \otimes m_\lambda)$.

In all cases above, formulas for $b \otimes m$ associated with a solution $(\hat{R}, \lambda, b \otimes m)$ are given by (2.15) evaluated at the ordered eigenvalues of $(U+\lambda n \otimes a)(U+\lambda a \otimes n)$.

These formulas can be calculated for $U = U_3$ of the form (2.2). Such formulas are given by Wechsler, Lieberman and Read [18] in this case. They are based on the "double-shear mechaism" and not on energy considerations as here. They are part of the crystallographic theory of martensite which is summarized by Wayman [17]. The formulas gave austenite/martensite interface planes within about $1\frac{1}{2}^\circ$ from members of the $\{110\}$ family of planes. The pairs of twin and interface planes which are associated with solutions of (2.29) are nearly 60° apart — the pairs of $\{110\}$ planes which are 90° apart are not associated with solutions. Also λ comes out close to $\frac{1}{3}$ or $\frac{2}{3}$ for all solutions. These results agree with observations.

3. LIMITED FINENESS

The calculation of the minimizing sequences described above suggests that the twinned martensite should be infinitely fine whereas the observed twin spacing is small but obviously nonzero. In fact, twin spacings vary over at least four orders of magnitude (1 nm to 10's of μm) in different alloys while in a given alloy the twin spacing in the austenite/finely twinned martensite interface is more or less constant.

It is tempting to attribute the limited fineness to surface energy on twin boundaries. That is, we can assign a constant energy per unit area to twin boundaries, which is not accounted for by ϕ, and add it to the bulk energy. As we make the twins finer and finer ($j \to \infty$) we add twin boundary area and therefore increase the total surface energy while decreasing the bulk energy. On the other hand, if we make j small we are stuck with a large bulk energy. This suggests that the introduction of a small surface energy can account for limited fineness without invalidating the overall conclusions we have reach based on bulk energies alone.

Further support for this idea comes from another calculation. It is widely observed that sufficiently small crystals do not contain the austenite/finely twinned martensite interface (Otsuka and Shimizu [12]). Consider a cube of side L divided by an austenite/finely twinned martensite interface. The bulk energy of the crystal goes like $L^2 j^{-1}$, j being the parameter in (2.24), while the twin boundary area is approximately const. $L^3 j$. If we regard the total energy as the bulk energy plus the surface energy (which we assume is proportional to the total twin boundary area) we get

$$E(j) = \text{const.}\, L^2 j^{-1} + \text{const.}\, L^3 j\,. \tag{3.1}$$

The function $E(j)$ is minimized as a function of $j > 0$ when

$$j = \text{const.}\, L^{-\frac{1}{2}}\,. \tag{3.2}$$

Hence, we obtain limited fineness from this calculation. Of interest with regard to the observation on small crystals is to look at the dependence of the twin spacing on L. The twin spacing is proportional to j^{-1} by (2.29), so that the twin spacing is proportional to $L^{\frac{1}{2}}$ by (3.2). Hence, if L is sufficiently small, the twin spacing exceeds the length of a side of the cube!

These rough calculations make the "surface energy penalization" look attractive, but other observations are hard to reconcile with it. For example, in some martensitic crystals it is not difficult to create a single twin boundary which cuts across a corner of the crystal. This can be done by lowering the temperature to some temperature below the transformation temperature and then by applying suitable loads. When these loads are removed, the deformation gradients on each side of the twin are uniform and at potential well minima. However, it seems that an arbitrarily small perturbation of the twin boundary — say moving it to a nearby location parallel to its original position but nearer to the corner of the crystal — will decrease its area. Such a perturbation can be done without changing the bulk energy. Furthermore, while surface energy need not be proportional to area, it is highly probable that it strictly increases with area in the situation described above. Hence, any such perturbation will strictly decrease the total energy, and it becomes difficult to understand how such a boundary could remain stationary from this point of view. One can disallow the perturbations mentioned above by a suitable definition of relative minimizer in the manner of Parry [13].

In general, much useful information comes out of the bulk theory, excluding surface energy. In addition to observed interface and twin orientations, preliminary calculations suggest that it predicts accurately the variation of transformation temperature with applied stress in tension and compression (see Burkart and Read [7] for such observations in InTl). We also think it will be successful in describing the possible overall deformations of single and polycrystals subject to a phase transformation.

4. CURVED AUSTENITE/MARTENSITE INTERFACES AND LAYERS WITHIN LAYERS

Below the transformation temperature the austenite disappears. The typical microstructure especially in polycrystals, is a twinned structure consisting of layers within layers as in Fig. 2(a) or even layers within layers within layers. Close scrutiny of these pictures suggest that impinging sets of layers of similar fineness do not meet at an exact interface, i.e. dark bands may impinge on either dark or light bands as in Fig. 2(a).

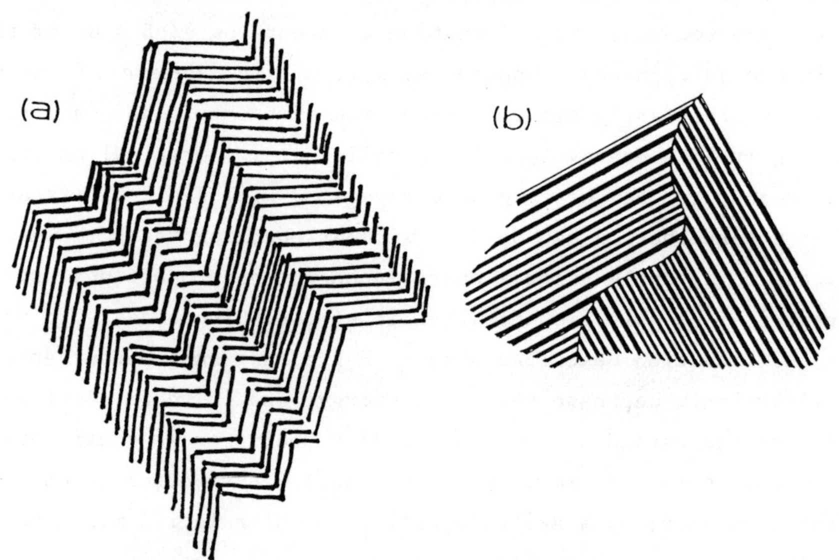

FIG. 2 Complex microstructures consisting of martensite only

The cover photograph of the Journal of Materials Science, Vol. 21-1 (1986), shows a beautiful photograph of these layers within layers. Also, in alloys that have been heated and cooled several times rapidly, curved martensite/martensite interfaces (Fig. 2(b) are occasionally formed. The relative proportion of the twin variants changes along the interface in this microstructure.

A possible way of understanding these microstructures is the following. For simplicity we consider cubic to tetragonal transformations. Below the transformation temperature there are only three symmetry related potential wells at say U_1, U_2, U_3 defined by (2.8) but with η_1 and η_2 slightly different from those at the transformation temperature (η_1 and η_2 depend weakly on θ for $\theta < \theta_0$). Twin connections are possible between these wells as listed in (2.20). Let F_1 and F_2 be twinned potential well minima:

$$F_2 - F_1 = a \otimes n . \tag{4.1}$$

We can create a laminate with average deformation gradient $G(\lambda)$ given by

$$G(\lambda) = \lambda F_2 + (1-\lambda) F_1 . \tag{4.2}$$

Consider for each $\lambda \in (0,1)$ the equation

$$RG(\lambda)\bar{R} = G(\lambda) + c_\lambda \otimes d_\lambda , \tag{4.3}$$

to be solved for a rotation R, for $\bar{R} \in P_c$ and for nonzero vectors c_λ and d_λ. This equation is just the twinning equation and has solutions (cf. Gurtin [9]) except possibly for very special twin families and isolated values of λ. Also,

$$RG(\lambda)\bar{R} = \lambda R F_2 \bar{R} + (1-\lambda) RF_1 \bar{R} , \tag{4.4}$$

and since by (4.1)

$$R F_2 \bar{R} - R F_1 \bar{R} = Ra \otimes \bar{R}^T n , \tag{4.5}$$

then $RG(\lambda)\bar{R}$ can be thought of as the average deformation of a fine laminate with gradients $RF_2\bar{R}$ and $RF_1\bar{R}$. Equation (4.3) is just the condition of compatibility for the two fine laminates of equal twin proportion λ, and it is clear we can obtain structures like that in Fig. 2a from minimizing sequences. Such structures often have the appearance of being twinned on the larger scale. Furthermore, by letting $G(\lambda)$ vary with λ along an interface with normal m_λ, we obtain sequences which look something like the curved martensite/martensite interface of Fig. 2(b).

Single crystal martensite/finely twinned martensite interfaces are also seen [6]. These can be obtained in the following way. Consider U_1, U_2 and U_3 as in (2.8) with

$$R_2 U_2 - R_1 U_1 = a \otimes n . \tag{4.6}$$

For a martensite/finely twinned martensite interface we must have,

$$\lambda R_2 U_2 + (1-\lambda) R_1 U_1 = U_3 + e \otimes f , \tag{4.7}$$

which can be rewritten,

$$U_1 U_3^{-1} + \lambda a \otimes U_3^{-1} n = R_1^T (1 + e \otimes U_3^{-1} f) . \tag{4.8}$$

But (4.8) is of the same form as (2.30) and therefore is governed by Theorem 2.2. Since $U_1 U_3^{-1}$ has an eigenvalue equal to 1, we turn to part II of Theorem 2.2. The scalar μ^* comes out

$$\mu^* = \frac{\eta_1^2}{\eta_2^2} \left[\left(\frac{\eta_2^2}{\eta_1^2} - 1 \right)^2 \right] , \tag{4.9}$$

which is positive if $\eta_1 \neq \eta_2$. Hence, (4.8) has solutions. The condition $\mu^* > |a|^2/4$ of Theorem 2.2, which guarantees solutions for all λ, becomes

$$2(\eta_1^2 + \eta_2^2) > \eta_1^2 \eta_2^2 \tag{4.10}$$

which is clearly satisfied for all the common alloys.

The two examples given above deliver average deformations $y = Fx$ with F on an "H" or a "T" in the set of deformation gradients. Δs, rank-one connections between each pair of three deformation gradients, are not possible between the potential wells at U_1, U_2 and U_3 if $\eta_1 \neq \eta_2$. The T and H calculations can be iterated by successive twinning in the manner of (4.3) and (4.4). It becomes clear that the convex hull of these potential wells is abundantly criss-crossed with lines representing deformation gradients that are limits of minimizing sequences.

It would be interesting to know all of the linear deformations that can be obtained from minimizing sequences with gradients at or near the potential well minima. Limitations on these overall deformations can be obtained from the set of weakly continuous functions (cf. Ball, Currie and Olver [1]). They definitely place nonobvious restrictions on the overall deformations in special cases.

5. QUESTIONS

The analysis of the austenite/martensite interface was highly motivated by pictures of microstructures. In the general case it would be nice to know whether such minimizing sequences are possible.

Many transformations satisfy assumptions similar to those introduced in Section 2. That is, there is a measured transformation strain matrix U_1 and a finite group of rotations $\mathcal{P}$. The free energy $\phi(F,\theta)$ at the transformation temperature $\theta=\theta_0$ has equipotential wells at

$$S = S_{\text{parent}} \cup S_{\text{trans}}\,, \tag{5.1}$$

where

$$\begin{aligned} S_{\text{parent}} &= \{F \in M^{3\times 3} : F^T F = 1,\ \det F > 0\} \\ S_{\text{trans}} &= \{F \in M^{3\times 3} : F^T F = \bar{R} U_1^2 \bar{R}^T,\ \bar{R} \in \mathcal{P},\ \det F > 0\}. \end{aligned} \tag{5.2}$$

Also, as mentioned in Section 2, the typical measured values of U_1 do not have an eigenvalue equal to 1, so no rank-one connections exist between S_{parent} and S_{trans} and therefore there are no continuous, piecewise differentiable deformations taking on values in both S_{parent} and S_{trans}. The austenite/martensite interfaces provide one way to accomplish the phase transformation but the necessary and sufficient conditions for their existence are quite restrictive. For example, by operating the basic equation (2.30) on a vector perpendicular to m and n, it is found that U_1 must have an eigenvalue $\geqslant 1$ and an eigenvalue $\leqslant 1$. Many transformation strains do not satisfy these conditions.

QUESTION 1. Given a function $\phi(\cdot, \theta_0)$ with suitable growth in its first argument, under what conditions on U_1 and P is it possible for each $\delta \in (0, \text{meas}\,\Omega)$ to have a minimizing sequence $y_k^\delta \in W^{1,\infty}(\Omega, \mathbb{R}^3)$ for the problem

$$\min \int_\Omega \phi(Dy, \theta_0)\, dx \tag{5.3}$$

satisfying

$$\lim_{k\to\infty} \text{meas}\, \{x \in \Omega : Dy_k^\delta(x) \in S_{\text{parent}}\} = \delta\, ? \tag{5.4}$$

With mild assumptions on ϕ, this is really a question about functions, since under appropriate conditions Dy_k^δ must lie near $S_{\text{parent}} \cup S_{\text{trans}}$ except on a set whose measure $\to 0$ as $k \to \infty$.

A special case of Question 1 is

QUESTION 2. What functions $y \in W^{1,\infty}(\Omega, \mathbb{R}^3)$ satisfy

$$Dy \in S_{\text{parent}} \cup S_{\text{trans}} \quad \text{a.e. ?} \tag{5.5}$$

This question is nontrivial because there exist functions $y \in W^{1,\infty}(\Omega, \mathbb{R}^3)$ satisfying

$$Dy \in \{F_1, F_2, \cdots, F_n\} \quad \text{a.e.} \tag{5.6}$$

where $F_1, \cdots, F_n$ are constant matrices such that $F_1 - F_i$ not a rank-one matrix for $i = 2, \cdots, n$. An example of such a function with $n = 5$ is given by Ball and James [2]. There exist such examples with $n = 4$ but not with $n = 3$ (Ball, James and Jodeit [3]). There may exist functions of the form (5.6) with no rank-one relations between any pair of the F_i's, i.e. the Sierpinski gasket has no proper interfaces. However, the calculation goes beyond just picking a geometric figure. Also, the role of the rotations in Questions 1 and 2 is unclear.

Answers to these questions would improve our understanding why some materials pass reversibly through phase transformations while others fracture or suffer plastic deformation.

ACKNOWLEDGEMENT

I wish to thank John Ball and Robin Knops for stimulating discussions and warm hospitality during visits to Heriot-Watt

University in the summers of 1985 and 1986. The research summarized here was supported by an S.E.R.C. visiting fellowship and by the U.S. National Science Foundation (MSM-8612420).

REFERENCES

1. J.M. Ball, J.C. Currie and P.J. Olver, "Null Lagrangians, weak continuity and variational problems of arbitrary order", *J. Functional Anal.* **41** (1981), pp.135-174.
2. J.M. Ball and R.D. James, "Fine phase mixtures as minimizers of energy", *Arch. Ration. Mech. Anal.* (to appear).
3. J.M. Ball, R.D. James and M. Jodiet, (in preparation).
4. Z.S. Basinski and M.A. Christian, "The cubic-tetragonal phase transformation in manganese copper alloys", *J. Inst. Metals* **80** (1951-52), pp.659-666.
5. Z.S. Basinski and M.A. Christian. "Crystallography of deformation by twin boundary movements in indium-thallium alloys", *Acta. Met.* **2** (1954), pp.101-116.
6. Z.A. Basinski and M.A. Christian, "Experiments on the martensitic transformation in single crystals of indium-thallium alloys. *Acta. Met.* **2** (1954), pp.148-166.
7. M.W. Burkart and T.A. Read, "Diffusionless phase change in the indium-thallium system", *Trans. AIME J. Metals* **197** (1953), pp.1516-1524.
8. J.W. Gibbs, "On the equilibrium of heterogeneous substances", in *The Scientific Papers of J. Willard Gibbs*, Vol. 1, Dover Publications, New York, 1961.
9. M.E. Gurtin, "Two-phase deformations of elastic solids", *Arch. Ration. Mech. Anal.* **84** (1983), pp.1-29.
10. R.D. James, "Displacive phase transformations in solids", *J. Mech. Phys. Solids* **34** (1986), pp.359-394.
11. S.W. Meeks and B.A. Auld, "Periodic domain walls and ferroelastic bubbles in neodymium pentaphosphate", *Appl. Phys. Lett.* **47** (1985), pp.102-104.
12. K. Otsuka and K. Shimizu, "Morphology and crystallography of thermoelastic γ' Cu-Al-Ni martensite", *Japanese J. Appl. Phys.* **8** (1969), pp.1196-1204.
13. G. Parry, "On shear bands in unloaded crystals", (to appear).
14. Yu.G. Reshetnyak, "On the stability of conformal mappings in multidimensional spaces", *Siberian Math. J.* **8** (1967), pp.69-85.
15. L. Tartar, "Étude des oscillations dans les equations aux derivées partielles nonlineares", in *Lecture Notes in Physics 195*, Springer-Verlag (1984), pp.384-412.
16. G. van Tendeloo, J. van Landuyt and S. Amelinckx, "The α-β phase transition in quartz and $AlPO_4$ as studied by electron microscopy and diffraction", *Phys. Stat. Sol.* **a 33** (1976), pp.723-735.

17. C.M. Wayman, *Introduction to the Crystallography of Martensitic Transformations*, MacMillan, 1964.
18. M.S. Wechsler, D.S. Lieberman and T.A. Read, "On the theory of the formation of martensite", *Trans. AIME J. Metals* (1953), pp.1503-1515.
19. L.C. Young, *Lectures on the Calculus of Variations and Optimal Control Theory*, Chelsea 1980.

15

STANDING WAVES OF NONLINEAR SCHRÖDINGER EQUATIONS: EXISTENCE AND STABILITY

CHRISTOPHER K. R. T. JONES

Abstract. Much of our understanding of the nonlinear Schrödinger equation is organised around standing waves. There are two quite distinct techniques that have been developed for attacking the central questions of existence and stability of these waves; these are the variational (PDE) and the geometric (ODE) approaches. In this paper, I describe the respective hierarchies of waves that each of these methods has produced together with the various known stability and instability results. The instability results that have grown out of the two different perspectives are quite distinct from each other, but ironically this suggests some connection between the two hierarchies. This connection is formulated as a conjecture.

1. INTRODUCTION

A general class of nonlinear Schrödinger type equations, that have received much attention recently, can be written as

$$i\phi_t = \Delta\phi + f(|\phi|)\phi \tag{1}$$

where $\phi \in \boldsymbol{C}$,

$$\Delta = \sum_{i=1}^{n} \frac{\partial^2}{\partial x_i^2}$$

with $x = (x_1, \cdots, x_n) \in \mathbb{R}^n$. The nonlinear term $f: \boldsymbol{R} \to \boldsymbol{R}$ is assumed to be, at least, smooth and satisfy $f(0) = 0$.

---

†Supported, in part, by the National Science Foundation under Grant #DMS 8501961.

DEFINITION. A standing wave for (1) is a solution which has the form

$$\phi(x,t) = e^{i\beta t}u(x) .$$

In the following, I shall assume that $\beta < 0$ and that u is real.

A standing wave satisfies the semilinear elliptic equation

$$\Delta u + f(|u|)\,u + \beta u = 0 , \tag{2}$$

and can be seen as a steady state of equation (1) in a rotating frame, with frequency β:

$$i\phi_t = \Delta\phi + f(|\phi|)\,\phi + \beta\phi . \tag{3}$$

The nonlinear Schrödinger equation arises in nonlinear optics and the standing waves are special solutions that, for physical reasons, form the centrepiece of the analysis. A step towards understanding the overall behaviour is therefore a systematic study of the existence and stability of standing waves.

The aim of this paper is to describe the known results that address this question and compare the different basic techniques. The two main approaches have been variational methods and geometric ODE arguments. Each approach has supplied a basic set of standing waves, together with a different property that indexes them and permits a partial classification. For general nonlinear terms f, it is not understood how these indices which emerge from the different techniques are related to each other. The paper will end with a conjecture whose verification would form a substantial bridge between them.

Each technique has made its own contribution to the stability question. Certain forms of the variational characterisation are intrinsically related to the invariants of the nonlinear Schrödinger equation and consequently carry stability information. There are a plethora of variational approaches and only some of these are illuminating about stability questions. The stability problem reveals interesting and deep connections between the dynamical problems (1) and the variational approach for standing waves, but these are not completely understood.

The dynamical systems point of view has produced a result on the instability of certain standing waves. A geometric framework is formulated for the eigenvalue problem associated to the linearisation of the equation (3) about the wave. A criterion can then be determined that forces the existence of a positive eigenvalue. Informally, this criterion is an index of the wave that says the wave is anomalous in the hierarchy. It is this idea that leads to the conjecture about the relationship between the two hierarchies.

2. EXISTENCE

The main example of a nonlinear term $f(|\phi|)$, which has motivated most of the results, is $f(|\phi|) = |\phi|^{\sigma}$ where $\sigma > 0$. The standing wave equation then becomes

$$\Delta u + |u|^{\sigma} u + \beta u = 0 . \tag{4}$$

The variational approach has produced infinitely many solutions of (4) under the assumption $0 < \sigma < 4/(n-2)$. This is contained in the important paper of Strauss [22]. There were many earlier results in specific space dimensions, see Nehari [15], Berger [5], Coffman [7], but to this author's knowledge Strauss's paper was the first to consider the problem in general space dimensions. Berestycki-Lions [3] proved an extension of this result in which the nonlinear term is permitted to grow at a rate less than one of the powers described above, but does not have to grow at any particular rate.

The condition $0 < \sigma < 4/(n-2)$ is used to obtain some compactness in the problem, via the Palais-Smale condition. The work of Pohozaev [17] shows that existence cannot be expected with higher powers.

In this paper I shall not attempt to enunciate the most general conditions under which each result holds. There are various different power growth restrictions, for large $|\sigma|$, that arise and each of these is a generalisation of a condition on σ in the pure power case. Indeed, the restrictions for general nonlinear terms are motivated by the special pure power case. In order to avoid overloading this paper with technical conditions I shall state theorems that apply to more general nonlinearities than the pure power with the hypothesis: "f satisfies certain growth conditions" and an indication, in parentheses, of the conditions on σ that are being generalised. With this understanding I shall state the main existence theorem achieved by the variational method.

THEOREM 1 (Strauss [22], Berestycki-Lions [3]). Under certain growth conditions on f $(0<\sigma<4/(n-2))$, (2) has infinitely many classical solutions in $L^2(\mathbb{R}^n)$.

The solutions found by all of these authors are actually radially symmetric. Although the variational method is a PDE technique and therefore, *a priori*, not restricted to finding solutions that are functions of a single variable, the condition of radial symmetry is imposed to guarantee compactness of certain Sobolev embeddings. It follows that the resulting standing waves must be radially symmetric. For the positive ones, the theorem of Nirenberg-Gidas-Ni [8] entails that they are radially symmetric. For non-positive waves, to this author's knowledge, there are not any existence results that produce real, non-radially symmetric waves, as long as the domain is all of $\mathbb{R}^n$ and $f(0)=0$, see Smoller [20] for related results.

The variational strategy is to find the waves as critical points of certain functionals. Let $H^1(\mathbb{R}^n)$ denote the Sobolev space of functions in $L^2(\mathbb{R}^n)$ whose first order weak derivatives are also in $L^2(\mathbb{R}^n)$. $H^1_r(\mathbb{R}^n)$ denotes the subspace of functions that are radially symmetric, set

$$M = \{u \in H^1_r(\mathbb{R}^n) : Q(u) = 0\} \tag{5}$$

where Q is some functional on $H_r^1(\mathbb{R}^n)$. For $k \in \mathbb{N}$, set

$$\Gamma_k = \{A \subset M : A \text{ is homeomorphic to } S^k \text{ if } k > 0 \text{ or } A = \text{point if } k = 0\}.$$

Critical points for another functional $J(u)$ are then sought on M. The strategy of Liusternik-Schnirelman theory is to show that the number

$$b_k = \inf_{A \in \Gamma_k} \sup_{u \in A} J(u) \tag{6}$$

is a critical value of J on Γ and therefore generates a weak solution of

$$J'(u) - \lambda Q'(u) = 0. \tag{7}$$

With appropriate choices of J and Q, the Lagrange multiplier λ can be set so that (7) becomes equation (2) in a weak form. Elliptic regularity is then invoked to bootstrap a classical solution of (7).

The compactness required to make Liusternik-Schnirelman theory work is supplied by the Palais-Smale condition through the growth condition and the imposed radial symmetry. The values of k in (6) should be thought of as indexing the hierarchy. It can usually be shown that b_k is non-decreasing in k and $b_k \to +\infty$.

The key is obviously the judicious choice of J and Q in each of the different variational approaches. In the work of Strauss [22], restricting to our particular form of the nonlinear term, the following choices are made:

$$\begin{aligned} J(u) &= \int_{\mathbb{R}^n} \left\{ \tfrac{1}{2} |\nabla u|^2 - \frac{1}{\sigma+2} |u|^{\sigma+2} \right\} dx \\ Q(u) &= \tfrac{1}{2} \int_{\mathbb{R}^n} |u|^2 \, dx - 1 . \end{aligned} \tag{8}$$

To achieve their more general theorem, Berestycki-Lions [3] use a different grouping of the functional and constraint:

$$\begin{aligned} J(u) &= \tfrac{1}{2} \int_{\mathbb{R}^n} |\nabla u|^2 \, dx \\ Q(u) &= \int_{\mathbb{R}^n} \left\{ \frac{1}{\sigma+2} |u|^{\sigma+2} + \frac{\beta}{2} |u|^2 \right\} dx - 1 . \end{aligned} \tag{9}$$

Again, I stress that these are the functionals used by these author's written here in the particular case of a power nonlinearity.

The geometric ODE approach has also produced a hierarchy of infinitely many standing waves. With this approach the waves are indexed by their nodal properties.

THEOREM 2 (Jones-Küpper [11]). Under certain growth conditions on f $(0<\sigma<4/(n-2))$ equation (2) has infinitely many classical, radially symmetric solutions $u(r)$ with $u(r)\to 0$ exponentially as $r\to\infty$. Moreover there is at least one with m zeroes, for each integer $m>0$.

Our approach is to consider the radial form of (2)

$$u_{rr}+\frac{n-1}{r}u_r+f(|u|)u+\beta u=0 \qquad (10)$$

and write it as a system of ODE's. This is nonautonomous and the dimension of the phase space has to be increased by one. The resulting equations are massaged by various transformations and compactifications. The problem is then reduced to that of finding a certain connecting orbit in a three dimensional phase space. The number of zeroes is related to a certain oscillation of the trajectories around an axis which corresponds to the trivial solution.

There is earlier work on the existence of these waves with specified nodal structure, see Ryder [18], Nehari [15] and Coffman [7]. Indeed there are results connecting the variational approach and nodal structure which depend on monotonicity type assumptions for the nonlinear term f, see Coffman [7] and also the more recent work of Heinz [10]. There should be some general result connecting the two approaches, but as can be seen from the above description the two approaches are fundamentally very different. One of the basic tenets of modern physics is that complexity, in the form of nodal structure, should be related to energy. Understanding the connections between these two approaches is therefore of fundamental importance.

3. STABILITY AND INSTABILITY BY THE VARIATIONAL APPROACH

Firstly we need an exact definition of the stability of a set of standing waves.

DEFINITION. A set S of standing waves is stable if for all $\varepsilon > 0$, there exists $\delta > 0$ such that

$$\inf_{u \in S} \|\phi_0(x) - u\|_{H^1(\mathbb{R}^n)} < \delta$$

implies that

$$\inf_{u \in S} \|\phi(x,t) - u\|_{H^1(\mathbb{R}^n)} < \varepsilon$$

for all $t \geqslant 0$, where $\phi(x,t)$ is a solution of (1) with $\phi(x,0) = \phi_0(x)$. Such a set S is unstable if it is not stable.

In this definition, the set S may, for instance, be the set

$$S_u = \{e^{i\theta} u(x+y) : \theta \in [0,2\pi],\quad y \in \mathbb{R}^n\}$$

where u is some fixed standing wave. This would allow mild instabilities to drift in the phase or the centering of the wave and still preserve stability of the set S_u; this is usually called orbital stability.

In the variational approach, the waves that occur as minima, i.e. $k = 0$ in the Liusternik-Schnirelman hierarchy, are good candidates for stability. Cazenave-Lions [6] proved such a theorem which they stated for the pure power case: $f(|u|) = |u|^{\sigma}$. A minimum in this context is usually called a ground state but this is obviously dependent on which variational characterisation is being used. The following definition is the one used by Cazenave-Lions [6].

DEFINITION. A ground state is a standing wave which solves the minimisation problem:

$$\inf \{J(u) : u \in H^1(\mathbb{R}^n) \text{ and } \|u\|_{L^2} = \mu\}$$

for some μ, where

$$J(u) = \int_{\mathbb{R}^n} \left\{ \tfrac{1}{2} |\nabla u|^2 - \frac{1}{\sigma+2} |u|^{\sigma+2} \right\} dx . \tag{11}$$

Notice that this is close to Strauss's variational approach. A ground state must be positive since $|u|$ will also be minimum and regularity forbids this if u is ever zero. By the theorem of Nirenberg-Gidas-Ni [8] it then follows that $u = u(|x|)$, i.e. u is radially symmetric.

THEOREM 3 (Cazenave-Lions [6]). If $0 < \sigma < 4/n$ then the set of all ground states is nonempty and stable.

The idea of the proof is that minima of a conserved energy should be stable but to make sense of this intuition uses the structure of Lions concentrated compactness. The non-emptiness part of the theorem is significant and actually fails when the power condition is not satisfied as then the "energy" is not bounded below. The range of powers for which stability holds is narrower than the range for existence since $4/n < 4/(n-2)$. The critical level $\sigma = 4/n$ is known from the work of Glassey [9] in which finite time blow-up was proved for some initial data.

Weinstein [23] has attacked the stability question from a slightly different point of view. He starts with another definition of a ground state.

DEFINITION. A ground state is a positive standing wave.

By the remarks made above, a ground state in the sense of Cazenave-Lions is one in Weinstein's sense. The converse is not necessarily true. Notice again that a ground state in Weinstein's sense must be radially symmetric by Nirenberg—Gidas—Ni [8]. Weinstein's theorem addresses the stability question in the context of a curve of standing waves. In this sense it has the flavour of a bifurcation theory result. Suppose then that we have a one parameter family of standing waves $u(x, \beta)$, where β is the frequency as in (2). Set $\mu(\beta) = ||u, \beta||_{L^2}$.

THEOREM 4 (Weinstein [23]). If there is a unique nondegenerate standing wave $u(x, \beta)$ for each $\beta < 0$ then

$$\frac{d}{d\beta}\, \mu(\beta)\bigg|_{\beta=\beta_0} > 0$$

implies that $u(x, \beta_0)$ is stable.

To interpret the theorem for the pure power case, the condition is satisfied exactly when $0 < \sigma < 4/n$. The uniqueness means that the family $u(x, \beta)$ gives the set of all ground states. These are then minima and the energy type arguments can be developed.

I turn now to instability results that are obtained by the variational approach. The first is an instability result of Berestycki-Cazenave [2] which applies to the supercritical case $(4/n < \sigma < 4/(n-2))$ and ground states of (2). However the original Cazenave-Lions definition for a ground state will not suffice here since the functional used is not bounded below for the super-critical case. Set

$$G(u) = \int_0^u f(|v|)\, v\, dv \tag{12}$$

and define the energy for (2) to be

$$E(u) = \int_{\mathbb{R}^n} \left\{ \tfrac{1}{2} |\nabla u|^2 - G(u) - \frac{\beta u^2}{2} \right\} dx. \tag{13}$$

DEFINITION. A ground state is a solution of (2) that minimises $E(u)$ amongst $u \in H^1(\mathbb{R}^n)$ solutions of (2) with $u \neq 0$.

THEOREM 5 (Berestycki-Cazenave [2]). If f satisfies certain growth conditions $(4/n < \sigma < 4/(n-2))$ then any ground state (in the sense of this definition) is unstable.

Berestycki and Cazenave in fact show that there are initial data lying arbitrarily close to a ground state that lead to a solution which blows up in finite time. The idea of the proof is to recast the existence in terms of a constrained functional and then apply a style of argument that dates back to Payne-Sattinger [16]. In this argument it is shown that the particular form of the constrained variational characterisation, together with some invariance properties of the equations, gives some estimates which force some solutions to blow up in finite time.

This kind of idea can be generalised to standing waves that are obtained at minimax values and this has been done by Berestycki-Lions [4]. Again a new variational formulation is used and their attendant standing waves obtained. For these results Berestycki-Lions restrict to the pure power case: $f(|\phi|) = |\phi|^{\sigma}$. The relevant functionals are the energy

$$E(u) = \int_{\mathbb{R}^n} \left\{ \frac{1}{2} |\nabla u|^2 - \frac{1}{\sigma+2} |u|^{\sigma+2} - \frac{\beta u^2}{2} \right\} dx \tag{14}$$

and a subsidiary functional

$$Q(u) = \int_{\mathbb{R}^n} \left\{ |\nabla u|^2 - \frac{n}{2} \frac{\sigma}{\sigma+2} |u|^{\sigma+2} \right\} dx \,. \tag{15}$$

Then set

$$M = \{u \in H_r^1(\mathbb{R}^n) : u \neq 0,\ Q(u) = 0\}$$

and Γ_k = set of $h(S^k)$ where $h : S^k \to M$ is an odd, one-to-one continuous map on the k sphere. The functional Q comes from a scaling property and it is known *a priori* that $Q(u) = 0$ for any standing wave.

DEFINITION. A k-excited state is a critical point of E on M with a critical value:

$$c_k = \inf_{A \in \Gamma_k} \sup_{u \in A} E(u) \,.$$

It is shown in Berestycki-Lions that if $4/n \leqslant \sigma < 4/(n-2)$ then there is a k-excited state for each $k > 0$, an integer.

THEOREM 6. (Berestycki-Lions [4]). If $4/n \leqslant \sigma < 4/(n-2)$ then any k-excited state (for any integer $k > 0$) is unstable.

In this case as well they obtain the stronger result of blow-up in finite time from a neighbourhood of such waves.

Another development of this kind of theorem is due to Strauss-Shatah [19]. To state their result as it applies to the nonlinear Schrödinger equation involves the introduction of another variational characterisation. Recall $G(u)$ from (12) and set

$$K(u) = \int_{\mathbb{R}^n} \left\{ \left(\tfrac{1}{2} - \tfrac{1}{n} \right) |\nabla u|^2 - \tfrac{1}{2} u^2 + G(u) \right\} dx . \qquad (16)$$

DEFINITION. A ground state is a standing wave which solves the minimisation problem

$$\inf \left\{ \int_{\mathbb{R}^n} |\nabla u|^2 : u \neq 0, \quad u \in H_r^1(\mathbb{R}^n) , \quad K(u) = 0 \right\} .$$

The condition $K(u) = 0$ is again one that is known to be satisfied at a standing wave from certain invariance properties. The result of Strauss and Shatah shares something of the flavour of Weinstein's work. We assume that we have a one-parameter family of standing waves: $u(x, \beta)$ and set again

$$\mu(\beta) = \|u(x, \beta)\|_{L^2} . \qquad (17)$$

THEOREM 7 (Shatah-Strauss [19]). If a ground state has

$$\frac{d}{d\beta} \mu(\beta) \Big|_{\beta = \beta_0} < 0$$

then it is unstable.

If the nonlinear term is restricted to the pure power case then the condition of the theorem becomes

$$4/n < \phi < 4/(n-2).$$

The Strauss-Shatah theorem is the complementing instability result to Weinstein's result (Theorem 4) in the same way that Berestycki-Cazenave (Theorem 5) complements Cazenave-Lions (Theorem 3).

There should be a generalisation of Theorem 7 to excited states. The work of Sternberg [21] is in this direction for the nonlinear Klein-Gordon equation.

4. INSTABILITY BY THE GEOMETRIC ODE APPROACH

The methods of ordinary differential equations have a chance of being applied because the standing waves under consideration are radially symmetric. It will be seen below that certain eigenvalue problems arise in studying linearised stability which lead to a system of ODE's. In particular some definitions in order to focus on particular standing waves in the hierarchy whose instability can be proved. Let $u(r)$ be some given standing wave.

DEFINITION. $Q=$ number of zeroes of $u(r)$, $P=$ Morse index of $u(r)$ relative to the energy $E(u)$ given in (13) on $H^1_r(\mathbb{R}^n)$.

In other words P is the number of positive eigenvalues of the operator

$$Lv = \Delta v + f'(|u|)v + f'(|u|)v + \beta v \tag{18}$$

on $H^1_r(\mathbb{R}^n)$. If the world were simple there would be an exact relationship between P and Q. There is a fundamental inequality that they satisfy which follows from some Sturm-Liouville considerations, see Jones [12].

LEMMA. For any standing wave $u=u(r)$, $P-Q \geqslant 1$.

Consider, again, the nonlinear Schrödinger equation in a rotating frame (3). I shall write (3) in real and imaginary parts, linearise at a radially symmetric standing wave $u=u(r)$ and restrict to radially symmetric perturbation $\phi = p + iq$. The equations are then

$$\begin{aligned} p_t &= q_{rr} + \frac{n-1}{r} q_r + f(|u|)q + \beta q \\ q_t &= -p_{rr} - \frac{n-1}{r} p_r - f(|u|)p - f'(|u|)|u|p - \beta p . \end{aligned} \tag{19}$$

Rewrite these as:

$$\begin{pmatrix} p \\ q \end{pmatrix}_t = \begin{pmatrix} 0 & L_- \\ -L_+ & 0 \end{pmatrix} \begin{pmatrix} p \\ q \end{pmatrix} = N \begin{pmatrix} p \\ q \end{pmatrix},$$

where

$$L_- = \frac{d^2}{dr^2} + \frac{n-1}{r}\frac{d}{dr} + f(|u|) + \beta$$

$$L_+ = \frac{d^2}{dr^2} + \frac{n-1}{r}\frac{d}{dr} + f(|u|) + f'(|u|)|u| + \beta .$$

The numbers P and Q have relevance to the operators L_+, L_-. By the definition of P we have that

$$P = \text{number of eigenvalues of } L_+ \text{ on } H_r^1(\mathbb{R}^n) \text{ in } \{\lambda > 0\} .$$

Notice that $L_- u = 0$ is exactly the standing wave equation and consequently u is an eigenfunction of L_- with eigenvalue zero. From Sturm-Liouville theory the number of zeroes of $u(r)$ on $r \in [0,\infty)$ is equal to the number of eigenvalues of L_- in $\{\lambda > 0\}$, therefore

$$Q = \text{number of eigenvalues of } L_- \text{ of } H_r^1(\mathbb{R}^n) \text{ in } \{\lambda > 0\} .$$

For the sake of studying instability, the eigenvalues of N, considered as an operator on $H_r^1 \times H_r^1$, need to be understood. The eigenvalue equations for N are

$$L_- q = \lambda p , \quad L_+ q = -\lambda q . \tag{21}$$

It is not obvious how the spectrum of N is related to that of the individual operators L_+ and L_-. It is a tantalising temptation to believe that the spectrum of N is built out of that of L_+ and L_- in some way but in general this seems to be complicated. The following instability theorem is a contribution to this end.

THEOREM 8. (Jones [12]). If $P-Q > 1$ then N has an eigenvalue in $\{\lambda > 0\}$.

This theorem is proved by (very) geometric ODE methods. Equation (2.1) is a system of coupled second order ODE's and as such it can be rewritten as a system of equations on $\boldsymbol{R}^4$ (since we can assume $\lambda \in \boldsymbol{R}$).

$$
\begin{aligned}
p' &= y \\
q' &= w \\
y' &= -\frac{n-1}{r}\, y - g(r)\, p - \lambda q \\
w' &= -\frac{n-1}{r}\, w - h(r)\, q + \lambda p\,,
\end{aligned} \tag{22}
$$

where

$$g(r) = f(|u|) + f'(|u|)\,|u| + \beta$$

$$h(r) = f(|u|) + \beta\,.$$

An eigenvalue of N corresponds to a value of λ at which (22) has a solution in $H^1_r(\mathbb{R}^n)$. This turns out to be equivalent to finding a bounded solution of (22). In Jones [12] this condition is reformed as a connection problem in the Grassmannian of two dimensional subspaces of $\boldsymbol{R}^4$.

The basis of this idea is that there is a two dimensional subspace of solutions satisfying the boundary condition at $r=0$. This can be viewed as a trajectory of the flow induced by the linear system (22) on this Grassmannian. The boundedness condition then translates into determining a certain behaviour of this trajectory as $r \to +\infty$. This is a connection problem that can be attacked by a shooting method. The strategy is that a solution of the problem is obtained when some relevant "index" of these trajectories changes as λ varies. This index must be topological and since we are measuring some property of trajectories it must be reflected in a nontrivial fundamental group. Unfortunately $\pi_1(G_{2,4}) \equiv \boldsymbol{Z}_2$ and therefore will not supply very much information. However, the equations (22) have a certain special structure and there is an invariant submanifold $\Lambda(2) \subset G_{2,4}$ in which the relevant trajectories live. This manifold is diffeomorphic to the space of Lagrangian planes and therefore $\pi_1(\Lambda(2)) \equiv \boldsymbol{Z}$, see Arnol'd [1]. Some winding of trajectories can therefore be measured, this is, in fact, the Maslov index. This winding changes from $P-Q-1$ at $\lambda=0$ to 0 when λ is very large positive. Therefore if $P-Q>1$ a positive eigenvalue is forced to exist.

As an application of this result, suppose that all standing waves with m zeroes are nondegenerate. Notice that, in fact, there must be at least two of them, since if u is a solution to (10) so is $-u$. If there are more then some are unstable.

THEOREM 9. (Jones [12]). If there are more than two radial standing waves with m zeroes then at least one has $P-Q > 1$ and hence is linearly unstable.

5. CONCLUSION

In the preceding sections I have described some instability results that arise from two fundamentally different points of view. A natural goal is to assimilate these into a wider perspective and thereby cast light on both the general structure of the nonlinear Schrödinger equation and the relationship between the variational and geometric methods.

There is evidence that the instability mechanism established by these two techniques are quite distinct and perhaps even complement each other. To illustrate this point, consider the variational instability theorems for ground states. In each of the different definitions of a ground state it was concluded that the standing wave in question is positive, thus $Q=0$. To qualify as a ground state, according to each of the definitions before the instability theorems, it must be the minimum of some constrained problem. In the Cazenave-Lions definition the energy is being minimised. Since P measures the number of directions in which the energy decreases it follows that at the minimum of such a constrained problem either $P=0$ or $P=1$. The uncertainty in the calculation of P is due to the constraint. However from the Lemma in Section 4, we know that $P-Q \geqslant 1$, so then $P=1$ and $P-Q=1$ at such a ground state. It follows that the instability results arising from the variational method cannot be deduced from Theorem 8.

If this situation is viewed from the bifurcation theory point of view it appears less strange. Theorem 8 says that the branches of solutions with $P-Q>1$ are inherently unstable. For a ground state $P-Q=1$ and the variational arguments show that the waves on this branch may or may not be stable. Indeed Theorems 4 and 7 connect the stability or instability to a condition that can be seen in terms of the direction of the bifurcation curve. The work of Maddocks [13] addresses the problem of bifurcation in variational problems and seems to corroborate this interpretation. The picture emerging from these considerations is that there are two different classes of standing waves: those with $P-Q>1$ which are definitely unstable and those with $P-Q=1$ whose stability depends on a subsidiary condition.

Consider again the hierarchy of standing waves supplied by the geometric ODE approach (Theorem 2). These are indexed by their number of zeroes. There are no results for the uniqueness of these waves with a prescribed number of zeroes, indeed the uniqueness question for positive solutions is difficult, see Serrin-McLeod [14]. However the minimal number of waves allowed for each number of zeroes is two (if $u(r)$ is a solution then so is $-u(r)$). I believe it can be shown that if this minimal situation actually holds and each wave is nondegenerate then $P-Q=1$ at each one of them. The deviation of $P-Q$ from 1 thus measures how anomalous the waves are in the geometric hierarchy.

In the variational hierarchy, the special waves are the ones that appear as "minimax" solutions, i.e. the k-excited states that are found by the Liusternik-Schnirelman approach. The minimal situation here is of two k-excited states for each k and presumably if both hierarchies are in their minimal configurations then they must coincide in a natural way. If multiplicities of solutions exist that disturb this configuration and introduce extra waves then it is still tempting to conjecture that the "special" ones maintain some correspondence. A natural formulation of this is the following:

CONJECTURE. At each minimax level there is an excited state with $P-Q=1$.

This actually begs the question as to which variational characterisation is being used and should more properly be stated as speculating the existence of a variational characterisation for which the conjecture holds. The proof of this conjecture would have some fascinating consequences. There is usually a natural relationship between the Morse index P and the Liusternik-Schnirelman level k, which should be $P=k+1$. If this can be shown to hold here then there would exist a k-excited state with $Q=k$, in other words with k zeroes. This would establish a very satisfying bridge between the two hierarchies.

Both from the geometric ODE and the variational PDE points of view there are some very important open problems. I shall state what I believe are the two most fundamental problems, one stemming from each approach. It is possible that they are actually manifestations of the same problem depending on the exact relationship of the two hierarchies.

In the geometric hierarchy it is not understood whether the solutions with $P-Q=1$ but $Q>0$ are stable or not. For instance if the minimal situation holds, as described above, then the solutions with Q zeroes have $P-Q=1$. The question is: are these stable? In this case if the nonlinear Schrödinger equation is viewed as a Hamiltonian system then the second variation

$$D^2H = \begin{pmatrix} L_+ & 0 \\ 0 & L_- \end{pmatrix}$$

is indefinite and, in fact, has $P+Q$ "positive" directions. It is not understood whether these cause instability or not and this may be a very subtle question involving such considerations as Arnol'd diffusion.

From the variational point of view the most glaring open

question is the stability of excited states in the subcritical ($\sigma < 4/n$) case. A good guess is that the non-excited states (those which do not appear as minimax solutions) are unstable (perhaps they have $P-Q>1$!) but the stability of the excited states is more subtle. If the conjecture stated above is correct, this is closely related to the problem given above for the geometric ODE case.

ACKNOWLEDGEMENTS

The author wishes to thank Professors Strauss and Weinstein for their helpful comments on this paper.

ADDENDUM

Since the writing of this paper, the author has received a preprint of a paper by Grillakis-Shatah-Strauss concerning the instability of standing waves by the variational approach. These new results significantly extend some of those mentioned in this paper. They also further corroborate the conjectures stated above as they prove instability under hypotheses that include the assumption that $P=1$ and $Q=0$. The work of Grillakis should also be mentioned in the context of existence results, He proves a generalisation of Theorem 2 above using a degree-theoretic argument. This now introduces a third technique into the arena. The relevant papers are:

M. Grillakis, "Nodal characterization and instability for bound states of the Klein-Gordon equation", Ph.D. Thesis, Brown University, 1986.

M. Grillakis, J. Shatah and W. Strauss, "Stability theory of solitary waves in the presence of symmetry, I", Preprint.

REFERENCES

1. V.I. Arnol'd. "Characteristic class entering in quantization conditions", *Functional Analysis and its Applications* 1 (1) (1967), pp. 1-13.
2. H. Berestycki and T. Cazenave, "Instabilité des états stationnaires dans les equations de Schrödinger et de Klein-Gordon nonlinéaires", *Comptes Rendus Acad. Sc. Paris*, **293**, Série 1, (1981), pp. 489-492.

3. H. Berestycki and P.-L. Lions, "Nonlinear scalar field equations, II, existence of infinitely many solutions", *Arch. Rational Mech. Anal.* **82**, (1983), pp.347-376.
4. H. Berestycki and P.-L. Lions, "Théorie des points critiques et instabilité des ondes stationnaires pour des équations de Schrödinger nonlinéaires", *Comptes Rendus Acad. Sci. Paris*, **300**, Série 1, (10), (1985), pp.319-322.
5. M.S. Berger, "On the existence and structure of stationary states for a nonlinear Klein-Gordon equation", *J. Func. Anal.* **9**, (1972), pp.249-261.
6. T. Cazenave and P.-L. Lions, "Orbital stability of standing waves for some nonlinear Schrödinger equation", *Comm. in Math. Phys.* **85**, (1982), pp.560-561.
7. C.V. Coffman, "Uniqueness of the ground state solution for $\Delta u - u + u^3 = 0$ and a variational characterisation of other solutions", *Arch. Rat. Mech. Anal.* **46**, (1973), pp.81-95.
8. B. Gidas, W.-M. Ni and L. Nirenberg, "Symmetry and related properties via the maximum principle, *Comm. Math. Phys.* **68**, (1979) pp.209-243.
9. R.T. Glassey, "On the blow-up of solutions to the Cauchy problem for nonlinear Schrödinger equations", *J. Math. Phys.* **18**, (1977), pp.257-260.
10. H.-P. Heinz, "Nodal properties and variational characterisations of solutions to nonlinear Sturm-Liouville problems", *J. Diff. Eqns.* **62**, (1986), pp.299-333.
11. C. Jones and T. Küpper, "On the infinitely many solutions of a semilinear elliptic equation", *SIAM J. Math. Anal.* **17** (4) (1986), pp.803-835.
12. C. Jones, "An instability mechanism for radially symmetric standing waves of a nonlinear Schrödinger equation", *J. of Differential Equations*, to appear.
13. J. Maddocks, *Stability and Folds*, preprint #174, Institute of Mathematics and its Applications, University of Minnesota.
14. K. McLeod and J. Serrin, "Uniqueness of solutions of semilinear Poisson equations", *Proc. Nat. Acad. Sci.* **78**, (1981) pp.6592-6595.
15. Z. Nehari, "On a nonlinear differential equation arising in nuclear physics", *Proc. Roy. Irish Acad.* **62**, (1963), pp.117-135.
16. L.E. Payne and D.H. Sattinger, "Saddle points and instability of nonlinear hyperbolic equations", *Israel J. of Mathematics* **22**, (1975), pp.273-303.
17. S.I. Pohozaev, "Eigenfunctions of the equation $\Delta u + \lambda f(u) = 0$", *Soviet Math. Dokl.* **5**, (1965), pp.1408-1411.
18. G.H. Ryder, "Boundary value problems for a class of nonlinear differential equations", *Pacific J. Math.* **22**, (1968), pp.477-503.
19. J. Shatah and W. Strauss, "Instability of nonlinear bound states", *Comm. Math. Phys.* **100**, (1985), pp.173-190.

20. J. Smoller, "Symmetry breaking for solutions of semilinear elliptic equations with general boundary conditions". Preprint.
21. N. Sternberg, "Blow up near higher modes of nonlinear wave equations", *Trans. AMS* **296** (1), 1986, pp.315-325.
22. W.A. Strauss, "Existence of solitary waves in higher dimensions", *Comm. Math. Phys.* **55**, (1977), pp.149-162.
23. M. Weinstein, "Lyapunov stability of ground states of nonlinear dispersive equations", *Comm. Pure and Applied Math.* **34**, (1986), pp.51-68.

16

REMARKS ABOUT EQUILIBRIUM CONIFUGRATIONS OF CRYSTALS

DAVID KINDERLEHRER

1. INTRODUCTION

The morphology of a crystal may show several phases and these may be altered with changes in its mechanical or thermal environment. For example, some crystals may be deformed to consist of several twin related phases. Indeed, our interest is to study defect structures in materials which have mobile phase boundaries whose existence, position, and orientation are sensitive to applied loads, temperature, and electromagnetic fields.

To set these phenomena into the context of thermoelasticity, Ericksen ([19] – [22]) has derived a stored energy density which exhibits invariance with respect to change of the crystallographic lattice basis of the material. Such a density is invariant with respect to an infinite discrete group as well as frame indifferent. A body governed by it is rendered highly unstable with respect to certain motions. For example, at a smooth local minimum of energy in a constant temperature heat bath, the Cauchy stress reduces to a pressure, cf. Ericksen [18]. So it seems unlikely that even setting homogeneous boundary conditions leads to a homogeneous extremal.

In this note we take up a direct method for finding and analysing equilibrium configurations under displacement loading conditions. We favour this approach as a means of surmounting the difficulties imposed by the defect structures on the stability of smooth solutions. We are able to illustrate how an important role in the thermodynamics of the crystal is played by its subenergy. This concept, related to the traditional free energy used by many workers, was introduced by Ericksen [23] and is based in part on a method of Flory [30]. Often the energy assumed by a configuration is its subenergy and its stress is a pressure determined by it. This may be reconciled with one common thermodynamic view where phase diagrams are expressed in terms of pressure, or specific volume, and temperature.

The equilibrium configuration is characterized by a minimizing sequence, rather than just its limit. A convenient way of expressing this is through Young's idea of a parametrized measure. With this measure, energy and stress may be calculated. Morphology may be ascertained by inspection of the linearized equation that the parametrized measure provides. We may regard this measure as an accounting device to summarize the properties of a minimizing sequence.

What emerges from these considerations is a coarse theory which in some manner accommodates information from a finer structure to yield macroscopic properties. A limit configuration found in this way may be a macroscopically homogeneous but infinitely twinned array of states of minimum energy. It would appear that the material seeks to assume the lowest energy available to it by suffering small, kinematically admissible shears. Of course, a physical crystal cannot be infinitely twinned, but what we are able to construct may offer an approximation to the actual body.

This paper is primarily a report on [12]. Our remarks here will be a summary of:

- a brief background of the theory
- the minimum energy calculation
- the existence of solutions and the notion of a parametrized measure solution

- the analysis of parametrized measure minima
- the analysis of the energy density at an equilibrium configuration
- brief examples related to phase transitions.

There are two facets to these questions as topics in the calculus of variations. One is to consider energy functions in order to seek those properties of material symmetry and continuity which permit us to determine minimum energy configurations. Fonseca [32], [33] has undertaken an analysis in this direction. Another is to restrict our attention to the elastic crystal to assess the behaviour of minimizing sequences and their various state functions. This is our topic here. The two methods are complementary and they agree in their calculation of the minimum energy available to a configuration.

It would be out of place for us to attempt to mention here the body of work devoted to the challenging questions which Ericksen's ideas have stimulated. We might suggest to the interested reader consultation of some work of Ericksen [18–29], James [36–41], Parry [52], and Pitteri [56–60]. Ball and James study fine twinning in [7]. We show how our ideas are consistent with theirs in Section 6. Mathematical analysis of the dead loading problem and related issues is due to Fonseca [31]. Additional questions are considered by Fonseca and Tartar [34]. One interesting feature of the theory is that it may be used to derive the relations of Müller's interaction theory for the ferroelectric transition in Rochelle salt, cf. [43]. An interesting earlier work about phase transitions is Tisza [64].

Given a bounded domain $\Omega\subset\mathbb{R}^n$ ($n=2$ or 3) with adequately smooth boundary $\partial\Omega$ and a stored energy density $W(F)$, we investigate deformations y of Ω satisfying

$$\delta \int_\Omega W(\nabla y)\, dx = 0 \tag{1.1}$$

or

$$\int_\Omega W(\nabla y)\, dx = \inf \int_\Omega W(\nabla v)\, dx \tag{1.2}$$

where the admissible variations belong to a class

$$A = \{v \in H^{1,\infty}(\Omega): \quad v = y_0 \text{ on } \partial\Omega\}$$

with y_0 prescribed [(1)].

The analytical difficulties encountered in the study of (1.1) or (1.2) are well publicized, by now, but may withstand a brief review. By an elastic crystal we understand a three dimensional lattice given by three independent lattice vectors $\{l_1, l_2, l_3\}$, written as a matrix $L = (l_1\, l_2\, l_3)$ with columns l_i. Owing to the frame indifference of the Helmholtz free energy Φ, we may express

$$\Phi = \Phi(L^T L). \tag{1.3}$$

The lattice basis is not unique. Indeed, for any matrix M having integer entries with $\det M = \pm 1$, the matrix

$$L' = LM$$

is another lattice basis, whence

$$\Phi(L^T L) = \Phi(M^T L^T LM) \quad \text{for} \quad M \in GL(\mathbb{Z}^3). \tag{1.4}$$

For a fixed L we define the elastic bulk energy of the crystal by

$$W(F) = \Phi(L^T CL), \quad C = F^T F, \quad \det F > 0. \tag{1.5}$$

Thus

$$W(F) = W(QFH) \quad \text{for} \quad \det F > 0, \quad Q^T Q = 1, \quad \det Q = 1, \quad H \in \mathbb{H}, \tag{1.6}$$

where $\mathbb{H} = LGL(\mathbb{Z}^3)L^{-1}$ is a conjugate group of $GL(\mathbb{Z}^3)$. We impose on W, or Φ, the conditions

$$W(F) \geqslant 0, \quad W(1) = 0, \quad \text{and} \quad \lim_{\det F \to 0} W(F) = \infty.$$

More generally, Φ may depend on temperature, polarization (when electromagnetic fields are active), or lattice shifts (for the case of complicated crystals).

For any F, $\det F > 0$, and $H = 1 + a \otimes n \in \mathbb{H}$, a direct calculation shows that

$$W(F(1 + \lambda a \otimes n)), \quad -\infty < \lambda < \infty, \tag{1.8}$$

is a periodic function of period 1. In particular this implies that there are points λ for which

$$C_{ijhk}\, a^i a^h n^j n^k < 0 ,\qquad C_{ijhk} = \frac{\partial^2 W}{\partial F_{ij}\, \partial F_{hk}} (F(1+\lambda a \otimes n)). \tag{1.9}$$

Thus W is not rank 1 convex, and in particular, neither quasiconvex nor such that the integral in (1.1) is lower semicontinuous. In fact the deformation satisfying (1.9) is infinitesimally unstable, whereas homogeneous deformations are minima of quasiconvex integrands by definition. A further contrast is afforded by the uniqueness result of Knops and Stuart [44].

The role of quasiconvexity, introduced by Morrey, cf. [50], in the investigation of isotropic elastic solids has been explored by Ball ([3], [4] for example). We call the reader's attention to the work of Ball and Murat [8] and Dacorogna [14] as well. Although not pertinent to elasticity, the work of Meyers [48] sheds some light on these issues.

Moreover, the periodicity of W shows that the functional of (1.1) does not grow for all large values of $|F|$. It is a dictum of the calculus of variations that rapid growth at infinity assists the finding of minima. Where growth is only linear, a minimizer need not be in an ordinary function space; its gradient may be a measure for example. This occurs in the classical theory of elastic-perfectly plastic deformation, c.f. Témam [63] or [35]. When there is no growth at all, it may be even more difficult to characterize solutions. Our choice here is Young measures. We shall strive to convince the reader that the parametrized measure arises here in a very natural way which also is rather distinct from its use in the study of hyperbolic conservation laws. Its implementation there was suggested by Tartar [62] and it was further developed by DiPerna, for example [15]. Articles in the volume [61] are devoted to this subject; Slemrod

[61] has written an expository one.

The notion of effective modulus, used in the theory of periodic structures, offers a point of comparison. The paper by Kohn and Milton [45] contains a very readable account of this. Given a tensor $C(x)$, $x \in Q$, a unit cube, its effective modulus tensor $\bar{C}$ is

$$\bar{C}A \cdot A = \inf_{H_0^1(Q)} \int_Q C(A + \nabla\zeta) \cdot (A + \nabla\zeta)\, dx .$$

If $C(x)$ is constant, then $\bar{C} = C$. In general $\bar{C}$ is the tensor of coefficients of the homogenized operator associated to C. Likewise, if $W(A)$ is a variational integrand, we may call

$$\bar{W}(A) = \inf_{H^{1,\infty}(Q)} \int_Q W(A + \nabla\zeta)\, dx$$

its effective modulus in some sense. When W is quasiconvex, $\bar{W} = W$. However, $\bar{W}(A)$ is usually complicated and difficult to compute. One of the features of our W, satisfying (1.5) – (1.7), is that $\bar{W}$ is the subenergy density which is a function of the determinant alone.

In the theory of effective moduli, one seeks to achieve the optimal coefficients of $\bar{C}$ in some sense by various constructions, cf. Milton [49] for example, or the papers in [45]. The question of whether or not there are optimal parametrized measure solutions for (1.1) or (1.2) has not yet been investigated. One connection between crystals and laminates has been explored in [13].

The illustration on p.151 of [49] shows a laminate construction which turns out to be similar to the construction used by us to find parametrized measure solutions. It is not dissimilar to what is seen under the microscope in the laboratory, say, of shape memory material untreated to render it a single crystal, cf. e.g., Barrett and Massalski [9], p.527.

2. MINIMUM ENERGY

The minimum energy available to a configuration turns out to be its subenergy, or related to it. For a given $W(F)$ this subenergy is defined to be

$$\phi(t) = \inf_{\det A = t} W(A) \tag{2.1}$$

For the moment let us suppose fixed a bounded open set Ω with piecewise smooth boundary $\partial\Omega$ and y_0 a smooth deformation of $\bar{\Omega}$ with $\det \nabla y_0 > 0$. Let us set

$$E_\Omega(y_0) = \operatorname{Inf}_{A(y_0)} \int_\Omega W(\nabla v)\, dx \tag{2.2}$$

$$A(y_0) = \{v \in H^{1,\infty}(\Omega) : \det \nabla v > 0 \text{ in } \Omega \text{ and } v = y_0 \text{ on } \partial\Omega\} . \tag{2.3}$$

We may state this result.[†]

Assume that $\phi(t)$ is convex and Ω and y_0 are as above. Then

$$E_\Omega(y_0) = \inf_{A(y_0)} \int_\Omega \phi(\det \nabla v)\, dx . \tag{2.4}$$

In the case where $y_0(x) = Fx$ is a homogeneous deformation and ϕ is convex, by Jensen's inequality and (2.1),

$$\begin{aligned} \int_\Omega W(\nabla v)\, dx &\geq \int_\Omega \phi(\det \nabla v)\, dx \\ &\geq \phi\left(|\Omega|^{-1} \int_\Omega \det \nabla v\, dx\right) |\Omega| \\ &\geq \phi\left(|\Omega|^{-1} \int_\Omega \det F dx\right) |\Omega| \\ &= \phi(\det F)\,|\Omega| . \end{aligned}$$

Thus from (2.4),

$$E_\Omega(Fx) = \phi(\det F)\,|\Omega| . \tag{2.5}$$

The idea of the proof of (2.4) is to show (2.5) first and then to approximate. There are two ideas involved, both elementary, so we might mention them.

Let $\Omega \subset \mathbb{R}^3$ be a bounded open domain with piecewise smooth

† see footnote on p.237.

boundary as before. Let

$$F_0,\ \det F_0 > 0\,,$$
$$B = 1 + a \otimes b\,,\ \det B = 1 + a\cdot b > 0\,, \text{ and}$$
$$\theta, 0 < \theta < 1\,,$$

be given and consider

$$F = F_0(1 + \theta a \otimes b) = (1-\theta)F_0 + \theta F_0 B\,. \tag{2.6}$$

We assume that $|b| = 1$. Let $\chi(t)$ be the characteristic function of the interval $(0,\theta) \subset (0,1)$ and extend it to be a 1-periodic function on $\mathbb{R}$. It is easy to check that

$$f^k(x) = \chi(kx\cdot b)\,, \quad k = 1,2,3,\cdots$$

satisfy

$$f^k \to \theta \ \text{ in } L^\infty(\mathbb{R}^3) \text{ weak} * \text{ as } k\to\infty\,. \tag{2.7}$$

Let us recall that this means that for any ball B,

$$\int_B f^k dx \to \theta \ \text{ as } k \to \infty\,,$$

or equivalently, that

$$\int_\Omega f^k \zeta\, dx \to \theta \int_\Omega \zeta\, dx \ \text{ as } k\to\infty \text{ for any } \zeta \in L^1(\Omega)\,.$$

Setting

$$u(y) = F_0\left(y + \int_0^r \chi(t)\,dt\, a\right), \quad r = y\cdot b\,,$$

and

$$u^k(x) = k^{-1} u(kx)\,, \tag{2.8}$$

it is clear that

$$u^k \to Fx \ \text{ in } H^{1,\infty}(\Omega) \text{ weak} *\,. \tag{2.9}$$

This means that

$$u^k \to Fx \ \text{ uniformly in } \bar{\Omega}$$

and

$$F^k = \nabla u^k = F_0(1 + f^k a \otimes b)$$
$$= (1-f^k)F_0 + f^k F_0 B \to F \ \text{ in } L^\infty(\Omega) \text{ weak} *\,.$$

Now

$$W(F^k) = (1-f^k)\,W(F_0) + f^k\, W(F_0 B)\,,$$

whence

$$W(F^k) \to (1-\theta)\,W(F_0) + \theta W(F_0 B) \ \text{ in } L^\infty(\Omega) \text{ weak} *\,.$$

A particular consequence of this is that whenever

$$F = (1-\theta)F_0 + \theta F_0 B, \quad B = 1 + a \otimes b ,$$

then

$$E_\Omega(Fx) \leqslant (1-\theta)W(F_0) + \theta W(F_0 B) . \tag{2.10}$$

If $W(F_0) = \phi(\det F_0)$ and $B = H = 1 + a \otimes n \in \mathbb{H}$, the symmetry group of W then

$$W(F_0 H) = W(F_0) = \phi(\det F_0) = \phi(\det F) ,$$

from which it follows that

$$E_\Omega(Fx) \leqslant |\Omega| \phi(\det F) .$$

In the special case $F_0 = 1$ and $W(F) \geqslant W(1) = 0$ for all F, then

$$E_\Omega((1 + \theta a \otimes n)x) = 0 .$$

To attempt a first motivation of our idea, observe that this minimum has been calculated via a minimizing sequence without regard to the numerical value of $W(1 + \theta a \otimes n)$.

This computation, although illustrative, is not sufficient to establish (2.5). There must be some iterative argument; care must be exercised with boundary conditions, and so forth. We found useful this fact about matrices: For any matrix A, $\det A > 0$, there are vectors a_i, n_i, $i = 1,2$, and a rotation Q such that

$$A = (\det A)^{1/3} Q(1 + a_2 \otimes n_2)(1 + a_1 \otimes n_1) . \qquad a_i \cdot n_i = 0, \; i = 1,2. \tag{2.11}$$

In general there need only be one iteration. A similar fact is proved in Fonseca [33].

Returning to the idea behind (2.10), let $\psi(A)$ be any continuous function on $\mathbf{M}$, the set of matrices with positive determinant. Then the sequence $(\psi(F^k))$ is bounded in $L^\infty(\Omega)$ and indeed

$$\psi(F^k) = (1 - f^k)\psi(F_0) + f^k \psi(F_0 B) \to (1-\theta)\psi(F_0) + \theta\psi(F_0 B) \text{ in } L^\infty(\Omega) \text{ weak}* .$$

Now we may by leap of imagination write

$$(1-\theta)\,\psi(F_0)+\theta\psi(F_0 B)=\int_{\mathbb{M}}\psi(A)\,d\nu \tag{2.12}$$

for

$$\nu = (1-\theta)\,\delta_{F_0}+\theta\,\delta_{F_0 B},$$

where δ_A denotes the Dirac delta measure at the matrix A. The measure ν is called a parametrized measure or Young measure [65] for the sequence (u^k).

Reconsidering the case where $B = H = 1 + a\otimes n \in \mathbb{H}$ and $W(F_0) = \phi(\det F_0)$ realizes the value of the subenergy, we find that (u^k) is a minimizing sequence and we may refer to ν as a parametrized measure minimum.

The formula (2.12) may be used to evaluate various state functions. For example, the Piola-Kirchhoff stress $S(F) = \frac{\partial W}{\partial F}(F)$ and the linear elasticity tensor $S'(F) = \partial S(F)/_{\partial F} = \partial^2 W(F)/_{\partial F^2}$ are given by

$$\bar{S} = \int_{\mathbb{M}} S(A)\,d\nu \quad\text{and}\quad \bar{S}' = \int_{\mathbb{M}} S'(A)\,d\nu\,. \tag{2.13}$$

Observe that F itself is given by

$$F = \int_{\mathbb{M}} A\,d\nu\,.$$

It is not clear which matrices F give rise to minimizing Young measures. In view of (2.11), a dense set in $\mathbb{M}$ is given by those of the form

$$F = QF_0\,(1+\theta_2\,a_2\otimes n_2)\,(1+\theta_1\,a_1\otimes n_1) \tag{2.14}$$

where

$$W(F_0) = \phi(\det F_0)\,,\quad \theta_i\in\mathbb{R}\,,\quad Q^TQ = 1\,,$$

and

$$H_i = 1 + a_i\otimes n_i \in \mathbb{H}\,.$$

The proof of this is analogous to the proof of (2.5) and the measure ν associated to F is

$$\begin{aligned}\nu = (1-\theta_1)\,(1-\theta_2)\,\delta_{QF_0} &+ \theta_1(1-\theta_2)\,\delta_{QF_0 H_1}\\ &+\theta_2(1-\theta_1)\,\delta_{QF_0 H_2}+\theta_1\,\theta_2\,\delta_{QF_0 H_2 H_1}.\end{aligned} \tag{2.15}$$

More generally if

$$H_i = 1 + a_i \otimes n_i \in \mathbb{H} \quad \text{and} \quad \theta_i \in \mathbb{R}, \quad i = 1, \cdots, N,$$

then

$$F = QF_0 \prod_1^N (1 + \theta_i a_i \otimes n_i)$$

determines a Young measure minimum given by a formula analogous to (3.15).

We might say a few words about the case when the subenergy ϕ is not assumed convex. In this situation (2.4) must be modified to

$$E_\Omega(y_0) = \inf_{A(y_0)} \int_\Omega \phi^{**}(\det \nabla v)\, dx \qquad (2.16)$$

where ϕ^{**} is the convex minorant of ϕ. There are several ways to show this. The easiest relies on the formula (2.10) in the case where a, b are not necessarily orthogonal and a result of Ball and Murat, [8], p.240. Another proof may be given using Mascolo and Schianchi [47] and Moser [51] under some additional hypotheses.

3. PROPERTIES OF MINIMUM ENERGY CONFIGURATIONS

Ericksen's deduction [18] that the Cauchy stress of a minimizing configuration is a constant pressure turns out to be valid under arbitrary deformation, at least in some sense. To discuss this, we might first agree on what we mean by a parametrized measure minimum and subsequently on what we mean by a stable minimum.

Let us say that a sequence (u^k) in $H^{1,\infty}(\Omega)$ determines the parametrized measure $(\nu_x)_{x \in \Omega}$ provided that there is a $y \in H^{1,\infty}(\Omega)$ such that

$$\begin{aligned} &u^k \to y \quad \text{in} \quad H^{1,\infty}(\Omega) \text{ weak*} \\ &F^k = \nabla u^k \text{ satisfy } \det F^k > 0, \end{aligned} \qquad (3.1)$$

$$E_\Omega(y) = \lim_{k \to \infty} \int_\Omega W(F^k)\, dx\,, \qquad (3.2)$$

and for any continuous function $\psi(A)$,

$$\psi(F^k) \to \bar{\psi} \quad \text{in} \quad L^\infty(\Omega) \text{ weak*} \quad \text{where} \quad \psi(x) = \int_{\mathbb{M}} \psi(A)\, d\nu_x\,. \qquad (3.3)$$

Since the $\|F^k\|_{L^\infty(\Omega)}$ are bounded, the support of any parametrized measure is compact.

Note that if $y(x)$, $x \in \Omega$, is a Lipschitz minimum, it is also a parametrized measure minimum with

$$\mu_x = \delta_{F(x)} .$$

However there may be sequences (u^k) convergent to y which determine measures (ν_x) different from (μ_x) above. This is an essential feature of the parametrized measure as a macroscopic device for recording properties of the smaller scale structure.

When ϕ is convex it is possible to check that $W(A)$ is $d\nu_x\, dx$ integrable and

$$E_\Omega(y) = \lim_{k \to \infty} \int_\Omega \phi(\det F^k)\, dx. \tag{3.4}$$

Moreover,

$$\begin{gathered} W(F(x)) = \phi(\det F(x)) \\ \operatorname{supp} \nu_x \subset \{A: \det A = \det F(x)\} \quad \text{a.e. in } \Omega. \end{gathered} \tag{3.5}$$

From (3.5), it follows easily that for a subsequence of k,

$$\det F^k \to \det F \quad \text{a.e. in } \Omega \text{ and in } L^p(\Omega), \quad 1 \leqslant p < \infty .$$

Since $\operatorname{adj} \nabla v$ is a weak* continuous function, the above implies that

$$(F^k)^{-1} \to F^{-1} \quad \text{in } L^\infty(\Omega) \text{ weak*} . \tag{3.6}$$

Most desirably, an equilibrium solution renders stationary the energy functional within a suitable class of disturbances. This is an attribute of stability which is not necessarily conferred on a state which realizes minimum energy. Much has been written about this, especially by Ball [4] , [5] , and it is fair to report that our difficulties here are connected to the growth of W as $\det F \to 0$. For example, that $(\nu_x)_{x\in\Omega}$ is a parametrized measure minimum ought to mean that

$$\int_\Omega \overline{W}\, dx \leqslant \int_\Omega \int_{\mathbb{M}} W(A + \nabla\zeta)\, d\nu_x\, dx , \quad \zeta \in H^{1,\infty}(\Omega) ,$$

at least for $\|\zeta\|_{H^{1,\infty}(\Omega)}$ sufficiently small. Unfortunately, since nothing has been established about $\det F$, it is not evident that there are any ζ for which the right hand side is finite.

Conditions for variation of domain are the easiest to control, so our criteria will be expressed in these terms. We say that a parametrized measure minimum is *stable* provided that for each $\zeta \in H^{1,\infty}(\Omega)$, there is an $\varepsilon_0 > 0$ such that

$$\int_\Omega \bar{W}\,dx \leqslant \int_\Omega \int_{\mathbb{M}} W(A(1+\varepsilon\nabla\zeta))\det(1+\varepsilon\nabla\zeta)^{-1}\,d\nu_x\,dx < \infty \quad \text{for } \varepsilon < \varepsilon_0 . \tag{3.7}$$

From this we may infer a familiar form of the equilibrium equation:

$$\int_\Omega \int_{\mathbb{M}} \{A^T S(A)\cdot\nabla\zeta - W(A)\operatorname{div}\zeta\}\,d\nu_x\,dx = 0 \quad \text{for } \zeta \in H^{1,\infty}(\Omega) \tag{3.8}$$

There are several conditions which guarantee that a minimum is stable. About the solution $y(x)$, with $F = \nabla y$, if

$$\det F \geqslant c > 0 \quad \text{in } \Omega, \tag{3.9}$$

then the measure $(\nu_x)_{x\in\Omega}$ is stable. This is true for any parametrized measured minimum whose underlying deformation gradient is F.

Also note that if

$$m(\det A)^{-\alpha} \leqslant W(A) \leqslant M(\det A)^{-\alpha} + C \tag{3.10}$$

holds for some $0 < m \leqslant M$, $\alpha > 0$, and $C \geqslant 0$, then any parametrized measure minimum is stable. In fact, (3.10) need only hold on the support of ν.

The energy density derived in Eftis, McDonald, and Arkilic [16],[17] may be written in the form

$$W(A) - g(\det A) \tag{3.11}$$

where $W(A)$ satisfies (3.10) but g is a convex function of $\det A$ which has the property

$$\lim_{t\to 0} g(t) = +\infty .$$

The minima of interest to them are usually homogeneous, so (3.9) applies. More generally, as our result below makes clear, a parametrized measure minimum for W determines a stationary point of (3.11).

Stable minima have the property which Ericksen observed. We state this result.

Assume that $(\nu_x)_{x \in \Omega}$ is a parametrized measure minimum in the sense of (3.1), (3.2), (3.3) which is stable in the sense of (3.7). Let the subenergy ϕ be strictly convex. Then $\det F$ and $\phi'(\det F)$ are constant in Ω and the Cauchy stress

$$T = \phi'(\det F)\, 1 \tag{3.12}$$

is a constant pressure.

There are some interpretations of this as well. Suppose our situation involves minimum energy configurations associated to $W(F, \theta)$, where θ is temperature. Then the entropy density at a minimum

$$\eta = -\frac{\partial W}{\partial \theta}(F, \theta) = -\frac{\partial \phi}{\partial \theta}(\det F, \theta) \tag{3.13}$$

and the specific heat, at constant volume say,

$$C_v = \theta^{-1} \frac{\partial \eta}{\partial \theta} \tag{3.14}$$

depend only on the specific volume, that is $\det F$, not on the details of the boundary conditions.

For example, a Curie point associated to a singularity of the specific heat depends only on the specific volume, even though it may indicate a change of shape of the crystal. This is consistent with the traditional view of thermodynamicists who frequently write phase diagrams in terms of specific volume, pressure, and temperature, cf. Pippard [55], p.136 *et seq.* for instance.

Many analytical questions now arise, for example regarding the form of a parametrized measure minimum. This is currently under investigation. For example, suppose that the measure ν corresponds to a homogeneous deformation $y = Fx$ and

$$\nu = \mu \otimes \delta_{F_0 H},$$

that is

$$\int_{\mathbb{M}} \psi(A)\, d\nu = \int_{SO(3)} \psi(QF_0 H)\, d\mu \,.$$

Then

$$F = \Big(\int_{SO(3)} Q\, d\mu \Big) F_0 H = PF_0 H$$

and by weak continuity,

$$\det F = \int_{SO(3)} \det(QF_0 H)\, d\mu = \det F_0$$

and

$$F^{-1} = \int_{\mathbb{M}} (QF_0 H)^{-1} d\nu = H^{-1} F_0^{-1} P^T .$$

From this

$$1 = \det FF^{-1} = \det P \det P^T = (\det P)^2 ,$$

so

$$\det P = 1 \,.$$

Also,

$$1 = FF^{-1} = PP^T ,$$

hence P is a rotation. Now μ is a probability measure on the sphere of radius $\sqrt{3}$ in $\mathbb{R}^9$, which is the boundary of a strictly convex set. It follows that μ is a delta measure at P, so

$$\nu = \delta_P \otimes \delta_{F_0 H} .$$

Other characterizations in this spirit are also possible.

4. PARAMETRIZED MEASURE EQUILIBRIA

Some of the preceding conclusions remain valid when the measure does not provide an absolute minimum, but this depends on what is granted about equilibrium configurations. Surely they ought to be local minima in some sense, and the extent to which they are will determine their properties. While dynamical considerations may offer the most satisfactory notions of metastability, as developed in Andrews and Ball [2] and Pego [53],[54] for example, these are not yet available for our three dimensional problem.

Maintaining our static viewpoint, an alternative approach

to the analysis of the measure is to investigate the behaviour of the energy density at an equilibrium configuration. The essential features of the argument are illustrated by considering measures which minimize with respect to small compactly supported perturbations. Under these circumstances a form of quasiconvexity may be established. Similar considerations in a different context were initiated independently by Ball [6]. When the admissible perturbations are sufficiently large, conclusions like (3.11)—(3.13) follow. Such may be seen as either a defect or an advantage of the theory.

We found this point of view to be a useful complement to the study of absolute minima, but in order to keep this presentation short, we shall omit the details.

5. SOME EXAMPLES RELATED TO PHASE TRANSITIONS

The phase transformations we have in mind involve the onset of a change of shape or other properties with little hysteresis as the crystal, or more general periodic structure, is slowly cycled through the transformation temperature. Let θ_0 denote the critical temperature and suppose that $\Phi(L^T(\theta)\,L(\theta),\theta)$ is arranged so that the reference configuration is the stable homogeneous parent phase for $\theta \geqslant \theta_0$; i.e.,

$$F = 1 \quad \text{for} \quad \theta \geqslant \theta_0 \tag{5.1}$$

is a minimum of $W(F,\theta)$.

The symmetry group of a homogeneous configuration with deformation gradient F is

$$L = \{H \in \mathbb{H} :\ H^T C H = C\}, \quad C = F^T F. \tag{5.2}$$

We assume that the symmetry group of (5.1) is constant for $\theta \geqslant \theta_0$,

$$L_1 = \{H \in \mathbb{H} :\ H^T H = 1\} \tag{5.3}$$

Below θ_0, there is a stable homogeneous daughter phase

$$F = F_0(\theta),\ C_0(\theta) = F_0(\theta)^T F_0(\theta)$$

whose symmetry group is properly contained in L_1. Other daughter phases with Cauchy-Green tensors C are related to $F_0(\theta)$ by symmetry, namely, in such a way that

$$C = H^T C_0 H \text{ for some } H \in L_1 . \tag{5.4}$$

The phases may coexist at critical temperature. These concepts are explained by Ericksen [21]. To simplify matters let us think that

$$\begin{aligned} W(1, \theta) &= 0 \text{ for } \theta \geqslant \theta_0 \\ W(F_0(\theta), \theta) &= 0 \text{ for } \theta \leqslant \theta_0 . \end{aligned} \tag{5.5}$$

Our sketch of the profile of W as a function of θ is the standard austenite/martensite one.

We may discuss the transition in terms of the extent of kinematic compatibility between parent and daughter phases.

CASE 1. *Second order transition*
$F_0(\theta_0) = 1$ *although* $F_0(\theta) \neq 1$ *for* $\theta < \theta_0$.

Examples of this are rare. For practical purposes, the ferroelectric transition in Rochelle salt is one. Here the linearized equation exists at the critical temperatures, however it necessarily exhibits certain degeneracies which are generalized Clausius-Clapeyron relations or Ehrenfest relations. These are the Müller relations, cf. Lines and Glass [46] or Tisza [64]. In [43], the ferroelectric transition is discussed in particular and it is shown that Agmon-Douglis-Nirenberg conditions [1] must fail for certain boundary value problems at critical temperature.

CASE 2. *Kinematically compatible first order transition*
$F_0(\theta_0) - 1$ *is rank* 1.

Although the starting point of James' discussion in [40], this analysis does not abide by the parent/daughter symmetry breakdown described above. The daughter phases are actually certain types of aggregate phases, perhaps the sort which we shall encounter in Case 3. One possible configuration is given in Figure 7, [40]. There is a possibility of piecewise affine, hence Lipschitz, solutions. At the critical temperature, the domain Ω may be decomposed into disjoint subdomains Ω_1, $i = 1, \cdots, n$, with

$$F = \chi_{\Omega_1} 1 + \sum_{2 \leqslant i \leqslant n} \chi_{\Omega_1} F_1 , \tag{5.6}$$

where, say, $F_2 = F_0(\theta_0)$ and the F_i, $i > 2$, are variants of F_2 with respect to some group action. State functions may be determined by evaluation directly on F. In other words, the parametrized measure is merely

$$\nu = \chi_{\Omega_1} \delta_1 + \sum_{2 \leqslant i \leqslant n} \chi_{\Omega_1} \delta_{F_1} .$$

We are not presently aware of any examples where this behaviour takes place in crystals. When $F_0 = 1 + a \otimes b$, then

$$C_0 = (1 + a \otimes b)(1 + b \otimes a)$$

has an eigenvector $p = a \wedge b$ of eigenvalue 1. It is not difficult to check that if the parent phase is a primitive cubic lattice, then the daughter phase is not likely to be a primitive tetragonal lattice. In this case, 1 would be an eigenvalue of C_0 with multiplicity two.

CASE 3. Kinematically incompatible parent and daughter phases: fine phase twinning

This is the subject addressed by Ball and James [7]. Here we wish to point out that our ideas are consistent with theirs. In fine twinning, a homogeneous laminate constructed of daughter phases is compatible with the parent phase across the parent/daughter interface. We refer to the parent phase as austenite and the daughter phase as martensite. The phenomena is rather widespread, indeed, typical of many austenite/martensite transitions. Indium-thallium [10], [11] is an example and the ferroelectric transition in Barium titanate is another [42]. The martensite, to render itself kinematically compatible with the

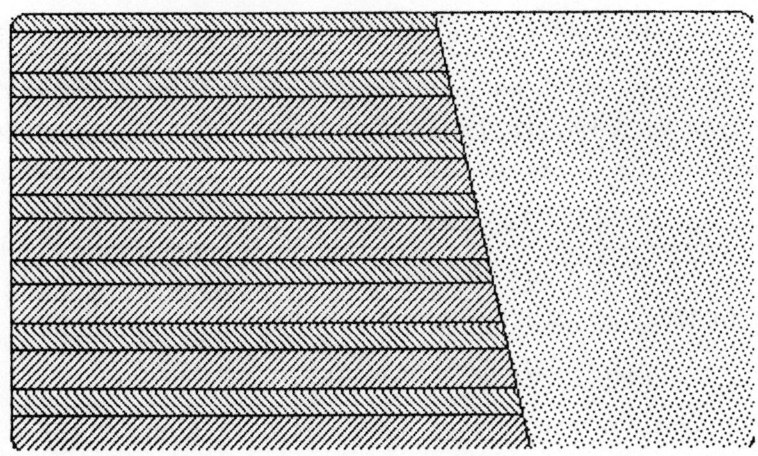

Fine twinning with martensite/austenite interface

austenite, appears to average itself among twin related phases.

Suppose we are give a plane $x \cdot m = 0$ in Ω to serve as the austenite/martensite interface and require that

$x \cdot m > 0$ in the austenite, and

$x \cdot m < 0$ in the martensite.

In experience, the possibilities for this plane will be limited by the symmetry considerations mentioned at the beginning of this section. We suppose that F_0 and $F_{00} = F_0(1 + a \otimes n)$, $1 + a \otimes n \in \mathbb{H}$, are daughter phases and that for some θ, $0 < \theta < 1$, the sequence (u^k) of (2.8) satisfies

$$u^k \to y \text{ in } H^{1,\infty}(\Omega) \text{ weak*} \tag{5.7}$$

where

$y(x)$ and x are kinematically compatible across $x \cdot m = 0$.

Thus

$$y(x) = (1 + b \otimes m)\, x = Fx \tag{5.8}$$

for some $b \in \mathbb{R}^3$. We may take $|m| = 1$.

Our first observation is that the sequence (u^k) may be slightly altered to form a minimizing sequence. Let

$$\eta(t) = \begin{cases} t+1 & -1 \leq t \leq 0 \\ 1 & t > 0 \end{cases} \tag{5.9}$$

Set $\eta^k(x) = \eta(kx\cdot m)$, $k = 1,2,3,\cdots$, and

$$v^k(x) = (1-\eta^k(x))\,u^j(x) + \eta^k(x)\,Fx \tag{5.10}$$

for a $j=j(k)$ which we shall specify now. Indeed, since $u^k \to y$ uniformly as $k\to\infty$, it is easy to check that we may choose a subsequence (u^j) of (u^k) such that

$$v^k \to \begin{cases} x & x\cdot m \geq 0 \\ Fx & x\cdot m < 0 \end{cases} \quad \text{in } H^{1,\infty}(\Omega) \text{ weak}* . \tag{5.11}$$

This implies, cf. Section 3,

$$\det \nabla v^k \to \det F \quad \text{a.e. in } \{x\cdot m < 0\} \cap \Omega ,$$

so that owing to the form of ∇v^k, $\det \nabla v^k \geq \frac{1}{2}\det F > 0$ for k sufficiently large. It follows that

$$\int_\Omega W(\nabla v^k)\,dx = \int_{\Omega\cap\{x\cdot m < -1/k\}} W(F^k)\,dx$$

$$+ \int_{\Omega\cap\{-1/k < x\cdot m < 0\}} W(\nabla v^k)\,dx + \int_{\Omega\cap\{x\cdot m > 0\}} W(1)\,dx$$

$$\to 0$$

as $k\to\infty$ since the first and last integrals vanish and the middle one may be estimated by

$$\left|\int_{\Omega\cap\{-1/k < x\cdot m < 0\}} W(\nabla v^k)\,dx\right| \leq \text{const.}/k .$$

Now the limit of the (u^k) is given by (2.6), so

$$F_0(1+\theta a\otimes n) = 1 + b\otimes m . \tag{5.12}$$

Let $p = m\wedge n$. The equation (5.12) implies that

$$F_0 p = p$$

so $F_0 - 1$ is of rank 2 and its range is in the plane spanned by m and n. We write

$$F_0 = 1 + c_0\otimes n + b_0\otimes m .$$

The same holds for F_{00} since

$$F_{00}(1+(\theta-1)a\otimes n) = 1 + b\otimes m ,$$

so

$$F_{00} = 1 + c_{00} \otimes n + b_{00} \otimes m .$$

Substituting in here the equation defining F_{00}, $F_{00} = F_0(1 + a \otimes n)$, we see immediately that $b_0 = b_{00} = b$.

Expressing u^k by

$$u^k(x) = F_0(x + k^{-1} w(kx)\, a) ,$$

where

$$k^{-1} w(kx) \to \theta x \cdot n \quad \text{uniformly} ,$$

one checks that

$$u^k(x) - y(x) = c_0 n \cdot x + k^{-1}\, w(kx)\, F_0 a ,$$

so that

$$0 = \lim (u^k(x) - y(x)) = c_0 n \cdot x + \theta x \cdot n\, F_0 a .$$

Consequently, $c_0 = -\theta F_0 a$. Also, $c_{00} = (1-\theta) F_0 a$. Setting $c = F_0 a$, we obtain the representation [7]

$$\begin{aligned} F_0 &= 1 - \theta c \otimes n + b \otimes m \\ F_{00} &= 1 + (1-\theta) c \otimes n + b \otimes m. \end{aligned} \tag{5.13}$$

With the representation (5.13) in hand, one may begin to consider the issues posed by the symmetries of the austenite and the martensite to determine, for example, the relationships among the vectors m and n and the proportion θ. This has been done in [7] to excellent agreement with experiment. A parametrized measure minimum is given by the measure

$$\nu = (1 - \chi_a) \{ (1-\theta)\, \delta_{F_0} + \theta \delta_{F_{00}} \} + \chi_a \delta_1 , \tag{5.14}$$

where χ_a is the characteristic function of $\{x \cdot m > 0\} \cap \Omega$.

(1) The space $H^{1,p}(\Omega)$ is the Sobolev space of functions in $L^p(\Omega)$ whose derivatives are in $L^p(\Omega)$, $1 \leqslant p \leqslant \infty$.

---

† See p.223. The conclusion (2.4) requires the technical assumption that

$$\inf_{A(y_0) \cap C^1(\bar{\Omega})} \int_\Omega \phi(\det \nabla v)\, dv = \inf_{A(y_0)} \int_\Omega \phi(\det \nabla v)\, dx .$$

ACKNOWLEDGEMENTS

The author would like to express his appreciation to J.L. Ericksen for his continuous assistance and interest in this work. He would also like to thank J.M. Ball, M. Chipot, I. Fonseca, and R. James for many useful discussions.

Portions of this research were accomplished while the author was visiting Heriot Watt University. He would like to thank J.M. Ball and R. Knops for their hospitality.

Work partially supported by NSF grant MCS 83-01345.

REFERENCES

1. S. Agmon, A. Douglas and Nirenberg, L. "Estimates near the boundary for solutions of elliptic partial differential equations satisfying general boundary conditions, II", *Comm. Pure Appl. Math.* **22**, (1964), pp.35-92.
2. G. Andrews and J. Ball, "Asymptotic behaviour and change of phase in one dimensional nonlinear viscoelasticity", *J. Diff. Eqns.* **44**, (1982), pp.306-341.
3. J.M. Ball, "Constitutive equations and existence theorems in nonlinear elastostatics", (ed. Knops, R.), Heriot-Watt Symp., I, Pitman (1976).
4. J.M. Ball, "Strict convexity, strong ellipticity, and regularity in the calculus of variations, *Math. Proc. Camb. Phil. Soc.* **87**, (1980), pp.501-513.
5. J.M. Ball, "Singular minimizers and their significance in elasticity". *Phase Transformations and Material Instabilities in Solids*, (ed. Gurtin, M.), Academic Press (1984), pp.1-20.
6. J.M. Ball, to appear
7. J.M. Ball and R. James, "Fine phase mixtures as minimizers of energy", *Arch. Rat. Mech. Anal.* to appear.
8. J.M. Ball and F. Murat, "$W^{1,p}$ — quasiconvexity and variational problems for multiple integrals", *J. Funct. Anal.* **58**, (1984), pp.225-253.
9. C. Barrett and T.B. Massalski, *Structure of Metals*, 3rd edition revised, (1980), Pergamon.
10. Z.S. Basinski and J.W. Christian, "Experiments on the martensitic transformations in single crystals of indium-thallium alloys", *Acta. Met.* **2**, (1954), pp.149-166.
11. M.W. Burkart and T.A. Read, "Diffusionless phase change in the indium-thallium system", *J. of Metals* 5, (1953) pp.1516-1524.
12. M. Chipot and D. Kinderlehrer, "Equilibrium configurations of crystals", to appear.
13. M. Chipot, D. Kinderlehrer and G. Vergara-Caffarelli, "The smoothness of linear laminates", IMA preprint 199, *Arch. Rat. Mech. Anal.* **96** (1986), pp.81-96.

14. B. Dacorogna, *Weak Continuity and Weak Lower semicontinuity of Nonlinear Functionals*, Springer Lecture Notes 922 (1982).
15. R. Di Perna, "Convergence of approximate solutions to conservation laws", *Arch. Rat. Mech. Anal.* **63**, (1977), pp.337-407.
16. J. Eftis, D.E. MacDonald and G.M. Arkilic, "Theoretical calculation of the pressure variation of second-order elastic coefficients for alkali, *Mater. Sci. Eng.* **7**, (1971), pp.141-150.
17. J. Eftis, D.E. MacDonald and G.M. Arkilic, "Theoretical calculation of plane wave speeds for alkali metals under pressure, *Mater. Sci. Eng.* **8**, (1971), pp.210-219.
18. J.L. Ericksen, "Loading devices and stability of equilibrium", in *Nonlinear Elasticity*, Academic Press (1973), pp.161-173.
19. J.L. Ericksen, "Special topics in elastostatics", *Adv. in Appl. Mechanics*, (ed. Yih, C.-S.) Academic Press 7, (1977) pp.189-243.
20. J.L. Ericksen, "On the symmetry of deformable crystals, *Arch. Rat. Mech. Anal.* **72**, (1979), pp.1-13.
21. J.L. Ericksen, "Some phase transitions in crystals", *Arch. Rat. Mech. Anal.* **73**, (1980), pp.99-124.
22. J.L. Ericksen, "Changes in symmetry in elastic crystals", IUTAM Symp. Finite Elasticity, (eds. Carlson, D.E. and Shield, R.T.) M. Nijhoff, (1981)m pp.167-177.
23. Ericksen, J.L. "Some simpler cases of the Gibbs phenomenon for thermoelastic solids", *J. of Thermal Stresses*, **4**, (1981), pp.13-30.
24. J.L. Ericksen, "Crystal lattices and sublattices", *Rend. Sem. Mat. Padova* **68**, (1982), pp.1-9
25. J.L. Ericksen, "Ill posed problems in thermoelasticity theory", in *Systems of Nonlinear Partial Differential Equations*, (ed. Ball, J.M.) D. Reidel, (1983), pp.71-95.
26. J.L. Ericksen, "The Cauchy and Born hypotheses for crystals", *Phase Transformations and Material Instabilities in Solids*, (ed. Gurtin, M.), Academic Press, (1984), pp.6178.
27. J.L. Ericksen, "Some surface defects in unstressed solids", *Arch. Rat. Mech. Anal.* **88**, (1985), pp.337-345.
28. J.L. Ericksen, "Stable equilibrium configurations of elastic crystals, *Arch. Rat. Mech. Anal.* **94**, pp.1-14.
29. J.L. Ericksen, "Twinning of crystals. I. Metastability and incompletely posed problems", in *IMA Vol. Math. Appl.* **3**, (eds. Antman, S., Ericksen, J.L., Kinderlehrer. D. and Müller, I.) Springer (1987), pp.77-94.
30. P.J. Flory, "Thermodynamic relations for high elastic polymers", *Trans. Faraday Soc.* **57**, (1961), pp.829-838.
31. I. Fonseca, "Variational methods for elastic crystals", (Thesis, Univ. of Minn.) *Arch. Rat. Mech. Anal.* (1985).
32. I. Fonseca,"Stability of elastic crystals", to appear.
33. I. Fonseca, "The lower quasiconvex envelope of the stored energy function of an elastic crystal", to appear.

34. I. Fonseca and L. Tartar, "The displacement problem for elastic crystals", to appear.
35. R. Hardt and D. Kinderlehrer, "Elastic plastic deformation", *Appl. Math. Optim.* **10**, (1983), pp.203-246.
36. R.D. James, "Mechanics of coherent phase transformations in solids, MRL Report, Brown U. Division of Engineering (1982).
37. R.D. James, "The arrangement of coherent phases in a loaded body", *Phase Transformations and Material Instabilities in Solids*, (ed. M. Gurtin), Academic Press, (1984) pp.79-98.
38. R.D. James, "Stress free joints and polycrystals", *Arch. Rat. Mech. Anal.* **86**, (1984), pp.13-37.
39. R.D. James, "Phase transformations and non-elliptic free energy", *New Perspectives in Thermodynamics* (ed. Serrin, J.) Springer (1986), pp.223-239.
40. R.D. James, "Displacive phase transformations in solids", *J. Mech. Phys. Solids* **34**, (1986), pp.359-394.
41. R.D. James, "The stability and metastability of quartz", Metastability and incompletely posed problems, *IMA Vol. Math. Appl.* **3**, (eds. Antman, S., Ericksen, J.L., Kinderlehrer, D. and Müller, I.) Springer (1987), pp.147-176.
42. F. Jona and G. Shirane, *Ferroelectric Crystals*, Pergamon, (1962).
43. D. Kinderlehrer, "Twinning in crystals II, Metastability and incompletely posed problems", *IMA Vol. Math. Appl.* **3**, (eds. Antman, S., Ericksen, J.L., Kinderlehrer, D. and Müller, I.) Springer (1987), pp.185-211.
44. R.J. Knops and C.A. Stuart, "Quasiconvexity and uniqueness of equilibrium solutions in nonlinear elasticity", *Arch. Rat. Mech. Anal.* **86**, pp.233-249.
45. R. Kohn and G. Milton, "On bounding the effective conductivity of anisotropic composites Homogenization and effective moduli of materials and media", *IMA Vol. Math. Appl.* (eds. Ericksen, J.L., Kinderlehrer, D., Kohn, R. and Lions, J.-L.) (1986), pp.97-125.
46. M.E. Lines and A.M. Glass, *Principles and Applications of Ferroelectrics and Related Materials*, Oxford (1977).
47. E. Mascolo and R. Schianchi, "Existence theorems for non-convex problems", *J. Math Pures et Appl.* **62**, (1983), pp.349-359.
48. N.G. Meyers, "Quasiconvexity and lower semi-continuity of multiple variational integrals of any order", *Trans. A.M.S.* **119**, (1965), pp.125-149.
49. Milton, G. "Modelling the properties of composites by laminates", Homogenization and effective moduli of materials and media, *IMA Vol. Appl. Math.* **1**, (eds. Ericksen, J.L., Kinderlehrer, D., Kohn, R. and Lions, J.-L.), (1986) pp.150-175.
50. C.B. Morrey, Jr., *Multiple Integrals in the Calculus of Variations*, Springer (1966).
51. J. Moser, "On the volume elements on a manifold", *Trans AMS*, **120**, (1965), pp.286-294.

52. G. Parry, "On phase transitions involving internal strain", *Int. J. Solids Structures*, **17**, (1981), pp.361-378.
53. R. Pego, "Phase transitions: stability and admissibility in one dimensional nonlinear viscoelasticity", *Arch. Rat. Mech. Anal.* **97**, (1987), pp.353-394.
54. R. Pego, "Phase transitions in one dimensional nonlinear viscoelasticity: admissibility and stability, Dynamical problems in Continuum Physics, *IMA Vol. Math. Appl.* 4, (eds. Bona, J., Dafermos, C., Ericksen, J.L. and Kinderlehrer, D.), (1987), pp.277-285.
55. A.B. Pippard, *The Elements of Classical Thermodynamics*, Cambridge, (1957).
56. M. Pitteri, "Reconciliation of local and global symmetries of crystals", *J. Elasticity* **14**, (1984), pp.175-190.
57. M. Pitteri, "On the kinematics of mechanical twinning in crystals", *Arch. Rat. Mech. Anal.* **88**, (1985), pp.25-57.
58. M. Pitteri, "On $\nu + 1$ lattices", *J. Elasticity*, **15**, (1985), pp.3-25.
59. M. Pitteri, "On type-2 twins in crystals", *Int. J. Plasticity*, 2, (1986), pp.99-106.
60. M. Pitteri, "A contribution to the description of natural states for elastic crystalline solids", Metastability and incompletely posed problems, *IMA Vol. Math. Appl.* **3**, (eds. Antman, S., Ericksen, J.L., Kinderlehrer, D., Müller, I.), Springer (1987), pp.295-310.
61. M. Slemrod, "Interrelationships among mechanics, numerica- analysis, compensated compactness, and oscillation theory, Oscillation Theory, Computation, and Methods of Compensated Compactness, *IMA Vol. Math. Appl.* **2**, (eds. Dafermos, C., Ericksen, J.L., Kinderlehrer. D. and Slemrod, M.), (1986).
62. L. Tartar, "Compensated compactness and applications to partial differential equations", in *Nonlinear Analysis and Mechanics*, Heriot-Watt Symp. IV, (ed. Knops, R.J.) Pitman, (1979), pp.136-212.
63. R. Témam, *Mathematical Problems in Plasticity*, Gauthier-Villers, (1985).
64. L. Tisza, *On the General Theory of Phase Transformations in Solids*, (eds. Smeluchowski, R., Mayer, J.E. and Weyl, W.A.), Wiley, (1951)., pp.1-35.
65. L.C. Young, *Lectures on Calculus of Variations and Optimal Control Theory*, W.B. Saunders (1969).

17

SOME REMARKS ON UNIQUENESS PROPERTIES

J. L. LIONS

INTRODUCTION

The following remarks are motivated by the question of *exact controllability* in distributed systems.

In fact, we have introduced (J.L. Lions, [1] — [3]) a systematic method to deal with this question *in linear problems*. This method is based on the construction of *Hilbert structures* on the initial data, these Hilbert structures being connected with *uniqueness properties* (we write HUM for the Hilbert Uniqueness Method).

For example, let us consider the wave equation

$$\frac{\partial^2 \phi}{\partial t^2} - \Delta\phi = 0 \quad \text{in} \quad \Omega \times (0\,, T)\,, \tag{1}$$

where Ω is a bounded open set of $\mathbb{R}^n$ with boundary Γ, with

$$\phi(0) = \phi^0\,, \quad \phi'(0) = \phi^1 \quad \text{in} \quad \Omega \tag{2}$$

(where $\phi(0)$ stands for $x \to \phi(x\,,0)$, $\phi'(0) = \frac{\partial \phi}{\partial t}(0)$) and subject to the boundary condition

$$\phi = 0 \quad \text{on} \quad \Gamma \times (0\,,T)\,. \tag{3}$$

We define a semi-norm on the data by setting

$$\|\{\phi^0, \phi^1\}\|_F = \left(\int_{\Sigma_0} \left(\frac{\partial \phi}{\partial \nu}\right)^2 d\Gamma\, dt\right)^{\frac{1}{2}} \tag{4}$$

where

$$\Sigma_0 \subset \Sigma = \Gamma \times (0,T). \tag{5}$$

We say that Σ_0 is a *uniqueness set* if every function ϕ which satisfies (1), (3) and

$$\frac{\partial \phi}{\partial \nu} = 0 \quad \text{on } \Sigma_0 \tag{6}$$

is identically zero (i.e. if we have a uniqueness theorem). [It is classical that such a property is indeed true for Σ_0 "suitable" and for T "large enough".]

Therefore if Σ_0 is a uniqueness set, (4) *defines a norm* and this is the beginning of HUM. One can then obtain exact controllability and stabilization properties in a systematic manner (cf. J.L. Lions, *loc cit.*). □

In a sense, with this method, we "identify" uniqueness theorems and exact controllability; hence we have a motivation to look for (apparently) new uniqueness results, for the sake of exact controllability.

We present here two such examples, dealing with the operator $\frac{\partial^2}{\partial t^2} + \Delta^2$.†

Let us here state the results in a formal fashion, which will be made precise in the text below. □

Let u be a solution of

$$u'' + \Delta^2 u = 0 \quad \text{in } \Omega \times]0,T[, \tag{7}$$

subject to

$$u = \frac{\partial u}{\partial \nu} = 0 \quad \text{on } \Gamma \times]0,T[\tag{8}$$

and subject to

$$\Delta u = 0 \quad \text{on } \Gamma_0 \times]0,T[. \tag{9}$$

Then for *a suitable* Γ_0 *and for* T *large enough*, say $T > T_0$,* it follows that $u = 0$.

REMARK 1.

Usual uniqueness results for (7),(8), are obtained by adding 2 boundary conditions on Γ, namely $\Delta u = \frac{\partial \Delta u}{\partial v} = 0$,

---

† Other examples, for various models of plates, will be found in J. Lagnese and J.L. Lions [1] and E. Zuazua [6].

* This will be made precise below.

with T arbitrarily small. The above situation (which seems new?) is therefore different. □

REMARK 2.

Many open questions remain. For instance,

(i) what are "the best" set of conditions on Γ_0 T_0 such that the above result holds true?

(ii) what happens if we *replace* (9) by

$$\frac{\partial \Delta u}{\partial \nu} = 0 \quad \text{on} \quad \Gamma_0 \times (0\,, T) \quad ? \quad \square \tag{10}$$

Let us give now — again in a formal fashion to be made precise below — the second example we have in mind.

Let u_1 and u_2 be solutions of

$$u_i'' + \Delta^2 u_i = 0 \quad \text{in} \quad \Omega \times]0\,, T[\,, \quad i = 1\,,2 \tag{11}$$

subject to

$$\left| \begin{aligned} & u_1 = \frac{\partial u_1}{\partial \nu} = 0 \quad \text{on} \quad \Gamma \times]0\,, T[\\ & u_2 = \Delta u_2 = 0 \quad \text{on} \quad \Gamma \times]0\,, T[\end{aligned} \right. \tag{12}$$

$$\frac{\partial \Delta u_2}{\partial \nu} = 0 \quad \text{on} \quad \Gamma_0 \times]0\,, T[\tag{13}$$

$$\Delta u_1 + \frac{\partial}{\partial t}\frac{\partial u_2}{\partial \nu} = 0 \quad \text{on} \quad \Gamma \times]0\,, T[\ . \tag{14}$$

<u>Then for</u> Γ_0 <u>suitable and</u> $T > T_0$, <u>one has</u>

$$u_1 = u_2 = 0\ . \quad \square$$

REMARK 3

A positive answer to the question of Remark 2,(ii), would imply $u_2 = 0$, hence $\Delta u_1 = 0$ on $\Gamma \times]0, T[$, which is more than necessary to imply $u_1 = 0$, by using the first example. □

To comply with the topics of this Symposium, we add, in Section 3, a *nonlinear* uniqueness theorem....

1. FIRST UNIQUENESS THEOREM

Let x^0 be arbitrarily chosen in $\mathbb{R}^n$. We shall set

$$m(x) = \{x_k - x_k^0\}$$

$$\Gamma(x^0) = \{x \mid x \in \Gamma\ ,\ m(x)\,\nu(x) \geqslant 0\} \tag{1.1}$$

where $\nu(x)$ = unit normal to Γ directed towards the exterior of Ω,

$$\Gamma_*(x^0) = \Gamma \setminus \Gamma(x^0)\ . \tag{1.2}$$

We introduce λ_0, μ_0 such that

$$\begin{aligned} |\nabla\phi| &\leqslant \frac{1}{\lambda_0}\ |\Delta\phi| \qquad \forall\ \phi \in H_0^2(\Omega)\ ,^{\dagger} \\ |\phi| &\leqslant \frac{1}{\mu_0}\ |\Delta\phi| \qquad \forall\ \phi \in H_0^2(\Omega)\ . \end{aligned} \tag{1.3}$$

[We have set

$$|\psi| = \left(\int_\Omega \psi^2\,dx\right)^{\frac{1}{2}}, \qquad |\nabla\phi| = \left(\int_\Omega \Sigma\left(\frac{\partial\phi}{\partial x_i}\right)^2 dx\right)^{\frac{1}{2}};$$

λ_0 and μ_0 are such that λ_0^2 and μ_0^2 are the first eigenvalues for

$$\Delta^2 w = -\lambda_0^2\,\Delta w\ ,\quad w \in H_0^2(\Omega), \qquad \Delta^2\rho = \mu_0^2\,\rho\ ,\quad \rho \in H_0^2(\Omega).]$$

We set

$$R(x^0) = \sup.\ m(x)\,\nu(x) \tag{1.4}$$

$$T(x^0) = \frac{R(x^0)}{\lambda_0} + \frac{|n-2|}{2\mu_0}. \tag{1.5}$$

We can now state

THEOREM 1.1. Let u be a solution of (1) – (3) of the Introduction, where

$$\Gamma_0 = \Gamma(x^0)\ ,\quad T_0 = T(x^0) \tag{1.6}$$

and we assume that

$$u \in L^\infty(0, T;\ H_0^2(\Omega))\ . \tag{1.7}$$

Then if $T > T_0$

$$u \equiv 0\ . \tag{1.8}$$

REMARK 1.1

One can have the same generalized set for $\Gamma(x^0)$ with different points x^0. For instance, if Ω is convex, then $\Gamma(x^0) = \Gamma$ for every x^0 inside Ω. Then "the" best choice of x^0 is such that $R(x^0) = \frac{1}{2}$ diameter Ω.

† We use standard Sobolev spaces:

$$H_0^2(\Omega) = \{\phi \mid D^\alpha\phi \in L^2(\Omega)\ ,\ |\alpha| \leqslant 2\ ,\ \phi = \frac{\partial\phi}{\partial\nu} = 0 \text{ on } \Gamma\}\ .$$

Again with Ω convex, $\Gamma(x^0)$ is "smaller" as x^0 gets "farther and farther", so that $T(x^0)$ increases as $\Gamma(x^0)$ "decreases" – according to common sense ! □

REMARK 1.2

It would be very interesting to obtain *counter examples*. For instance take for Ω *a ball*, and for Γ_0 a set *strictly contained* in a half sphere of Γ. Is the conclusion of Theorem 1.1 false *no matter how large we take* T ? □

REMARK 1.3

Komornik [7] has proven that one can take

$$T(x^0) = \frac{R(x^0)}{\lambda_0} \quad \text{if} \quad n \geqslant 2 ,$$

in (1.5). □

PROOF OF THEOREM 1.1

STEP 1. Let us consider a solution ϕ of

$$\left|\begin{array}{l} \phi'' + \Delta^2\phi = 0 \quad \text{in} \quad \Omega \times]0,T[, \\ \phi = \dfrac{\partial\phi}{\partial\nu} = 0 \quad \text{on} \quad \Gamma \times]0,T[, \\ \phi(0) = \phi^0,\ \phi'(0) = \phi^1 ,\quad \phi^0 \in H_0^2(\Omega) ,\ \phi^1 \in L^2(\Omega) . \end{array}\right. \tag{1.9}$$

We are going to show that, given $T > T_0$, there is a constant $c > 0$ such that

$$\int_{\Gamma(x^0)\times(0,T)} (\Delta\phi)^2\, d\Gamma\, dt \geqslant c\left[\|\phi^0\|^2_{H_0^2(\Omega)} + \|\phi^1\|^2_{L^2(\Omega)} \right] . \tag{1.10}$$

Clearly (1.10) *implies* Theorem 1.1.

STEP 2. *An Identity*

We multiply (1.9) by $m_k \dfrac{\partial\phi}{\partial x_k}$, where we use the summation convention for repeated indices. We shall set

$$(\phi,\psi) = \int \phi\psi\, dx ,$$

$$\iint \phi = \iint_{\Omega\times(0,T)} \phi\, dx\, dt , \qquad \int \phi = \int_{\Gamma\times(0,T)} \phi\, d\Gamma\, dt .$$

We also set

$$X = \left(\phi'(t)\ ,\ m_k \frac{\partial\phi(t)}{\partial X_k}\right)\Big|_0^T \tag{1.11}$$

$$Y = (\phi'(t)\ ,\ \phi(t))\Big|_0^T\ . \tag{1.12}$$

We obtain

$$X - \iint \phi' m_k \frac{\partial\phi'}{\partial x_k} + \int \frac{\partial\Delta\phi}{\partial\nu} m_k \frac{\partial\phi}{\partial x_k}$$

$$- \int \Delta\phi \frac{\partial}{\partial\nu}\left(m_k \frac{\partial\phi}{\partial x_k}\right) + \iint \Delta\phi\, \Delta\left(m_k \frac{\partial\phi}{\partial x_k}\right) = 0\ ,$$

i.e.

$$X - \iint \frac{m_k}{2} \frac{\partial}{\partial x_k} (\phi')^2 - \int (\Delta\phi)^2\ m_k\ \nu_k +$$

$$+ \iint \frac{m_k}{2} \frac{\partial}{\partial x_k} (\Delta\phi)^2 + 2\iint (\Delta\phi)^2 = 0\ . \tag{1.13}$$

We integrate by parts in (1.13); it becomes

$$X + \frac{n}{2}\iint \phi'^2 - (\Delta\phi)^2 - \frac{1}{2}\int m_k\, \nu_k (\Delta\phi)^2 + \iint (\Delta\phi)^2 = 0\ . \tag{1.14}$$

We rewrite (1.14) as follows:

$$X + \frac{n-2}{2}\iint \phi'^2 - (\Delta\phi)^2 + \iint \phi'^2 + (\Delta\phi)^2 - \frac{1}{2}\int m_k\, \nu_k (\Delta\phi)^2 = 0\ . \tag{1.15}$$

But (conservation of energy)

$$\frac{1}{2}\left[\ |\phi'(t)|^2 + |\Delta\phi(t)|^2 = \frac{1}{2}\left[\ |\phi^1|^2 + |\Delta\phi^0|^2\right] = E_0 \right. \tag{1.16}$$

and

$$\iint \phi'^2 - (\Delta\phi)^2 = Y\ . \tag{1.17}$$

Using (1.16), (1.17) in (1.15) gives

$$X + \frac{n-2}{2} Y + 2TE_0 - \frac{1}{2}\int (m_k\, \nu_k)\, (\Delta\phi)^2 = 0\ . \tag{1.18}$$

This is the identity we had in mind.

STEP 3. *Estimates*.

We divide the surface integral in (1.18) in two parts:

$$2TE_0 + \int_{\Gamma_*(x^0)\times(0,T)} \left(-\frac{1}{2} m_k\, \nu_k\right) (\Delta\phi)^2 =$$

$$= \int_{\Gamma(x^0)\times(0,T)} \left(\frac{m_k \nu_k}{2}\right)(\Delta\phi)^2 - X - \frac{n-2}{2} Y .$$

It follows that

$$2TE_0 \leqslant \frac{R(x^0)}{2} \int_{\Gamma(x^0)\times(0,T)} (\Delta\phi)^2 + |X| + \frac{|n-2|}{2} |Y| . \tag{1.19}$$

We now estimate $|X|$ and $|Y|$. We observe that[†]

$$\left|\left(\phi'(t), m_k \frac{\partial\phi(t)}{\partial x_k}\right)\right| \leqslant R(x^0)\, |\phi'(t)|\,|\nabla\phi(t)| \leqslant$$
$$\leqslant \frac{R(x^0)}{\lambda_0} |\phi'(t)|\,|\nabla\phi(t)| \leqslant \frac{R(x^0)}{\lambda_0} E_0 ,$$

hence

$$|X| \leqslant \frac{2R(x^0)}{\lambda_0} E_0 . \tag{1.20}$$

We have also

$$|(\phi'(t), \phi(t))| \leqslant \frac{1}{\mu_0} |\phi'(t)|\,|\Delta\phi(t)| \leqslant \frac{1}{\mu_0} E_0 ,$$

hence

$$|Y| \leqslant \frac{2}{\mu_0} E_0 . \tag{1.21}$$

Using (1.20) and (1.21) in (1.19) gives

$$2(T - T(x^0)) E_0 \leqslant \frac{R(x^0)}{2} \int_{\Gamma(x^0)\times(0,T)} (\Delta\phi)^2 . \tag{1.22}$$

Hence (1.10) follows. □

REMARK 1.4

The preceding proof of (1.10) is an extension of the proof of L.O. Ho [1] for *hyperbolic* equations. □

2. SIMULTANEOUS UNIQUENESS THEOREM

We use again the notation of Section 1 and we introduce ν_0 by

$$|\phi| \leqslant \frac{1}{\nu_0} |\nabla\phi| \quad \forall\ \phi \in H_0^1(\Omega) . \tag{2.1}$$

We set

$$T_*(x^0) = R(x^0)\left(\frac{1}{\nu_0} + \max\left(\frac{1}{\lambda_0}, \frac{1}{\nu_0}\right)\right) + \frac{1}{2}\max\left(\frac{|n-2|}{\mu_0}, \frac{n}{\nu_0^2}\right) . \tag{2.2}$$

---

† One improves $T(x^0)$ (cf. Komornik [7]) by estimating $|X + (n-2)/2\ Y|$ instead of estimating $|X|$ and $|Y|$ separately.

We then have

THEOREM 2.1. Let u_i, $i = 1,2$, be solutions of (11) – (14) of the Introduction, with $\Gamma_0 = \Gamma(x^0)$, $T > T_*(x^0)$. Then assuming that

$$u_1 \in L^\infty(0,T;\ H_0^2(\Omega)),\quad u_2 \in L^\infty(0,T;\ H^3(\Omega)) \tag{2.3}$$

one has

$$u_1 = u_2 = 0.$$

The proof is again divided in 3 steps.

STEP 1. Let ϕ_i, $i = 1,2$, be defined as the solutions of

$$\left|\begin{aligned} &\phi_i'' + \Delta^2 \phi_i = 0 \\ &\phi_i(0) = \phi_i^0,\quad \phi_i'(0) = \phi_i^1 \quad,\ i = 1,2, \\ &\phi_1 = \frac{\partial \phi_1}{\partial \nu} = 0 \quad \text{on} \quad \Gamma \times]0,T[\ , \\ &\phi_2 = \Delta\phi_2 = 0 \quad \text{on} \quad \Gamma \times]0,T[\ . \end{aligned}\right. \tag{2.4}$$

We assume *different* regularity hypotheses on the initial data for ϕ_1 and for ϕ_2:

$$\left|\begin{aligned} &\phi_1^0 \in H_0^2(\Omega),\quad \phi_1^1 \in L^2(\Omega), \\ &\phi_2^0 \in H^3(\Omega),\quad \phi_2^0 = \Delta\phi_2^0 = 0 \quad \text{on}\ \Gamma, \\ &\phi_2^1 \in H_0^1(\Omega). \end{aligned}\right. \tag{2.5}$$

We set

$$\left|\begin{aligned} &E_{01} = \tfrac{1}{2}\,[|\Delta\phi_1^0|^2 + |\phi_1^1|^2], \\ &E_{02} = \tfrac{1}{2}\,[|\nabla\Delta\phi_2^0|^2 + |\nabla\phi_2^1|^2]. \end{aligned}\right. \tag{2.6}$$

We have (multiply $\phi_2'' + \Delta^2\phi_2 = 0$ by $-\Delta\phi_2'$):

$$\tfrac{1}{2}[|\nabla\Delta\phi_2(t)|^2 + |\nabla\phi_2'(t)|^2] = E_{02} \tag{2.7}$$

so that

$$\phi_2 \in L^\infty(0,T;\ H^3(\Omega)),\quad \phi_2' \in L^\infty(0,T;\ H_0^1(\Omega)).^\dagger \tag{2.8}$$

---

† Actually ϕ_2 (resp. ϕ_2') is *continuous* from $[0,T] \to H^3(\Omega)$ (resp. $H_2^1(\Omega)$).

We shall set

$$E_0 = E_{01} + E_{02} \,. \tag{2.9}$$

We shall prove below that for $T > T_*(x^0)$ there is a constant c such that

$$\int_{\Gamma\times(0,T)} \left(\Delta\phi_1 + \frac{\partial\phi_2'}{\partial\nu}\right)^2 + \int_{\Gamma(x^0)\times(0,T)} \left(\frac{\partial\Delta\phi_2}{\partial\nu}\right)^2 \geqslant c\,E_0 \tag{2.10}$$

which verifies the result.

STEP 2. *Some Identities.*

We introduce

$$X_1 = (\phi_1'(t),\, m\nabla\phi_1(t))\big|_0^T,\quad Y_1 = (\phi_1'(t),\, \phi_1(t))\big|_0^T \tag{2.11}$$

(as in Section 1, adding the index "1")

$$X_2 = (\phi_2'(t), -m\nabla\Delta\phi_2(t))\big|_0^T,\quad Y_2 = (\phi_2'(t), -\Delta\phi_2(t))\big|_0^T. \tag{2.12}$$

We rewrite (1.18) as

$$X_1 + \frac{n-2}{2}\,Y_1 + 2TE_{01} - \tfrac{1}{2}\int (m\nu)\,(\Delta\phi_1)^2 = 0\,. \tag{2.13}$$

We then multiply the equation $\phi_2'' + \Delta^2\phi_2 = 0$ by

$$-m_k\,\frac{\partial}{\partial x_k}\,\Delta\phi_2\,. \tag{2.14}$$

We obtain

$$\begin{aligned} X_2 + \iint \phi_2'\, m\nabla\Delta\phi_2' - \int \frac{\partial\Delta\phi_2}{\partial\nu}\, m\nabla\Delta\phi_2 \\ + \iint \frac{\partial\Delta\phi_2}{\partial x_j}\,\frac{\partial}{\partial x_j}\left(m_k\,\frac{\partial\Delta\phi_2}{\partial x_k}\right) = 0\,. \end{aligned} \tag{2.15}$$

The second term in (2.15) equals

$$\begin{aligned} -\iint \frac{\partial}{\partial x_j}\,(m_k\,\phi_2')\,\frac{\partial^2\phi_2'}{\partial x_j\,\partial x_k} &= -\iint \frac{m_k}{2}\,\frac{\partial}{\partial x_k}\,|\nabla\phi_2'|^2 - \iint \phi_2'\,\Delta\phi_2' \\ &= -\int \frac{m\nu}{2}\left(\frac{\partial\phi_2'}{\partial\nu}\right)^2 + \left(\frac{n}{2}+1\right)\iint |\nabla\phi_2'|^2 \end{aligned} \tag{2.16}$$

The last term in (2.15) equals

$$\iint \tfrac{1}{2} m_k \frac{\partial}{\partial x_k} |\nabla\Delta\phi_2|^2 + \iint |\nabla\Delta\phi_2|^2 = \tfrac{1}{2}\int m\nu |\nabla\Delta\phi_2|^2 - \tag{2.17}$$

$$-\frac{n}{2}\iint |\nabla\Delta\phi_2|^2 + \iint |\nabla\Delta\phi_2|^2 .$$

Using (2.16)(2.17) in (2.15) gives

$$X_2 - \int \frac{m\nu}{2}\left(\frac{\partial\phi_2'}{\partial\nu}\right)^2 - \int \frac{m\nu}{2}\left(\frac{\partial\Delta\phi_2}{\partial\nu}\right)^2 + \frac{n}{2}\iint |\nabla\Delta\phi_2|^2 - |\nabla\Delta\phi_2|^2 -$$

$$+ \iint |\nabla\phi_2'|^2 + |\nabla\Delta\phi_2|^2 = 0 . \tag{2.18}$$

But one has (multiply $\phi_2'' + \Delta^2\phi_2 = 0$ by $-\Delta\phi_2$)

$$\iint |\nabla\phi_2'|^2 - |\nabla\Delta\phi_2|^2 = Y_2 ,$$

hence

$$X_2 + \frac{n}{2} Y_2 + 2TE_{02} - \tfrac{1}{2}\int (m\nu)\left[\left(\frac{\partial\phi_2'}{\partial\nu}\right)^2 + \left(\frac{\partial\Delta\phi_2}{\partial\nu}\right)^2\right] = 0 . \tag{2.19}$$

It follows from (2.13) and (2.19) that

$$2TE_0 \leqslant C + \frac{R(x^0)}{2}\int_{\Gamma(x^0)\times(0,T)} (\Delta\phi_1)^2 + \left(\frac{\partial\phi_2'}{\partial\nu}\right)^2$$

$$+ \frac{R(x^0)}{2}\int_{\Gamma(x^0)\times(0,T)} \left(\frac{\partial\Delta\phi_2}{\partial\nu}\right)^2 , \tag{2.20}$$

where

$$C = |X_1| + |X_2| + \frac{|n-2|}{2}|Y_1| + \frac{n}{2}|Y_2| .^{\dagger} \tag{2.21}$$

We now use another identity. We multiply the equation for ϕ_1 (resp. ϕ_2) by ϕ_2' (resp. ϕ_1'). We obtain

$$\frac{d}{dt}(\phi_1', \phi_2') - \int_\Gamma \Delta\phi_1 \frac{\partial\phi_2'}{\partial\nu} + \frac{d}{dt}(\Delta\phi_1, \Delta\phi_2) = 0 ,$$

hence

$$\int_{\Gamma\times(0,T)} \Delta\phi_1 \frac{\partial\phi_2'}{\partial\nu} = Z \tag{2.22}$$

where

$$Z = (\phi_1'(t), \phi_2'(t)) + (\Delta\phi_1(t), \Delta\phi_2(t))\Big|_0^T . \tag{2.23}$$

---

† One can improve the estimates by not separating X_i and Y_i.

Since we have in (2.22) an integral extended over all of Γ, we use a weakened form of (2.20)

$$2TE_0 \leqslant C + \frac{R(x^0)}{2} \int_{\Gamma\times(0,T)} (\Delta\phi_1)^2 + \left(\frac{\partial\phi'_2}{\partial\nu}\right)^2 + $$

$$+ \frac{R(x^0)}{2} \int_{\Gamma(x^0)\times(0,T)} \left(\frac{\partial\Delta\phi_2}{\partial\nu}\right)^2 = \text{(using (2.22))} =$$

$$= C + \frac{R(x^0)}{2} \int_{\Gamma\times(0,T)} (\Delta\phi_1)^2 + \left(\frac{\partial\phi'_2}{\partial\nu}\right)^2 + 2\Delta\phi_1 \frac{\partial\phi'_2}{\partial\nu} -$$

$$- R(x^0)\, Z + \frac{R(x^0)}{2} \int_{\Gamma(x^0)\times(0,T)} \left(\frac{\partial\Delta\phi_2}{\partial\nu}\right)^2 . \qquad (2.24)$$

Hence

$$2TE_0 \leqslant C + R(x^0)\,|Z| + \frac{R(x^0)}{2} \int_{\Gamma\times(0,T)} \left(\Delta\phi_1 + \frac{\partial\phi'_2}{\partial\nu}\right)^2 + \qquad (2.25)$$

$$+ \frac{R(x^0)}{2} \int_{\Gamma(x^0)\times(0,T)} \left(\frac{\partial\Delta\phi_2}{\partial\nu}\right)^2 .$$

STEP 3. *Estimates*

We already know from Section 1 that

$$|X_1| \leqslant \frac{2R(x^0)}{\lambda_0} E_{01} , \quad |Y_1| \leqslant \frac{2}{\mu_0} E_{01} . \qquad (2.26)$$

We observe that

$$|(\phi'_2(t), -m\nabla\Delta\phi_2(t))| \leqslant R(x^0) \frac{1}{\nu_0} |\nabla\phi'_2(t)|\,|\nabla\Delta\phi_2(t)|$$

$$\leqslant \frac{R(x^0)}{\nu_0} E_{02} , \ ,$$

hence

$$|X_2| \leqslant \frac{2R(x^0)}{\nu_0} E_{02} . \qquad (2.27)$$

Next

$$|(\phi'_2(t), -\Delta\phi_2(t))| \leqslant \frac{1}{\nu_0^2} |\nabla\phi'_2(t)|\,|\nabla\Delta\phi_2(t)| \leqslant \frac{1}{\nu_0^2} E_{02} ,$$

hence

$$|Y_2| \leqslant \frac{2}{\nu_0^2} E_{02} . \qquad (2.28)$$

We notice now that

$$|(\phi_1'(t),\phi_2'(t)) + (\Delta\phi_1(t),\Delta\phi_2(t))| \leq \frac{1}{\nu_0}|\phi_1'(t)||\nabla\phi_2'(t)| + $$

$$+\frac{1}{\nu_0}|\Delta\phi_1(t)||\nabla\Delta\phi_2(t)| \leq \frac{1}{\nu_0}E_0 . \tag{2.29}$$

Then $|Z| \leq \frac{2}{\nu_0}E_0$ and

$$\begin{aligned} C + R(x^0)|Z| &\leq 2R(x^0)\left(\frac{1}{\lambda_0}E_{01} + \frac{1}{\nu_0}E_{02}\right) + \\ &+ \frac{|n-2|}{\mu_0}E_{01} + \frac{n}{\nu_0^2}|E_{02}| + \frac{2R(x^0)}{\nu_0}E_0 \\ &\leq 2\left[R(x^0)\left(\frac{1}{\nu_0} + \max\left(\frac{1}{\lambda_0},\frac{1}{\nu_0}\right)\right) + X\left(\frac{|n-2|}{2\mu_0},\frac{n}{2\nu_0^2}\right)\right]E_0 \\ &= 2T_*(x^0)E_0 , \end{aligned}$$

hence (2.10) follows. □

3. A NONLINEAR UNIQUENESS THEOREM

We end these remarks by giving a (somewhat artificial!) nonlinear example.

THEOREM 3.1. Let u be a solution of

$$u'' + \Delta^2 u + u^3 = 0 \text{ in } \Omega\times]0,T[, \tag{3.1}$$

satisfying

$$u \in L^\infty(0,T;\ H_0^2(\Omega) \cap L^4(\Omega)),^{\dagger} \tag{3.2}$$

and subject to

$$u = \frac{\partial u}{\partial \nu} = 0 \text{ on } \Gamma\times(0,T) \tag{3.3}$$

and to

$$\Delta u = 0 \text{ on } \Gamma(x^0)\times(0,T). \tag{3.4}$$

Assume that

$$T > T_{**}(x^0) = \frac{2}{k}\frac{R(x^0)}{\lambda_0} + \frac{n-k}{k\mu_0} \tag{3.5}$$

where

$$k = \min\left(2,\frac{n}{3}\right) = \frac{n}{3}, \text{ assuming } n \leq 6 . \tag{3.6}$$

Then

$$u \equiv 0. \tag{3.7}$$

---

† $H_0^2(\Omega) \cap L^4(\Omega) = H_0^2(\Omega)$ if $n \leq 8$.

PROOF. We consider the solution ϕ of

$$\left|\begin{array}{l} \phi'' + \Delta^2\phi + \phi^3 = 0 \text{ in } \Omega \times]0,T[, \\ \phi(0) = \phi^0 \in H_0^2(\Omega) , \ \phi'(0) = \phi^1 \in L^2(\Omega) , \\ \phi = \dfrac{\partial\phi}{\partial\nu} = 0 \text{ on } \Gamma \times (0,T). \end{array}\right. \tag{3.8}$$

We use the same notation as in Section 1. Multiplying by $m\nabla\phi$, we obtain (cf. (1.14))

$$X + \frac{n}{2}\iint \phi'^2 - (\Delta\phi)^2 - \frac{1}{2}\int (m\nu)(\Delta\phi)^2 + \iint (\Delta\phi)^2 + \iint \frac{m_k}{4}\frac{\partial}{\partial x_k}(\phi^4) = 0 \tag{3.9}$$

i.e.

$$X + \frac{n}{2}\iint \phi'^2 - (\Delta\phi)^2 - \frac{1}{2}\int (m\nu)(\Delta\phi)^2 + \iint (\Delta\phi)^2 - \frac{n}{4}\iint \phi^4 = 0. \tag{3.10}$$

We observe that

$$\frac{1}{2}|\phi'(t)|^2 + \frac{1}{2}|\Delta\phi(t)|^2 + \frac{1}{4}\int_\Omega \phi(t)^4\,dx = E_0$$
$$E_0 = \frac{1}{2}|\phi^1|^2 + \frac{1}{2}|\Delta\phi^0|^2 + \frac{1}{4}\int_\Omega (\phi^0)^4\,dx \tag{3.11}$$

and that

$$\iint \phi'^2 - (\Delta\phi)^2 - \phi^4 = Y. \tag{3.12}$$

We rewrite (3.10) as follows:

$$X + \frac{n-k}{2}\iint \phi'^2 - (\Delta\phi)^2 - \phi^4 + \frac{k}{2}\iint \phi'^2 + (\Delta\phi)^2 + \frac{1}{2}\phi^4 + (2-k)\iint(\Delta\phi)^2 + \frac{n-3k}{4}\iint \phi^4 - \int \frac{m\nu}{2}(\Delta\phi)^2 = 0 ,$$

i.e.

$$X + \frac{n-k}{2}Y + kTE_0 + (2-k)\iint(\Delta\phi)^2 + \frac{n-3k}{4}\iint \phi^4 - \int_{\Gamma_x(x^0)\times(0,T)} \frac{m\nu}{2}(\Delta\phi)^2 = \int_{\Gamma(x^0)\times(0,T)} \frac{m\nu}{2}(\Delta\phi)^2 \tag{3.13}$$

With the choice (3.6) for k, (3.13) implies

$$kTE_0 \leqslant \frac{R(x^0)}{2} \int_{\Gamma(x^0)\times(0,T)} (\Delta\phi)^2 + |X| + \frac{n-k}{2}|Y| .^{\dagger} \tag{3.14}$$

Hence

$$\frac{R(x^0)}{2} \int_{\Gamma(x^0)\times(0,T)} (\Delta\phi)^2 \geqslant k(T-T_{**}(x^0))\,E_0 . \tag{3.15}$$

Theorem 3.1 follows, since we can define $u(0)=u^0 \in H_0^2(\Omega)$, $u'(0)=u^1 \in L^2(\Omega)$, and since (3.1),(3.3) with initial data in $H_0^2(\Omega)$ is well set if $n \leqslant 6$. □

REFERENCES

1. L.F. Ho, "Observabilité frontière de l'équation des ondes", *C.R.A.S. Series* **302**, (1986).
2. J. Lagnese and J.L. Lions, "Exact controllability and stabilization for various models of plates". (to appear).
3. J.L. Lions, "Contrôlabilité exacte des systèmes distribués", *C.R.A.S. Paris* **302**, (1986), pp.471-475.
4. J.L. Lions, "Exact controllability, stabilization and perturbations for distributed systems", J. von Neumann Lecture, SIAM Boston, (1986), *SIAM Review*, March 1988.
5. J.L. Lions, "Contrôlabilité exacte, stabilization et perturbations pour les systèmes distribués", Lecture Notes of the College de France, (3 vols.), Notes by E. Zuazua. cf. also the Appendix by Bardos and Lebeau. (to appear).
6. E. Zuazua, *CRAS Series*, (1986).
7. V. Komornik, *CRAS Series*, (1987).

† One can here also improve the estimates by non-separating X and Y.

18

ON A CERTAIN VARIATIONAL PROBLEM OF PHASE EQUILIBRIUM

K. A. LURIE *and* A. V. CHERKAEV

1. INTRODUCTION

We consider a problem of equilibrium of two elastic phases occupying some domain Ω of space and maintained at the temperature T of phase transition. Along the boundary of Ω, the displacement vector u of the body is prescribed. The value of energy U stored within the body in the state of phase equilibrium is determined by the boundary conditions; its density depends on strain in its own way for each phase, the strain itself varying from one point to another. In the state of equilibrium, the phases are distributed all over the body so as to minimise its total energy U under some prescribed value S_0 of entropy (the Gibbs' principle):

$$U = \int_\Omega \left[v(e) + TS \right] dx \triangleq \int_\Omega w(e,S)\, dx = \min_{\substack{e = \operatorname{def} u;\, S \\ u|_{\partial\Omega} = u_0}} \tag{1}$$

$$\int_\Omega S dx = S_0$$

In what follows, we will assume that the entropy density $S(x)$ is constant within each phase:

$$S(x) = \chi_1(x)\, S_1 + \chi_2(x)\, S_2$$

$$\chi_i(x) = 1\,, \quad x \in \Omega_i \tag{2}$$

$$\chi_i(x) = 0\,, \quad x \notin \Omega i\,, \quad i = 1,2.$$

Here, Ω_1, Ω_2 denote the domains occupied by the first and the second phase, respectively, and $\chi_1(x), \chi_2(x)$ denote the characteristic functions of those domains. Under the aforementioned assumptions, the assignment of the overall entropy S_0, $S_0 \in (S_1, S_2)$ is equivalent to that of the total amounts of the given phases. The equilibrium value of the total energy equals

$$I = \inf_{\substack{e = \operatorname{def} u \\ u|_{\partial S} = u_0}} \int_\Omega \left[\chi_1 w_1(e) + \chi_2 w_2(e)\right] dx$$

$$w_i(e) \triangleq v_i(e) + TS_i, \qquad i = 1, 2. \tag{3}$$

The infimum operation performed with respect to χ for some fixed value e of strain results in the following representation

$$I = \inf_{\substack{e = \operatorname{def} u \\ u|_{\partial S} = u_0}} \int_\Omega w(e)\, dx \tag{4}$$

where

$$w(e) = \min\{w_1(e), w_2(e)\}. \tag{5}$$

The variational problem (4) thus possesses a non-convex integrand (5) except for the trivial case $w_1(e) < w_2(e)$, $\forall e$, when there arises no phase transition. For this reason, the functional (4) is generally not semi-continuous with respect to e, and does not attain its lower bound [1]. More specifically, the minimizing sequences $\{e^s\}$ include functions which display more and more rapid oscillations as the number s increases. Because the strain suffers jumps across the phase boundaries, its rapid oscillation is associated with rapid alternation of regions occupied by various phases. This type of distribution of materials represents the sole mechanism able to provide oscillatory behaviour of strain if the external loads and the form of a body are assumed smooth. We conclude that the faster is the alternation of domains occupied by the given materials, the lower is the overall energy value stored in the body, this type of phase distribution materializing the idea of formation of microstructure within a body subject to a phase transition, see [7]. The domain structures of that type are known to arise in martensite transformations and, specifically, in crystals subject to mechanical twinning [2].

2. THE RELAXED PROBLEM

Consider now the variational problem (4), (5) in more detail. We will pass to a "relaxed" problem, i.e. we will be looking for a pointwise transformation of the integrand (5) which possesses the following properties:

(i) The solution to the relaxed problem exists and the relaxed functional attains its minimum within some set of smooth functions $e(x)$.

(ii) The infimum of U in the initial problem equals its minimum in the relaxed problem.

(iii) The relaxed problem describes the behaviour of the strain averaged over some physically small volume provided that the phases constitute a microstructure possessing minimal total energy.

At the same time, we will determine the form of micro-inclusions realizing the infimum of the total energy.

Any element $e(x)$ of the set E of admissible functions belongs to some $L_p(\Omega)$ and is subject to the differential constraints [3]

$$\mathrm{Ink}\, e(x) \triangleq \nabla\times(\nabla\times e)^T \equiv 0 \qquad \forall e \in E \tag{6}$$

expressing the necessary conditions of existence of the displacement vector u. This condition is substantial because it implies continuity of displacement across the boundary line dividing the neighbouring phases. Note that the restriction (6) is associated only with phase transitions in solids: for ideal fluids it ceases to be important since the particles are then allowed to move freely along the phase boundaries.

In the absence of (6), the infimum of U coincides with the minimal value of the relaxed functional

$$U_c = \int_\Omega w_c(e)\, dx \tag{7}$$

where

$$w_c(e) = \min_{\substack{m_1, m_2 \geq 0 \\ m_1 + m_2 = 1}} \; \min_{\substack{e_1, e_2 \\ m_1 e_1 + m_2 e_2 = e}} [m_1 w_1(e_1) + m_2 w_2(e_2)] \quad \forall e_1, e_2 \tag{8}$$

In other words, there holds the inequality

$$\min_{e \in L_p(\Omega)} U(e) = \min_{e \in L_p(\Omega)} U_c(e) . \tag{9}$$

The function w_c determined by (8) represents the lower convex envelope of w, i.e. the largest convex function not exceeding w (Fig. 1).

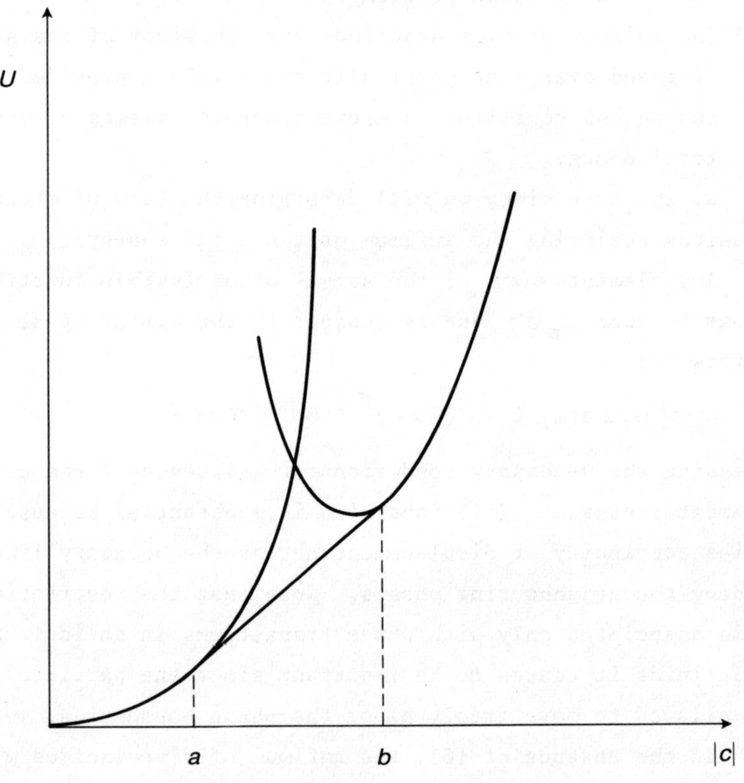

FIG. 1

When w_c is substituted for w, the functional $U_c(e)$ becomes semicontinuous, which implies existence of the extremal element. At those points where the values of w_c and w are the same, there should be placed a pure constituent; wherever $w_c(e) < w(e)$, there appears a mixture of both constituents whose relative amount (concentration) is variable. The line on Fig. 1 corresponds to the latter case; it expresses the well known Maxwell's rule: the derivative of energy with respect to e is constant. Note that in the absence of (6), the values of $w_c(e)$ depend only on the concentration of each phase in the mixture and do not depend on its microstructure at all.

Returning to phase transitions in solids, observe that the condition (6) implies, instead of (9), only the inequality

$$\inf_{e \in E} U(e) \geq \min_{e \in E} \int_\Omega w_c(e)\, dx\,. \tag{10}$$

This is because we do not take the compatibility conditions (6) into account when we construct the lower convex envelope, and the functional is minimized on some broader set of admissible functions: its minimum does not exceed therefore the lower bound provided by the initially admitted class of functions.

Bearing (6) in mind, we construct, instead of a convex envelope, the so-called quasiconvex envelope w_q of w; for this, there holds the inequality [4,5,8]

$$\inf_{e \in E} \int_\Omega w(e)\, dx = \min_{e \in E} \int_\Omega w_q(e)\, dx\,. \tag{11}$$

There is no standard technique for constructing $w_q(e)$ for any given function $w(e)$; we know, however, that a lower bound for $w_q(e)$ is given by the function $w_p(e)$,

$$w_c \leq w_p \leq w_q \leq w \tag{12}$$

constructed according to the rule [5]:

$$w_p(e) = \max_{\alpha_i > 0} \left\{ \left[w(e) + \sum_i \alpha_i \phi_i(e) \right]_c - \sum_i \alpha_i \phi_i(e) \right\}. \tag{13}$$

Here $\phi_i(e)$ $(i = 1, 2, \cdots)$ is a set of quasiconvex (but non-convex) functions which exist due to (6) and can be constructed explicitly (see [4,5]).

The inequality (12) shows that the lower bound of the functional U for w_p (and all the more for w_q) cannot be lower than its value for w_c and cannot be greater than its value for w_q. According to (11) and (12), the value of the functional for w_p implies a lower bound for $\inf U(e)$; the exact form of this bound may be checked by immediate calculations performed for some special kinds of microstructure ([5]). Note that the lower bound will this time be dependent on the microstructure, i.e. on the form of microinclusions, which is because we now take into account the compatibility conditions induced by (6).

3. AN EXAMPLE

As an illustration, consider an example where the energy density of constituents is given by the formulae

$$w_1(e) = e \cdot\cdot D \cdot\cdot e \qquad \text{First phase}, \tag{14}$$

$$w_2(e) = e \cdot\cdot D \cdot\cdot e + \varepsilon \cdot\cdot e + \gamma \qquad \text{Second phase}. \tag{15}$$

Here ε denotes a tensor of the 2nd rank associated with the proper deformation $D^{-1} \cdot\cdot \varepsilon$ of the second phase, and $\gamma = \text{const}$. The elastic constants of the constituents are assumed the same. We will restrict ourselves in what follows to the case of small plane deformations $e = (\nabla + \nabla^T)\, u$:

$$e_{xx} = \frac{\partial u_1}{\partial x}, \quad e_{yy} = \frac{\partial u_2}{\partial y}, \quad e_{xy} = \frac{1}{2}\left(\frac{\partial u_1}{\partial y} + \frac{\partial u_2}{\partial x}\right).$$

The isotropic tensor D allows the following representation

$$D \triangleq D(\kappa, \mu) = \kappa a_1 a_1 + \mu (a_2 a_2 + a_3 a_3).$$

Here κ, μ denote, respectively, the dilatation and shear moduli, $a_1 = (1/\sqrt{2})(ii + jj)$, $a_2 = (1/\sqrt{2})(ii - jj)$, $a_3 = (1/\sqrt{2})\ (ij + ji)$, represent a triple of eigentensors, and i, j denote orthogonal unit vectors.

The quasiconvex function $\phi(e)$ is given by [5]

$$\phi = -\det e = -\frac{\partial u_1}{\partial x}\frac{\partial u_2}{\partial y} + \frac{1}{4}\left(\frac{\partial u_1}{\partial y} + \frac{\partial u_2}{\partial x}\right)^2 = -\frac{1}{2}\, e \cdot\cdot\, D(1,-1) \cdot\cdot\, e. \tag{16}$$

The expression for $w_p(e)$ can now be obtained by elimination of $w_1(e)$, $w_2(e)$ and ϕ from (13) with subsequent use of (8):

$$w_p(e) = \max_{\alpha>0}\left\{ \min_{\substack{m_1,\, m_2 \geqslant 0\\ m_1+m_2=1}} \; \min_{\substack{e_1, e_2\\ m_1e_1+m_2e_2=e}} \left[m_1 A_1(\alpha_1 e_1) + m_2 A_2(\alpha_2 e_2)\right] - \alpha e \cdot\cdot\, D(1,-1) \cdot\cdot\, e \right\}. \tag{17}$$

Here

$$A_1(\alpha, e) \triangleq e \cdot\cdot\, D(\kappa+\alpha, \mu-\alpha) \cdot\cdot\, e,$$

$$A_2(\alpha, e) \triangleq e \cdot\cdot\, D(\kappa+\alpha, \mu-\alpha) \cdot\cdot\, e + \varepsilon \cdot\cdot\, e + \gamma.$$

The stationary values e_{10}, e_{20} of strains e_1 and e_2 are equal to (provided that $0 \leqslant \alpha < \mu$, the latter inequality implying positive definiteness of $D(\kappa+\alpha, \mu-\alpha)$)

$$e_{10} = e + \frac{m_2}{2} D^{-1}(\kappa+\alpha, \mu-\alpha) \cdot\cdot\, \varepsilon,$$

$$e_{20} = e - \frac{m_1}{2} D^{-1}(\kappa+\alpha\ \ \mu-\alpha) \cdot\cdot\, \varepsilon.$$

We have now

$$w_p(e) = \max_{\mu>\alpha\geqslant 0} \; \min_{\substack{m_1,\, m_2\geqslant 0\\ m_1+m_2=1}} A(\alpha, m),$$

where

$$A(\alpha, m) = e \cdot\cdot\, D(\kappa, \mu) \cdot\cdot\, e - \frac{1}{4} m_1 m_2 \varepsilon \cdot\cdot\, D^{-1}(\kappa+\alpha, \mu-\alpha) \cdot\cdot\, \varepsilon + m_2(\gamma + \varepsilon \cdot\cdot\, e).$$

It can be seen that $A(\alpha, m)$ is convex-concave with respect to m and α, and therefore

$$w_p(e) = \min_{\substack{m_1,\, m_2\geqslant 0\\ m_1+m_2=1}} B(m), \tag{18}$$

where $$B(m) = \max_{\mu>\alpha\geqslant 0} A(m, \alpha).$$

$$B(m) = \begin{cases} e \cdot\cdot D \cdot\cdot e - \frac{1}{2}\frac{m_1 m_2}{+\mu}\lambda_2^2 + m_2(\gamma + \varepsilon \cdot\cdot e) & \text{if } \det(D^{-1} \cdot\cdot \varepsilon) \geqslant 0 \\ e \cdot\cdot D \cdot\cdot e - \frac{1}{4} m_1 m_2 \varepsilon \cdot\cdot D^{-1} \cdot\cdot \varepsilon + m_2(\gamma + \varepsilon \cdot\cdot e) & \\ & \text{if } \det(D^{-1} \cdot\cdot \varepsilon) \leqslant 0. \end{cases} \tag{19}$$

Here λ_2 denotes the largest (in absolute value) eigenvalue of ε.

4. LAYERED MICROSTRUCTURES

We will now show that the formulae (18), (19) for $w_p(e)$ at the same time provided the exact expression for $w_a(e)$. To this end, we construct some special (layered) microstructures assembled of the constituents (14) and (15) taken, respectively, in proportions m_1 and m_2; this construction will demonstrate that the energy density of microstructure is given by (19) which means coincidence of w_p and w_q. In fact, the value of energy density evaluated for some specific microstructure provides the upper estimate for w_q which leads to the desired result in view of (12).

The stress in the medium equals

$$\begin{aligned} \sigma^1 &= 2D \cdot\cdot e^1 && \text{for the first phase ,} \\ \sigma^2 &= 2D \cdot\cdot e^2 + \varepsilon && \text{for the second phase.} \end{aligned} \tag{20}$$

Here and below, the upper index denotes the number of the phase. Consider a layered microstructure assembled of the given constituents; let n, t be the unit normal and tangent vectors to the layers, and m_1, m_2 $(m_1 + m_2 = 1)$ the concentrations of phases in the mixture.

Along the phase boundary, the following compatibility conditions hold

$$\begin{aligned} (e^1 - e^2) \cdot\cdot tt &= 0 \\ (\sigma^1 - \sigma^2) \cdot\cdot nn &= 0 \\ (\sigma^1 - \sigma^2) \cdot\cdot (nt + tn) &= 0 . \end{aligned} \tag{21}$$

The mean strain e equals

$$m_1 e^1 + m_2 e^2 = e . \tag{22}$$

The energy density of such a microstructure will be calculated assuming that the vector n is placed along the proper direction of the tensor ε. For that case

$$\varepsilon \cdot\cdot (nt + tn) = 0$$

and

$$(\sigma^1 - \sigma^2) \cdot\cdot (nt + tn) = 2(e^1 - e^2) \cdot\cdot D \cdot\cdot (nt + tn) = 0.$$

In view of the isotropy of D we conclude that

$$(e^1 - e^2) \cdot\cdot (nt + tn) = 0\,. \tag{23}$$

The second equation (21) shows in view of (20) that

$$e^1_{nn} - e^2_{nn} = \frac{1}{\kappa + \mu}\, \varepsilon_{nn} \tag{24}$$

(here we also took into account the first equation (21)).

The convolution of (22) with the nn-tensor leads to the relationship

$$m_1 e^1_{nn} + m_2 e^2_{nn} = e_{nn}\,. \tag{25}$$

Equations (24), (25), (20), (23) and $(21)_1$ show that

$$e^1 = e + \frac{m_2}{\kappa + \mu}\, \varepsilon_{nn}\, nn\,, \qquad e^2 = e - \frac{m_1}{\kappa + \mu}\, \varepsilon_{nn}\, nn\,.$$

The mean energy density $\tilde{B}(m)$ equals

$$\tilde{B}(m) = m_1 e^1 \cdot\cdot D \cdot\cdot e^1 + m_2 (e^2 \cdot\cdot D \cdot\cdot e^2 + \varepsilon \cdot\cdot e^2 + \gamma) =$$

$$= e \cdot\cdot D \cdot\cdot e + m_2 (\gamma + \varepsilon \cdot\cdot e) - \frac{m_1 m_2}{2(\kappa + \mu)}\, \varepsilon^2_{nn}\,.$$

This expression coincides with that appearing in equation $(19)_1$ if the normal n to the layers is placed along that proper axis of ε which corresponds to the largest (in absolute value) eigenvalue of this tensor.

To illustrate the exact character of equation $(19)_2$, again consider the layered microstructure placing it this time so that $tt \cdot\cdot D^{-1} \cdot\cdot \varepsilon = 0$ (this is possible since $\det(D^{-1} \cdot\cdot \varepsilon) \leqslant 0$). For such orientation, $(\sigma^1 - \sigma^2) \cdot\cdot tt = 0$. In fact, equation $(21)_1$ together with (20) shows that

$$0 = (e^1 - e^2) \cdot\cdot tt = \tfrac{1}{2}(\sigma^1 - \sigma^2) \cdot\cdot D^{-1} \cdot\cdot tt + \tfrac{1}{2}\varepsilon \cdot\cdot D^{-1} \cdot\cdot tt.$$

The last term in the right hand side vanishes due to the accepted choice of axes, and from vanishing of the second term, in view of isotropy of D and of equation $(21)_2$ it follows that $(\sigma^1-\sigma^2)\cdot\cdot tt = 0$. Together with (21) this means that the tensor σ is itself continuous: $\sigma^1=\sigma^2=\sigma$.

Multiplying now (22) by $2D$ and making use of (20), we get

$$\sigma = 2D\cdot\cdot e+m_2\varepsilon .$$

The energy density now equals

$$\tilde{B}(m) = m_1 e^1\cdot\cdot D\cdot\cdot e^1+m_2(e^2\cdot\cdot D\cdot\cdot e^2+\varepsilon\cdot\cdot e^2+\gamma) =$$
$$= e\cdot\cdot D\cdot\cdot e-\frac{m_1m_2}{4}\varepsilon\cdot\cdot D^{-1}\cdot\cdot\varepsilon+m_2(\varepsilon\cdot\cdot e+\gamma).$$

This expression coincides with that in the second line of (19).

5. THE QUASICONVEX HULL

We describe now the quasiconvex hull $w_q(e)$. To this end, we substitute the expression (19) for $B(m)$ in (18). The final expression for $w_q(e)$ depends on the proper deformation $D^{-1}\cdot\cdot\varepsilon$. If $\det(D^{-1}\cdot\cdot\varepsilon)>0$, $(19)_1$, then

$$w_q(e) = \begin{cases} w_1(e) & \text{if } \dfrac{\kappa+\mu}{\lambda_2}(\gamma+\varepsilon\cdot\cdot e) > \frac{1}{2}, \\ w_2(e) & \text{if } \dfrac{\kappa+\mu}{\lambda_2}(\gamma+\varepsilon\cdot\cdot e) < -\frac{1}{2}, \\ e\cdot\cdot\left(D-\dfrac{\varepsilon\otimes\varepsilon}{2\lambda_2^2}\right)\cdot\cdot e+\frac{1}{2}\varepsilon\cdot\cdot e+\text{const.} & \text{otherwise.} \end{cases}$$

Here, the symbol $\otimes$ denotes a dyadic product. The last line corresponds to the laminates described earlier in Section 4, the concentration m_1 of the first phase in the laminate equals $\frac{1}{2}+\lambda_2^{-1}(\kappa+\mu)(\gamma+\varepsilon\cdot\cdot e)$. Within a zone occupied by the microstructure, the following constitutive equation holds:

$$\sigma = \tilde{D}\cdot\cdot e+\tfrac{1}{2}\varepsilon . \tag{26}$$

The tensor

$$\tilde{D} = 2D-\frac{\kappa+\mu}{\lambda_2^2}e\otimes\varepsilon$$

is not positive definite; it equals

$$\tilde{D} = (\kappa+\mu)\left[1-\left(\frac{\lambda_1}{\lambda_2}\right)^2\right]tttt + \left[\kappa-\mu-(\kappa+\mu)\frac{\lambda_1}{\lambda_2}\right](nntt + ttnn) + \\ + \mu(nt+tn)(nt+tn). \qquad (27)$$

One of the eigenvalues of this fourth rank tensor is negative, and two others are positive. Note that the Legendre-Hadamard condition $tt \cdot\cdot\, \tilde{D} \cdot\cdot\, tt \geqslant 0,\ \forall t$ is fulfilled which guarantees ellipticity of the corresponding variational problem [6].

If $\det(D^{-1} \cdot\cdot\, \varepsilon) < 0$, then $w_q(e)$ coincides with the convex hull $w_c(e)$ of $w(e)$.

6. CONCLUSIONS

The results obtained here allow us to give a correct statement of the problem of determination of strain within the body subject to a phase transition. At the same time we will find the function $\chi_1(x)$, i.e. we specify which parts of the body are occupied by each phase and which part of it is filled in by the microstructure.

This analysis shows that layered microstructures may arise as energetically optimal in phase transitions of the first kind. Though we simplified our consideration by the assumption of isotropy of the constituents, one may expect that the laminar character of microstructure will remain valid also for more general cases. One may also refer to the case of two phases possessing different moduli of elasticity in the absence of proper deformation. For this case, optimal microstructure is provided by laminates, too, but this time by "layers of layers", i.e. laminates of certain rank. The experimental data show that laminates are actually observed as domain structures associated with martensite transformations.

ACKNOWLEDGEMENT

The authors wish to thank Professors J.M. Ball and M.A. Grinfel'd for fruitful discussions which stimulated this work.

REFERENCES

1. L.C. Young, *Lectures on the Calculus of Variations and Optimal Control Theory*, W.B. Saunders Company, Philadelphia, London, Toronto, 1969.
2. A.L. Roitburd. "Theory of formation of heterophase structure under phase transitions in a solid state", (in Russian) *Uspekhi Fizicheskikh Nauk* (*Soviet Physics: Uspekhi*), **113**(1), (1974), pp.69-104.
3. A.I. Lurie, "Theory of Elasticity", (in Russian), *Nauka*, Moscow, (1970).
4. B. Dacorogna, "Weak continuity and weak lower semicontinuity of nonlinear functionals", *Lecture Notes in Mathematics* **922**, Springer-Verlag, Berlin Heidelberg, New York, 1982.
5. K.A. Lurie and A.V. Cherkaev, "A method for construction of exact bounds for effective constants of composites", (in Russian), in *Collection: Problems of Nonlinear Mechanics of Continua*, Valgus, Tallinn, 1985.
6. J.M. Ball, "Convexity conditions and existence theorems in nonlinear elasticity", *Arch. Ration. Mech. Anal.* **63** (4), (1977), pp.337-403.
7. J.M. Ball and R.D. James, "Fine phase mixtures as minimizers of energy", *Arch. Rat. Mech. Anal.*, (to appear).
8. E. Acerbi and N. Fusco, "Semicontinuity problems in the calculus of variations", *Arch. Rat. Mech. Anal.* **86**, (1984), pp.125-145.

19

SOME REMARKS ON NON-CONVEX PROBLEMS

ELVIRA MASCOLO

Consider a functional of the calculus of variations

$$F(u) = \int_G f(x, u, Du)\, dx$$

where G is an open subset of $\mathbb{R}^N$ with sufficiently regular boundary ∂G and $f:(x,s,p) \in G \times \mathbb{R} \times \mathbb{R}^N \longrightarrow \mathbb{R}$ is a Carathéodory function such that:

$$\lambda_1 |p|^\alpha - \lambda_2 \leq f(x, s, p) \leq \Lambda_1 + \Lambda_2 |S|^\alpha + \Lambda_3 |p|^\alpha , \qquad (0.1)$$

where $\alpha > 1$. Consider the problem:

$$\text{Inf}\,\{F(v)\,,\ v \in H^{1,\alpha}(G)\,,\ v = u_0 \ \text{ on } \ \partial G\} \qquad (0.2)$$

with

$$u_0 \in H^{1,\alpha}(G) \cap L^\infty(G)\,.$$

It is well known that if f is not convex with respect to p, there are no general results of existence for solutions of (0.2). In fact, the classical "direct methods" do not apply, since F is not a semicontinuous functional.

However, the existence of minima for non-convex problems is closely related to the problem of finding equilibrium configurations in nonlinear elasticity ([3], [4], [5] etc.). In this connection, it is very interesting to find at least sufficient conditions on f to get the existence of solutions.

Marcellini in [10] shows that, in dimension one, the convexity of f with respect to p is not necessary for the existence of solutions. Other results of existence for non-convex problems in dimension one are due to Ericksen [5], Aubert and Tahraoui [1].

Starting from some duality arguments, introduced in [10], the study of the existence in dimensions higher than one has been developed by Mascolo and Schianchi [11] – [14].

In these papers the main assumption on f, which depends on $(x,p) \in G \times \mathbb{R}^N$, is the affinity of f^{**}, the lower convex envelope of f with respect to p, in the set in which $f^{**} < f$. We remark explicitly that this condition is always satisfied in one dimension.

In this paper we first consider the problem:

$$\operatorname{Inf}\left\{\int_G f(x,Dv)\,dx, \quad v \in H_0^{1,2}(G) + u_0\right\} \tag{0.3}$$

with f satisfying (0.1), with $\alpha = 2$. In [14] we proved that (0.3) has at least one solution if f^{**} is affine in the bounded open sets

$$K(x) = \left\{p \in \mathbb{R}^N :\ f^{**}(x,p) < f(x,p)\right\}, \quad x \in G$$

i.e. there exist $(m+1)$ functions defined in G, m_i, $i=1,\cdots,N$ and q such that

$$f^{**}(x,p) = \sum_{i=1}^{N} m_i(x)\,p_i + q(x), \quad p \in K(x)$$

and $m_i \in C^1(\bar{G})$ with

$$\operatorname{meas}\left\{x \in G :\ \sum_{i=1}^{N} D_i m_i(x) = 0\right\} = 0 .$$

Here we prove a theorem of existence without the last technical assumption on m_i. This result is obtained by using a suitable approximation method.

Secondly, we prove a theorem of existence of minima of a non-convex functional with an integrand which depends also on u.

The main ideas contained in the present paper can be traced back to the following example in dimension one, which was pointed out to me by G. Buttazzo: consider

$$\text{Inf}\left\{\int_0^1 (1-v'^2)^2\,dx + \int_0^1 g(v)\,dx,\quad v(0)=v(1)=0\right\},\qquad (0.4)$$

where $g(s)=s^+=\max(0,s)$. It is easy to check that a solution of problem (0.4) is the infimum of the set of solutions of the following problem:

$$\text{Inf}\left\{\int_0^1 (1-v'^2)^2\,dx,\quad v(0)=v(1)=0\right\}.\qquad (0.5)$$

If we consider $g(s)=s^-=\min(0,s)$, a solution of (0.4) is the supremum of the set of solutions of (0.5).

Consider now the following problem in G:

$$\text{Inf}\left\{\int_G f(Dv)\,dx + \int_G g(x,v)\,dx,\quad v\in H^{1,\alpha}(G)+u_0\right\}\qquad (0.6)$$

where $f+g$ satisfies (0.1). Moreover, g is a monotone function of s for a.e. $x\in G$ and f satisfies the condition of affinity on the set

$$K=\{p\in\mathbb{R}^N : f(p)\neq f^{**}(p)\}.$$

In particular, K can be a union of open bounded sets and f^{**} affine in each of them. We obtain results close to those for the one dimensional case. If g is a decreasing function, a solution of (0.6) is the supremum of a certain subset S of the convex problem:

$$\text{Inf}\left\{\int_G f^{**}(Dv)\,dx + \int_G g(x,v)\,dx,\quad v\in H^{1,\alpha}(S)+u_0\right\}.\qquad (0.7)$$

If g is an increasing function, a solution of (0.6) is the infimum of the set S. Moreover, if g is strictly monotone, every solution of (0.7) is also a solution of (0.6).

We recall that problems of the type (0.6) have been studied by Aubert and Tahraoui in [2] using quite different methods and by obtaining different results. To get results of existence for problems of (0.3) and (0.6), we need the relaxed problem of (0.3) and the convex problem (0.7) to have solutions in $C^{0,1}_{\text{loc}}(G)$.

If f is supposed to be strictly convex and regular for p sufficiently large (see (AI) of Section 1, and (HI) of Section 2) and g satisfies a suitable condition (see (H2) of Section 2), we can use the same arguments of regularity as in [14] and we obtain solutions in $C^{0,1}_{\text{loc}}(G)$.

1. Let $f \in C^0(\bar{G} \times \mathbb{R}^N)$ be such that

$$c_1|p|^2 - c_2 \leqslant f(x,p) \leqslant c_3|p|^2 + c_4, \quad p \in \mathbb{R}^N, \quad x \in G \tag{1.1}$$

where $c_i \geqslant 0$. Consider the problem

$$\operatorname{Inf}\left\{F(v) = \int_G f(x, Dv)\, dx, \quad v \in H_0^{1,2}(G) + u_0\right\}, \tag{1.2}$$

with

$$u_0 \in H^{1,2}(G) \cap L^\infty(G).$$

Let f^{**} be the lower convex envelope of f with respect to p and consider the related relaxed problem of (1.2), i.e.

$$\operatorname{Inf}\left\{F^{**}(v) = \int_G f^{**}(x, Dv)\, dx, \quad \hat{v} \in H_0^{1,2}(G) + u_0\right\}. \tag{1.3}$$

In particular we suppose that $f(x,\cdot) \in C^{0,1}_{\text{loc}}(G)$ uniformly with respect to p, so f^{**} has the same property.

Since F^{**} is a convex functional, it is lower semicontinuous in the weak topology of $H^{1,2}$. Then, via direct methods, and by (1.1), we have that (1.3) has at least one solution.

However, as F^{**} is convex, but not strictly convex, and we did not make any hypotheses of differentiability on f, (1.3) is not covered by general existence results in $C^{0,1}(G)$. Nevertheless, in [14] we give some sufficient conditions on f, in order to obtain locally Lipschitz continuous solutions of (1.3). (In this connection we recall also the results of [7] for Neumann boundary conditions.)

More precisely, defining for $r > 0$, $B_r = \{p \in \mathbb{R}^N, \ |p| < r\}$, we assume[†]

(A1) <u>There exists</u> $R > 0$ <u>such that</u> $f \in C^2(\bar{G} \times (\mathbb{R}^N - B_R))$ <u>and</u>

[†] For a function $h = h(x,p)$ we denote by $h_x(x,p)$ and $h_p(x,p)$ the vector gradient of h with respect to x and the vector gradient of h with respect to p.
For a function $g = g(x)$ we denote Dg the vector gradient of g, $Dg = (D_1 g, D_2 g, \cdots, D_N g)$ with $D_i = \partial/\partial x_i$.

$$|f_p(x,p)|\ ,\ \ |f_{px}(x,p)| \leqslant \mu(1+|p|)\ ,\ \ |f_{pp}(x,p)| \leqslant \mu$$

<u>for</u> $p \in \mathbb{R}^N - B_R$ <u>and</u> $x \in G$. <u>Moreover, there exist</u> $\lambda, \Lambda > 0$ <u>such that</u>

$$\lambda|\xi|^2 \leqslant f_{p_i p_j}(x,p)\,\xi_i\xi_j \leqslant \Lambda|\xi|^2\ ,\ \ x \in \bar{G}\ ,\ \ p \in \mathbb{R}^N - B_R\ ,\ \ \xi \in \mathbb{R}^N - \{0\}.$$

It is easy to check that if f verifies (A1) then since $f = f^{**}$ for $p \in \mathbb{R}^N - B_R$, f^{**} also verifies (A1).

The following regularity result holds (see Theorem I.2 of [14]):

THEOREM 1.1: <u>Assume that</u> f <u>satisfies (1.1) and (A1). Then, every solution of (1.3) is in</u> $C^{0,1}_{\mathrm{loc}}(G)$.

In order to find solutions of the non-convex problems (1.2) we need an additional assumption on f. Actually, it is a condition of affinity on f^{**} in the set in which f is strictly greater than f^{**}. Define, for $x \in G$

$$K(x) = \{p \in \mathbb{R}^N : f(x,p) > f^{**}(x,p)\}\ ,$$

and suppose that:

(A2) <u>For a.e.</u> $x \in G$, $K(x)$ <u>is a connected bounded open subset of</u> $\mathbb{R}^N$ <u>and there exist</u> $(N+1)$ <u>functions</u> m_i $i = 1, \cdots, N$, <u>and</u> q <u>in</u> $C^{0,1}(G)$, <u>such that</u>

$$f^{**}(x,p) = \sum_{i=1}^{N} m_i(x)p_i + q(x), \tag{1.4}$$

<u>for</u> $p \in K(x)$ <u>and</u> $x \in G$.

(A3) <u>The functions</u> m_i <u>in (1.4) are in</u> $C^1(\bar{G})$ <u>and</u>

$$\mathrm{meas}\left\{x \in G : \sum_{i=1}^{N} D_i m_i(x) = 0\right\} = 0$$

In [14] (see Theorem 2.2) the following existence theorem is given:

THEOREM 1.2: Assume (1.1) and (A1)−(A3). Then every solution u of (1.3) is such that $Du(x) \in \mathbb{R}^N - K(x)$ for a.e. $x \in G$, so u is a solution of (1.2), too.

Our purpose is now to prove a theorem of existence for (1.2) without the assumption (A3). We prove:

THEOREM 1.3: Assume (1.1), (A1) and (A2). Moreover, suppose that f is strictly convex in $\mathbb{R}^N - K(x)$ for a.e. $x \in G$. Then problem (1.2) has at least one solution.

We give a rough idea of the proof. We show that there exists a solution u of (1.3) with $Du(x) \in \mathbb{R}^N - K(x)$ a.e. in G so that u is a solution of (1.2). For this, we approximate the function f^{**} with a suitable sequence of functions f^J, satisfying (A1), (A2) and (A3). Then, we consider a sequence of problems (P_J) as in (1.3) with f^{**} replaced by f^J. For each solution u_J of (P_J) we apply Theorem 1.2 and then $Du_J(x) \in \mathbb{R}^N - K(x)$ a.e. in G. The main step is to prove that (u_J) converges in $H^{1,2}$ to a solution $\bar{u}$ of (1.3), which satisfies $D\bar{u}(x) \in \mathbb{R}^N - K(x)$ a.e. in G.

PROOF OF THEOREM 2.2: Let $\boldsymbol{m} = (m_1, \cdots, m_N)$, with m_i the functions in (1.4). Consider the sequence of polynomials $\boldsymbol{m}^J = (m_1^J, \cdots, m_N^J)$, $J \in \mathbb{N}$, of order greater than one, which converges to $\boldsymbol{m}$ in $(C^{0,1}(G))^N$. For each J,

$$\operatorname{div} \boldsymbol{m}^J = \sum_{i=1}^{N} D_i m_i^J(x) \neq 0$$

a.e. in G. Consider, for $(x,p) \in G \times \mathbb{R}^N$, the sequence of functions

$$\begin{aligned} f^J(x,p) &= f^{**}(x,p) + (\boldsymbol{m} - \boldsymbol{m}^J) \times p \\ &= f^{**}(x,p) + \sum_{i=1}^{N} (m_i - m_i^J)\, p_i \,^{\dagger} \end{aligned}$$

---

† For $\boldsymbol{x} = (x_1, \cdots, x_N)$ and $y = (y_1, \cdots, y_N)$ we denote

$$\boldsymbol{x} \times \boldsymbol{y} = \sum_{i=1}^{N} x_i y_i .$$

It is easy to verify that f^J converges uniformly to f^{**} in $G \times C$, with C a compact subset of $\mathbb{R}^N$. In fact, for $p \in C$, we have

$$|f^J(x,p) - f^{**}(x,p)| \leqslant \sum_{i=1}^{N} |m_i^J(x) - m_i(x)|\,|p_i|, \quad x \in G.$$

Consider the sequence of problems:

$$(P_J) \quad \operatorname{Inf}\left\{\int_G f^J(x,Du)\,dx, \quad v \in H^{1,2}(G) + u_0\right\}.$$

Since f and thus f^{**} verify (1.1) and (A1), each f^J satisfies the same assumptions with constants independent of J.

By Theorem 1.1, every solution u_J of (P_J) is in $C^{0,1}_{\mathrm{loc}}(G)$. Moreover, since $f^J(x,p)$ satisfies (A2) and (A3), by applying Theorem 1.2, we get, for all $J \in \mathbb{N}$:

$$Du_J(x) \in \mathbb{R}^N - K(x) \quad \text{for a.e. } x \in G. \tag{1.5}$$

We first show that (u_J) is a minimizing sequence for (1.3). By (1.1) and the dominated convergence theorem, we have for all $v \in H^{1,2}(G) + u_0$,

$$\lim_J \int_G f^J(x,Dv)\,dx = \int_G f^{**}(x,Dv)\,dx. \tag{1.6}$$

Moreover, by (1.1), since u_J is a solution of (P_J), we obtain:

$$\tilde{c}_1 \int_G |Du_J|^2\,dx \leqslant \int_G f^J(x,Du_J)\,dx + \tilde{c}_2 \leqslant$$
$$\leqslant \int_G f^J(x,Dv)\,dx + \tilde{c}_2.$$

The preceding inequalities and (1.6) imply that (u_J) is bounded in $H^{1,2}(G)$ and so a subsequence converges weakly in $H^{1,2}(G)$ to $\bar{u}$. The function $\bar{u}$ is a solution of (1.3). In fact, from the semicontinuity of F^{**}, we have:

$$\int_G f^{**}(x,D\bar{u})\,dx \leqslant \lim_J \int_G f^{**}(x,Du_J)\,dx. \tag{1.7}$$

From (1.5) we have

$$f^{**}(x,Du_J) = f^J(x,Du_J) - (\boldsymbol{m} - \boldsymbol{m}^J) \times Du_J, \qquad x \in G. \tag{1.8}$$

Then, since (u_J) is bounded in $H^{1,2}(G)$ and $(\boldsymbol{m}^J)$ converges to $\boldsymbol{m}$ in $(L^2(G))^N$, from (1.6), (1.7) and (1.8), we get:

$$\int_G f^{**}(x,D\bar{u})\,dx \leq \lim_J \int_G f^J(x,Du_J)\,dx - \sum_{i=1}^N \int_G (m_i^J - m_i)\,D_i u_J\,dx$$
$$\leq \lim_J \int_G f^J(x,Du^J)\,dx \leq \lim_J \int_G f^J(x,Dv)\,dx$$
$$\leq \int_G f^{**}(x,Dv)\,dx\,, \quad \text{for all} \quad v \in H^{1,2}(G) + u_0\,.$$

We now show that (u_J) converges to $\bar{u}$ in the strong topology of $H^{1,2}(G)$. Consider the Euler equations of problem (P_k) and (P_h), $k \neq h$:

$$\int_G f_p^k(x,Du_k) \times D\psi\,dx = 0$$
$$\int_G f_p^h(x,Du_h) \times D\psi\,dx = 0\,,$$

for all $\psi \in H_0^{1,2}(G)$.

Since $u_k = u_h = u_0$ on ∂G we have in particular:

$$\int_G (f_p^k(x,Du_k) - f_p^h(x,Du_h)) \times (Du_k - Du_h)\,dx = 0\,. \tag{1.9}$$

Since, by (1.5)

$$f_p^J(x,Du_J) = f_p(x,Du_J)) + (\boldsymbol{m}^J - \boldsymbol{m})\,, \quad \text{for all } J\,,$$

we have:

$$\int_G (f_p(x,Du_k) - f_p(x,Du_h)) \times (Du_k - Du_h)\,dx + \int_G (\boldsymbol{m}^k - \boldsymbol{m}^h) \times (Du_k - Du_h)\,dx = 0. \tag{1.10}$$

From the strict convexity of f with respect to p in $\mathbb{R}^N - K(x)$ and by the inequality $ab \leq \varepsilon a^2 + \varepsilon^{-1} b^2$ $(a,b > 0)$, we get:

$$c\int_G |Du_k - Du_h|\,dx \leq \frac{1}{\varepsilon}\int_G |\boldsymbol{m}^h - \boldsymbol{m}^k|^2\,dx + \varepsilon\int_G |Du_h - Du_k|^2\,dx.$$

This implies that (Du_J) converges in L^2 to a function w. Because (u_J) converges weakly in $H^{1,2}(G)$ to $\bar{u}$, we have $D\bar{u} = w$.

Therefore (u_J) converges to $\bar{u}$ in $H^{1,2}(G)$ and so, by passing to a subsequence, Du_j converges to Du a.e. in G. From (1.5) we get that $D\bar{u}(x) \in \mathbb{R}^N - K(x)$ a.e. in G, and thus $\bar{u}$ is a solution of (1.2).

2. Let G be as in Section 1. Let $f : \mathbb{R}^N \to \mathbb{R}_+$ be in $C^{0,1}_{\text{loc}}(\mathbb{R}^N)$ and let $g : \bar{G} \times \mathbb{R} \to \mathbb{R}_+$ be a continuous function such that:

$$\nu|p|^\alpha - C(|s|^\alpha + 1) \leqslant f(p) + g(x,s) \leqslant |p|^\alpha + |s|^\alpha + c_2 ,$$

with $\alpha > 1$, $c_1, c_2 > 0$ and c_1/ν strictly less than the constant in the Sobolev-Poincaré inequality.

Consider the following functional

$$F(v) = \int_G f(Dv)\,dx + \int_G g(x,v)\,dx ,$$

We are interested in finding solutions of the problem:

$$\operatorname{Inf}\left\{F(v) ,\quad v \in H_0^{1,\alpha}(G) + u_0\right\}, \tag{2.2}$$

where $u_0 \in H^{1,\alpha}(G) \cap L^\infty(G)$, without any assumption of convexity on f and g.

Denoting by f^{**} the lower convex envelope of f, consider the functional:

$$\bar{F}(v) = \int_G f^{**}(Dv)\,dx + \int_G g(x,v)\,dz ,$$

and the problem:

$$\operatorname{Inf}\left\{\bar{F}(v) ,\quad v \in H_0^{1,\alpha}(G) + u_0\right\}. \tag{2.3}$$

Since the functional $\bar{F}$ is lower semicontinuous in the weak topology of $H^{1,\alpha}$, by applying the direct method we have that (2.3) has at least one solution. Moreover, from Theorem 3.2 of Chapter V of [8] there exists $M > 0$ depending on α, u_0 and the constants in (1.1) such that every solution u of (2.3) satisfies

$$\|u\|_{L^\infty(G)} \leqslant M . \tag{2.4}$$

However, because $\bar{F}$ is convex, but not strictly convex, and since, as before, we did not make any differentiability hypotheses on f

and g , the existence of solutions in $C^{0,1}$ for (1.3) is not covered by general existence results. In order to have solutions of (1.3) in $C^{0,1}_{\mathrm{loc}}(G)$, we make similar assumptions and we use analogous arguments to those in [14].

We suppose that

(H1) <u>there exists</u> $R>0$ <u>such that</u> $f\in C^2(\mathbb{R}^N - B_R)$ <u>and for</u> $p\in\mathbb{R}^N - B_R$:

$$|f_p(p)| \leqslant \mu(1+|p|)^{\alpha-1}\ , \quad |f_{pp}(p)| \leqslant \mu\ . \tag{2.5}$$

<u>Moreover, there exist</u> $\lambda,\Lambda>0$ <u>such that</u>

$$\lambda(1+|p|)^{\alpha-2}|\xi|^2 \leqslant f_{p_i p_j}(p)\,\xi_i\,\xi_j \leqslant \Lambda|\xi|^2\,(1+|p|)^{\alpha-2} \tag{2.6}$$

<u>for</u> $p\in\mathbb{R}^N - B_R$ <u>and</u> $\xi\in\mathbb{R}^N - \{0\}$.

(H2) $g(\cdot\,,s)\in C^{0,1}(\bar{G})$ <u>for</u> s <u>varying in a compact set of</u> $\mathbb{R}$ <u>and</u> $g(x,\cdot)\in C^{0,1}_{\mathrm{loc}}(\mathbb{R})$ <u>uniformly with respect to</u> $x\in\bar{G}$. <u>For example, let</u> $g(x,s)=a(x)\,h(s)$, <u>with</u> $a\in C^{0,1}(\bar{G})$ <u>and</u> $h\in C^{0,1}_{\mathrm{loc}}(\mathbb{R})$.

As before, if f satisfies (H1), then since $f=f^{**}$ in $\mathbb{R}^N - B_R$, f^{**} also satisfies (H1).

THEOREM 2.1: <u>Suppose that</u> f <u>satisfies (2.1) and (H1). Moreover, suppose</u> g <u>satisfies (H2). Then there exists at least one solution of (2.3) in</u> $C^{0,1}_{\mathrm{loc}}(G)$.

PROOF: As in [14], we approximate f^{**} by a sequence of functions f_n of class C^2, satisfying (2.5) and (2.6) for all $p\in\mathbb{R}^N$ and such that $f_n=f^{**}$ in $\mathbb{R}^N - B_R$. Consider the sequence of problems:

$$\mathrm{Inf}\left\{\int_G f_n\,(Dv)\,dx + \frac{1}{n}\int_G |Dv|^\alpha\,dx + \int_G g(x,v)\,dx\ ,\quad v\in H^{1,\alpha}(G)+u_0\right\}.$$

We apply to the unique solution u_n of the above problem the classical arguments of regularity (see Chapter 4 of [8]).

Because each f_n satisfies (2.1) and (H1) with the same constants when $|p|$ is sufficiently large, by a suitable truncation method, we obtain:

$$\sup_{G_0} |Du_n| \leqslant c(G_0) , \qquad \forall\ n \in \mathbb{N} \text{ and } G_0 \subset\subset G. \qquad (2.6)$$

For more details see the proof of Proposition 1.1 of [14] and of Theorem 1.3 of [6]. It is easy to check that the presence of the term $\int_G g(x,u)\,dx$ does not give any trouble. In fact, because u_n satisfies (2.4) for all $n \in \mathbb{N}$, the assumption (H2) implies that $g_s(x,u_n)$ is bounded uniformly with respect to $x \in G$ and $n \in \mathbb{N}$.

Since each f_n satisfies (2.1) with the same constants, we have that (u_n) is bounded in the norm of $H^{1,\alpha}$. Thus a subsequence, again denoted (u_n), converges weakly in $H^{1,\alpha}(G)$, to a function u. From (2.6) and the Sobolev embedding theorem, (u_n) converges uniformly to u in $G_0 \subset\subset G$, and then $u \in C^{0,1}_{\text{loc}}(G)$. Semicontinuity arguments show that u is a solution of (2.3).

REMARK

If g is a convex function of s, by proceeding as in the last part of the proof of Theorem 1.2 of [14], we obtain that every solution of (2.3) is in $C^{0,1}_{\text{loc}}(G)$.

Let u now be a solution of (1.4) in $C^{0,1}_{\text{loc}}(G)$. Let $G_n \subset\subset G$ with $G_n \nearrow G$ and $G = \bigcup_n G_n$. Write $M_h = \|Du\|_{\infty, G_n}$.

Let $L_h \in \mathbb{R}_+$ be a sequence converging to infinity and such that $L_h \geqslant M_h$, for $h \in \mathbb{N}$.

Define

$$L = \left\{ v \in H^{1,\alpha}(G) \cap C^{0,1}_{\text{loc}}(G) : \ \|Dv\|_{\infty, G_n} \leqslant L_h , \quad v = u_0 \text{ on } \partial G \right\} .$$

Since $u \in L$, $L \neq \emptyset$. Consider the problem:

$$\operatorname{Inf} \{ \bar{F}(v) \ , \ v \in L \} . \qquad (2.7)$$

The function u is a solution of (2.7); then, denoting by

$$S = \{ w \in L, \ w \text{ is a solution of } (2.7) \}$$

we have that S is not empty.

Denote:

$$\begin{aligned}\bar{u}(x) &= \sup\{w(x)\ ,\ w\in S\}\ ,\\ \bar{\bar{u}}(x) &= \inf\{w(x)\ ,\ w\in S\}\ .\end{aligned} \tag{2.8}$$

PROPOSITION 2.2: <u>The functions $\bar{u}$ and $\bar{\bar{u}}$ are in S. Moreover, they are solutions of (2.3).</u>

PROOF: We prove that $\bar{u}$ defined in $(2.8)_1$ is an element of S. By proceeding as in Lemma 1 of [12], we prove that the supremum of a finite number of elements of S is still in S.

Fixing $h\in\mathbb{N}$, consider G_h. It is easy to check that there exists a sequence of elements of S, u_m^h, which converges uniformly to $\bar{u}$ in G_h. This implies that $\bar{u}\in C^{0,1}_{\mathrm{loc}}(G_h)$ and $\|D\bar{u}\|_{L^\infty,G_h}\leq L_n$.

Consider now the sequence $u_n=u_n^n$, obtained from u_n^h by a process of diagonalization. This sequence converges pointwise in G. In fact, since $G_h\nearrow G$

$$|u_n^h(x)-u_n^k(x)|<\varepsilon \quad \forall\ x\in G_k \text{ and } \forall\ k<h.$$

Let $x\in G$. There exists $h\in\mathbb{N}$ such that $x\in G_h$ and then for $n>h$ we have that

$$|u_n^n(x)-\bar{u}(x)|\leq|u_n^n(x)-u_n^h(x)|+|u_n^h(x)-\bar{u}(x)|<2\varepsilon.$$

Since $u_n\in S$, from (2.1) we have that the norms of u_n in $H^{1,\alpha}$ are bounded uniformly with respect to n, and thus a subsequence, again denoted (u_n), converges weakly in $H^{1,\alpha}$ to a function w. Then, possibly passing to a further subsequence, (u_n) converges a.e. in G to w. Consequently $w=\bar{u}$. So $\bar{u}\in H^{1,\alpha}(G)$. We now prove that $\bar{u}$ is a solution of (2.6). In fact, from the semicontinuity of $\bar{F}$, we have:

$$\begin{aligned}\int_G f^{**}(D\bar{u})+\int_G g(x,\bar{u})\,dx &\leq \lim_n \int_G f^{**}(Du_n)dx+\int_G g(x,u_n)\,dx\\ &= \operatorname{Inf}\{\bar{F}(v)\ ,\ v\in L\}\ .\end{aligned}$$

The above inequalities show also that $\bar{F}(u)=\bar{F}(\bar{u})$ and so $\bar{u}$ is a solution of (2.3). With the same arguments we prove that $\bar{\bar{u}}\in S$.

As in Section 1, in order to show the existence of solutions for the non-convex problem (1.2) we need a condition of affinity on f^{**}, in the set in which f is strictly greater than f^{**}.

Define

$$K = \{p \in \mathbb{R}^N, \quad f^{**}(p) < f(p)\}.$$

Since f satisfies (2.1) and (H1), K is a bounded subset of $\mathbb{R}^N$. Assume:

(H3) K <u>is the union of a finite number of open subsets of</u> $\mathbb{R}^N$,

$$K = \bigcup_{J=1}^{e} K^J$$

<u>and there exist</u> m_i^J, q^J <u>for</u> $i = 1,\cdots,N$ <u>and</u> $J = 1,\cdots,e$ <u>such that for</u> $p \in K^J$

$$f^{**}(p) = \sum_{i=1}^{N} m_i^J p_i + q^J . \tag{2.9}$$

(H4) <u>For a.e.</u> $x \in G$, $g(x,s)$ <u>is a monotone function of</u> s.

We state now our main theorem:

THEOREM 2.3: <u>Suppose (2.1) and (H1) – (H4) hold. Then Problem (2.2) has at least one solution.</u>

PROOF: Since K is a bounded subset of $\mathbb{R}^N$, there exists $L > 0$ such that $B_L \supset K$. On the other hand, since $L_h \nearrow +\infty$, we have $L_h > L$ for h large enough.

Let g be a decreasing function with respect to s. We prove that, in this case, the function $\bar{u}$ defined in $(2.8)_1$, is a solution of (2.2). Because $\bar{u}$ is a solution of (2.3), it is sufficient to show that $D\bar{u}(x) \in \mathbb{R}^N - K$, a.e. in G. In fact we obtain

$$F(\bar{u}) = \bar{F}(u) \leqslant \bar{F}(v) \leqslant F(v) \ , \quad \forall \ v \in H^{1,\alpha}(G) + u_0 \, .$$

Suppose that there exists $x_0 \in G$, at which $\bar{u}$ is differentiable and such that

$$D\bar{u}(x_0) \in K^J \quad \text{and} \quad |D\bar{u}(x_0)| < L, \quad \text{for some } j . \tag{2.10}$$

We have that $x_0 \in G_h$ for some h. From Lemma 4 of Appendix 2 of

[9], there exists $\psi \in C_0^1$ such that:

$$D\psi\_(x_0) = D\bar{u}(x_0), \qquad \psi\_(x_0) = \bar{u}(x_0), \tag{2.11}$$
$$\psi\_(y) < \bar{u}(y), \qquad \forall\ y \in B_r(x_0) - \{x_0\}\ ^{\dagger}$$
$$\text{for some } r > 0 .$$

Since K^J is open, there exists $0 < \delta < r$ such that, for $x \in B_\delta(x_0) = B_\delta \subset G_h$ we have

$$D\psi\_(x) \in K^J \quad \text{and} \quad |D\psi\_(x)| < L < L_h .$$

Let now $\xi \in C_0^\infty(B_\delta)$, $\xi(x_0) = 1$ and $0 \leqslant \xi(x) \leqslant 1$. For ε small enough, we have

$$D(\psi\_ + \varepsilon\xi)(x) \in K^J \text{ and } |D(\psi\_ + \varepsilon\xi)| < L, \quad x \in B_\delta . \tag{2.12}$$

Define the function

$$\chi(x) = \max\{\psi\_(x) + \varepsilon\,\xi(x), \bar{u}(x)\}, \quad x \in G .$$

Let $A \subset B_\delta$ be the open subset such that $\chi = \psi\_ + \varepsilon\xi$ in A and $\chi = \bar{u}$ in G—A. Consider:

$$\bar{F}(\bar{u}) - \bar{F}(\chi) = \int_A [f^{**}(D\bar{u}) - f^{**}(D\chi)]\,dx + \\ + \int_A [g(x,\bar{u}(x)) - g(x,\chi(x))]\,dx .$$

Since, from (2.9), because K^J is open, we obtain

$$f^{**}(p) \geqslant \sum_{i=1}^N m_i^J p_i + q, \quad \forall\ p \in \mathbb{R}^N ,$$

and in particular from $(2.12)_1$

$$f^{**}(D\chi) = \sum_{i=1}^N m_i^J D_i\chi(x) + q, \quad \text{in } A ,$$

we obtain :

† We denote $B_r(x_0)$ the ball of radius $r > 0$ and centre x_0.

$$\int_A [f^{**}(D\bar{u}) - f^{**}(D\chi)]\,dx = \\ = \sum_{i=1}^{N} \int_A m_i^J (D_i\bar{u} - D_i\chi)\,dx = \sum_{i=1}^{N} \int_{B_\delta} m_i^J (D_i\bar{u} - D_i\chi)\,dx\,.$$

Integrating by parts, because $\bar{u} = \chi$ on ∂B_δ, we have:

$$\int_A [f^{**}(D\bar{u}) - f^{**}(D\chi)]\,dx = 0\,.$$

On the other hand, since $\bar{u}(x) < \chi(x)$ for $x \in A$ and g is a decreasing function, we have:

$$g(x, \bar{u}(x)) \geq g(x, \chi(x)) \quad \text{for} \quad x \in A\,.$$

finally, we obtain:

$$\bar{F}(\bar{u}) \geq \bar{F}(\chi)\,.$$

So χ is an element of S, and since $\chi > \bar{u}$ in A, we get a contradiction. Thus, there are no x_0 satisfying (2.10).

From the choice of L, we have that $D\bar{u}(x) \in \mathbb{R}^N - K$ a.e. in G, and $\bar{u}$ is a solution of (2.2).

If g is an increasing function of s, with analogous arguments to the above, we show that the function $\bar{\bar{u}}$ defined in $(2.8)_2$ is a solution of problem (2.2). In particular, we show that there are no $x_0 \in G$ at which $\bar{\bar{u}}$ is differentiable such that:

$$D\bar{\bar{u}}(x_0) \in K^J \quad \text{and} \quad |D\bar{\bar{u}}(x_0)| < L\,, \quad \text{for some } j.$$

In this case, we consider the function $\psi_+ \in C_0^1$, existing by Lemma 4 of Appendix 2 of [9] such that

$$D\psi_+(x_0) = D\bar{\bar{u}}(x_0)\,, \qquad \psi_+(x_0) = \bar{\bar{u}}(x_0)$$

$$\psi_+(y) > \bar{\bar{u}}(y) \qquad \forall\ y \in B_r(x_0) - \{x_0\} \\ \text{for some } r > 0\,.$$

There exists $0 < \delta < r$ such that for $x \in B_\delta(x_0) = B_\delta \subset G_h$ we have:

$$D\psi_+(x) \in K^J \quad \text{and} \quad |D\psi_+(x)| < L < L_h\,.$$

Let $\phi \in C_0^\infty(B_\delta)$, $\phi(x_0) = 1$ and $0 \leq \phi \leq 1$, and define

$$\chi'(x) = \min\{\psi_+(x) - \tau\phi(x)\ ,\ \bar{\bar{u}}(x)\}\ ,$$

As before, for τ small enough, we show that $\chi' \in S$. Since $\chi'(x_0) < \bar{\bar{u}}(x_0)$, we get a contradiction to the definition of $\bar{\bar{u}}$. Thus we have that $D\bar{\bar{u}}(x) \in \mathbb{R}^N - K$ a.e. in G and that $\bar{\bar{u}}$ is a solution of (2.2).

If g is a strictly monotone function of s, we can state a more precise result.

THEOREM 2.4: <u>Suppose that f satisfies (2.1) and (H1) – (H3). If g is a strictly monotone function of s, every solution of (2.3) which is in $C^{0,1}_{loc}(G)$ is a solution of (2.2).</u>

PROOF: Suppose that g is strictly decreasing in s. Let $u \in C^{0,1}_{loc}(G)$ be a solution of (2.3), existing by Theorem 2.1. Suppose that there exists $x_0 \in G$ such that $Du(x_0) \in K^J$ for j. Let $\psi_- \in C^1_0$ be the function satisfying (2.11) with $\bar{u}$ replaced by u.

Then there exists $\delta > 0$ such that $D\psi_-(x) \in K^J$, for $x \in B_\delta$. Let ξ be as above. For ε small we have that

$$D\psi_-(x) + \varepsilon D\xi(x) \in K^J \quad \text{for} \quad x \in B_\delta\ .$$

Define:

$$\bar{\chi}(x) = \max\{\psi_-(x) + \varepsilon\xi(x)\ ,\ u(x)\}\ .$$

Let $A \subset B_\delta$ be such that $u = \bar{\chi}$ on ∂A and $u < \bar{\chi}$ in A. Now, because g is strictly decreasing,

$$g(x, u(x)) > g(x, \bar{\chi}(x)) \quad \text{for} \quad x \in A\ .$$

Consequently, by proceeding as in the proof of the previous theorem, we get

$$\bar{F}(u) > \bar{F}(\bar{\chi})$$

which is impossible because u is a solution of (2.3). Thus $Du(x) \in \mathbb{R}^N - K$ a.e. in G and u is a solution of (2.2).

REMARK

Consider the problem:

$$\operatorname{Inf}\left\{\int_G f(x,Dv)\,dx + \int_G g(x,v)\,dx \ , \ v \in H_0^{1,\alpha}\ G + u_0\right\} \quad (2.13)$$

with f satisfying the affinity condition (A2), g a strictly monotone function of s and $f+g$ satisfying (0.1).

In order to extend the previous existence results to problem (2.13) we need the convex problem:

$$\operatorname{Inf}\left\{\int_G f^{**}(x,Dv)\,dx + \int_G g(x,v)\,dx, \quad v \in H_0^{1,\alpha}(G) + u_0\right\}, \quad (2.14)$$

to have at least one solution in $C^{0,1}_{\mathrm{loc}}(G)$. Suppose that f satisfies some conditions similar to those in (A1) i.e. there exists $R > 0$ such that $f \in C^2(\bar{G} \times (\mathbb{R}^N - B_R))$ and for $x \in \bar{G}$ and $p \in \mathbb{R}^N - B_R$

$$|f_p(x,p)| \ , \ |f_{px}(x,p)| \leqslant \mu(1+|p|)^{\alpha-1}, \ |f_{pp}(x,p)| \leqslant \mu .$$

Moreover, there exists λ and Λ such that, for $x \in \bar{G}$ and $p \in \mathbb{R}^N - B_R$

$$\lambda|\xi|^2(1+|p|)^{\alpha-2} \leqslant f_{p_i p_j}(x,p)\,\xi_i\,\xi_j \leqslant \Lambda|\xi|^2(1+|p|)^{\alpha-2} ,$$

$$\xi \in \mathbb{R}^N - \{0\} .$$

If f satisfies (A2) and g satisfies (H2), by proceeding as in Theorem 2.1, we prove that (2.14) has at least one solution in $C^{0,1}_{\mathrm{loc}}(G)$. Then, by proceeding as in Theorem 2.4, we obtain that (2.13) has at least one solution.

ACKNOWLEDGEMENT

The author wishes to thank G. Buttazzo for useful discussions on the subject of this paper.

REFERENCES

1. G. Aubert and R. Tahraoui, "Théorèmes d'existence en calcul des variations", *J. Diff. Equations* **33**, (1), (1979), pp.75-100.
2. G. Aubert and R. Tahraoui, "Théorèmes d'existence en optimization non-convexes", *Appl. Analysis* **18**, (1984), pp.75-100.
3. J.M. Ball, "Constitutive inequalities and existence theorems in nonlinear elastostatics", Heriot-Watt Symposium, vol. 1, Pitman (1976).
4. J.M. Ball, "Convexity conditions and existence theorems in nonlinear elasticity", *Arch. Rat. Mech. Anal.* **63**, (1977), pp.337-403.

5. J.L. Ericksen, "Equilibrium of bars", *J. of Elasticity* **5**, (3-4), (1975) pp.191-201.
6. M. Giaquinta and G. Modica, "Regolarita lipschitziana per le soluzioni di alcuni problemi di minimo con vincolo", *Ann. Mat. Pura e Appl.* CVI, (1975), pp.95-117.
7. D. Kinderlehrer and E. Mascolo, "Local minima for non-convex problems", to appear.
8. O. Ladyzhenskaya and N. Ural'tseva, *Linear and Quasilinear Elliptic Equations*, Academic Press, (1968)
9. P.L. Lions, *Generalized Solution of Hamilton-Jacobi Equations*, Pitman (1982).
10. P. Marcellini, "Alcune osservazioni sull'esistenza del minimo di integrali del calcolo delle variazioni senza ipotesi di convessita", *Rend. Mat.* **13**, (1980), pp.271-281.
11. E. Mascolo and R. Schianchi, "Existence theorem for non-convex problems", *J. Math. Pures et Appl.* **62**, (1983), pp.349-359.
12. E. Mascolo and R. Schianchi, "Un théorème d'existence pour des problèmes du calcul des variations non-convexes", CRAS, Paris 297 (1983).
13. E. Mascolo and R. Schianchi, "Non-convex problems of the calculus of variations", *Nonlinear Analysis* **9**, (1985), pp.371-379.
14. E. Mascolo and R. Schianchi, "Existence theorems in the calculus of variations", to appear in *J. Diff. Equations*.

20

EXPERIMENTAL AND THEORETICAL ASPECTS OF CELLULAR AND DENDRITIC SOLIDIFICATION

D. G. McCARTNEY *and* J. D. HUNT

1. INTRODUCTION

The formation of a solid phase from the liquid phase, when the diffusion of both heat and solute are occurring, is one of the most frequently observed transformations in materials processing. For example large steel and aluminium alloy ingots are produced by continuous casting, rapid solidification is being increasingly used to manufacture a variety of novel metastable metallurgical alloys, and Czochralski crystal growth is used to produce large bulk single crystals of silicon and gallium arsenide for the semiconductor industry.

Let us begin by considering two different physical situations in which solid is growing into liquid. In the first case, as shown schematically in Fig. 1(a), a single solid sphere is growing in an undercooled liquid bath with heat flow away from the solid into the liquid phase. In general the sphere becomes unstable as it grows and typically six mutually perpendicular perturbations develop into what are termed isolated dendrites. This is discussed in more detail in Section 1.1. In the second case, as shown in Fig. 1(b), a planar solid-liquid interface is growing into the liquid phase and heat is being extracted through the solid. Under certain conditions the planar front becomes unstable and finger-like protrusions of solid develop. Depending on the growth conditions these protrusions develop into either

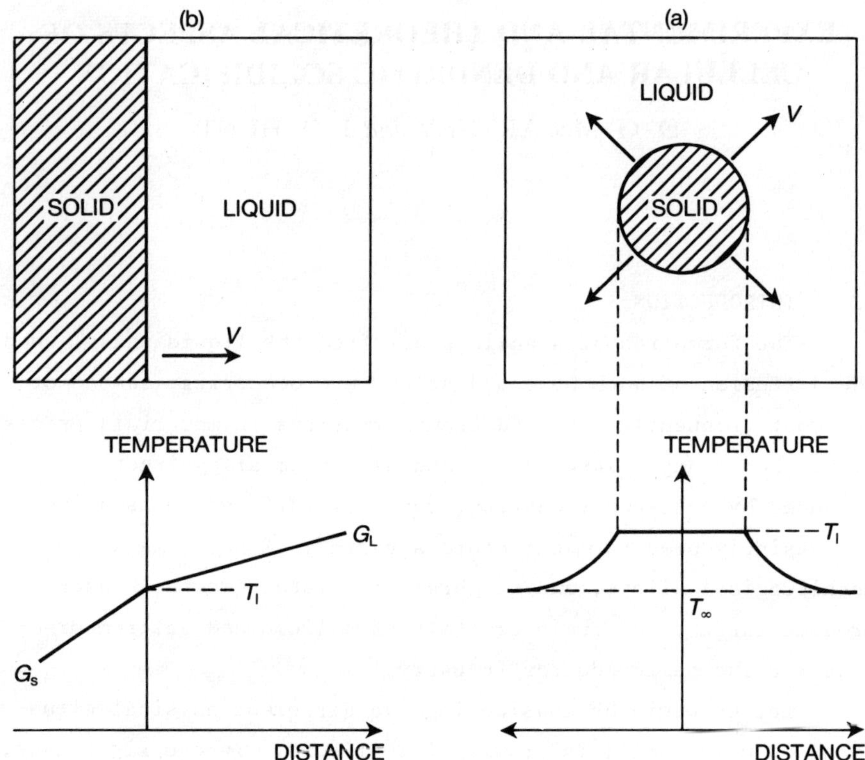

FIG. 1 Solid-liquid interface morphology and temperature distribution. In case (a) a pure solid sphere is growing radially into an undercooled bath. In case (b) a pure material is freezing with a planar front and heat is being extracted through the solid. T_I is the interface temperature, T_∞ is the bath temperature in (a), G_L and G_S are the liquid and solid temperature gradients respectively in (b).

cellular or dendritic arrays. The planar front instability is described in greater detail in Section 1.2 and the remainder of this paper is concerned with the growth of the arrays of cells or dendrites which subsequently develop. It is these cellular and dendritic arrays which are very frequently observed during alloy solidification in a wide range of casting processes [1,2]. Understanding their growth behaviour is thus of considerable technological importance.

1.1 *Isolated Dendrite Growth.* There has been considerable experimental and theoretical work on understanding the growth of isolated dendrites over the past few years. Glicksman and co-workers [3-5] have carried out extensive experimental studies on the growth of isolated dendrites in transparent, low melting-point, organic materials. There have, in addition, been numerous theoretical treatments of the problem [6-14] and reference [15] provides a succinct review.

Mullins and Sekerka [16] were the first to examine the problem of the instability of a growing sphere leading to the formation of isolated dendrites. The isolated dendrite growth problem is a free boundary one in that the shape of the solid-liquid interface is not known *a priori* but must be obtained as part of the overall solution. A feature of nearly all the analyses [6-11] is that they produce results which take the form of pairs of values of axial growth velocity, V, and tip radius, ρ, as a function of bath undercooling ΔT. ($\Delta T = T_M - T_\infty$ where T_M is the equilibrium freezing point of the material and T_∞ is the temperature of the liquid far from the dendrite.) It is found, however, that a unique value of (V, ρ) is not obtained for a given ΔT. Some means must therefore be used to determine uniquely the (V, ρ) pair which the system chooses in practice. (Experimentally it is found that for a given ΔT a unique (V, ρ) pair is selected.) The simplest procedure is to attempt to invoke an optimization criterion, where possible. This typically states that the dendrite will choose to grow at the maximum possible velocity for a given bath undercooling. Such a criterion is rather unsatisfactory, however, and recent work [12,14] has been concerned with applying a stability analysis to the dendrite tip to determine its operating point. This latter approach has been remarkably successful in producing results in close agreement with experimental observations [15].

1.2 *Instability of a Planar Front.* The conditions under which a planar front is unstable in a binary alloy are most simply predicted using the constitutional supercooling approach first put forward by Chalmers [17,18]. Consider an alloy of bulk composition C_∞ being unidirectionally solidified at steady state with an imposed growth velocity V in a liquid temperature gradient G_L. Figure 2 shows qualitatively how the driving force for instability of the plane front develops. For no constitutional supercooling, and hence planar front stability, the inequality

$$\frac{G_L}{V} > -\frac{m}{D_L}\left(\frac{1-k}{k}\right) C_\infty$$

must hold, where m is the slope of the liquidus line on the phase diagram [1,2], k is the distribution coefficient [1,2] and D_L is the diffusivity of the solute in the liquid alloy [1,2].

A major inadequacy of the constitutional supercooling criterion is that it is based on a thermodynamic argument relating to the presence or absence of a driving force for instability. Clearly a dynamic stability analysis is required and this problem was first analyzed by Mullins and Sekerka in the classic paper [19]. This linear stability analysis showed that the stability condition can be very different from that predicted by constitutional supercooling. Recently Trivedi and Kurz [20] have extended the Mullins and Sekerka analysis to examine planar front stability under conditions of rapid solidification. Numerical work on the early development of instabilities on a planar front has been carried out by Ungar and Brown [21], Ungar *et al.* [22], and McFadden and Coriell [23].

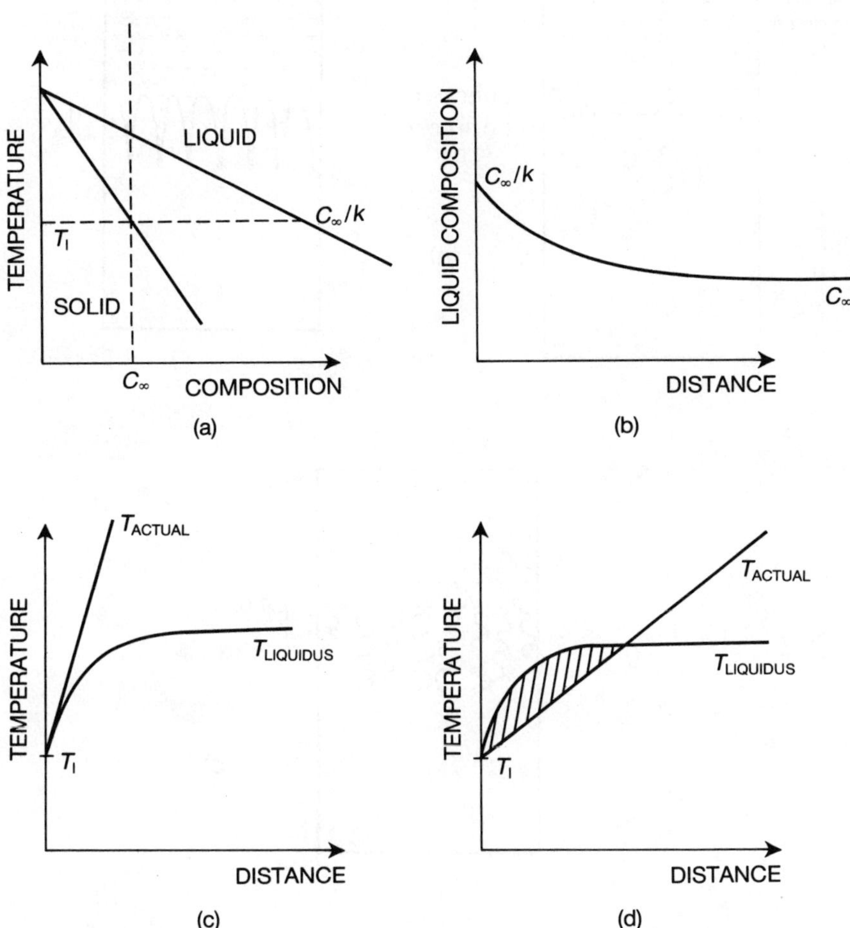

FIG. 2 Constitutional supercooling in alloy solidification. (a) phase diagram; (b) solute enriched layer ahead of freezing front; (c) no constitutional supercooling, stable interface; (d) constitutionally supercooled layer, unstable interface. T_I is the interface temperature, C_∞ the bulk alloy composition and k the distribution coefficient of the phase diagram

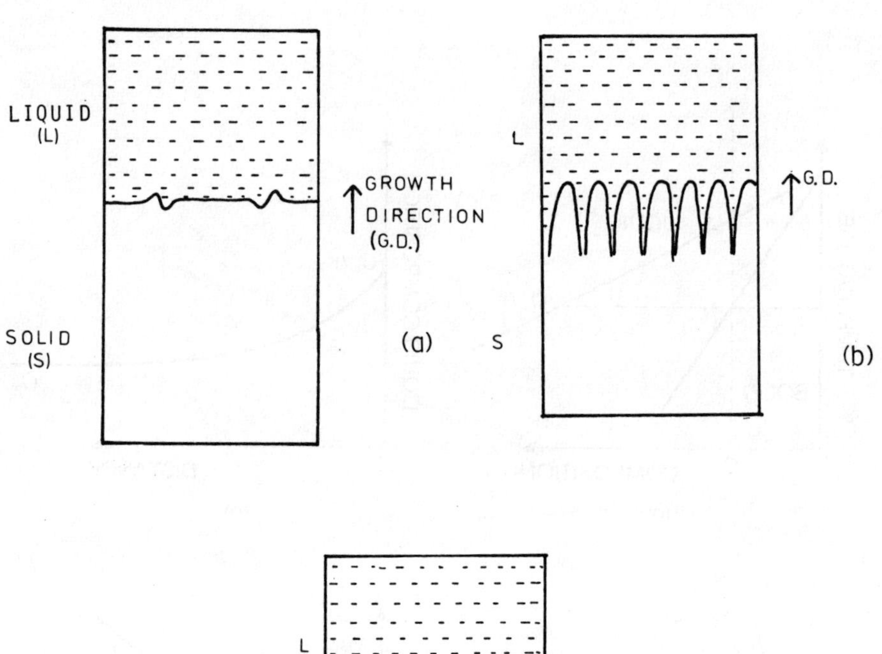

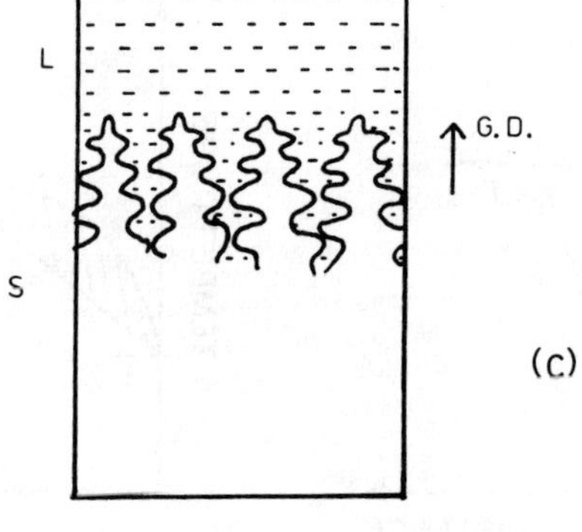

FIG. 3 Schematic illustration of the growth morphologies which develop when a planar front becomes unstable. (a) Initial perturbations when planar front is just unstable; (b) cellular array which forms when the growth conditions are close to those for stability; (c) dendritic array which forms over a wide range of growth conditions.

1.3 *Morphological Development of Planar Front Instabilities.* When a planar interface becomes unstable the initial growth morphology is highly irregular and takes the form of nodes and depressions on the solid-liquid interface [24,25]. As the growth conditions move away from those required to maintain a planar front a fairly regular hexagonal network of rods of circular cross-section can develop — these are termed cells [25]. At even lower values of G_L/V the cells develop side branches and are termed dendrites. These dendrites, in certain common alloy systems, develop cruciform-like cross-sections which are no longer arranged on a hexagonal array [25,26]. The main stalks of dendrites are termed primary arms and the side branches are referred to as secondary arms. Higher order branches can also be observed in certain circumstances. The development of these structures is shown by the schematic longitudinal sections of Fig. 3. Figure 4 illustrates the development of dendrites from cells. It should be noted that dendrites grow in preferred crystallographic directions determined by the crystal structure of the solid.

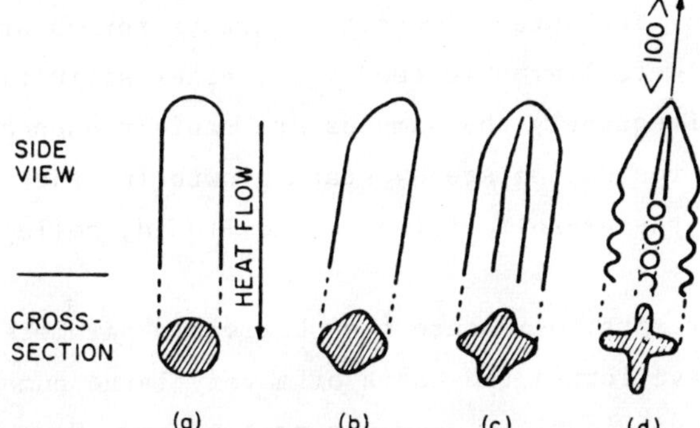

FIG. 4 Schematic illustration of the development of a fully branched dendrite (d) from a smooth sided cell (a). Changes (a) to (d) occur as G_L/V moves away from the critical value for planar stability. Cells (a) always grow along the direction of heat flow. Dendrites (d) grow in preferred crystallographic directions (denoted by $\langle 100 \rangle$) and may not grow parallel to heat flow.

2. EXPERIMENTAL ASPECTS OF CELLULAR AND DENDRITIC SOLIDIFICATION

Controlled unidirectional growth is used to study the characteristics of cellular and dendritic arrays. Generally it is accepted that if the growth conditions remain constant the composition of the cell/dendrite tips, the spacing at the primary arms, and the cell/dendrite tip radii will be constant with time. It is these parameters which are used to fully characterize the array and which are measured as a function of the experimental variables namely V, G_L and C_∞ (C_∞ is the bulk alloy composition, V the growth rate and G_L the liquid temperature gradient).

Unidirectional growth experiments have been carried out using both high melting point metallic alloys and low melting-point, transparent, organic materials which freeze like metals. In general the organic materials are solidified in thin cells, which results in the formation of two-dimensional arrays. These arrays can be observed during growth by carrying out the experiments on the stage of an optical microscope [27,28]. In metallic alloys unidirectional growth is achieved by pulling a long rod-shaped specimen out of a furnace and into a cold zone at a fixed velocity. The three-dimensional arrays formed are observed using standard metallographic techniques after solidification is complete. Frequently the samples are rapidly quenched to room temperatures during steady state growth in order to preserve the shape of the as-solidifying, solid-liquid, cellular or dendritic front.

Attempts to measure the observable parameters of arrayed growth have formed the basis of a very large number of experimental investigations over the past years. However, dendrite tip radii are rarely measured in opaque metallic alloy systems and nearly all of the experimental work has concentrated on measuring tip temperatures, compositions and primary arm spacings. The secondary arm spacing is also measured experimentally, but this parameter changes along the length of the primary stalk and is found to be proportional to $(\theta_s)^{0.3-0.4}$, where θ_s is the time spent in the semi-solid state.

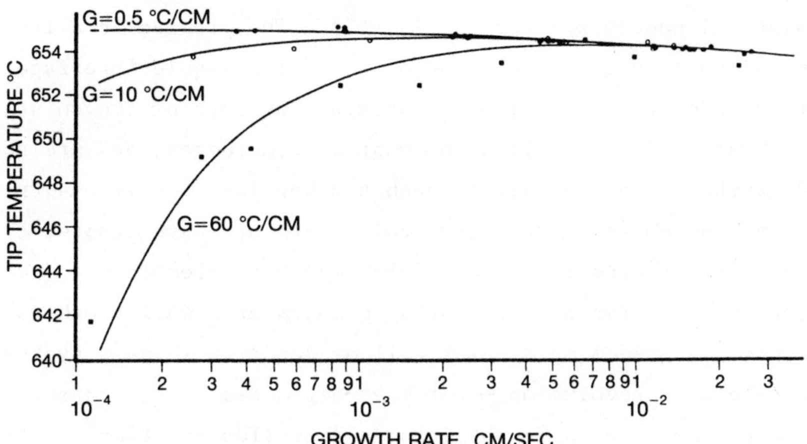

FIG. 5 Plot of dendrite tip temperature versus growth rate. The points are experimental values [29] and the full lines are the predicted values of the Burden and Hunt model [37]. Experimental points and calculated values are for the three different temperature gradients indicated.

There is now general agreement as to the behaviour of tip compositions and temperatures over a reasonably wide range of growth conditions. References 28-30 describe the most complete work in this area and Fig. 5 illustrates the variation of tip temperature with growth rate and temperature gradient in an Al-2 wt % Cu alloy. However, there is still considerable controversy in the literature regarding the relationship between primary arm spacing and growth variables. References 26 and 31-34 relate to some relatively recent experimental results on a variety of alloy systems. Most authors agree that the relationship between primary spacing, λ, and the growth variables is of the form

$$\lambda = G_L{}^{a}\ V^{b}\ C_\infty{}^{c}$$

but the exact values of a b and c are disputed. The work of McCartney and Hunt [26] shows that $a \simeq -0.5$, $b \simeq -0.25$ and $c \simeq 0.33$.

3. ANALYTICAL MODELS OF ARRAY GROWTH

Because of the importance of array growth in solidification processes [1,2] there have been a number of approximate analytical models put forward [34-42]. The problem is a free boundary one in that the shape of the solid-liquid interface is not known *a priori* but must be obtained as part of the solution. Both heat and solute diffusion must be considered, and are coupled at the phase boundary through the boundary conditions there. Some of the approximate analytical treatments have ignored the exact cell or dendrite shape [36-38,42] but tried to incorporate interactions between neighbouring primary arms whilst others [39-41] have assumed a particular cell or dendrite shape — typically a parabola of revolution — but have neglected the interaction between primary arms in obtaining a diffusion solution. A frequent assumption has also been that of a constant temperature gradient in both solid and liquid, and only solutions for the solute diffusion problem have been obtained.

The models of Burden and Hunt [37] and Trivedi [40] predict tip temperatures which are in good agreement with experimental work carried out by burden and Hunt [29] on Aℓ-Cu alloys as can be seen from Figs. 5 and 6. Hunt [38] and Kurz and Fisher [41] have developed models to predict the variation of primary arm spacing, λ, with growth conditions G_L, V and C_∞. In the dendritic growth regime both models predict a relationship of the form

$$\lambda \propto G_L^{-0\cdot5} \; V^{-2\cdot25} \; C_\infty^{0\cdot25}$$

but the constant of proportionality is different in the two cases. The predicted variation of λ becomes more complex in the cellular regime and reference 43 discusses the behaviour in some detail.

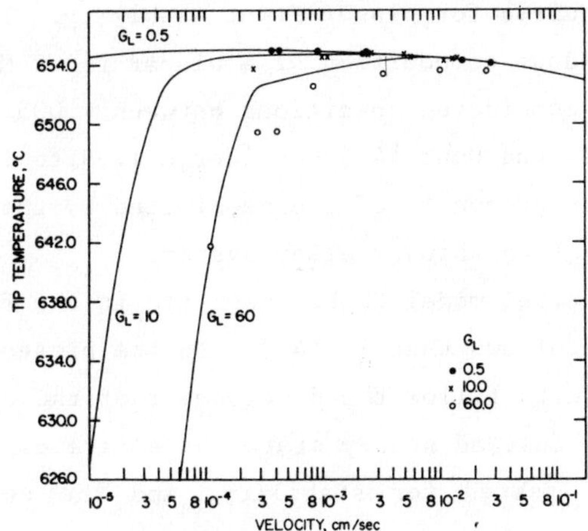

FIG. 6 Plot of dendrite tip temperature versus growth rate. The points are experimental values [29] at three different temperature gradients. The full lines are the tip temperature, velocity relationships predicted by the Trivedi model [40].

4. A NUMERICAL MODEL OF ARRAY GROWTH

4.1 *Introduction.* Clearly the analytical models described in the previous section are approximate and do not treat correctly the physics and mathematics of the array problem. The present authors believe that it is essential to consider the array to describe directional cellular and dendritic solidification correctly. Clearly the array must be considered when attempts are made to predict cell or dendrite spacings.

Ungar and Brown [21], Ungar *et al.* [22] and McFadden and Corriell [23] have all used numerical techniques to study array formation very close to planar front stability. In these models the cell groove depth was of similar dimensions to the cell spacing. McCartney and Hunt [44], however, modelled cells growing under steady state conditions and having very deep grooves which terminated well outside the domain of the numerical model. That work [44] was therefore substantially different from the other numerical studies in that the growth conditions were much further from those required for planar front stability. If C_∞^* is the critical breakdown composition of a planar front McFadden and Corriell [23] considered conditions between 1.005 and 1.13 C_∞^* while McCartney and Hunt [44] considered 1.39 to 13.6 C_∞^*. The depth of a deep groove is of course limited by the thermodynamic phase diagram of the binary alloy system.

The numerical model to be described in the following sections closely follows that in [44]. In the present work no attempt is made to follow the development of the cell from the planar front. Instead steady state shapes are calculated, these must later be tested for stability, and the cell groove is assumed to go back to a eutectic temperature which is outside the domain being modelled.

4.2 *The Problem.* Only a brief resumé of the numerical technique will be described here as it has already been described in detail in [44]. A cell from an array is assumed to be a solid growing within a cylinder. The approximation involved is illustrated in Fig. 7. The initial part of the problem is to calculate the steady-state cell shape. Since the shape does not change with time, it is convenient to transform the axes to axes moving with the cell tip.

The thermal and solute diffusion equation with coordinates moving in the z direction (Fig, 8), at the steady state velocity V, must be solved for composition and temperature in both the solid and liquid phases. These equations are

$$\nabla^2 U_i + \frac{V}{D_i}\frac{\partial U_i}{\partial z} = 0, \tag{1}$$

where U_i is either composition C or temperature T, in the liquid or solid (C_L, C_S, T_L or T_S see Appendix 1) and D_i is the relevant diffusivity. At the solid-liquid interface the usual interface conditions must be satisfied. The symbols used are listed in the symbol table (Appendix 1). The composition of the solid C_S is related to that in the liquid C_L by:

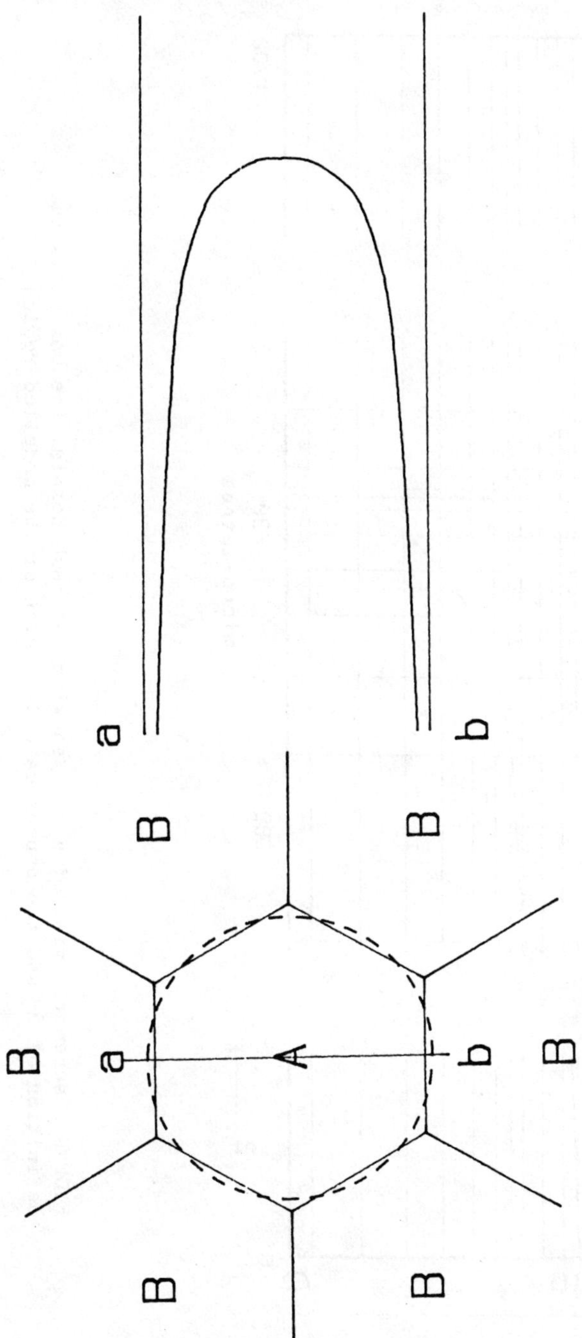

FIG. 7 Shows the approximations involved in assuming the cell has radial symmetry

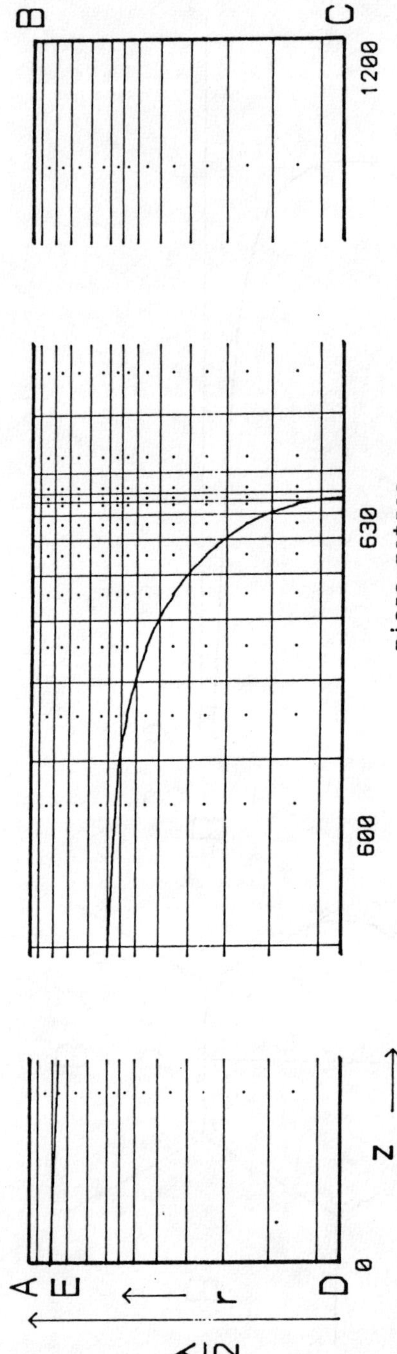

FIG. 8 A schematic view of a cell showing the mesh points, the box walls the definition of λ and the approximate dimension of the modelled region.

$$C_S = kC_L \tag{2}$$

and to conserve solute at the interface

$$V_n(1-k)C_L = D_S \frac{\partial C_S}{\partial n} - D_L \frac{\partial C_L}{\partial n} . \tag{3}$$

For heat

$$T_L = T_S \tag{4}$$

and

$$V_n L = K_S \frac{\partial T_S}{\partial n} - K_L \frac{\partial T_L}{\partial n} .$$

The temperature of the interface must be consistent with the composition curvature and kinetic mobility at each point. This may be written as an undercooling equation

$$T_0 - T_I = \Delta T_I = m(C_\infty - C_L) + \theta(R_1 + 1/R_2) + V_n/\mu \tag{6}$$

where m is the liquidus slope, θ the curvature undercooling constant, R_1 and R_2 the principal radii of curvature and μ the kinetic coefficient. When evaluated at the tip ΔT_I is termed the tip undercoating.

The far field conditions are

$$C_L = C_\infty \tag{7}$$

$$T_L = T_{L\infty} \tag{8}$$

as $z \to \infty$. (In practice a gradient condition a small distance ahead of the cell was used instead of equation (8), see later.)

A solution of the problem consists of solving equations (1), satisfying equations (7) and (8) for an interface shape such that equations (2)-(6) are simultaneously satisfied all along the solid liquid interface. Such a problem is difficult because the shape is not known *a priori* but must be obtained from the analysis.

4.3 *Method.* Since radial symmetry has been assumed the three-dimensional problem effectively becomes two dimensional. A rectangular array of points is set up on the diametric plane r, z (Fig. 8). Box walls are erected equidistant between the points. An equation for each point is obtained by considering the heat or solute crossing four walls surrounding the point. The problem is a steady state problem so that the sum (either heat or solute) crossing the walls equals zero. For the walls perpendicular to r, only a diffusive flux is present. However, for the walls perpendicular to the z direction, a flux due to the moving coordinates must also be included. The equation of a general point for either heat or solute is

$$A_E\left(D_i\frac{\partial U_i}{\partial z}+VU_i\right)_E-A_W\left(D_i\frac{\partial U_i}{\partial z}+VU_i\right)_W$$
$$+A_N\left(D_i\frac{\partial U_i}{\partial r}\right)_N-A_S\left(D_i\frac{\partial U_i}{\partial r}\right)_S=0\,, \tag{9}$$

where the subscripts E,W,N,S refer to east, west, north and south of the box, and A is the area of the relevant wall.

Because of the difficulty in handling curved phase boundaries with a uniform mesh, a non-uniform mesh was chosen such that mesh points lay on the solid-liquid interface (Fig. 8). Using this method the equations for the interface points can be obtained as before except that here two walls are solid and two liquid (Fig. 8). This satisfies equations (3) and (5) in a simple but accurate fashion.

In the previous work [44] the composition and composition gradient (or temperature and temperature gradient) at the walls were assumed to be that given by a linear interpolation between adjacent points. This is a reasonably good approximation provided the box size is small compared with D_i/V, where D_i is the relevant diffusivity. A better approximation for a moving box system is to use an exponential interpolation of the form

$$U_i = F + G\exp(-Vz/D_i) \tag{10}$$

where F and G are constants obtained from the composition or temperature of adjacent points. This interpolation was used in the present work so that solute diffusion in the solid could be included. The exponential interpolation has very little or no effect on heat flow and on diffusion of solute in the liquid.

The boundary conditions of the domain were treated such as before. The flux of heat or solute across the symmetry boundaries AB and CD (Fig. 8) are zero. Across BC the composition fluxes were obtained by assuming that the composition varies exponentially to a composition of C_∞ at infinity. This is the solution to equation (1) when there is no variation in the r direction (as must occur well ahead of the cell tip). A similar solution could be used for temperature. However, in practice during a directional growth experiment, a gradient is imposed near the cell tips rather than a temperature at infinity. For this reason a temperature gradient was imposed along BC.

Provided the modelled groove depth is large enough compared with the spacing, the temperature and liquid compositions vary very little with r at the base of the groove. At $z = 0$ (AD, Fig. 8) the temperature was assumed constant and initially put equal to zero. Later the temperature scale was adjusted to connect the temperature and composition fields.

The boundary condition for solute in the liquid at $z = 0$ (AE, Fig. 8) was

$$\frac{\partial T}{\partial z} = m \frac{\partial C_i}{\partial z}. \tag{11}$$

This is necessary to satisfy equation (6) when $\partial C_i/\partial r$ and $(\partial T/\partial r) \to 0$.

The boundary condition for solute in the solid at $z = 0$ (ED, Fig. 2) was assumed to be

$$\frac{\partial C_s}{\partial z} = 0 . \tag{12}$$

This is only strictly valid when $D_S \to 0$. However, in this work $D_S \simeq 10^{-3} D_L$, so that no large errors are to be expected from this approximation.

The method used to calculate the principal radii of curvature from the interface points is identical to that described in reference [44]. An analytical curve approximating the final shape was fitted to three adjacent interface points. One principal radius of curvature of the centre point could then be calculated and the slope at that point used to obtain the other principal radius of curvature.

In principle, it is possible to set up an equation for the composition and temperature at each mesh point and an additional interface equation (6), for each of the points on the solid-liquid interface. These could then be solved to give the temperature and composition at each point as well as the interface position. The equations, however, are nonlinear and very large arrays and extensive computing would be needed.

In the present work and in that reported previously [44], advantage is taken of the fact that once the interface shape is fixed the equations for the composition and temperature at each mesh point are all linear. The procedure used to obtain a solution was thus to assume an interface shape. The temperature field was then simply calculated using Gaussian elimination. Knowing the temperature gradient in the groove, the composition could be calculated in a similar fashion.

This procedure satisfies all the equations with the exception of the undercooling equation (6). This equation must be simultaneously satisfied at each of the interface points by suitably modifying the cell shape.

The temperature scale was adjusted as mentioned previously so that there was no error in equation (6) for the interface point nearest $z=0$. The temperature error e, is defined from equation (6) as

$$e = -m\,(C_\infty - C_L) - \theta\,(1/R_1 + 1/R_2) - V_n/\mu + T_0 - T_I\,. \qquad (13)$$

In the previous work [44] an intuitive technique was used to adjust the interface shape to give a zero error at each point. The interface was moved in the z direction by an amount proportional to the error. After about one hundred iterations, errors of less than 10^{-6} K could be produced. As well as shifting the points, it is necessary to rearrange the points on the solid liquid interface after each iteration. This ensures that points do not become too close. Unfortunately, the procedure used in the earlier work led to shoulders of the cell being poorly defined by the interface points (Fig. 9). It was found that when the shoulders were well defined (Fig. 10), the iteration procedure did not converge and the undercooling errors slowly oscillated. This was because changing the position of one point effects the errors on each of the other points.

In the present work, the problem of eliminating the undercooling error was overcome by perturbing the position of the interface in a number of independent ways. In the simplest form, if there were p interface points, each interface point, other than the first, was shifted separately by a small amount in the z direction. The change on the undercooling error at that point and all the other points was obtained from the numerical calculation. This gave the terms

$$e_{ij} = \frac{\delta e_j}{\delta e_i} \tag{14}$$

(that is the change in the error of point j due to a shift δz in point i). These error gradients were then used to set up linear equations

$$\sum_{i=2}^{p} e_{ij}\,\delta z_i = -e_j \tag{15}$$

which could be solved to give all the $(p-1)$ δz_i terms or a combined shift which should eliminate the error at all the interface points. In practice a fraction of the calculated shift was used to maintain stability. Under favourable circumstances, it was found that this procedure gave errors of less than 10^{-6} K after only 3 or 4 iterations.

Typical results with well defined shoulders are shown in Fig. 10. The tip undercoolings at large spacings were different from those obtained in the earlier numerical work. This is because of the poorly defined shoulders of the cells which were used. It was found that provided sufficient points were present

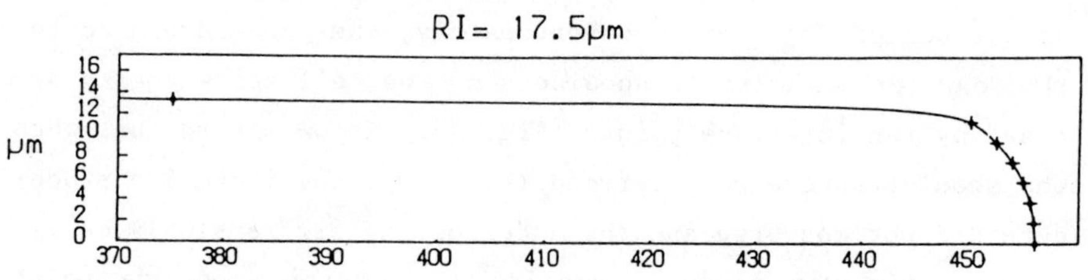

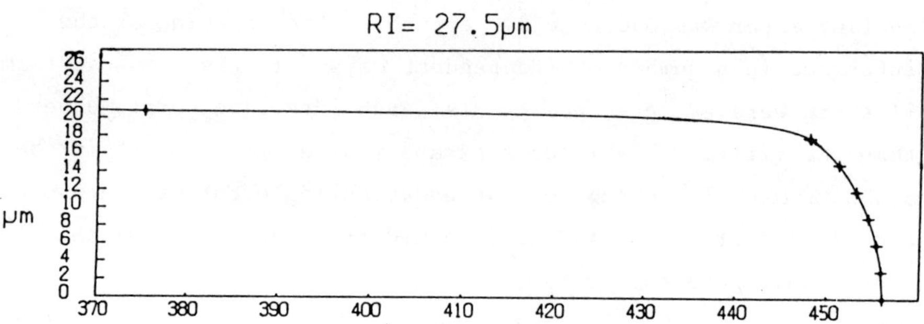

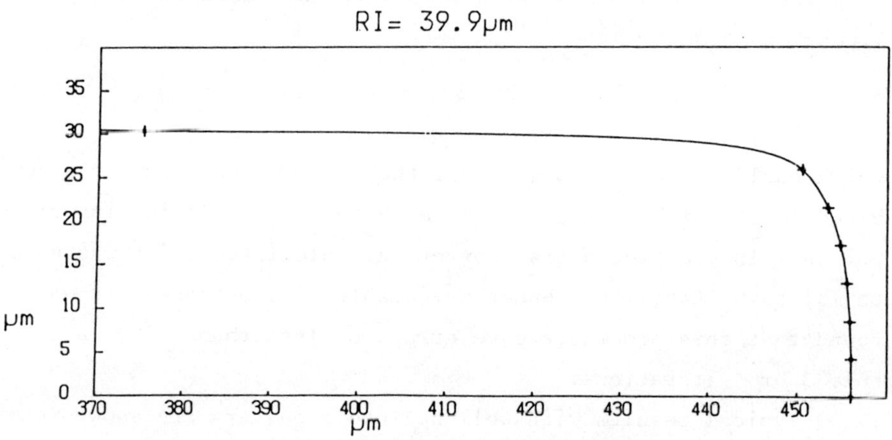

FIG. 9 Cellular shapes calculated using poorly defined shoulders [44] ($V = 8\mu m/s$)

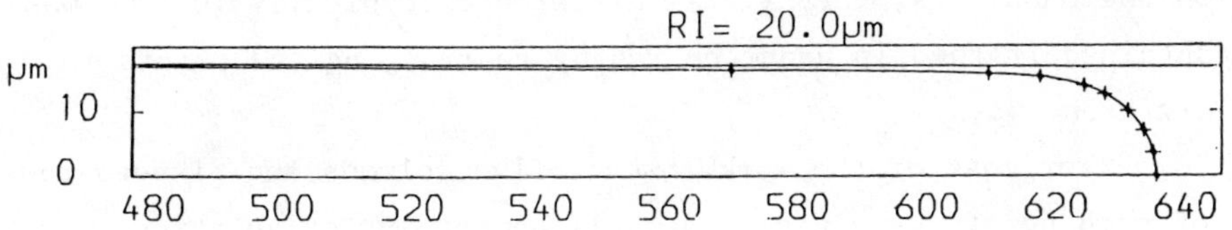

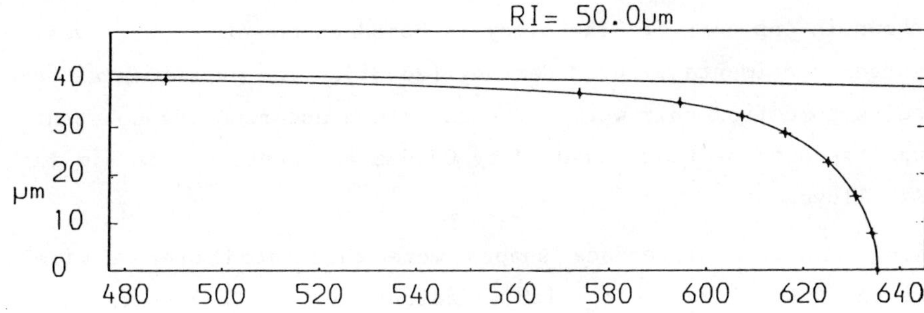

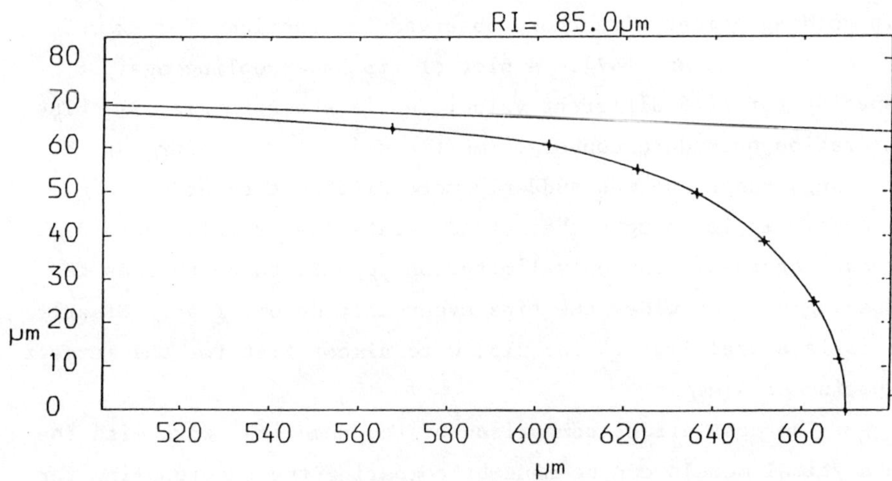

FIG. 10 Typical cell tip shapes; the modified Scheil shape is the top line ($V = 8\mu m/s$)

on the shoulders, very little difference (typically 10^{-2} K) was obtained in the tip undercooling by rearranging the points in different ways.

For most of the work, twenty-five columns and sixteen rows of mesh points were used. A small number of calculations were carried out using double the number of rows and columns. The results were not significantly different.

The values of the constants used in the calculation are shown in Appendix 1. The alloy is based on an Aℓ alloy investigated experimentally by McCartney [26,34]. The experimental results come from this work. The curvature undercooling constant was taken to be that measured by Gunduz and Hunt [45] in similar Aℓ alloys.

4.4. *Results*. Interface shapes were calculated over a wide range of spacings, velocities, gradients and surface energies. Examples of the tip region are shown in Fig. 10. This figure also shows the Scheil shape [1,2] modified to allow for diffusion in the z direction [46]. The cell shapes are qualitatively in good agreement with those observed in practice, for example reference [26] and [47]. A plot of tip undercooling against spacing for five different velocities is shown in Fig. 11. The iteration procedure converged most easily at the centre of the spacings range and was suddenly more difficult towards the extremes. It is thought that steady state shapes exist outside those obtained. The only limitation appears to be that as the spacing becomes wider the tips eventually become flat. Clearly this is a real limit. The tips were almost flat for the largest spacing of 4 μm/s.

The most direct comparison of the numerical work with the analytical models can be made by comparing the undercooling for a particular tip radius. The undercoolings were calculated using Burden and Hunt [37] equation (17) (ΔT_{BH}), Kurz and Fisher [41] equation (7) (ΔT_{KF}), and Trivedi [40] equation (33) (ΔT_T).

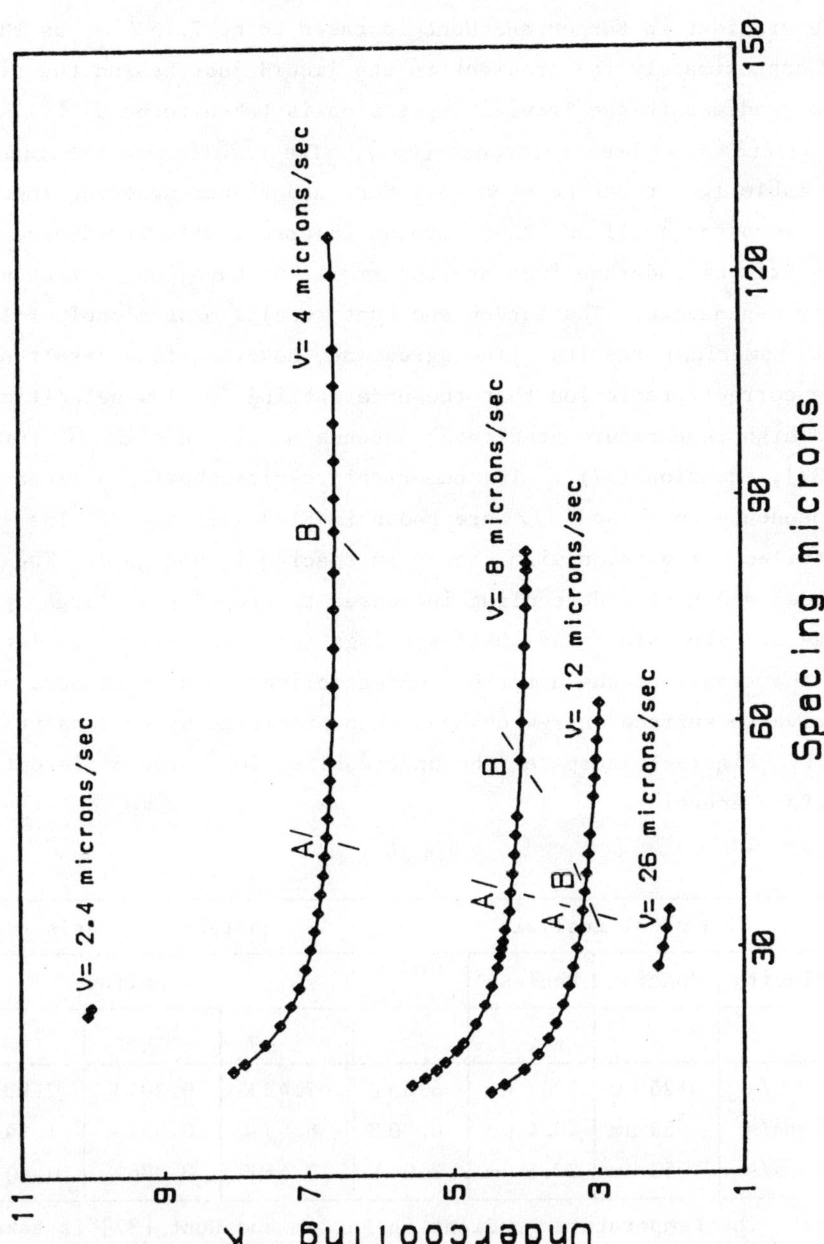

FIG. 11 Plots of tip undercooling against spacing for five different velocities. Points A represent the experimentally measured spacings [26] and points B those calculated using the present stability analysis (Figs. 17-19)

The gradient in Burden and Hunt is taken to be 7.75 K/mm as this is approximately the gradient in the liquid just behind the tip. The gradient in the Trivedi expression is taken to be $G = (K_L G_L + K_S G_S)/(K_L + K_S)$ (see reference [40]). The results are tabulated in Table 1. It can be seen that Kurz and Fisher undercoolings are much too small and they have an incorrect velocity dependence. The Trivedi undercoolings are too small but have the correct velocity dependence. The Burden and Hunt results most closely follow the numerical results. The agreement, however, is a result of the correct prediction that the undercooling (at low velocities and high temperature gradients) depends mainly on a GD_L/V term ([37], equation (17)). The numerical results showing a linear dependence on G and $1/V$ are shown in Figs. 12 and 13. The detailed correlation with change in spacing is not good. The Burden and Hunt undercooling increases too rapidly at large spacings and stoo slowly as small spacings (see Fig. 14). The latter is the result of the numerical undercoolings being much more sensitive to surface energy changes than predicted by the analytical work. Figure 15 compares the undercooling for three different surface energies.

TABLE 1

Numerical analysis				Analytical models		
Velocity	Spacing	Radius	Under cooling	Undercooling		
V	λ	R_1	ΔT	ΔT_{BH}	ΔT_{KF}	ΔT_T
4 μm/s	125 μm	220 μm	6.65 K	7.42 K	0.304 K	2.93 K
4 μm/s	50 μm	40.4 μm	6.70 K	6.99 K	0.051 K	1.94 K
12 μm/s	50 μm	22.5 μm	3.06 K	2.45 K	0.086 K	0.90 K

Note: The temperature gradient in Burden and Hunt [37] is taken to be $G = 7.75$ K/mm and in Trivedi [40] is taken to be $G = (K_L G_L + K_S G_S)/(K_L + K_S)$.

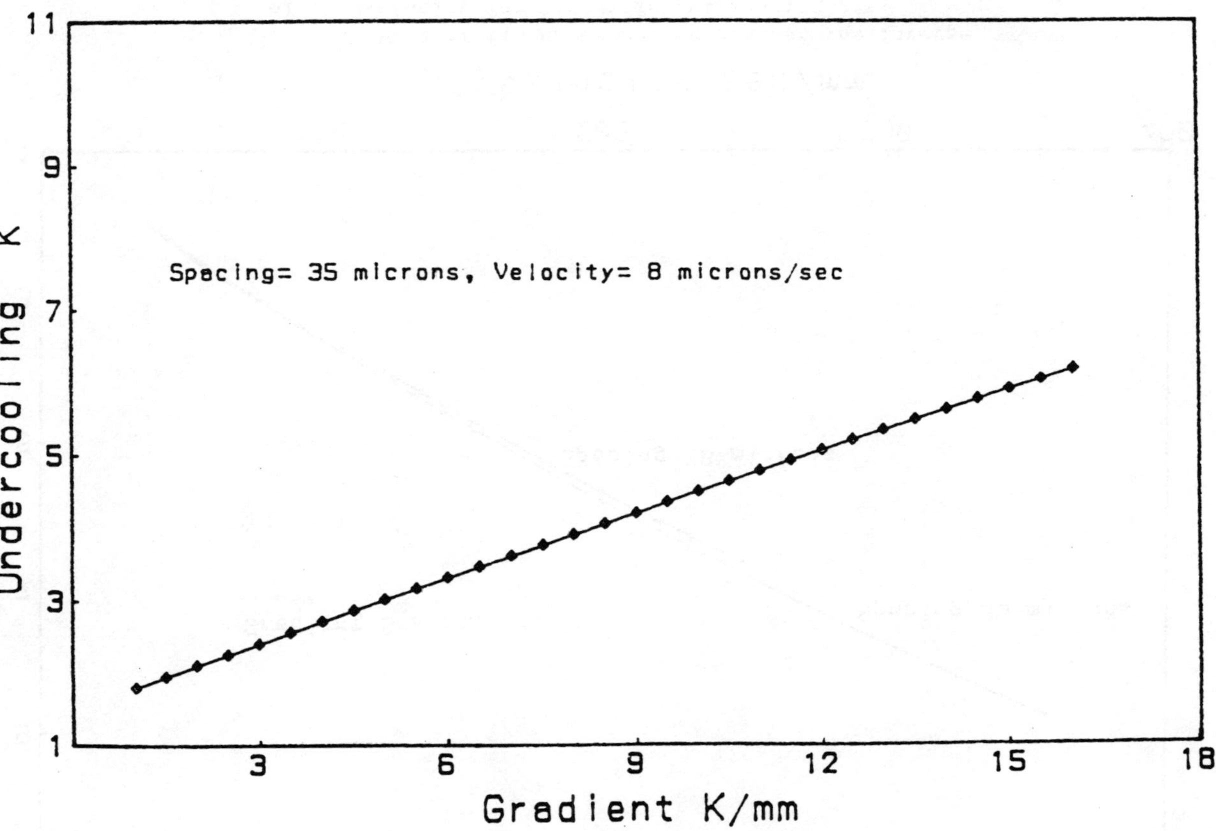

FIG. 12 Plot of tip undercooling against temperature gradient in the liquid

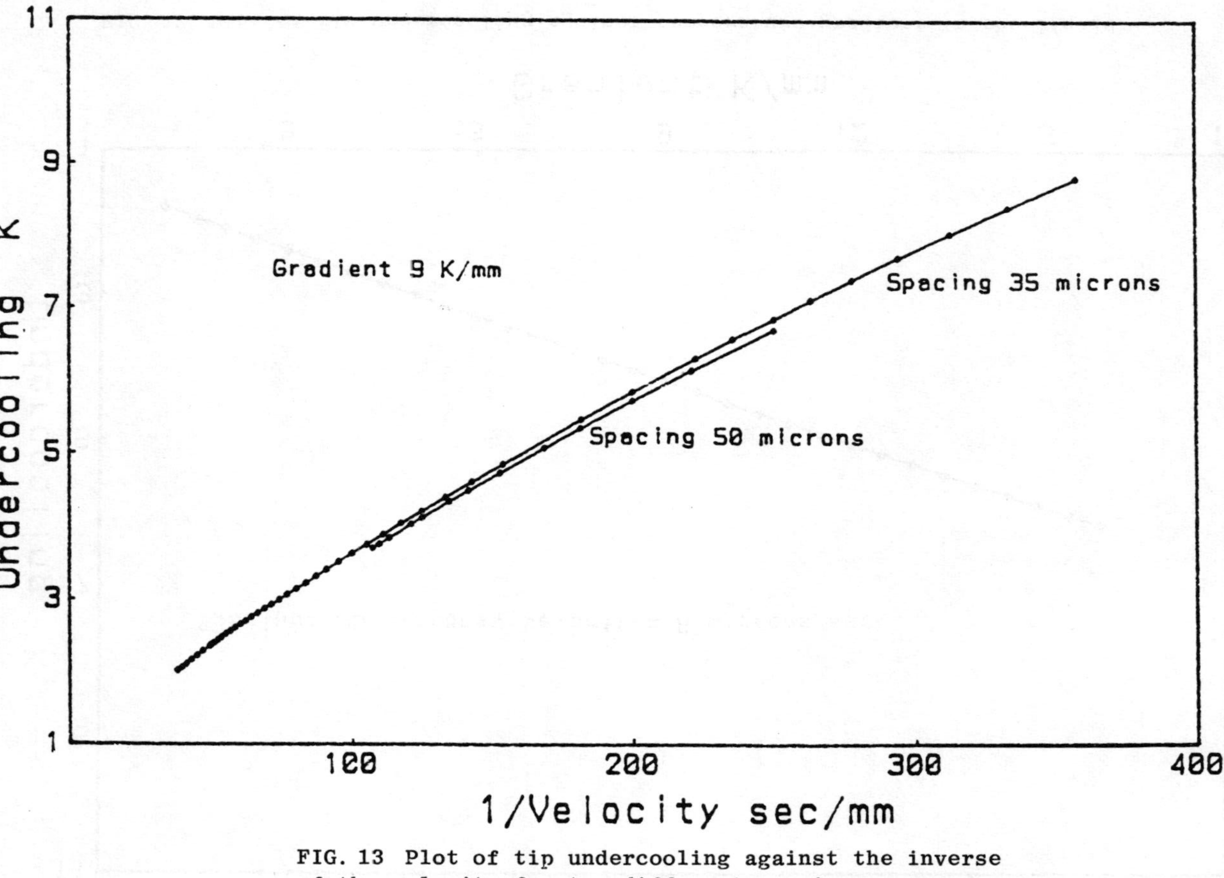

FIG. 13 Plot of tip undercooling against the inverse of the velocity for two different spacings

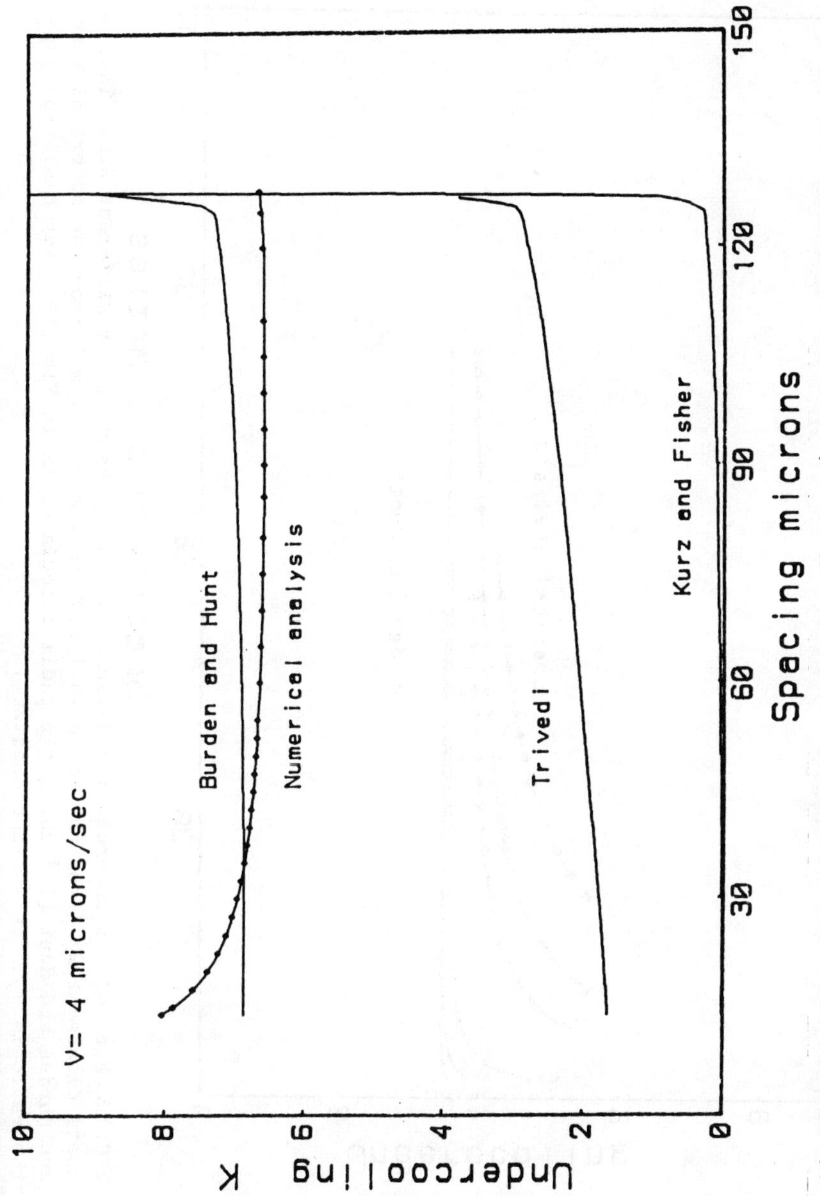

FIG. 14 Plot of tip undercooling against spacing ($V = 4\,\mu$ms) from the numerical work and the predictions made by references [37, 40 and 41]. The tip radii used in the analytical expressions correspond to the radii found in the numerical work (note the radius $\rightarrow \infty$ as $\lambda \rightarrow 128\,\mu$m)

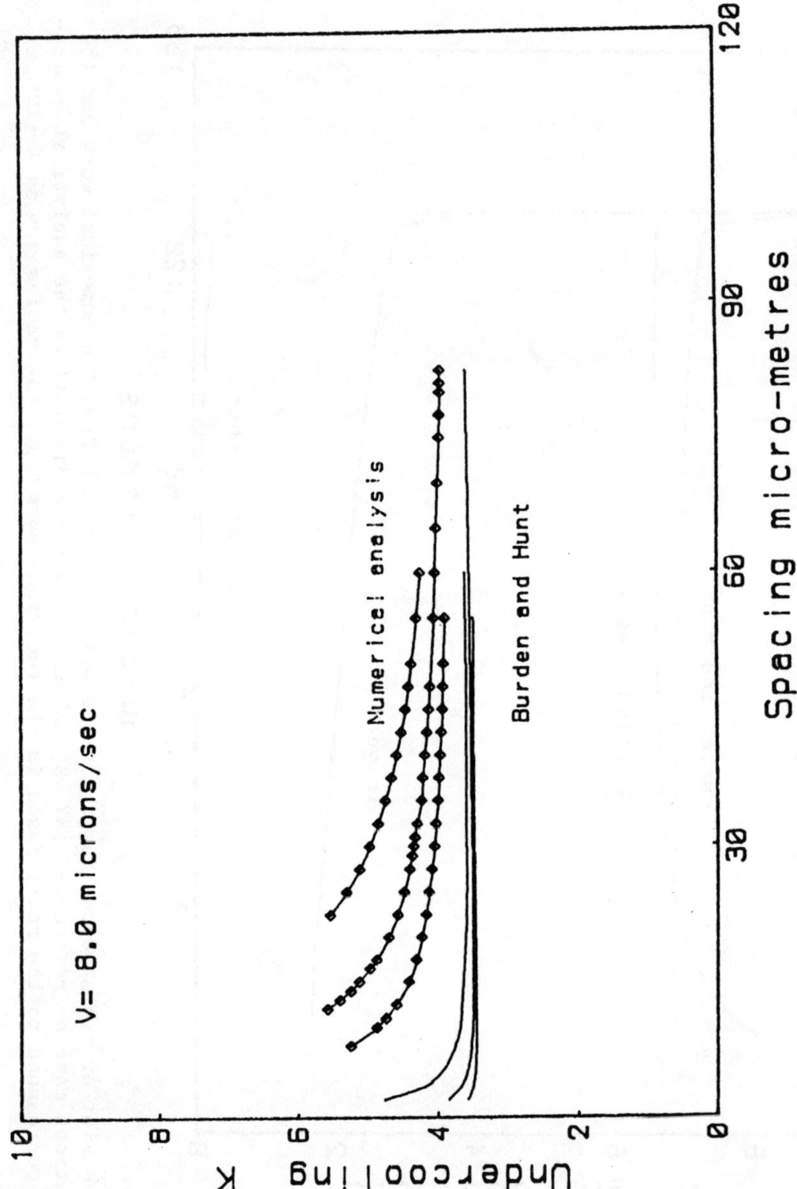

FIG. 15 Plot of tip undercooling against spacing ($V = 8\mu m/s$) for different Gibbs Thomson coefficients [1,2] (from the top down 1.0, 0.28 and 0.1 K.μm.) from the numerical work and Burden and Hunt [37] using tip radii calculated as in Fig. 14. (at small spacings the ratio is λ/R is assumed constant)

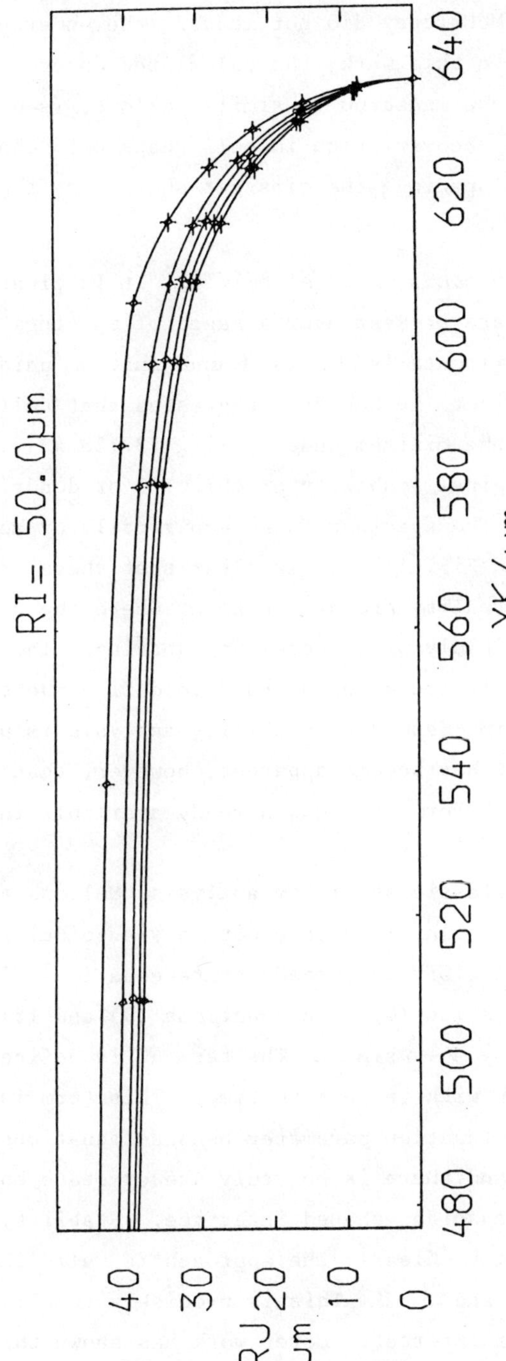

FIG. 16 Shows the effect of varying velocity from 4→12 μm on the shape of the cell tip. The tip radius of curvature decreases with increasing velocity

Although McCartney did not measure the undercoolings in the alloy modelled in this work, the calculated undercoolings are consistent with those measured in similar alloys, (see Figs. 5 and 6). Figure 16 shows the variation in cell shape with change in velocity at a fixed spacing; the tips get sharper as the velocity increases.

4.5 *The growth condition – stability.* It is clear from Fig. 11 that solutions are present over a range of spacings for a fixed velocity. Experimentally it is found that a unique spacing exists in practice. It has been suggested that cells or dendrites grow at the minimum undercooling [37,38,42], and more recently by marginal stability of the cell or dendrite tip [12-14, 40, 41, 43]. The spacings found empirically by McCartney [26, 34] are shown in Fig. 11. It is clear that these are near the minimum but in reality are at the point where the undercoolings begin to rise rapidly with decreasing spacing. The minimum is so shallow that it could not be said to define the spacing.

A fully time-dependent stability analysis is planned in future work. It has become apparent, however, that some preliminary stability information was already available in the present work.

In their classic stability analysis, Mullins and Sekerka [19], perturbed a flat interface with a sinusoidal perturbation $\delta \sin \omega x$. They satisfy the steady state equations (1) and the undercooling equation (6). In equations (3) and (5) the velocity V is replaced by $V + \dot{\delta} \sin \omega x$. The term $\dot{\delta}$ is defined as the derivitive of δ with respect to time. This term is, however, effectively a relaxation parameter because equations (1) are not time dependent and there is no truly steady state solution for an arbitrary sinusoidal shaped interface. Stability is then given when $\dot{\delta}/\delta < 0$. Clearly the approach is valid in the limit when both $\dot{\delta} \rightarrow 0$ and $\delta \rightarrow 0$. This is precisely the limiting condition which is of interest. Later work has shown the approach to be fully justified, for example reference [23].

The present authors suggest that the other interface condition, the undercooling equation (6) in the form of equation (13), could have been relaxed. Instead of writing equation (13) as $e=0$ it could be assumed that $e=\delta e \sin \dot{\omega} r$. If this is done and V is kept constant the ratio $\delta e/\delta$ gives an identical numerator to that given for $\dot{\delta}/\delta$ in the Mullins and Sekerka [19] expression for a planar interface. It is the numerator which determines the sign and the most unstable wavelength of $\dot{\delta}/\delta$. Stability would then be given by $\delta e/\delta < 0$. There does not appear to be any fundamental reason that the continuity equations (3) and (5) should be relaxed in preference to the undercooling equation (6) although there is perhaps a good physical reason.

In the present numerical work the continuity equations cannot easily be relaxed. In fact, it is the undercooling equation which was relaxed to obtain self-consistent solutions. To obtain preliminary stability information it is therefore suggested that stability of a cell can be examined by perturbing a steady state shape and examining the sign of $\delta e/\delta$, the undercooling error.

Steady state shapes were perturbed by applying a perturbation to interface shape of the form

$$z = z(r)_0 + \delta \cos(2\pi r/\lambda_P) \tag{17}$$

where $z(r)_0$ is the steady state cell shape and λ_P is the wavelength of the perturbation. The amplitude δ was typically 0.1μ (for a spacing $\simeq 30\mu$). Because of the slope of the interface in the cell groove this form of perturbation effectively only perturbs the tip.

A problem arises in connecting the temperature field to the composition field. In the steady state analysis, the temperature of the first point is assumed to correspond to the temperature calculated from the composition and curvature of that point. The interface shape is then adjusted so that no error exists at all the other points. The choice of the zero error point is more difficult for the perturbed shape. It is arbitrarily suggested that the zero point should be at the base of the cell groove furthest away from the perturbed tip and that the temperature and composition fields should remain constant at this point. The composition can be made constant by replacing the gradient condition equations (11) and (12) with a composition condition (the steady state value). Alternatively the shape of the perturbation can be modified in the cell groove to give a zero error for the first few points. Both of these procedures gave tip stability curves such as those shown in Figs. 17-19. These figures show plots of $\delta e/\delta$ against the perturbation wavelength (divided by the cell spacing) for different cell spacings. It can be seen that the results are much as would be expected. Below a particular spacing the cell tips are completely stable. It is interesting to note that the most unstable wavelength, (when just marginally stable), occurs when the ratio of wavelength to spacing is about one for each of the three velocities investigated. Initially the perturbation wavelength was divided by the tip radius since in the isolated dendrite stability analysis [14] it was predicted that the most unstable wavelength occurs at a wavelength equal to the tip radius of curvature. In the present work, however, the marginally stable ratio λ_P/R varies from about $3.7 \rightarrow 2.0$. Thus it appears in this preliminary work that it is the spacing rather than the tip radius which may be the more important dimension in the array.

The limits of stability of a single cell tip are plotted on Fig. 11 and should be compared with the experimentally measured cell spacings. The marginal stability spacings vary much more rapidly with change in velocity. Empirically the cellular structure begins to become dendritic at about 12 μm/s. Thus it appears that the marginal stability condition for a single cell, at least in this analysis, does not predict the cell spacing but may well show the breakdown from a cellular to a dendritic structure.

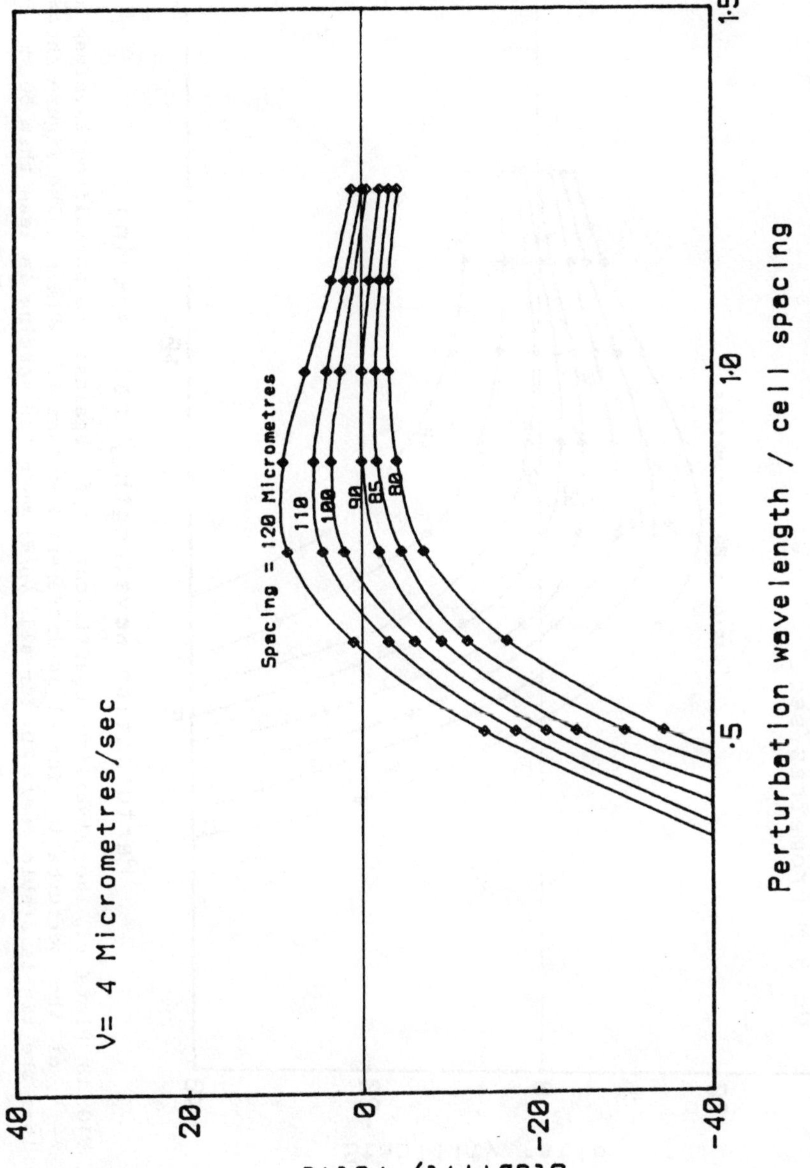

FIG. 17 Plots of the stability coefficient $\delta e/\delta$. against the normalised wavelength λ_p/λ, of the perturbation for six different spacings ($V = 4\mu m/s$). The figure shows that the tip is stable ($\delta e/\delta < 0$ for all λ_p/λ) when the spacing is less than 90 μm

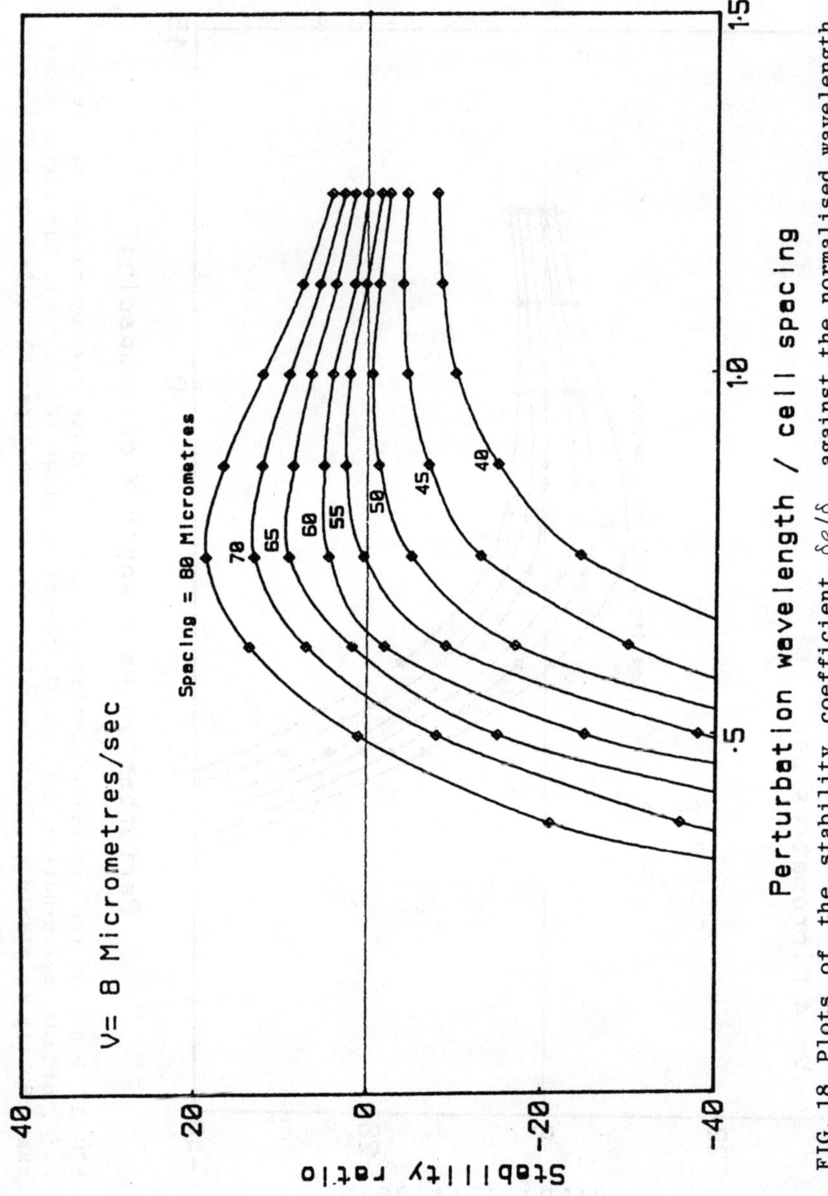

FIG. 18 Plots of the stability coefficient $\delta e/\delta$, against the normalised wavelength λ_p/λ, of the perturbation for eight different spacings ($V = 8\mu$m/s). The figure shows that the tip is stable ($\delta e/\delta < 0$ for all λ_p/λ) when the spacing is less than 50 μm

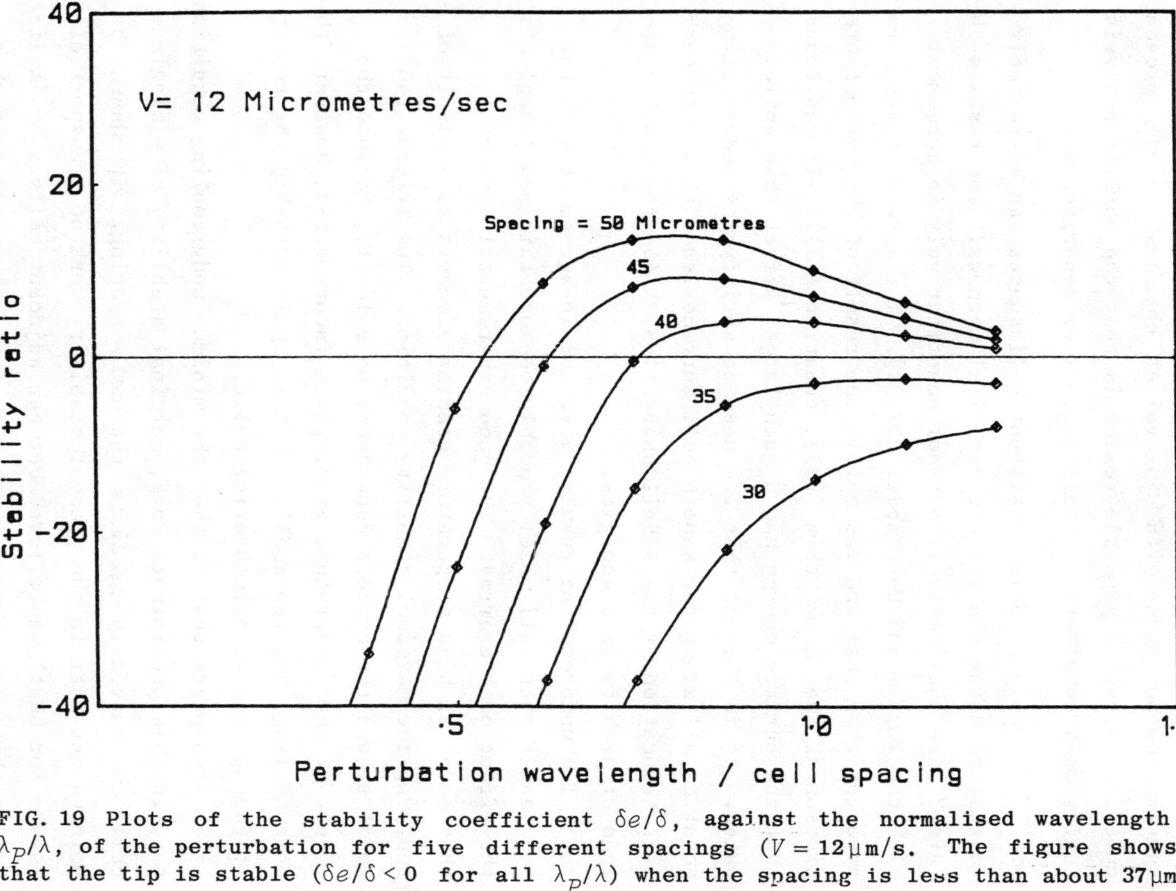

FIG. 19 Plots of the stability coefficient $\delta e/\delta$, against the normalised wavelength λ_p/λ, of the perturbation for five different spacings ($V = 12\,\mu$m/s. The figure shows that the tip is stable ($\delta e/\delta < 0$ for all λ_p/λ) when the spacing is less than about $37\,\mu$m

The marginal stability condition used by other authors [41] predicts much larger spacings than found experimentally; also the spacings do not agree with the values obtained in the present numerical work. A probable reason is that the earlier analysis really only considers an isolated cell or dendrite.

4.6 *Discussion.* Self-consistent cell shapes can be reliably calculated using the present numerical methods. The results obtained were qualitatively but not quantitatively in agreement with the Burden and Hunt model [37,38]. The agreement was more apparent than real and was mainly the result of the correct prediction of the GD_L/V term ([37], equation (17)). It was found that the surface energy had a much larger effect than anticipated. It is thought that the need to satisfy the undercooling equation all along the smooth cell surface leads to a much earlier truncation of the Scheil shape [1-3, 8, 46] for small spacings or high surface energies.

The undercooling results were not in agreement with the Kurz and Fisher [41] model probably because this model neglects the effect of the array. The good experimental agreement of this model at high temperature gradients occurs as a result of applying the marginal stability condition. The Trivedi [40] model showed the correct dependence on velocity, because the effect of the temperature gradient is incorporated, however, the undercoolings are too small. This is again probably the result of the neglect of neighbouring cells.

It appears that neither the minimum undercooling condition nor, in this preliminary work, marginal stability of a single cell tip correctly predicts the cell spacing. It should be pointed out that in the present stability work no transport of solute (or heat) was allowed between adjacent cells. By implication this means that each member of the array was perturbed in an identical way. Clearly this is too limiting a condition. What is necessary is for the macroscopic cellular interface to be perturbed in such a way that transport is allowed between the cells. It seems probable that this will give a small range of stable cell spacings. Practically, [26,34] it is found that there is a range of stable spacings present on a specimen. In some areas the spacing is sufficiently narrow for a cell to be overgrown by the surrounding cells, while in other regions the spacing is decreased by a cell tip splitting mechanism. An attempt is being made to examine the stability of such a multiple cell system.

5. SUMMARY AND CONCLUSIONS

In this paper the distinctly different phenomena of (a) isolated dendritic growth in an undercooled liquid bath and (b) arrayed cellular and dendritic growth in a positive temperature gradient with heat extraction through the solid, have been described. A very brief review of the growth of isolated dendrites has been presented in which both experimental observations and theoretical models have been discussed. Experimental observations on the growth of cellular and dendritic arrays have been briefly outlined, and simple approximate analytical models for such growth have been described.

The free boundary problem of arrayed growth has been detailed together with the deficiencies in the approximate analytical treatments. A numerical finite difference model of arrayed growth, developed by the present authors, has been presented and the results have been briefly compared both with experimental observations and with previous approximate analytical models.

It is concluded from this work that smooth steady state shapes can be calculated using the numerical technique described. Only one cell shape is found for a fixed spacing, velocity, gradient and alloy composition. From the results it is clear that neither a minimum undercooling nor a marginal stability (as applied in this analysis) growth criterion correctly predicts the experimentally observed cellular spacings. It is not unexpected that marginal stability of a single cell in the array fails to correctly predict the cellular spacing. It is probable that one must consider perturbing the array of cells. Such an analysis is currently being undertaken and will be reported in a subsequent publication.

APPENDIX

List of symbols and values used in the calculations unless otherwise stated.

Symbol	Meaning	Value	Units
C_S	solid solute composition		wt%
C_L	liquid solute composition		wt%
C_∞	bulk alloy composition	0.48	wt%
D_i	thermal or solute diffusivity (D_L, D_S, α_L or α_S)		m^2s^{-1}
D_L	solute diffusivity in the liquid	3.5×10^{-9}	m^2s^{-1}
D_S	solute diffusivity in the solid	1×10^{-11}	m^2s^{-1}
e	the temperature error defined in equation (12)		K
e_j	the temperature error for interface point j		K
G	temperature gradient in bulk liquid	9000	Km^{-1}
k	solute distribution coefficient	0.18	
K_L	thermal conductivity in liquid	105	$Wm^{-1}K^{-1}$
K_S	thermal conductivity in solid	180	$Wm^{-1}K^{-1}$
L	Latent heat of fusion per unit volume	10.04×10^8	Jm^{-3}
m	the slope of the liquidus	−5.5	K/wt%
n	distance normal to the interface		m
r	distance from cell centre (see Fig.8)		m
R_1, R_2	the principal radii of curvature		m
T_L	temperature in the liquid		K

Symbol	Meaning	Value	Units
T_0	liquidus temperature for alloy of composition C_∞		K
T_S	temperature in the solid		K
U_i	C_S, C_L, T_L or T_S		
V	steady state velocity in z direction		ms^{-1}
V_n	velocity normal to the interface		ms^{-1}
z	distance along axis of cell		m
α_L	thermal diffusivity in liquid	4.2×10^{-5}	m^2s^{-1}
α_S	thermal diffusivity in solid	7×10^{-5}	m^2s^{-1}
ΔT_i	tip undercooling		K
θ	curvature undercooling constant	2.8×10^{-7}	Km
δ	amplitude of perturbation		m
δe	amplitude of the temperature error		K
λ	cell spacing (see Fig. 8)		m
λ_p	wavelength of the perturbation		m
μ	kinetic coefficient	1×10^{-1}	$ms^{-1}K^{-1}$

REFERENCES

1. M.C. Fleming, *Solidification Processing*, McGraw-Hill, New York, 1974.
2. W. Kurz and D.J. Fisher, "Fundamentals of solidification", *Trans. Tech. Publications*, Switzerland, 1984.
3. M.E. Glicksman, R.J. Schaeffer and J.D. Ayers, *Metall. Trans.* **7A**, 1976, p.1747,
4. S.C. Huang and M.E. Glicksman, *Acta Metall.* **29**, 1981, p.701.
5. M. Chopra, Ph.D. Thesis, Renselaer Polytechnic Institute, Troy, NY, 1983.
6. G.P. Ivantsov, *Dokl. Akad. Nauk SSSR*, **58**, 1947, p.567.
7. M.E. Glicksman and R.J. Schaefer, *J. Cryst. Growth* **1**, 1967, p.297.
8. M.E. Glicksman and R.J. Schaefer, *J. Cryst. Growth* **2**, 1968, p.239.
9. R. Trivedi, *Acta Metall.* **22**, 1970, p.287.
10. G.E. Mash and M.E. Glicksman, *Acta Metall.* **22**, 1974, p.1283.
11. G. Horvay and J.W. Cahn, *Acta Metall.* **9**, 1961, p.695.
12. J.S. Langer and H. Muller-Krumbhaar, *Acta Metall.* **26**, 1979, p.1681.
13. J.S. Langer and H. Muller-Krumbhaar, *Acta Metall.* **26**, 1979, p.1689.
14. J.S. Langer and H. Muller-Krumbhaar, *Acta Metall.* **26**, 1979, p.1697.
15. M.E. Glicksman, *Mat. Sci. and Eng.* **65**, 1984, p.45.
16. W.W. Mullins and R.F. Sekerka, *J. Appl. Phys.*, **34**, 1963, p.323.
17. J.W. Rutter and B. Chalmers, *Can. J. Phys.* **31**, 1953,
18. W.A. Tiller, K.A. Jackson, J.W. Rutter and B. Chalmers, *Acta Metall.* **1**, 1953, p.428,
19. W.W. Mullins and R.F. Sekerka, *J. Appl. Phys.* **35**, 1964, p.444.
20. R. Trivedi and W. Kurz, *Acta Metall*, **34**, 1986, p.1663
21. L.H. Ungar and R.A. Brown, *Phys. Rev.* **B31**, 1985, p.5931.
22. L.H. Ungar, M.J. Bennett and R.A. Brown, *Phys. Rev.* **B31**, 1985, p.5923.
23. G.B. McFadden and S.R. Corriell, *Physica* **12D**, 1984, p.253.
24. K.A. Jackson and J.D. Hunt, *Acta Metall.* **13**, 1965, p.1212.
25. L.R. Morris and W.C. Winegard, *J. Cryst. Growth* **5**, 1969, p.36.
26. D.G. McCartney and J.D. Hunt, *Acta Metall.* **29**, 1981, p.1851.
27. K.A. Jackson and J.D. Hunt, *Acta Metall.* **13**, 1965, p.1212.
28. K. Somboonsuk, J.T. Mason and R. Trivedi, *Metall. Trans.* **15A**, 1984, p.967.
29. M.H. Burden and J.D. Hunt, *J. Cryst. Growth* **22**, 1974, p.99.
30. I. Jin and G.R. Purdy, *J. Cryst. Growth* **23**, 1974, p.37.
31. K.P. Young and D.H. Kirkwood, *Metall. Trans.* **6A**, 1975, p.197.
32. C.M. Klaren, J.D. Verhoeven and R. Trivedi, *Metall. Trans.* **11A**, 1980, p.1853.

33. J.T. Mason, J.D. Verhoeven and R. Trivedi, *J. Cryst. Growth* **59**, 1982, p.516.
34. D.G. McCartney, D. Phil. Thesis, University of Oxford, Oxford, 1981.
35. G.F. Bolling and W.A. Tiller, *J. App. Phys.* **31**, 1960, p.2040.
36. P.K. Rohatgi and C.M. Adams, *Trans. TMS-AIME* **239**, 1967, p.1737.
37. M.H. Burden and J.D. Hunt, *J. Cryst. Growth* **22**, 1974, p.109.
38. J.D. Hunt, *Solidification and Casting of Metals*, The Metal Society, London, 1979, p.3.
39. I. Jin and G.R. Purdy, *J. Cryst. Growth* **23**, 1974, p.29.
40. R. Trivedi, *J. Cryst. Growth* **49** , 1980, p.219.
41. W. Kurz and D.J. Fisher, *Acta. Metall.* **29**, 1981, p.11.
42. V. Laxmann, *Acta Metall.* **33**, 1985, p.1037.
43. R. Trivedi and K. Somboonsuk, *Mat. Sci. and Eng.* **65**, 1984, p.65.
44. D.G. McCartney and J.D. Hunt, *Metall. Trans.* **15A**, 1984, p.983.
45. M. Gunduz and J.D. Hunt, *Acta Metall.* **33**, 1985, p.1651.
46. T.F. Bower, H.D. Brody and M.C. Flemings, *Trans. TMS-AIME*, **236**, 1966, p.624.
47. J.D. Hunt and K.A. Jackson, *Acta Metall.* **13**, 1965, p.1212.

21

ON THE VISCOUS CAHN–HILLIARD EQUATION

A. NOVICK-COHEN

Abstract. Two equations are derived and analyzed which model the dynamics of viscous first order phase transitions. When viscous and gradient energy terms are included, our derivation yields a viscous Cahn-Hilliard equation

$$c_t = \Delta(f(c) + \nu c_t - K\Delta c). \tag{1}$$

Here $c(x,t)$ is a concentration, $f(c)$ is an intrinsic chemical potential, which typically is non-monotone, and ν and K are respectively coefficients of viscosity and gradient energy. In the limit $K \to 0$, equation (1) reduces to a viscous diffusion equation

$$c_t = \Delta(f(c) + \nu c_t) \tag{2}$$

and in the limit $\nu \to 0$, (1) reduces to the well-known Cahn-Hilliard equation [4]

$$c_t = \Delta(f(c) - K\Delta c). \tag{3}$$

Equations (1)-(3) can be viewed as viscous and/or gradient energy regularizations of the backwards-forwards heat equation. Early stages in the evolution of (1) follow closely the behaviour of equation (2) for small K, and in this manner phase separated transients can be seen to develop rapidly. During the late stages, the phase separated transients probably evolve and decay in much the same way as is believed to occur in the Cahn-Hilliard equation, although such a similarity has yet to be proved. The

---

† Research sponsored by the Air Force Office of Scientific Research, Air Force Systems Command, USAF, under Contract/ Grant No. AFSOR-86-0179.

newly derived equations are of lower order than the Cahn-Hilliard equation and to some extent lend themselves to simpler analysis. In what follows, the derivation of equations (1) and (2) is outlined and several results are given. Full details are to appear in Novick-Cohen and Pego [1, 2, 3].

INTRODUCTION

There are many two component systems in which phase separation can be induced by rapid cooling ("quick quenching") the system. Thus, if a two component system of appropriate composition which is spatially uniform at temperature T_1 is rapidly cooled to a second sufficiently lower temperature T_2, then the cooled system will separate itself oût into regions of higher and lower concentration. A rough description of the behaviour of such systems can be obtained by energy minimization arguments. The claim would be that there exists some critical temperature T_c, so that for $T > T_c$ the appropriate quantity, the free energy, is single welled whereas for $T < T_c$ the free energy is double welled. Referring to Fig. 1, a system of average concentration c^* which was spatially uniform at temperature T_1, when cooled to temperature T_2 would find it energetically preferrable to subdivide itself into two systems, one at concentration c_A and one at concentration c_B.

A naive attempt to extrapolate the dynamics of phase separation from the energy minimization principle would lead to the assumption that the mass flux J is proportional to the gradient of $\tilde{\mu}$, the intrinsic chemical potential,

$$J \propto \nabla \tilde{\mu}$$

where

$$\tilde{\mu} = \frac{\partial}{\partial c} F(c, T) (= f_T(c)).$$

For $T > T_c$, $f_T(c)$ is monotone. For $T < T_c$, $f_T(c)$ is nonmonotone, i.e. there is an interval (α, β) such that for $c \in (\alpha, \beta)$, $f_T'(c) < 0$. The interval (α, β) is called the subspinodal region.

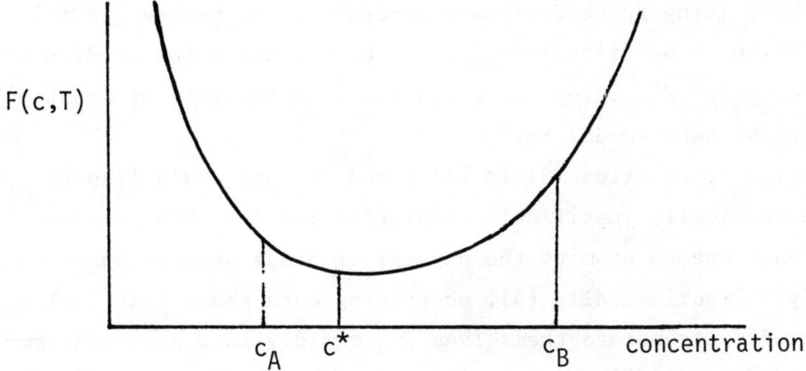

Fig. 1(a). The free energy at some temperature $T_1 > T_c$

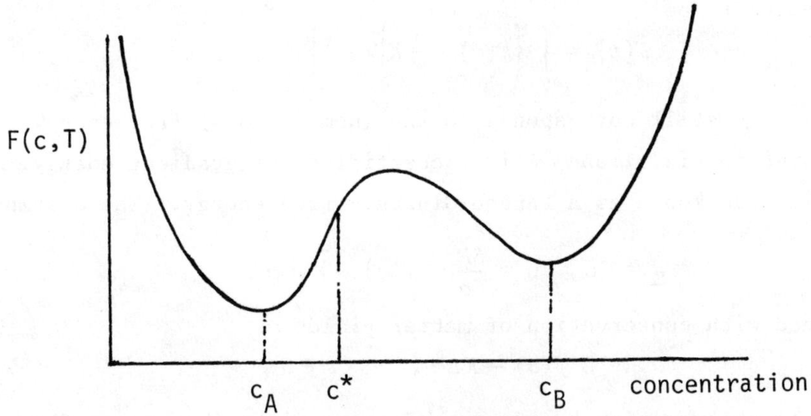

Fig. 1(b). The free energy at some temperature $T_2 < T_c$

Coupling the above assumption with mass conservation yields the backwards-forwards heat equation

$$c_t = \Delta f_T(c). \tag{3}$$

Thus the driving force for phase separation can be seen in this description to be diffusion against the concentration gradient, or "backwards" diffusion which results from the loss of monotonicity in the subspinodal region.

Clearly equation (3) is ill-posed and one would like to find a physically justifiable regularization for this equation. This leads one to examine the physics of phase separation more closely. Experimentally [5], scattering data shows that cooled systems first separate themselves out rapidly into some configuration of alternately high and low concentration. Then, slowly, the system sorts itself out, and larger and larger regions become dominated by a single phase. This points to the importance of gradient energy effects.

Cahn and Hilliard [4] obtained a regularization of equation (3) by including energy contributions in their definition of the free energy. They defined the relevant free energy of the system to be

$$\tilde{F}(t) = \int_V \left\{F(c) + \tfrac{1}{2}K|\nabla c|^2\right\} dV, \tag{4}$$

where $F(c)$ still corresponds to the (homogeneous) free energy portrayed in Fig. 1 and K is a coefficient of gradient energy. $\tilde{F}(t)$ is also known as a Landau-Ginzburg free energy. The assumption

$$\underline{J} \propto \nabla\tilde{\mu}, \quad \tilde{\mu} = \frac{\delta\tilde{F}}{\delta c} = f(c) - K\nabla^2 c,$$

combined with conservation of matter yields

$$c_t = \Delta\, f(c) - K\Delta c, \qquad x \in \Omega. \tag{5}$$

Sensible accompanying boundary conditions might then be no flux

$$\underline{n}\cdot\underline{J} = 0 \quad \text{and} \quad \underline{n}\cdot\nabla c = 0, \qquad x \in \partial\Omega. \tag{6}$$

The physical justification of the first or no flux boundary condition is clearer than the justification of the second or natural

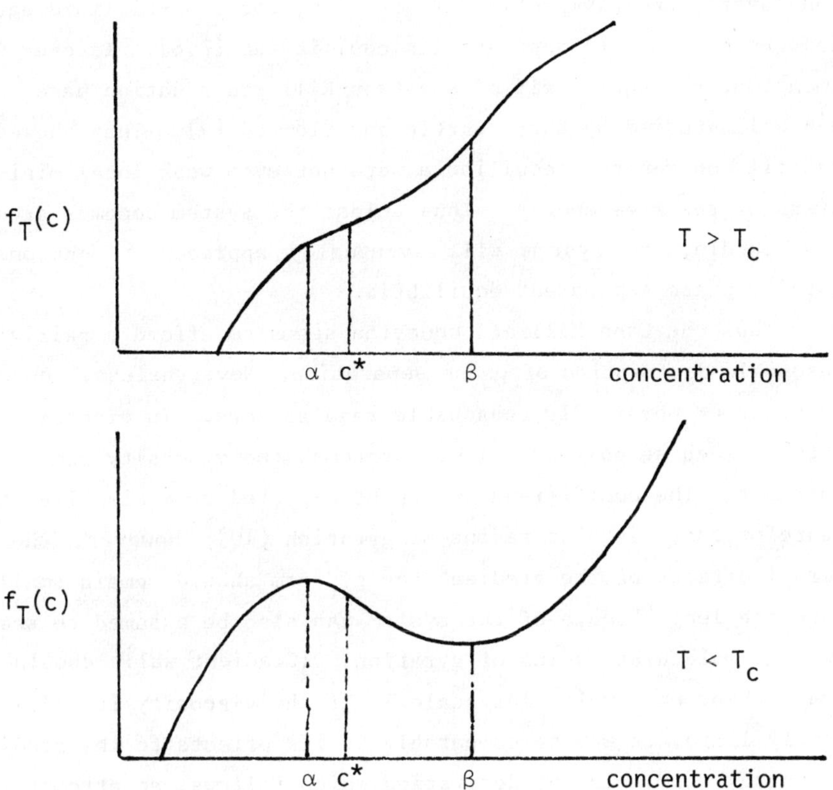

Fig. 2. Typical diagrams for $f_T(c) = (\partial/\partial c)\, F(c, T)$ for $T > T_c$ above, and for $T < T_c$ below. The region $\alpha < c < \beta$ in which $f_T'(c) < 0$ is known as the subspinodal region

boundary condition. Here Ω is taken to be some bounded domain in $\mathbb{R}^n$.

Numerically [1,6], the Cahn-Hilliard equation is known to develop phase separated transients characterized by alternating regions of high and low concentration separated by thin transition layers, and asymptotically as $t\to\infty$, the Cahn-Hilliard equation can be shown to approach its equilibrium [7,8]. In one dimension, the equilibria of the Cahn-Hilliard equation have been well studied by Carr, Gurtin and Slemrod [9]. They showed that all non-monotone equilibria were not even weak local minimizers of the free energy. Thus unless the system becomes stuck at a saddle, the system will eventually approach a monotone "totally phase separated" equilibria.

Thus the Cahn-Hilliard equation seems to afford a fairly reasonable description of phase separation. Nevertheless, there may be other physically reasonable regularizers. In viscous systems, such as polymer-polymer systems, the viscosity can be important. The coefficient K can be expected to scale like the square of the molecular radius of gyration [10]; however, the overall effects of the gradient energy term should remain small since the length scale of the system can also be assumed to scale like the molecular radius of gyration. (Gradient walls should remain thin, on a molecular scale.) If the viscosity is sufficiently large, it may be comparable in its effects to the gradient energy term. In the derivation which follows, an attempt is made to incorporate both viscous and gradient energy effects coherently.

DERIVATION. Our system will be assumed to consist of two chemical species located within some bounded region, $\Omega\subset\mathbb{R}^n$. We shall assume our system to satisfy, additionally, the following hypotheses:

(1) The two chemical species will both be assumed to be long molecules.

(2) Each of the two species will have similar physical properties, but below some critical temperature, there should be a miscibility gap (or double welled region) in the phase diagram relating the two species. A system satisfying this hypothesis could be, for example, polybutadiene and deutonated polybutadiene.

(3) Our system should be viscous in the neighbourhood of the critical temperature. This could reflect a statement about the existence of a glass transition temperature in the proximity of the critical temperature.

(4) The phase transition should take place in a creep flow regime. In particular, most of the viscous dissipation in the system should result from motion generated by the phase separation itself and not from a large amplitude underlying flow field.

A mixture theory approach will be employed in order to describe the dynamics of our system (see [12]). Thus there should be four equations — conservation of mass equations for the two densities, ρ_1 and ρ_2, and two conservation of momentum equations written in terms of the individual velocities $\underline{v}_1$ and $\underline{v}_2$. In order to complete our model, we must make a number of constitutive assumptions. If we write the total stress tensor as

$$t_{ij} = \sum_{\alpha=1}^{2} (t_{ij}^{\alpha} - \rho_\alpha (v_i^\alpha - v_i)(v_j^\alpha - v_j))$$

where $\underline{v} = (\rho_1 \underline{v}_1 + \rho_2 \underline{v}_2)/(\rho_1 + \rho_2)$ is the barycentric velocity, then the individual stress tensors will be assumed to be Newtonian both in the barycentric velocity and in the relative momentum, i.e.

$$t_{ij}^{\alpha} = -p^{\alpha}(\rho, c)\,\delta_{ij} + \tfrac{1}{2}\mu(v_{i,j} + v_{j,i}) + \tfrac{1}{2}\bar{\mu}(J_{i,j}^{\alpha} + J_{j,i}^{\alpha})$$

where $\underline{J}^{\alpha} = \rho\alpha(\underline{v}^{\alpha} - \underline{v})$ and where p^{α} is the partial pressure of component α. Here the notation $v_{i,j} = (\partial/\partial x_j)\, v_i$, etc. is being employed. Note that since $J^1 + J^2 = 0$, t_{ij} will only contain classical Newtonian viscosity. The momentum production vector m_1^1 will be taken as

$$m_i^1 = M(v_i^1 - v_i^2) + (p^1_{,i} - \rho_1 \mu^2_{,i}) ,$$

where $\tilde{\mu}$ is the intrinsic chemical potential and $M(<0)$ is a drag coefficient.

We now reduce our equations to dimensionless form by re-writing them in terms of the following dimensionless quantities $P_\rho = \bar{p}_\rho/\bar{v}^2$, $P_c = \bar{p}_c \bar{c}/\bar{\rho}\bar{v}^2$, $v = \mu/\bar{\rho}\bar{v}$, $\bar{v} = \mu/\bar{\rho}\bar{v}$, $\bar{M} = M\bar{t}/\bar{c}\bar{\rho}$, $U_\rho = \bar{\mu}_\rho \bar{\rho}/\bar{v}^2$ and $U_c = \mu_c \bar{c}/\bar{v}^2$. Here $\bar{p}_\rho = (\partial/\partial\rho)(p^1(\bar{\rho},\bar{c}) + p^2(\bar{\rho},\bar{c}))$, $\bar{\mu}_\rho = (\partial/\partial\rho)\tilde{\mu}(\bar{\rho},\bar{c})$, etc., where $\bar{c}$ and $\bar{\rho}$ are respectively the average concentration and the average density ($c = \rho_1/\rho$ and $\rho = \rho_1 + \rho_2$) and where $\bar{v}$ and $\bar{t}$ are representative velocity and time scales.

Let $\varepsilon = -M^{-1}$ be a small parameter. We incorporate hypothesis (1—4) into our model by making the following scaling hypotheses: $p_c = 0(\varepsilon^2)$, $p_\rho = 0(\varepsilon)$, $\tilde{\mu}_\rho = 0(\varepsilon^2)$, $\tilde{\mu}_c = 0(\varepsilon)$, $\bar{v} = 0(\varepsilon)$, $\bar{t} = 0(1)$, $\bar{\rho} = 0(1)$. This yields the relative size of the dimensionless quantities, $P_\rho = 0(\varepsilon^{-1})$, $P_c = 0(1)$, $U_\rho = 0(1)$, $U_c = 0(\varepsilon^{-1})$ and $\bar{M} = 0(\varepsilon^{-1})$. To lowest order the equations governing our system are

$$\rho_t + (\rho v_j)_{,j} = 0$$

$$\rho(c_t + v_j c_{,j}) + J_{i,i} = 0$$

$$-(p_1 + p_2)_{,j}\delta_{ij} + 2\mu(v_{i,j} \quad v_{j,i})_{,j} = 0$$

$$\rho_1 \tilde{\mu}_{\,i} = \alpha[\tfrac{1}{2}\mu(v_{i,j} + v_{j,i}) + \bar{\mu}(J_{i,j} + J_{j,i})] - \beta[v_i^1 - v_i^2]$$

where $\alpha = (U_c \bar{c})^{-1}\varepsilon$ and $\beta = -(U_c\bar{c})^{-1}\bar{M}\varepsilon^2$.

When our system is one dimensional and when the above equations are written in Lagrangian (first component) coordinates, v_x and J may be eliminated. Two equations remain,

$$c_t = [c(1-c)\,\rho[\rho_1\,\rho(\tilde{\mu}_x) + \alpha\rho(\mu(\rho_t/\rho)) + 2\bar{\mu}(\rho c_t)_x]_x$$

$$p - p_0 = 2\alpha(-\rho_t/\rho) , \qquad (7)$$

where x is actually the Lagrangian variable

$$x' = \int^{x} \rho(s,t)\, ds\,.$$

Note that since to lowest order $p = p(\rho)$, the second equation is uncoupled from the first and allows the particular solution $\rho_t = 0$. When $\rho_t = 0$, the first equation is of the form

$$c_t = [A(c,\rho)\,\tilde{\mu}_x + B(c,\rho)[C(c,\rho)\,c_t]_x]_x \tag{8}$$

where A, B, and C are positive.

Equation (8) motivates the consideration in one dimension of the model equation

$$c_t = \tilde{\mu}_{xx} + \nu c_{txx}$$

or analogously in higher dimensions

$$c_t = \Delta[\tilde{\mu} + \nu c_t].$$

We can now obtain equations (1) and (2) by choosing the chemical potential appropriately. If we take $\tilde{\mu} = f(c)$ we obtain

$$c_t = \Delta[f(c) + \nu c_t] \tag{9}$$

and if we take $\tilde{\mu} = f(c) - K\Delta c$, then

$$c_t = \Delta[f(c) + \nu c_t - K\Delta c]\,. \tag{10}$$

If we consider equations (9) and (10) to be defined on some domain $\Omega \subset \mathbb{R}^n$, $n \leqslant 3$, then plausible boundary conditions for (9) are

$$n \cdot \nabla[f(c) + \nu c_t] = 0 \quad \text{on} \quad \partial\Omega \tag{11}$$

i.e. no flux, and plausible boundary conditions for (10) are

$$n \cdot \nabla[f(c) + \nu c_t - K\Delta c] = 0 \quad \text{and} \quad n \cdot \nabla c = 0 \quad \text{on} \quad \partial\Omega \tag{12}$$

i.e. no flux and natural boundary conditions, where $\Omega \subset \mathbb{R}^n$, $n \leqslant 3$.

Some Rigorous Results. In what follows, let us assume that

(i) $\Omega \subset \mathbb{R}^n$, $n \leqslant 3$ is a bounded domain with a smooth boundary, and that

(ii) $f(c)$ is locally Lipschitz.

(*)

Note that equation (9) is a singular limit of equation (10). The Landau-Ginzburg free energy

$$\tilde{F}(c) = \int_\Omega \{F(c) + \tfrac{1}{2}K|\nabla c|^2\}\, dV$$

acts as a Liapounov functional both for the original Cahn-Hilliard equation and for equation (10). Setting $K=0$, $\tilde{F}(c)$ is also a Liapounov functional for equation (9). However, equations (9) and (10) differ with regard to regularity and asymptotic behaviour.

The invertibility of the operator $\omega = (1-\nu\Delta)f$, $n\cdot\nabla d = 0$, $\omega \in L_\infty$, allows us to rewrite equations (9) and (10) as

$$c_t = (1-\nu\Delta)^{-1}\,\Delta f(c)$$
$$c_t = (1-\nu\Delta)^{-1}\,\Delta\left(f(c) - \frac{K}{\nu}c\right) + \frac{K}{\nu}\,\Delta c\,.$$

Written in this form, (9) can be seen to be a nonlocal ODE and (10) is essentially a reaction-diffusion equation with a nonlocal reaction term. The regularity properties of the two equations behave accordingly.

THEOREM 1. Under assumptions (*), for arbitrary initial data $c_0(x) \in L_\infty$ for any $T>0$, equation (9) has a unique solution $c(x,t)$ with $c(x,t) \in C^1([0,T), L^\infty(\Omega))$ and $c(x,0) = c_0(x)$. If $c_0(\cdot)$ is continuous at x, then $c(\cdot,t)$ is continuous at x. If $c_0(\cdot)$ is not continuous at x, then $c(\cdot,t)$ is not continuous at x.

And likewise,

THEOREM 2. Under assumptions (*), for arbitrary initial data $c_0(x) \in L_\infty$, equation (10) has a unique global classical solution.

The asymptotic behaviour of equation (9) is partially described by the following theorem.

THEOREM 3. In addition to the assumptions (*), let us suppose that $f(c)$ is monotone increasing on $I = \{c < c < \beta\}$. (Here II is the subspinodal region, see Fig. 2.) Let $c(x,t)$ be a solution to equation (9) with $c(x,0) = c_0(x) \in L^\infty$, and let

$$\mu(t)_{\text{unstable}} = \{\text{meas}(x) \mid c(x,t) \in \text{II}\}.$$

Then $\lim_{t\to\infty} \mu(t)_{\text{unstable}}$ exists and either

(i) $c(x,t)$ converges pointwise a.e. to some $c_\infty(x)$ such that $f(c_\infty(x)) = f_\infty$ for some f_∞ and $c_\infty(x) \in \text{I}$, or

(ii) $\lim_{t\to\infty} \mu(t)_{\text{unstable}} > 0$.

Possibility (ii) can be eliminated directly if $f(c)$ is required to satisfy certain additional non-degeneracy requirements [11]. Alternatively, note that possibility (ii) implies that a finite amount of concentration remains in the unstable or subspinodal region. This is dynamically improbable since any such configuration would be linearly unstable. This is further supported by the following principle of phase invariance.

THEOREM 4. Let us assume as in Theorem 3 that the assumptions (*) hold and that $f(c)$ is monotone increasing on $I = \{c<\alpha\} \cap \{c>\beta\}$ and that $f(c)$ is monotone decreasing on $\text{II} = \{\alpha<c<\beta\}$ (see Fig. 2). Let $f_e \in (f(\beta), f(\alpha))$ be arbitrary. Suppose now that $\|c_0\|_{L^\infty} \leqslant M_0$ and that for $\varepsilon_0, \varepsilon_1$ positive and sufficiently small,

$$\text{meas}\{x \in \Omega \mid |c_0(x) - c_e(x)| > \varepsilon_1\} < \varepsilon_0$$

where $c_e(x)$ is any possible steady state satisfying $f(c_e(x)) = f_e$ and $c_e(x) \subset I$. It then follows that

$$\text{if } c_0(x) < \alpha \text{ then } c(x,t) < \alpha \text{ for all } t > 0$$

$$\text{if } c_0(x) > \beta \text{ then } c(x,t) > \beta \text{ for all } t > 0.$$

Thus if the initial data is near some steady state which is a mixture of two phases neither of which lie in the subspinodal region, then the phase does not change at any point.

There are similarities between the asymptotic behaviour as $t \to \infty$ of equation (10) and the asymptotic behaviour of the Cahn-Hilliard equation. By studying the compactness of the ω-limit set in H^1, the solutions of (10) can be shown to converge to some equilibrium solution of (10) as $t \to \infty$. By our previous remarks on the nature of the equilibrium becomes caught at a saddle, it will tend eventually towards a monotone (totally phase separated) equilibrium state. On the other hand, from Theorems 1 and 3 we can see that solutions evolving according to equation (9) are not being driven towards smooth monotone solutions. The following comparison Theorem allows us to compare the early evolution of the two equations.

THEOREM 5. In addition to assumptions (*), let us suppose that $f(c) \in C^3$ and $c_0(x) \in C^4$. If $2 \leqslant n \leqslant 3$, let us assume also that $n \cdot \nabla c_0 = 0$ for $x \in \partial\Omega$. Define $\gamma = K/\nu$. Suppose that $u(x,t)$ is the solution to the equations

$$
\begin{aligned}
u_t &= \Delta[f(u) + \nu u_t] && x \in \Omega \\
\underline{n} \cdot \nabla[f(u) + \nu u_t] &= 0 && x \in \partial\Omega \\
u(x,0) &= c_0(x) && x \in \Omega
\end{aligned}
$$

and suppose that $v(x,t)$ is the solution to the equations

$$
\begin{aligned}
v_t &= \Delta[f(v) + \nu v_t - K\Delta v] && x \in \Omega \\
n \cdot \nabla[f(v) + \nu v_t - K\Delta v] &= 0 && x \in \partial\Omega \\
n \cdot \nabla v &= 0 && x \in \partial\Omega \\
v(x,0) &= c_0(x) && x \in \Omega .
\end{aligned}
$$

Then for any $0 < \nu < \nu_0$ for some ν_0 and for any $T > 0$, there exists a $\bar{\gamma}$ such that for any $0 < \gamma < \bar{\gamma}$ and for any $0 < t < T$

$$\|u - v\|_\infty \leqslant \gamma^{1/2} G_1 \quad \text{if} \quad n = 1$$

$$\|u - v\|_\infty \leqslant \gamma^{1/4} G_2 \quad \text{if} \quad n = 2 \text{ or } n = 3$$

where $G_i = G_i(T, \nu_0)$.

Combined with an understanding of the rate at which equation (9) approaches its discontinuous phase separated equilibria, Theorem 5 could be useful in determining the rate at which phase separated transients are produced in equation (10).

CONCLUSION

A new equation is developed to describe the dynamics of viscous first order phase transitions which take into account viscous stresses arising from the relative fluxes of the two components. The derivation also yields two new regularizations of the backwards-forwards heat equation. Written in inverted form

$$c_t = (1 - \nu\Delta)^{-1}\Delta\left(f(c) - \frac{K}{\nu}c\right) + \frac{K}{\nu}\Delta c$$

the new equation can be seen to be a concentration conservative analogue of the reaction-diffusion equation. By comparing the early evolution of the viscous Cahn-Hilliard equation with the phase separating viscous diffusion equation, and by studying the evolution towards equilibrium at large time, the viscous Cahn-Hilliard equation can be seen to describe a system which form phase separated transients which eventually decay. This can arguably be claimed to also be true of the original Cahn-Hilliard equation — however, for the viscous Cahn-Hilliard equation, there is a particularly simple description of the early evolution (Theorems 3 and 5). At present, there is no comparable description of the early stages of evolution of the Cahn-Hilliard equation. In fact, there may well prove to be significant differences in the early evolution of these two equations. In real applications these differences could be critical, since other external effects (e.g., drying, rinsing, ...) may come into play before the transients have time to slowly evolve and decay.

REFERENCES

1. A. Novick-Cohen and R.L. Pego, "Viscosity and the dynamics of phase decomposition", in preparation.
2. A. Novick-Cohen and R.L. Pego. "On the viscous Cahn-Hilliard equation", in preparation.
3. A. Novick-Cohen and R.L. Pego, "Analysis of a viscous diffusion equation", in preparation.
4. J.W. Cahn and J.E. Hilliard, "Free energy of a nonuniform system. I. Interfacial free energy", *J. Chem. Phys.* **28** (1958) pp.258-367.

5. J.O. Gunton, M. San-Miguel and P.S. Sahni, in *Phase transitions and Critical Phenomena*, Academic Press (1983).
6. C.M. Elliott and D.A. French, "Numerical studies of the Cahn-Hilliard equation for phase separation", Technical Report Series, Center for Applied Mathematics, Purdue University # 29 (1986).
7. B. Nicolaenko and B. Scheurer, "Low dimensional behaviour of the pattern formation of the Cahn-Hilliard equation", *Physica D*, (1985).
8. S. Zheng, "Asymptotic behaviour of the solutions to the Cahn-Hilliard equations", Technical Report Series, Center for Applied Mathematics, Purdue University #25 (1986).
9. J. Carr, M.E. Gurtin and M. Slemrod, "Structural phase transitions on a finite interval", *Arch. Rat. Mech. Anal.* **86** (1984), pp.317-351.
10. P. Debye, "Angular dissymmetry of the critical opalescence in liquid mixtures", *J. Chem. Phys.* **31** (1959), p.680.
11. G. Andrews and J.M. Ball, "Asymptotic behaviour and changes of phase in one-dimensional nonlinear viscoelasticity", *J. Diff. Eq.* **44** (1982), pp.306-341.
12. I. Müller, *Thermodynamics*, Pitman, Boston, 1985.

22

SOME ASYMPTOTIC PROBLEMS OF LINEAR ELASTICITY

O. A. OLEINIK

1. THE LINEAR ELASTICITY SYSTEM

We consider the behaviour at infinity of solutions of the linear elasticity system with finite energy integrals. We first prove some generalizations of the Korn inequality.

Set

$$S_{a,b} = \{x : a < |x| < b\}, \quad S_b = \{x : |x| < b\}, \quad x = (x_1, \cdots, x_n),$$

$$E(u) \equiv \sum_{k,j=1}^{n} \left(\frac{\partial u_j}{\partial x_k} + \frac{\partial u_k}{\partial x_j} \right)^2, \qquad u = (u_1, \cdots, u_n).$$

LEMMA 1. Let $u = (u_1, \cdots, u_n)$, $u_j = u_j(x_1, \cdots, x_n)$ $u_j \in L^2(S_1)$, $\partial u_j / \partial x_k \in L^2(S_1)$, $j, k = 1, \cdots, n$. Then there exist constants $\delta > 0$ and $C > 0$ such that

$$\int_{S_1} \sum_{j=1}^{n} |\operatorname{grad} u_j|^2 \, dx \leq C \left(\int_{S_1} E(u) \, dx + \int_{S_{1-\delta}} |u|^2 \, dx \right), \tag{1}$$

where the constants C and δ do not depend on u.

PROOF. From the classical Korn inequality we get

$$\int_{S_1} \sum_{j=1}^{n} |\operatorname{grad} u_j|^2 \, dx \leq C_1 \left(\int_{S_1} E(u) \, dx + \int_{S_1} |u|^2 \, dx \right), \tag{2}$$

where the constant C_1 does not depend on u. We first prove that for any $\varepsilon = \text{constant.} > 0$ there exist constants $\delta(\varepsilon) \in (0,1)$ and $C > 0$ such that

$$\int_{S_1} |u|^2\, dx \leq \varepsilon \int_{S_1} \sum_{j=1}^{n} |\operatorname{grad} u_j|^2\, dx + C \int_{S_{1-\delta(\varepsilon)}} |u|^2\, dx\,. \tag{3}$$

We can now suppose that $u_j \in C^\infty(\mathbb{R}^n)$. Let δ be an arbitrary positive number, $\delta \leq \frac{1}{4}$, and let $(\tau, \omega) = (\tau, \omega_1, \cdots, \omega_{n-1})$ be polar coordinates in $\mathbb{R}^n$, $\tau = |x|$, $0 \leq \eta \leq \frac{1}{2}$. Then for $1-\eta \leq \tau \leq 1$ we get:

$$|u_j(\tau,\omega)|^2 \leq 2\,|u_j(1-\eta,\omega)|^2 + 2\eta \int_{\frac{1}{2}}^{1} \left|\frac{\partial u_j}{\partial \tau}\right|^2 d\tau\,. \tag{4}$$

According to the mean value theorem there is a η^* such that

$$\int_{|x|=1-\eta^*} |u_j|^2\, d\sigma = \frac{1}{\delta} \int_{S_{1-2\delta,1-\delta}} |u_j|^2\, dx\,, \quad \delta \leq \eta^* \leq 2\delta\,. \tag{5}$$

Setting $\eta = \eta^*$ in (4) and integrating the resulting inequality over the sphere σ_1 with radius 1 and centre at the origin, we find that

$$\int_{\sigma_1} |u_j(\tau,\omega)|^2\, d\sigma_1 \leq 2 \int_{\sigma_1} |u_j(1-\eta^*,\omega)|^2\, d\sigma_1 + 4\delta 2^{n-1} \int_{S_{\frac{1}{2},1}} |\operatorname{grad} u_j|^2\, \partial x\,. \tag{6}$$

Using (5) and integrating (6) with respect to τ from $1-\eta^*$ to 1 we get

$$\int_{S_{1-\eta^*,1}} |u_j|^2\, dx \leq 2\delta^{-1}(1-\eta^*)^{1-n}\eta^* \int_{S_{1-2\delta,1-\delta}} |u_j|^2\, dx + 4\delta 2^{n-1}\eta^* \int_{S_1} |\operatorname{grad} u_j|^2\, dx\,. \tag{7}$$

Since $1-\eta^* > \frac{1}{2}$, it follows from (7) that

$$\int_{S_{1-\delta,1}} |u_j|^2\, dx \leq 2^{n+1} \int_{S_{1-\delta}} |u_j|^2\, dx + \delta 2^{n+1} \int_{S_1} |\operatorname{grad} u_j|^2\, dx\,. \tag{8}$$

If we take $2^{n+1}\delta = \varepsilon$, we get from (8) that

$$\int_{S_1} |u_j|^2\, dx \leqslant 2^{n+1} \int_{S_{1-\delta}} |u_j|^2\, dx + \varepsilon \int_{S_1} |\operatorname{grad} u_j|^2 dx, \quad j = 1,\cdots,n.$$

Summing up these inequalities with respect to j from 1 to n, we obtain (3). From (3) and (2), taking $\varepsilon = \frac{1}{2}$, we find for some $\delta \in (0,1)$ that (1) is valid. Lemma 1 is proved.

LEMMA 2. Let $u = (u_1,\cdots,u_n)$, $u_j = u_j(x_1,\cdots,x_n)$, $u_j \in L^2_{\text{loc}}(\mathbb{R}^n)$, $\partial u_j/\partial x_k \in L^2_{\text{loc}}(\mathbb{R}^n)$, $k,j = 1,\cdots,n$,

$$\int_{|x|<\lambda} \sum_{k,j=1}^{n} \left(\frac{\partial u_j}{\partial x_k} + \frac{\partial u_k}{\partial x_j}\right)^2 dx \leqslant C_1(1+\lambda)^{m_1},$$

for any $\lambda > 0$, where the constants C_1, m_1 do not depend on λ. Then there exist constants C_2, m_2 such that for any $\lambda > 0$

$$\int_{|x|<\lambda} \sum_{j=1}^{n} |\operatorname{grad} u_j|^2\, dx \leqslant C_2(1+\lambda)^{m_2},$$

where C_2 does not depend on λ; m_2 depends on m_1 and n only.

PROOF. Let $\bar{u} = (\bar{u}_1,\cdots,\bar{u}_n)$, $\bar{u}_j$ the mean value of u_j over $S_{1-\delta}$, δ the constant from Lemma 1. Then according to Lemma 1 we have

$$\int_{S_1} \sum_{j=1}^{n} |\operatorname{grad} u_j|^2 \leqslant C_3\left(\int_{S_1} E(u)\, dx + \int_{S_{1-\delta}} |u-\bar{u}|^2\, dx\right).$$

Applying the Poincaré inequality to estimate the last integral we get

$$\int_{S_1} \sum_{j=1}^{n} |\operatorname{grad} u_j|^2\, dx \leqslant C_4\left(\int_{S_1} E(u)\, dx + \int_{S_{1-\delta}} \sum_{j=1}^{n} |\operatorname{grad} u_j|^2\, dx\right)$$

$$C_4, C_3 = \text{const}.$$

After the change of variables $x' = \lambda x$, $\lambda = \text{const} > 1$, one can get from the last inequality

$$\int_{S_\lambda} \sum_{j=1}^{n} |\operatorname{grad} u_j|^2\, dx \leqslant C_4\left(\int_{S_\lambda} E(u)\, dx + \int_{S_{\lambda(1-\delta)}} \sum_{j=1}^{n} |\operatorname{grad} u_j|^2\, dx\right)$$

with $C_4 > 1$. Applying this inequality successively $k-1$ times to estimate the last integral of its right hand side, we obtain

$$\int_{S_\lambda} \sum_{j=1}^{n} |\operatorname{grad} u_j|^2 \, dx \leqslant C_4^k k \int_{S_\lambda} E(u)\, dx + C_4^k \int_{S_{\lambda(1-\delta)^k}} \sum_{j=1}^{n} |\operatorname{grad} u_j|^2 \, dx \tag{9}$$

Let us choose k so large that $\lambda(1-\delta)^k < 1$. Set $k = [a \ln \lambda] + 1$, where $a = \text{const.} > 0$, $[\mu]$ is the integer part of μ, and

$$1 + a \ln (1-\delta) < 0 .$$

Then $\lambda(1-\delta)^k < 1$ for $\lambda > 1$ and it follows from (9) that

$$\int_{S_\lambda} \sum_{j=1}^{n} |\operatorname{grad} u_j|^2 \, dx \leqslant C_5 (1+\lambda)^{m_2} ,$$

where the constants C_6, m_2 do not depend on λ. Lemma 2 is proved.

THEOREM 1. Assume that $u(x) = (u_1(x), \cdots, u_n(x))$ is a solution of the linear elasticity system with constant coefficients

$$\sum_{i,j,k=1}^{n} a_{hk}^{ji} = \frac{\partial^2 u_k}{\partial x_i \partial x_j} = f_h , \qquad h = 1, \cdots, n , \tag{10}$$

where a_{kh}^{ij} are constants,

$$a_{kh}^{ij} = a_{hk}^{ji} = a_{kj}^{ih} ,$$

$$\chi_1 \sum_{k,i=1}^{n} |\eta_k^i|^2 \leqslant \sum_{i,j,k,h=1}^{n} a_{kh}^{ij} \eta_k^i \eta_h^j \leqslant \chi_2 \sum_{k,i=1}^{n} |\eta_k^i|^2 ,$$

χ_1, χ_2 are positive constants, $f_h \in C^\infty(\mathbb{R}^n)$, $f_h \equiv 0$ for $|x| > l$, $h = 1, \cdots, n$; $\{\eta_k^i\}$ is any symmetric matrix, and also

$$\int_{\mathbb{R}^n} E(u)\, dx \equiv \int_{\mathbb{R}^n} \sum_{k,j=1}^{n} \left(\frac{\partial u_k}{\partial x_j} + \frac{\partial u_j}{\partial x_k}\right)^2 dx < \infty .$$

Then for $|x| > 1$

$$u_j(x) = \sum_{k=1}^{n} C_{jk} x_k + C_j$$

$$+ \sum_{\omega\leqslant|\alpha|\leqslant h} \sum_{l=1}^{n} \mathcal{D}^{\alpha}\Gamma_{lj}(x)\, C_{\alpha}^{l} + u_{jh}(x)\,, \tag{11}$$

where $\|C_{jk}\|$ is a skew-symmetric matrix, C_{jk}, C_j are constants, $j,k=1,\cdots,n$; $\omega=0$ for $n>2$ and $\omega=1$ for $n=2$; $\Gamma(x)=\|\Gamma_{lj}(x)\|$ is a fundamental matrix of solutions of system (10) such that

$$\Gamma(x) = p_0 \ln|x| + \Phi_1(x)\,,$$

if $n=2$, and

$$\Gamma(x) = \Phi_2(x)|x|^{2-n}\,,$$

if $n\geqslant 3$. Here p_0 is an $n\times n$-matrix, its elements are constants, the elements of the matrix $\Phi_1(x)$ and the matrix $\Phi_2(x)$ are homogeneous functions of order zero, C_α are constant vectors, $\alpha=(\alpha_1,\cdots,\alpha_n)$, $|\alpha|=\alpha_1+\cdots+\alpha_n$, $\mathcal{D}^{\alpha} = \partial^{|\alpha|}/\partial x_1^{\alpha_1}\cdots\partial x_n^{\alpha_n}$,

$$|\mathcal{D}^{\beta} u_{jh}(x)| \leqslant M_{\beta h}|x|^{2-n-|\beta|-h-1} \qquad \text{for} \quad |x|>1\,,$$

for any multi-index β and h, and $M_{\beta h}$ are constants.

The proof of Theorem 1 is based on Lemma 2 and the following theorems for elliptic systems with constant coefficients.

THEOREM 2. Consider an elliptic system of the form

$$\sum_{k=1}^{N} B_{kj}(\mathcal{D}_x)\, u_k = f_j\,, \quad j=1,\cdots,N\,, \quad x=(x_1,\cdots,x_n), \tag{12}$$

where $B_{kj}(\mathcal{D}_x)$ are homogeneous differential operators of order $2m$ with complex constant coefficients,

$$K_1|\xi|^{2mN} \leqslant \left|\det\|B_{kj}(\xi)\|\right| \leqslant K_2|\xi|^{2mN}$$

for any $\xi\in\mathbb{R}^n$, K_1, K_2 are positive constants. Assume that $f_j\in C_0^\infty(\mathbb{R}^n)$ and that $u=(u_1,\cdots,u_N)$ is a solution of the system (12) in $\mathbb{R}^n$ such that

$$|u(x)| \leqslant C(1+|x|)^q\,,$$

where $C=\text{const.}>0$ and q is a constant, $q\geqslant 0$. Then for $|x|>1$

$$u(x) = P_q(x) + \sum_{h_1\leqslant|\alpha|\leqslant h} \mathcal{D}^{\alpha}\Gamma(x)\, C_\alpha + u^h(x)\,,$$

where $C_\alpha = (C_1^\alpha, \cdots, C_N^\alpha)$, $C_j^\alpha = \text{const}$, $j = 1, \cdots, N$, $h_1 = 2m - n - q$, $\Gamma(x)$ is a fundamental matrix of solutions of system (12) such that

$$\Gamma(x) = p(x) \ln |x| + \Phi_1(x)|x|^{2m-n} ,$$

if $2m-n \geqslant 0$ and n is even, and

$$\Gamma(x) = \Phi_2(x)|x|^{2m-n} ,$$

if $2m-n < 0$ or n is odd, (see [1],[2]); $p(x)$ is a matrix its elements being polynomials of order $2m-n$, and the elements of the matrix $\Phi_1(x)$ and the matrix $\Phi_2(x)$ are homogeneous functions of order zero,

$$|\mathcal{D}^\beta u^h(x)| \leqslant M_{\beta h}(x)^{2m-n-h-1-|\beta|} , \quad M_{\beta h} = \text{const},$$

for $|x| > 1$ and for any multi-index β; $h \geqslant 0$ if n is odd or $2m-n < 0$; $h \geqslant 2m-n$ if $2m-n \geqslant 0$ and n is even;

$$P_q(x) = (P_1^q(x), \cdots, P_N^q(x)),$$

where $P_j^q(x)$ is a polynomial of order less than or equal to q.

THEOREM 3. Assume that $f_j \in C_0^\infty(\mathbb{R}^n)$, $j = 1, \cdots, N$,

$$u(x) = (u_1(x), \cdots, u_N(x))$$

is a solution of system (12) in $\mathbb{R}^n$ and

$$\int_{\mathbb{R}^n} \sum_{j=1}^{N} \sum_{|\alpha|=m} (1 + |x|)^{2\sigma} |\mathcal{D}^\alpha u_j|^2 \, dx < \infty$$

with $\sigma = \text{const.} \leqslant 0$. Then for $|x| > 1$

$$u(x) = P(x) + \sum_{\omega \leqslant |\alpha| \leqslant h} \mathcal{D}^\alpha \Gamma(x) C_\alpha + u^h(x) ,$$

where $P(x) = (P_1(x), \cdots, P_N(x))$, $P_j(x)$ is a polynomial of order less than $m_0 = \max\ (m, m - n/2 - \sigma)$, $j = 1, \cdots, N$; $C_\alpha = (C_1^\alpha, \cdots, C_N^\alpha)$, $C_j^\alpha = \text{const}$, $\Gamma(x)$ is the fundamental matrix of solutions of the system (12), described in Theorem 2, $\omega = m - n/2 + \sigma$, h is any integer if n is odd or $2m-n<0$, and $h \geqslant 2m-n$ if $2m-n \geqslant 0$, n is even,

$$|\mathcal{D}^p u^h(x)| \leqslant M_{\beta h}|x|^{2m-n-h-1-|\beta|}, \quad M_{\beta h} = \text{const},$$

for any multi-index β and $|x|>1$.

The proofs of Theorems 2 and 3 are given in [3].

Thus we have the asymptotic expansions in the neighbourhood of infinity of solutions of the linear elasticity system (10) when one of the following conditions is satisfied

(1) $$|u(x)| \leqslant C(1+|x|)^q, \quad C, q = \text{const}, \quad q \geqslant 0;$$

(2) $$\int_{\mathbb{R}^n} (1+|x|)^{2\sigma} \sum_{|\alpha|=1} |\mathcal{D}^\alpha u|^2\, dx < \infty, \quad \sigma = \text{const}, \quad \sigma \leqslant 0;$$

(3) $$\int_{\mathbb{R}^n} E(u)\, dx < \infty;$$

(4) $$\int_{\mathbb{R}^n} (1+|x|)^{2\sigma} E(u)\, dx < \infty, \quad \sigma = \text{const}, \quad \sigma \leqslant 0;$$

(5) $$\int_{|x|<\lambda} E(u)\, dx \leqslant C_1 (1+\lambda)^{m_1}, \quad C_1, m_1 = \text{const}, \; m_1 \geqslant 0, \quad \lambda = \text{const},$$

for any λ.

Let us now consider the linear elasticity system with variable coefficients

$$\sum_{k,i,j=1}^{n} \frac{\partial}{\partial x_i}\left(a_{kh}^{ij}(x)\frac{\partial u_k}{\partial x_j}\right) = f_h, \quad h = 1, \cdots, n, \tag{13}$$

where for any $x \in \mathbb{R}^n$

$$a_{kh}^{ij}(x) = a_{hk}^{ji}(x) = a_{kj}^{ih}(x),$$

$$\chi_1 \sum_{k,j=1}^{n} |\eta_k^j|^2 \leqslant \sum_{i,j,k=1}^{n} a_{kh}^{ij}(x)\, \eta_k^i \eta_h^j \leqslant \chi_2 \sum_{k,j=1}^{n} |\eta_k^j|^2,$$

χ_1, χ_2 are positive constants, $\|\eta_k^j\|$ is any symmetric matrix, $a_{kh}^{ij}, f_h \in C^\infty(\mathbb{R}^n)$, $f_h = 0$ for $|x| > l$, $l = \text{const}$, $h = 1, \cdots, n$. The following generalization of the Korn inequality is used to study the behaviour at infinity of solutions of the system (13).

LEMMA 3. Let $u = (u_1, \cdots, u_n)$, $u_j \in L^2_{\text{loc}}(\mathbb{R}^n)$,

$$\frac{\partial u_j}{\partial x_k} \in L^2_{\text{loc}}(\mathbb{R}^n), \quad k, j = 1, \cdots, n,$$

$$\int_{\mathbb{R}^n} E(u)\, dx < \infty .$$

Then

$$u(x) = P(x) + u^*(x),$$

where $P(x) = Cx + C_0$, C is a skew-symmetric $n \times n$-matrix,

$$C = \|C_{kj}\|, \quad C_{kj} = \text{const}, \quad C_0 = (C_0^1, \cdots, C_0^n),$$

$$C_0^j = \text{const}, \quad k, j = 1, \cdots, n,$$

$$\int_{\mathbb{R}^n} |\text{grad}\, u^*|\, dx \leqslant C_1 \int_{\mathbb{R}^n} E(u)\, dx ,$$

and the constant C_1 does not depend on u.

The proof of Lemma 3 is based on Lemma 2.

THEOREM 4. Let $u = (u_1, \cdots, u_n)$ be a solution of the system (13), $f_k \in C_0^\infty(\mathbb{R}^n)$, and every coefficient $a_{kh}^{ij}(x)$ of the system (13) have a limit as $|x| \to \infty$, $n > 2$. Assume that

$$\int_{\mathbb{R}^n} E(u)\, dx < \infty .$$

Then

$$u(x) = Cx + C_0 + u^* ,$$

where $C = \|C_{kj}\|$ is a skew-symmetric matrix, $C_{kj} = \text{const}$, C_0 is a constant vector,

$$|u^*(x)| = C_1 |x|^{2-n+\varepsilon}, \quad |x| > 1,$$

for any $\varepsilon > 0$, and the constant C_1 depends on ε.

For the proof of Theorem 4, Lemma 3 is used.

Problems connected with the asymptotic behaviour at infinity of solutions of elliptic equations and systems form the subject of numerous papers. In the case of second order elliptic equations the behaviour of solutions at infinity was studied in [4] and [5]. An asymptotic expansion at infinity for a class of solutions of elliptic systems with constant coefficients is given in [6]. This problem is considered also for some elliptic systems in [7], for polyharmonic equations in [8], (see also [9]).

2. THE BIHARMONIC EQUATION

Let us consider the Dirichlet problem for the equation

$$\Delta\Delta u = f + \sum_{j=1}^{2} \frac{\partial f_j}{\partial x_j} + \sum_{ij=1}^{2} \frac{\partial^2 f_{ij}}{\partial x_i\, \partial x_j} \tag{14}$$

in a bounded domain $\Omega \subset R^2$ with the boundary conditions

$$u\Big|_{\partial\Omega} = 0\,, \quad \operatorname{grad} u\Big|_{\partial\Omega} = 0\,, \tag{15}$$

where $f \in L^s(\Omega)$, $f_j \in L^p(\Omega)$, $j = 1,2$, $f_{ij} \in L^q(\Omega)$, $i,j = 1,2$, $s \geqslant 1$, $p > 1$, $q > 2$.

We denote by $H_0^2(\Omega)$ the completion of functions

$$u(x) \in C_0^\infty(\Omega)$$

with respect to the norm

$$\|u\|_2 = \Big(\int_\Omega \sum_{|\alpha| \leqslant 2} |D_u^\alpha|^2\, dx\Big)^{\frac{1}{2}}.$$

The function $u(x)$ is called a weak solution of the problem (14), (15) if $u \in H_0^2(\Omega)$ and for any $v \in H_0^2(\Omega)$ the following integral identity holds

$$\int_\Omega E(u,v)\, dx = \int_\Omega fv\, dx - \int_\Omega \sum_{j=1}^{2} f_j \frac{\partial v}{\partial x_j} dx + {} $$

$$+ \sum_{i,j=1}^{2} \int_\Omega f_{ij} \frac{\partial^2 v}{\partial x_i\, \partial x_j}\, dx\,,$$

where

$$E(u,v) \equiv \sum_{i,j=1}^{n} \frac{\partial^2 u}{\partial x_i \, \partial x_j} \frac{\partial^2 v}{\partial x_i \, \partial x_j}$$

Assume that 0 is the origin and $0 \in \partial\Omega$, $\partial\Omega$ denoting the boundary of Ω. Set

$$\sigma_t = \partial\Omega \cap \{x : |x| = t\}, \quad S_t = \Omega \cap \{x : |x| = t\}, \quad \Omega_t = \Omega \cap \{x : |x| < t\}.$$

Let ω be a constant such that $\pi \leqslant \omega \leqslant 2\pi$. Consider the equation for δ:

$$\sin^2(\omega\delta) = \delta^2 \sin^2 \omega, \tag{16}$$

where ω is a parameter. It is easy to see that equation (16) has a unique solution $\delta(\omega)$ for $\pi \leqslant \omega \leqslant 2\pi$ such that $0 < \omega\delta(\omega) \leqslant \pi$. One can easily see that $\delta(2\pi) = \frac{1}{2}$. We will suppose that $1.24\,\pi \leqslant \omega \leqslant 2\pi$ such that $0 \leqslant \omega\delta(\omega) \leqslant \pi$. In this case $\frac{1}{2} \leqslant \delta(\omega) \leqslant \frac{3}{4}$.

THEOREM 5. Let $1.24\,\pi \leqslant \omega \leqslant 2\pi$ and suppose that the length of every arc which belongs to S_t does not exceed ωt, and that σ_t is not empty if S_t is not empty. Then for a weak solution $u(x)$ of the problems (14), (15) the following inequalities hold

$$\int_{\Omega_t} E(u,u)\,dx \leqslant C\,t^{2\delta(\omega)} \Big[\|f\|^2_{L^s(\Omega)} + \sum_{j=1}^{2} \|f_j\|^2_{L^p(\Omega)} + \sum_{i,j=1}^{2} \|f_{ij}\|^2_{L^q(\Omega)} \Big], \tag{17}$$

$$|u(x)|^2 \leqslant C\,|x|^{2+2\delta(\omega)} \Big[\|f\|^2_{L^s(\Omega)} + \sum_{j=1}^{2} \|f_j\|^2_{L^p(\Omega)} + \sum_{i,j=1}^{2} \|f_{ij}\|^2_{L^q(\Omega)} \Big], \tag{18}$$

where $\delta(\omega)$ satisfies equation (16), $0 < \omega\delta(\omega) \leqslant \pi$, the constant C depends on ω, p, q, s and the diameter of the domain Ω, and $s \geqslant 1$, $p > 2/(2-\delta(\omega))$, $q > 2/(1-\delta(\omega))$.

THEOREM 6. Assume that the conditions of Theorem 5 are valid and that in addition for any $x^0 \in \partial\Omega$ with $|x^0| \leqslant \rho_1$, $\rho_1 = \mathrm{const} > 0$, the circle $|x - x^0| = \rho$ for $\rho \leqslant |x^0|/2$ has a nonempty intersection with $\partial\Omega$. Then for $|x| < \rho_1/2$ the inequality

$$|\operatorname{grad} u(x)|^2 \leqslant C|x|^{2\delta(\omega)} \left\{ \|f\|^2_{L^s(\Omega)} + \sum_{j=1}^{2} \|f_i\|^2_{L^p(\Omega)} + \sum_{i,j=1}^{2} \|f_{ij}\|^2_{L^q(\Omega)} \right\}, \quad (19)$$

holds, where the constant C depends on ω, s, p, q and the diameter of Ω only.

THEOREM 7. Assume that for every $x^0 \in \partial\Omega$ the circle $|x - x^0| = \rho$ for $\rho \leqslant t_0$, $t_0 = \mathrm{const} > 0$, has a nonempty intersection with $\partial\Omega$ and the length of every arc which belongs to $\Omega \cap \{x : |x - x^0| = \rho\}$ does not exceed $\omega\rho$. Let

$$f \in L^s(\Omega), \quad f_j \in L^p(\Omega), \quad j = 1, 2, \quad f_{ij} \in L^q(\Omega),$$

$$i, j = 1, 2; \quad s \geqslant 1, \quad \rho > 2/(2 - \delta(\omega)), \quad q > 2/(1 - \delta(\omega)).$$

Then the solution of the problem (14),(15) belongs to the Hölder class $C^{1+\delta(\omega)}(\Omega)$.

Theorems 5-7 are proved in [10],[11].

The estimates (17)-(19) are best possible. That is the constant $\delta(\omega)$ can not be replaced by $\delta(\omega) + \varepsilon$ with $\varepsilon = \mathrm{const} > 0$. This is proved in [10],[11] for $1.24\pi \leqslant \omega \leqslant 2\pi$. For $\omega \leqslant 1.24\pi$ it is an open problem to find best possible estimates of the form (17)-(19). Some estimates for this case are obtained in [12].

In the particular case $\omega = 2\pi$ from Theorems 5 and 6, we have

$$|u(x)| \leqslant C_1 |x|^{\frac{3}{2}}, \quad |\operatorname{grad} u(x)| \leqslant C_2 |x|^{\frac{1}{2}}, \quad C_1\, C_2 = \mathrm{const},$$

under only one condition, that the intersection of the circle $|x - x^0| = t$ with $\partial\Omega$ is not empty for $|x^0| < t_0$ and $t < |x^0|/2$,

$t_0 = \text{const} \geqslant 0$.

For physical problems it is important to estimate second derivatives of the solution $u(x)$ of the problem (14),(15) in a neighbourhood of an irregular point of $\partial\Omega$.

THEOREM 8. Let 0 be the origin, $0 \in \partial\Omega$. Suppose that the following conditions are satisfied:

(1) the intersection of $\partial\Omega$ with the circle $|x| = t$ for $t \leqslant T$, $T = \text{const} > 0$, is not empty;

(2) there exists $\beta = \text{const} > 0$, $\beta < 1$, such that for any $x^0 \in \partial\Omega$ and $|x^0| < \frac{1}{2}T$, $x^0 \neq 0$, the intersection of $\partial\Omega$ with the disk $|x - x^0| < \beta|x^0|$ contains a curve S_{x^0} whose end-points belong to the boundary of the disk, $x^0 \in S_{x^0}$; the curve S_{x^0} has the form

$$x_1 = \phi_2(x_2) \quad \text{or} \quad x_2 = \phi_1(x_1),$$

where

$$|\phi'_j(x_j)| \leqslant C_1, \quad |\phi''_j(x_j)| \leqslant C_2|x^0|^{-1}, \quad |\phi'''_j(x_j)| \leqslant C_3|x^0|^{-2}, \qquad j = 1,2,$$

and the constants C_1, C_2, C_3 do not depend on x^0; there are no other points of $\partial\Omega$ in a neighbourhood of S_{x^0} in the disk $|x - x^0| < \beta|x^0|$ and at least one of the two domains bounded by S_{x^0} and by a part of the circle $|x - x^0| = \beta|x^0|$ belongs to Ω.

Then there exists a constant C_4, which does not depend on u, f_1, f_2 and such that

$$\left|\frac{\partial^2 u(x)}{\partial x_i\, \partial x_j}\right| \leqslant C_4|x|^{-\frac{1}{2}}\left(\int_\Omega \sum_{j=1}^{2} |f_j|^p \, dx\right)^{1/p}, \quad i,j = 1,2, \tag{20}$$

for $|x| < \frac{1}{4}T$, $x \in \Omega$, $x \neq 0$. Estimate (20) is the best possible. The function $u(x)$ is a weak solution of the equation

$$\Delta\Delta u = \sum_{j=1}^{2} \frac{\partial f_j}{\partial x_j}$$

in Ω with the boundary conditions (15) and $f_j \in L^p(\Omega)$, $p>2$.

The proof of Theorem 8 is based on the Theorem 5 and *a priori* estimates for solutions of elliptic equations given in [13] and [14].

Theorems similar to Theorems 5–7 are also proved for solutions of the Dirichlet problem for the Karman system and for the Navier-Stokes system in non-smooth two-dimensional domains (see [11]).

A precise Saint-Venant principle can be proved for solutions of the problems (14),(15).

Assume that the origin 0 belongs to $\partial\Omega$ and any circle $|x|=t$ for $t \leqslant T$ has non-empty intersection with $\partial\Omega$ and the length of every arc which belongs to $S_t = \Omega \cap \{x: |x|=t\}$ does not exceed ωt; $f = f_j = f_{ij} = 0$ in $\Omega_T = \Omega \cap \{x: |x| < T\}$, $T = \text{const} > 0$. Then for the weak solution $u(x)$ of the problems (14),(15) we have

$$\int_{\Omega_t} E(u,u)\,dx \leqslant C\,\frac{t^{2\delta(\omega)}}{T^{2\delta(\omega)}} \int_{\Omega_T} E(u,u)\,dx\,, \tag{21}$$

where $1.24\,\pi \leqslant \omega \leqslant \pi$, $\delta(\omega)$ is a solution of the equation (16), $0 < \omega\delta(\omega) \leqslant \pi$, and the constant C does not depend on t, T, u. One can not replace $2\delta(\omega)$ by $2\delta(\omega)+\varepsilon$ in the inequality (21), $\varepsilon = \text{const} > 0$. In this sense the Saint-Venant type estimate (21) is precise.

The estimate (21) can be proved by the method used in Theorem 5 of [10].

Singular boundary points of a different type for the biharmonic equation are studied in [15],[16]. Estimates for the derivatives of any order near a singular point of the boundary for solutions of the elasticity system are given in [17]. The proof of the Saint-Venant's principle can be found in [18].

REFERENCES

1. J.B. Lopatinsky, "Fundamental solutions of a system of differential equations of elliptic type", *Ukr. Mat. Journal* **3** (1), (1951), pp.3-38.
2. L. Bers, F. John and M. Schechter, *Partial Differential Equations*, Interscience, New York, 1964.
3. V.A. Kondratiev and O.A. Oleinik, "On the behaviour at infinity of solutions of elliptic systems with finite energy integral", *Archive Rat. Mech. and Analysis*, (to appear).
4. J. Serrin and H. Weinberger, "Isolated singularities of linear elliptic equations", *Amer. Math. Journal* **88** (1), (1966), pp.258-272.
5. V.A. Kondratiev and O.A. Oleinik, "Sur un probleme de E. Sanchez-Palencia", *C.R. Acad. Sci. Paris* **299** (15), ser.1, (1984), pp.745-748.
6. B.R. Weinberg, "On solutions of elliptic equations with constant coefficients and a right hand side growing at the infinity", Vestnik of the Moscow University, *Math. Mach.*, ser. 1, (1), (1968), pp.41-48.
7. J.B. Lopatinsky, "The behaviour at infinity of solutions of systems of differential equations of the elliptic type", *Dokl. A.N. Ukr. SSR*, No.9, (1959), pp.931-935.
8. S.L. Sobolev, *Introduction into the Theory of Cubature Formulas*, Moscow, Nauka, 1974.
9. V.A. Kondratiev and O.A. Oleinik, "On periodic in the time solutions of a second order parabolic equations", Vestnik of the Moscow University, *Math. Mech.*, ser. 1 (4), (1985), pp.38-47.
10. V.A. Kondratiev, J. Kopacek, D.M. Lekveishvili and O.A. Oleinik, "Sharp estimates in Hölder spaces and the precise Saint-Venant principle for solutions of the biharmonic equation", *Trudy Mat. Institute in Steklov*, **166** (1984), pp.91-106.
11. V A Kondratiev and O.A. Oleinik, "Sharp estimates in Hölder spaces for weak solutions of the biharmonic equation, the Navier-Stokes system and the Karman system in non-smooth two-dimensional domains", *Vestnik Mosc. Univ. Mat. Mech.*, ser. 1 (6), (1983), pp.22-39.
12. O.A. Oleinik, V.A. Kondratiev and J. Kopacek, "On asymptotic properties of solutions of the biharmonic equation", *Differen. Uravnenia*, **17** (10), (1981), pp.1886-1899.
13. S. Agmon, A. Douglis and L. Nirenberg, "Estimates near the boundary for solutions of elliptic partial differential equations satisfying general boundary conditions, I". *Comm. Pure Appl. Math.* 12 (1959), pp.623-727.
14. C.B. Morrey, *Multiple Integrals in the Calculus of Variations*, Springer-Verlag, 1966.

15. O.A. Oleinik, "On some mathematical problems of elasticity", Proceedings of the Conference dedicated to Prof. M. Picone, May, 1985, *Acad. Naz. die Lincei*, Roma, *Atti dei convegni Lincei*, **77**, (1986), pp.259-273.
16. V.A. Kondratiev and O.A. Oleinik, "On the smoothness of weak solutions of the Dirichlet problem for the biharmonic equation in domains with non-regular boundary", in *Nonlinear Partial Differential Equations and their Applications*, College de France Seminar, No. 8, **7**, pp.180-199.
17. J. Kopacek and O.A. Oleinik, "On the behaviour of solutions of the elasticity system in a neighbourhood of irregular points of the boundary and infinity", *Transactions of the Moscow Math. Society*, **43** (1981).
18. O.A. Oleinik and G.A. Yosifian, "On the asymptotic behaviour at infinity of solutions in linear elasticity", *Archive Rat. Mech. and Analysis*, **78**(1), (1982), pp.29-53.

23

PHASE MIXTURES IN NONLINEAR VISCOELASTICITY IN ONE DIMENSION

ROBERT L. PEGO

1. INTRODUCTION

I will consider a model for one dimensional motion of a viscoelastic material with a rate type viscosity of a simple sort. Let $u(x,t)$ denote the displacement at time t of the material at reference point x. Denote the strain by $w = u_x$, the velocity by $v = u_t$. Assume the stress S is given by the constitutive relation $S = \sigma(u_x) + \mu u_{xt}$, where $\mu > 0$ is a constant. The equations of motion for a unit length of material held fixed at one end and subject to a prescribed force P at the other end take the form

$$\begin{aligned} w_t &= v_x && \text{for} \quad 0 < x < 1 \\ v_t &= (\sigma(w) + \mu v_x)_x && \phantom{\text{for}} \quad t > 0 \\ v(0,t) &= 0 && \text{for} \quad t > 0 \\ (\sigma(w) + \mu v_x)(1,t) &= P && \end{aligned} \tag{1.1}$$

with initial conditions

$$\begin{aligned} w(x,0) &= w_0(x) \\ v(x,0) &= v_0(x) \end{aligned} \qquad \text{for} \quad 0 < x < 1 \tag{1.2}$$

Following Ericksen [4] and others, let me assume that the function $w \to \sigma(w)$ is nonmonotonic. For simplicity assume $\sigma(w)$ has the shape indicated in Fig. 1, where $\sigma(w)$ is increasing for

$w < \alpha$ and $w > \beta$, and decreasing for $\alpha < w < \beta$. The material is said to be in the α-phase (resp. β-phase) when $w < \alpha$ (resp. $w > \beta$).

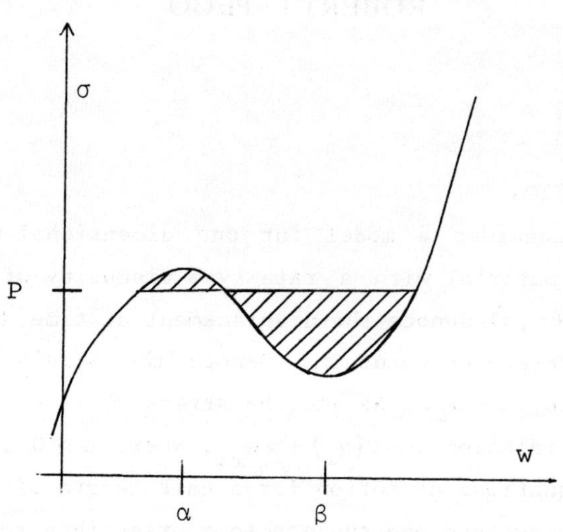

FIG. 1

A stationary solution $(w_s, v_s)(x)$ must satisfy

$$\begin{aligned} \sigma(w_s) &= P \\ v_s &= 0 \end{aligned} \qquad \text{for } 0 < x < 1 . \tag{1.3}$$

If attention is restricted to homogeneous stationary solutions, then as P is raised from a low value to a high one and decreased again, one expects to observe sudden transitions from the α-phase to the β-phase and back again, in a hysteresis loop. Here I am more interested in the many nonhomogeneous stationary solutions possible when P is fixed between $\sigma(\alpha)$ and $\sigma(\beta)$. Ignore the interval $\alpha < w < \beta$ for a moment. For each measurable subset of $(0,1)$, there is a stationary solution which lies in the α-phase on that subset and in the β-phase on its complement in $(0,1)$.

Traditional energy criteria for assessing the stability of phase mixtures, originating from Gibbs' rules for phase equilibrium, assert that just one of these solutions is stable, except when P is at the *Maxwell line*, where the shaded regions in Fig. 1 have equal areas. In that case, all are stable. In any case, the remaining solutions are often termed "metastable" [8].

This assessment differs markedly from that obtained by a recent analysis [10] of the long-time dynamics of solutions of (1.1)-(1.2). New results on several issues are obtained in [10]. Below, I will discuss results concerning: strong convergence of each solution to some stationary solution as $t \to \infty$, dynamic stability of all the "metastable" stationary solutions discussed above, and, regularity of solutions with discontinuous strain w. These results paint a rather comprehensive picture of long-time behaviour in (1.1)-(1.2). Implications concerning admissibility criteria for weak solutions of the purely elastic system,

$$w_t = v_x, \quad v_t = \sigma(w)_x \tag{1.4}$$

are discussed in the last section below.

2. EXISTENCE AND REGULARITY

Assume that $\sigma(w)$ is defined on a (large) interval $[A,B]$ containing $[\alpha,\beta]$ with $\sigma(A) < \sigma(\beta)$ and $\sigma(\alpha) < \sigma(B)$. Assume $\sigma(w)$ is Lipshitz continuous on $[A,B]$.

Concerning the initial data (1.2), assume that

$$w_0 \in L^\infty(0,1) \ , \quad v_0 \in L^2(0,1) \ . \tag{2.1}$$

These assumptions are close to those of Andrews and Ball [2], who establish global existence for weak solutions of (1.1)-(1.2). Existence for classical solutions was proved by Dafermos [3], who considered the more general constitutive relation $S = \sigma(u_x, u_{xt})$.

A simpler treatment of local existence theory under the assumptions above was given in [10] based on a transformation of (1.1). From now on, fix $\mu = 1$. Setting

$$V(x,t) = \int_1^x v(y,t)\,dy, \quad W = w - V. \tag{2.2}$$

one finds that in terms of V and W, (1.1) becomes

$$\begin{aligned} V_t &= V_{xx} + \sigma(V+W) - P \\ W_t &= \quad\quad - \sigma(V+W) + P \end{aligned} \quad \text{for} \quad \begin{aligned} 0 < x < 1 \\ t > 0 \end{aligned} \tag{2.3}$$

$$\begin{aligned} V_x(0,t) &= 0 \\ V(1,t) &= 0 \end{aligned} \quad \text{for } t > 0.$$

Standard existence theory for semilinear parabolic systems as in [6] yields strong solutions, and rather sharp regularity results may be obtained. Global existence is proved with the techniques of [1], [2].

In the theorem below, $W_P(w)$ denotes an antiderivative of the function $w \to \sigma(w) - P$ such that

$$\min_{[A,B]} W_P(w) = 0.$$

$C^{j,\nu}$ denotes the Banach space of functions on $[0,1]$ whose derivatives up to order j are Holder-continuous with exponent ν, and $C^{j,\beta}_{\text{loc}}((0,\infty),X)$ denotes the space of functions on $(0,\infty)$ with values in the Banach space X whose derivatives up to order j are locally Holder-continuous with exponent β. $W^{1,\infty}$ is the space of Lipschitz continuous functions on $(0,1)$.

THEOREM 2.1. Let $E>0$ be such that if $0 \leqslant w \leqslant 2E$, $\sigma(A+w) - P < 0$ and $\sigma(B-w) - P > 0$.

Assume, in addition to (2.1), that

$$w_0(x) \in [A + 2E, B - 2E] \text{ a.e.} \tag{2.4}$$

$$\int_0^1 \tfrac{1}{2} v_0^2(x) + W_P(w_0(x))\,dx < \tfrac{1}{2} E^2 .$$

Then a unique solution for (1.1)-(1.2) exists with, for any

$0 < \nu < \frac{1}{2}$ and for some $\beta > 0$,

$$w \in C([0,\infty),L^{\infty}) \cap C^{1,\beta}_{\text{loc}}((0,\infty),L^{\infty})$$

$$v \in C([0,\infty),L^{2}) \cap C((0,\infty),W^{1,\infty}) \cap C^{1,\beta}_{\text{loc}}((0,\infty)C^{0,\nu})$$

$$S \in C^{0,\beta}_{\text{loc}}((0,\infty),C^{1,\nu}) .$$

Further, there is a set $\Omega \subset [0,1]$ of Lebesgue measure 1, such that for $x \in \Omega$, the function $t \to W(x,t)$ given from (2.2) is a classical C^1 solution on $[0,\infty)$ of the ODE

$$\frac{dW}{dt} = -\sigma(W + V(x,t)) + P . \tag{2.5}$$

Lastly, if $\sigma(w)$ is defined and is locally Lipschitz continuous on $\mathbb{R}$, then the condition (2.4) is not needed.

Note that the total stress $S = \sigma(w) + v_x$ has an interesting spatial regularity property for $t > 0$: although w and v_x need lie only in L^{∞}, S lies in $C^{1,\nu}$ for any $\nu < \frac{1}{2}$ (even when σ is monotone). One may almost say that the momentum equation in (1.1) holds in a classical sense. However, the dissipative mechanism in (1.1) does *not* smooth the strain w:

PROPOSITION 2.2. Make the assumptions of Theorem 2.1. Then for any $T > 0$ and $x_0 \in (0,1)$, x_0 is a point of continuity of the function $x \to w(x,T)$ if and only if x_0 is a point of continuity of w_0.

The proof is very easy, based on the ODE (2.5) and the continuity of $V(x,t)$. Thus, any (essential) discontinuities in the initial data w_0 persist without moving for all time. Hoff and Smoller [7] have proved a result of this type.

3. CONVERGENCE TO EQUILIBRIUM

A question left outstanding by Andrews and Ball [2] was whether each solution of (1.1)-(1.2) converges strongly to some stationary state. The affirmative answer below is proved in [10].

THEOREM 3.1. Under the hypotheses of Theorem 2.1, there exists some function $w_\infty \in L^\infty(0,1)$ with $\sigma(w_\infty(x)) = P$ a.e., so that as $t \to \infty$, the solution of (1.1)-(1.2) satisfies

$$\begin{aligned} w(x,t) &\to w_\infty(x) && \text{bounded a.e.,} \\ v(x,t) &\to 0 && \text{uniformly ,} \\ v_x(x,t) &\to 0 && \text{boundedly a.e. ,} \\ \sigma(w) + v_x &\to P && \text{in the norm of } C^{1,\nu} \text{ for any } \nu < \tfrac{1}{2} . \end{aligned}$$

I will outline the proof here. Start with the energy identity

$$\begin{aligned} \int_0^1 (\tfrac{1}{2} v^2 + W_P(w))(x,t)\, dx + \int_0^t \int_0^1 v_x^2(x,T)\, dx\, dT \\ = \int_0^1 (\tfrac{1}{2} v_0^2 + W_P(w_0))(x)\, dx . \end{aligned} \tag{3.1}$$

Following Andrews and Ball [2], one obtains the *a priori* estimates

$$\sup_x |V(x,t)| \leqslant \|v(\cdot,t)\|_{L^2} \leqslant E , \tag{3.2}$$

$$W(x,t) \in [A+E, B-E] \quad \text{for } x \in \Omega , \tag{3.3}$$

for all $t>0$. The result (3.3) expresses the fact that the interval $[A=B, B-E]$ is positively invariant for W in (2.5) under the estimate (3.2) and the hypotheses of Theorem 3.1.

From (3.1), one may then show that

$$v_x \to 0 \quad \text{in } L^2(0,1). \tag{3.4}$$

I omit the technicalities. But then it follows that

$$V(x,t) \to 0 \text{ uniformly} . \tag{3.5}$$

For a given $x \in \Omega$, the hypothesis that $W(x,t)$ does not converge as $t \to \infty$ now leads to a contradiction. I thank J. Ball for the argument for this below: let

$$W_- = \liminf_{t\to\infty} W(x,t) , \quad W_+ = \limsup_{t\to\infty} W(x,t)$$

If $W_- < W_+$, then for any $W_0 \in (W_-, W_+)$ one may choose two sequencies $\{t_n^-\}$, $\{t_n^+\}$ such that

$$W(x, t_n^{\pm}) = W_0 ,$$

$$\frac{d}{dt} W(x, t_n^-) \leqslant 0 \leqslant \frac{d}{dt} W(x, t_n^+)$$

for all n. Using (2.5),(3.5) and taking $n \to \infty$, one finds

$$\sigma(W_0) = P \quad \text{for any} \quad W_0 \in [W_-, W_+] . \tag{3.6}$$

For $\sigma(w)$ as in Fig. 1, it follows

$$W_- = W_+ = \lim_{t \to \infty} W(x, t) .$$

But a short additional argument shows that $W_- = W_+$ even in degenerate cases where $\sigma = P$ is possible on an interval.

4. STABILITY OF MIXTURES

The stability of any stationary solution $w = w_s(x)$, $v = 0$ containing a mixture of α- and β- phases is asserted in the theorem below. The phases need not have the same energy densities, that is, perhaps $W_P(w_s(x))$ is not constant. Such a solution is not a strong minimizer of the stored energy functional

$$I(u) = \int_0^1 W_P(u_x(x))\, dx .$$

Instead, it is a weak relative minimizer. Below, let $\sigma(\beta) < P < \sigma(\alpha)$ and let $w_\alpha, w_\beta, w_\gamma$ satisfy

$$P = \sigma(w_\alpha) = \sigma(w_\beta) = \sigma(w_\gamma)$$

$$w_\alpha < \alpha < w_\gamma < \beta < w_\beta .$$

THEOREM 4.1. Make the assumptions of Theorem 2.1. If $\varepsilon > 0$ is sufficiently small, if

$$w_s(x) \in \{w_\alpha, w_\beta\} \quad \text{for a.e. } x \text{ in } (0, 1),$$

if a set Ω_ε exists such that the Lebesgue measure of Ω_ε is greater than $1 - \varepsilon$ and

$$|w_0(x) - w_s(x)| < \varepsilon \quad \text{for} \quad x \in \Omega_\varepsilon ,$$

and if

$$\int_0^1 v_0(x)^2\, dx < \varepsilon^2 ,$$

then

$$w_\infty(x) = w_s(x) \quad \text{for a.e. } x \text{ in } \Omega_\varepsilon ,$$

where

$$w_\infty(x) = \lim_{t\to\infty} w(x,t) \qquad \text{a.e.}$$

from Theorem 3.1.

In this theorem, $w_s(x)$ can have the form

$$w_s(x) = w_\alpha \chi(x) + w_\beta(1-\chi(x)) \tag{4.1}$$

where $\chi(x)$ is the characteristic function of any measurable subset of $(0,1)$. I stress that for any $\varepsilon > 0$, the hypotheses of Theorem 4.1 hold for continuous initial data w_0 in an open set of $C[0,1]$. Such data yield a solution with continuous strain w whose asymptotic state $w_\infty(x) \in \{w_\alpha, w_\beta, w_\gamma\}$ since $\sigma(w_\infty(x)) = P$ a.e. The conclusions of the theorem force w_∞ to be discontinuous if

$$\varepsilon < \int_0^1 \chi(x)\,dx < 1-\varepsilon .$$

Thus $w(x,t)$ does not converge uniformly in general; the same may be said of $\sigma(w(x,t))$ and $v_x(x,t)$. Instead, as $t\to\infty$, layers of transition between the phases persist and steepen indefinitely without moving much. The behaviour of the solution in these layers may be studied using equation (2.5) [10].

The proof of Theorem 4.1 hinges on the observation that if one has an *a priori* estimate for the kinetic energy

$$K(t) = \int_0^1 \tfrac{1}{2} v(x,t)^2\,dx < \tfrac{1}{2}\delta^2 \tag{4.2}$$

for a sufficiently small δ, the intervals $[A+E,\alpha]$, $[\beta, B-E]$ are positively invariant for W in (2.5), since (4.2) implies $|V| < \delta$, and $\sigma(\alpha) > P > \sigma(\beta)$.

One may say the *phases are invariant pointwise if the* *total* *kinetic energy is small.* Because there is a unique stationary state in each phase, the theorem follows easily from (4.2) and Theorem 3.1.

To derive the estimate (4.2) from the energy estimate (3.1), one needs an effective lower bound for the stored energy

$$I(t) = \int_0^1 W_P(w(x,t))\,dx.$$

When P is not at the Maxwell line, one may have

$$W_P(w_\alpha) > 0 = W_P(w_\beta)$$

for example, so it is unreasonable to expect that $I(0)$ can be made small enough to yield (4.2). In this case, one constructs a lower bound using the function

$$W_L(w) = \begin{cases} W_P(w_\alpha) & \text{if} \quad w < \alpha + \delta \\ W_P(w_\beta) = 0 & \text{otherwise} . \end{cases}$$

Define

$$I_L = \int_0^1 W_L(W(x,0))\,dx .$$

Then if ε is small, $I(0) \geqslant I_L$, and

$$K(0) + I(0) - I_L < \tfrac{1}{2}\delta^2 .$$

Now if (4.2) holds for $0 \leqslant t \leqslant T$, the phases remain invariant, which implies

$$I(t) \geqslant I_L \quad \text{for} \quad 0 \leqslant t \leqslant T, \tag{4.3}$$

because if $W(x,t) < \alpha$ then $W_P(w(x,t)) \geqslant W_P(w_\alpha)$. The energy estimate then yields

$$K(t) + I(t) - I_L \leqslant K(0) + I(0) - I_L < \tfrac{1}{2}\delta^2 .$$

Thus (4.3) implies (4.2) and vice versa. Then a continuity argument [10] shows that (4.2) and (4.3) hold for a maximal time $T = +\infty$.

5. ADMISSIBILITY

The simplest initial value problem for the inviscid equations (1.4) is the Riemann problem: Find a centered-wave weak solution of the form $(w,v)(x/t)$ with initial data

$$(w,v)(x,0) = \begin{cases} (w_-, v_-) & \text{for } x < 0 \\ (w_+, v_+) & \text{for } x > 0 . \end{cases} \tag{5.1}$$

James [9] has shown that this problem can have infinitely many solutions. In order to select the physically relevant ones, one must be able to decide which simple travelling wave solutions of (1.4) of the form

$$(w,v)(x,t) = \begin{cases} (w_-, v_-) & \text{for } x < st \\ (w_+, v_+) & \text{for } x > st \end{cases} \tag{5.2}$$

are to be regarded as physically relevant, or admissible.

One traditional approach to this issue is to regard a wave of the form (5.2) as admissible if it corresponds to a travelling wave of a system which includes some effects neglected in the system (1.4) but which may be physically relevant in zones of rapid transition. While other mechanisms must also be relevant, here I consider only the effect of a rate-type viscosity dependence of stress.

There is no reason to confine the considerations to the simple form in (1.1) however. In the general form taken by Dafermos [3], the stress satisfies $S = S(w, w_t)$. Writing

$$\sigma(w) = S(w,0), \quad \Lambda(w, w_t) = S(w, w_t) - S(w,0),$$

I consider systems of the form

$$w_t = v_x, \quad v_t = (\sigma(w) + \Lambda(w, w_t))_x. \tag{5.3}$$

Assume that Λ is smooth, $\Lambda(a,0) = 0$, and $\Lambda_b(a,b) > 0$ for all a and all real b.

If $(w,v)(x-st)$ is a travelling wave solution of (5.3), which corresponds to the wave (5.2) in the sense that

$$\lim_{x\to-\infty} (w,v)(x) = (w_-, v_-), \quad \lim_{x\to+\infty} (w,v)(x) = (w_+, v_+), \tag{5.4}$$

it follows that $(w,v)(x)$ must satisfy

$$\begin{aligned} -\Lambda(w, -sw') &= \sigma(w) - \sigma(w_-) - s^2(w - w_-) \\ -s(w - w_-) &= v - v_- \quad \text{for } x \in \mathbb{R}. \end{aligned} \tag{5.5}$$

Now it is not hard to verify that if $s \neq 0$, the differential equation in (5.5) has a solution with the limits (5.4) if and only if

$$\operatorname{sgn} s(w_+ - w_-) = \operatorname{sgn}(\sigma(w) - \sigma(w_-) - s^2(w - w_-) \quad \text{for all } w \text{ between } w_- \text{ and } w_+ . \tag{5.6}$$

This condition (5.6) is called the *chord condition*, since it means that the chord connecting the points $(w_-, \sigma(w_-))$ $(w_+, \sigma(w_+))$ does not intersect the graph of $\sigma(w)$, and that w_-, w_+ are ordered so that $s(w_+ - w_-) > 0$ if the chord lies above the graph, and $s(w_+ - w_-) < 0$ if the chord lies below.

In the remaining case, $s = 0$, equations (5.5) degenerate to

$$\sigma(w) = \sigma(w_-) , \quad v = v_- .$$

But if the wave (5.2) is a stationary solution of (1.4), then it satisfies these equations; it corresponds to itself. More generally, (1.4) and (5.3) have the same stationary solutions. In this case the corresponding viscous wave is not continuous. But the stability results described in Section 4 above, indicate that such waves can be stable asymptotic limits for smooth solutions in the problem (1.1)-(1.2). So, it is reasonable to regard them as admissible.

To summarize, we have the *viscoelastic admissibility* criterion for waves of the form (5.2) which are weak solutions of (1.4):

If $s \neq 0$, the wave is admissible if the chord condition (5.6) holds. (5.7)

If $s = 0$, the wave is admissible unconditionally.

Curiously enough, this criterion was used in 1982 by Shearer [13], who demonstrated existence and uniqueness of solutions to the Riemann problem when discontinuities are required to satisfy (5.7), assuming

$$\sigma''(w) < 0 \quad \text{for } w < \alpha, \ \sigma''(w) > 0 \quad \text{for } w > \beta .$$

Shearer seems to have admitted the waves with $s = 0$ on an *ad hoc* basis. But the conclusion is that the viscoelastic admissibility criterion does yield a self-consistent selection principle for solutions of the Riemann problem.

Paradoxically, however, uniqueness for the Riemann problem

may not be desirable from the physical point of view. The equations of motion of an isothermal viscous gas with a van der Waals equation of state have the form (1.1), where w is the specific volume and the pressure is

$$-\sigma(w) = \frac{R\theta}{w-b} - \frac{a}{w^2}$$

when the viscosity coefficient $\tilde{\mu}$ depends on density by $\tilde{\mu}(w) = \mu w$. But for liquid-gas phase transitions the admissibility criterion (5.7) and the stability results of Section 4 are unphysical. Phases do coexist at rest only when the pressure level is at the Maxwell line. Here the viscosity-capillarity criterion of Slemrod (see [5]) yields qualitatively better results: it permits coexistence at rest only at a Maxwell line, and also admits a family of moving two-phase waves which transform metastable states to stable ones. Shearer [14] has recently shown that for some pairs of states near the Maxwell line, solutions to the Riemann problem whose discontinuities are admissible by the viscosity-capillarity criterion are not unique. I have argued [11] that this result is physically meaningful, the mechanism being a likely loss of stability of metastable states in the limit of vanishing viscosity and capillarity, with a consequent sensitivity to nucleation and growth phenomena. Indeed, in [11] Shearer's analysis is extended to show that for homogeneous metastable rest states near the Maxwell line, two admissible solutions are possible. One is a continuation of the rest state, while the other is a two-phase solution in which the domain of the new phase grows from nothing.

One might say that nature has some difficulty deciding between the possibilities as well. Experimentally, metastable liquid states cannot be maintained indefinitely: a superheated liquid, apparently in equilibrium, suddenly changes phase after a random lifetime which fits Poisson statistics [15]. Physicists attribute this behaviour to nucleation and growth due to the random formation of clusters. It is hard to see how well-posed continuum theories can usefully model such behaviour. But curiously, Glimm's scheme, which involves a random choice, yields approximations for (1.4) which in some simple examples exhibit phase lifetimes having Poisson statistics, asymptotically as the mesh size vanishes [12].

All this is not to say that the viscoelastic model (1.1) with the behaviour described here is perfectly useless physically, though its utility is likely to be limited. In condensed media, one can sometimes observe coexisting phases of different energy densities. Models such as (1.1) may have a role in understanding the obstacles to achieving phase equilibrium in such systems.

ACKNOWLEDGEMENTS

This work was partially supported by the National Science Foundation under grant DMS-84-01614, and by the U.K. Science and Engineering Research Council.

REFERENCES

1. G. Andrews, "On the existence of solutions to the equation $u_{tt} = u_{xxt} + \sigma(u_x)_x$", *J. Diff. Eqns.* **35** (1980), pp.200-231.
2. G. Andrews and J.M. Ball, "Asymptotic behaviour and changes of phase in one-dimensional nonlinear viscoelasticity", *J. Diff. Eqns.* **44** (1982), pp.306-341.
3. C. Dafermos, "The mixed initial-boundary value problem for the equations of nonlinear one-dimensional viscoelasticity", *J. Diff. Eqns.* **6** (1969), pp.71-86.
4. J.L. Ericksen, "Equilibrium of bars", *J. Elasticity* **5** (1975), pp.191-201.
5. R. Hagan and M. Slemrod, "The viscosity-capillarity admissibility criterion for shocks and phase transitions", *Arch. Rat. Mech. Anal.* **83** (1983), pp.333-361.
6. D. Henry, *Geometric Theory of Semilinear Parabolic Equations*, Lecture Notes in Math. **840**, Springer, New York, 1981.
7. D. Hoff and J. Smoller, "Solutions in the large for certain nonlinear parabolic systems", *Anal. Non-Lin.* **2** (1985), pp.213-235.
8. R.D. James, "Coexistent phases in the one-dimensional static theory of elastic bars", *Arch. Rat. Mech. Anal.* **72** (1980), pp.99-140.

9. R.D. James, "The propagation of phase boundaries in elastic bars", *Arch. Rat. Mech. Anal.* **73** (1980), pp.125-158
10. R.L. Pego, "Phase transitions in one-dimensional nonlinear viscoelasticity: admissibility and stability", *Arch. Rat. Mech. Anal.* **97** (1987), pp.353-394.
11. R.L. Pego, "On the viscosity-capillarity admissibility criterion for phase transitions", in preparation.
12. R.L. Pego and D. Serre, "Instabilities in Glimm's scheme for two systems of mixed type. in preparation.
13. M. Shearer, "The Riemann problem for a class of conservation laws of mixed type", *J. Diff. Eqns.* **46** (1982), pp.426-443.
14. M. Shearer, "Nonuniqueness of admissible solutions of Riemann initial value problems for a system of conservation laws of mixed type", *Arch. Rat. Mech. Anal.* **93** (1986), pp.45-59.
15. V.P. Skirpov, *Metastable Liquids*, Wiley, New York, 1974.

24

STATISTICAL MECHANICS AND THE KINETICS OF PHASE SEPARATION

O. PENROSE

1. INTRODUCTION

When an alloy such as aluminium-zinc (Aℓ-Zn) is cooled the two components, previously perfectly mixed, show a tendency to separate. This tendency is well understood from the equilibrium theory of such alloys, according to which the equilibrium state is a single mixed phase at temperatures above the critical temperature T_c, but comprises two spatially separated phases of different compositions below T_c. Not quite so well understood, however, are the details of the process by which this phase separation takes place (when it takes place at all). It is the purpose of this article to summarize the theory of such processes, with particular reference to problems in it which may be of interest to mathematicians.

When a physical substance separates out into two phases, the two phases are characterized by different values of at least one local physical quantity. Such a physical quantity is called an *order parameter*. In the alloy system just mentioned, a suitable order parameter would be the concentration of one of the components, that is, the number of (say) aluminium atoms per unit volume. Another example is the ferromagnetic phase transition in iron for which a suitable order parameter is the magnetization per unit volume; this example illustrates the possibility of an order parameter which is not a scalar.

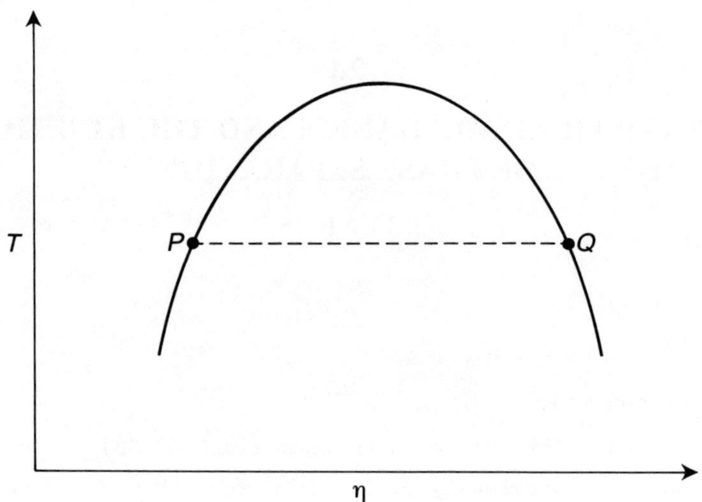

A useful tool for discussing such systems is the *phase diagram* illustrated in Fig. 1. The points outside the curve correspond to values of the temperature T and order parameter η for which the equilibrium state is a uniform phase. The central region inside the curve corresponds to values of T and η for which such an equilibrium state is not possible; instead, the equilibrium state corresponding to a point in this region (the *coexistence region*) consists of two phases with different values for the order parameter, existing side by side (like water and ice in a glass of iced water). These two co-existing phases correspond to two points on opposite branches of the curve shown (which is therefore called the *coexistence curve*) with equal values of T, for example the points P and Q in Fig. 1.

An important problem in the kinetics of phase transitions is to model the behaviour of a system which starts in equilibrium at some high temperature (and therefore in a uniform phase); which is suddenly cooled, or *quenched*, without changing the value of the order parameter, to a point (η, T) inside the coexistence region; and which is subsequently maintained at that temperature. Since the uniform phase at the new temperature is not an equilibrium state, equilibrium theory predicts that the system will eventually separate into two coexisting phases on the coexistence curve, joined by a horizontal chord in the phase diagram through (η, T); but we must turn to kinetics for a description of the way this separation takes place and how long it takes.

Two factors have a profound effect on the phase separation process. One is the nature of the microscopic dynamical process by which the system evolves, and in particular the conservation laws affecting this process. Especially important is the question of whether the order parameter itself obeys a conservation law or not. If the order parameter is the density or concentration of some substance, such as one of the components in an alloy, then the dynamical process cannot change the total amount of this substance in the system; the most it can do is to move the substance from one place to another. In such a case, we may speak of a *conserved order parameter* (COP). But if the order parameter is, for example, the magnetization density in a ferromagnet, there is no such restriction. In such cases, we may speak of a *non-conserved order parameter* (NCOP). In a common terminology due to Hohenberg and Halperin [30], the NCOP case is called 'Model A' and the COP case 'Model B'.

Other conservation laws may also be important in some cases, and impose limitations on the rate at which the order parameter can change even when it does not itself obey a conservation law. In particular, because energy is a conserved quantity, we often have to consider heat conduction as an important ingredient of the process being considered; but in metallic systems such as alloys and ferromagnets this effect can be ignored because metals conduct heat very well. This article concentrates on the simplest case, where these other conservation laws can be ignored.

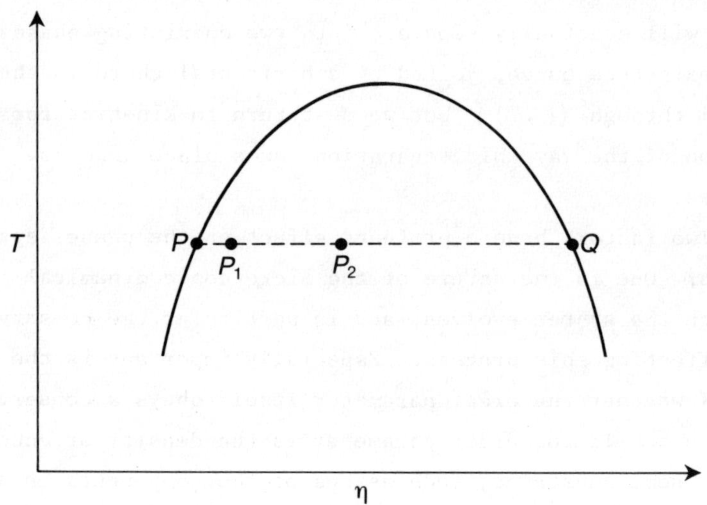

The second factor having a profound effect on the kinetics of phase separation is what part of the coexistence region the system arrives in after the quench. If it arrives at a point close to the coexistence curve, say the point P_1 in Fig. 2, we can speak of a *shallow quench*. In this case the phase separation is likely to proceed by *nucleation* followed by *growth*; that is, the first stage in the process is the formation of small roughly spherical nuclei inside which the system behaves similarly to the opposite phase — for example, in Fig. 2, the one corresponding to point Q — and the second stage is the growth of these nuclei to macroscopic size. Both these processes can be very slow and in particular the nucleation process may be so slow that it can be ignored; in this case the uniform phase P_1 is called a *metastable state*. A system in a metastable state behaves in many respects as though it were in a true equilibrium state.

If, on the other hand, the system arrives after the quench at a point such as P_2, somewhere in the middle of the coexistence region, we may speak of a *deep quench*. In this case the initial uniform state of the system will be unstable rather than metastable, and the subsequent evolution (a process known as *spinodal decomposition*) is likely to go considerably more quickly. As the two phases separate out the regions they occupy will be very complicated in shape, more like a sponge than a collection of spherical nuclei of one phase embedded in an otherwise continuous background of the opposite phase.

These complex phenomena have been studied both empirically and theoretically. The empirical studies are of two kinds: experiments with real alloys, and computer simulations of simplified theoretical models. The theoretical studies are also of two kinds: some based on mathematical models which are *macroscopic* in the sense of not trying to represent individual atoms or molecules, and some based on *microscopic* models which do include representations of individual atoms or molecules. The distinction is similar to the one between using the equations of gas dynamics and using Boltzmann's equation. One of the fundamental theoretical problems in the subject, therefore, is to relate the macroscopic and microscopic descriptions as rigorously as possible, so as to put the macroscopic description on a firm foundation; solving this problem would enable us to determine unambiguously the correct equations in the macroscopic description and give methods of calculating the coefficients and other invariant features of these equations in terms of the properties of atoms or molecules.

But this fundamental problem is, of course, only one of the mathematical problems in the subject. Even purely macroscopic models lead to a host of difficult and interesting mathematical problems, as the rest of this article should show.

2. MACROSCOPIC THEORIES

The simplest way to model the phase separation process is to assume that the space occupied by the system is partitioned into two regions with a common boundary, each region corresponding to a particular phase. A rule is then formulated describing the way the boundary surface moves.

A theory of this type is the Lifshitz-Slyozov theory [1] of the growth of spherical nuclei or 'grains' in a shallow quench with COP dynamics. This theory does not describe the nucleation process by which the nuclei are formed in the first place; its concern is what happens after they have formed. The theory takes into account the fact that the values of the order parameter on the two sides of a surface of discontinuity will depend on the curvature of that surface. In consequence, the order parameter will vary from place to place in the phase surrounding the grains. This non-uniformity causes the substance whose concentration the order parameter measures to diffuse, in such a way that the larger grains grow and the smaller ones shrink.

The equation used by Lifshitz and Slyozov to model this process can be written

$$\frac{\partial f}{\partial t} + \frac{\partial}{\partial s}(vf) = 0 \tag{2.1}$$

where $f(s,t)\,ds$ is the number of grains per unit volume with sizes (i.e. volumes) between s and $s+ds$ at time t, and $v(s,t)$ is the rate of growth of a grain of size s under the conditions prevailing at time t. Their formula for v can be written

$$v = v_0\left[ys^{1/3} - 1\right] \tag{2.2}$$

where v_0 is a constant and $y(t)$ depends linearly on $\eta_\infty(t)$, the average value of the order parameter in the region outside the grains at time t. Since the integral of the order parameter over the whole system is conserved (i.e. independent of time) and the integral of the order parameter over the regions inside the grains is proportional to their total volume $\int_0^\infty f(s,t)\,s\,ds$,

η_∞ obeys a relation of the form

$$\frac{d}{dt}\left[\eta_\infty(t) + (k - \eta_\infty(t))\int_0^\infty f(s,t)\,s\,ds\right] = 0 \tag{2.3}$$

where k is the (constant) value of the order parameter inside the grains. From this relation and the initial conditions we can determine $\eta_\infty(t)$ and hence $y(t)$.

For the limiting case where η_∞ is very small, Lifshitz and Slyozov derive an asymptotic (large-t) solution for these equations in self-similar form

$$f(s,t) = t^{-2}\,\phi(s/t) \tag{2.4}$$

$$y(t) = At^{-1/3} \tag{2.5}$$

where A is a constant and $\phi(x)$ satisfies

$$-2\phi(x) - x\frac{d\phi(x)}{dx} + \frac{d}{dx}\{v_0\,[Ax^{1/3} - 1]\,\phi(x)\} = 0\,, \tag{2.6}$$

the constant A being chosen so that $\phi(x)$ and its first derivative are continuous. They argue that this solution will represent the actual asymptotic behaviour, regardless of initial conditions. This contention is partly but not completely borne out by experiments and computer simulations (see Penrose and Buhagiar [10]). A mathematically rigorous investigation of the asymptotic behaviour of the Lifshitz-Slyozov equations would be a useful advance in this area.

The analogous model for deep quenches is much more difficult to analyse, since the shape of the boundary between the two phases is much more complicated — like the surface of a sponge rather than a collection of spheres. For NCOP dynamics, the evolution law proposed by Lifshitz [2] is that the surface moves in the direction perpendicular to itself with a speed proportional to the net curvature of the surface. For COP dynamics the velocity of the surface would depend instead on the rate at which material is delivered to the surface by diffusion, so that the evolution law would involve the solution of a diffusion equation in a region of very complicated shape. But even for the relatively simple case of NCOP dynamics no rigorous results have been obtained, and the non-rigorous ones amount to little more than dimensional analysis (Lifshitz [2], Allen and Cahn [21]) and empirical scaling laws (Phani *et al.* [3]). The 'dimensional' argument is that if R is a typical length describing the sponge-like surface, then dR/dt will be proportional to the mean curvature of the surface, which in turn is roughly proportional to $1/R$; thus we have $dR/dt = \text{const}/R$, so that R^2 varies linearly with time.

The outstanding mathematical problem in this area is to find a tractable way of describing and analysing complicated sponge-like surfaces. A hopeful sign is the progress which has been made recently with the theory of random surfaces in other contexts (see Fröhlich [4]).

3. FIELD-THEORETIC MODELS

The term 'field-theoretic models' is used here to mean that the order parameter is assumed to vary continuously with position; instead of surfaces of discontinuity, as in the models considered in the preceding section, we now have thin regions in which the order parameter varies rapidly in space. The equations for this model are derived from the following formula for the free energy F, which incorporates the assumption that a nonuniformity in the order parameter η gives rise to a term in the Helmholtz free energy density which is proportional to the square of the gradient of η:

$$F = \int_{\Omega} [f(\eta(\boldsymbol{x})) + \tfrac{1}{2}\,\epsilon\,(\nabla\eta)^2]\,d^3\boldsymbol{x}\,. \tag{3.1}$$

Here Ω is the region occupied by the system, $\eta(\boldsymbol{x})$ is the order parameter, which is a function of position $\boldsymbol{x}$, ϵ is a constant and $f(\eta)$ is the free energy of a system in which η is independent of $\boldsymbol{x}$. The hypothesis made is that, in view of the tendency of free energy towards a minimum (in a constant-temperature system), there is a generalized force at each point tending to decrease $\eta(\boldsymbol{x})$ and having a value equal to the functional derivative

$$\frac{\delta F}{\delta \eta(\boldsymbol{x})} = f'(\eta) - \varepsilon \nabla^2 \eta . \tag{3.2}$$

In the case of nonconserving dynamics, the equation of motion is obtained by assuming that $\eta(\boldsymbol{x})$ responds by decreasing at a rate proportional to this generalized force :

$$\frac{\partial \eta}{\partial t} = -K \frac{\delta F}{\delta \eta(\boldsymbol{x})} = K[\varepsilon \nabla^2 \eta - f'(\eta)] \qquad \text{(NCOP)} \tag{3.3}$$

where K is a constant. It is natural to take a Neumann condition on the boundary of Ω:

$$\frac{\partial \eta}{\partial n} = 0 \quad \text{on} \quad \partial\Omega . \tag{3.4}$$

This has the consequence that F, as defined in (3.1), is a Lyapunov functional:

$$\frac{dF}{dt} \leqslant 0 \tag{3.5}$$

with equality only when $\partial\eta/\partial t = 0$ throughout Ω.

For conserving dynamics the standard method of deriving an equation of motion is to assume that the flux of order parameter (denoted here by $\boldsymbol{J}(\boldsymbol{x})$) is proportional to the gradient of the generalized force, so that

$$\frac{\partial \eta}{\partial t} = -\operatorname{div} \boldsymbol{J} \tag{3.6}$$

where

$$\boldsymbol{J} = -D \operatorname{grad} \frac{\delta F}{\delta \eta} \tag{3.7}$$

and D is a constant related to the diffusion constant. The resulting equation of motion, due to Cahn and Hilliard ([5,6]; Cahn [32,33]), is

$$\frac{\partial \eta}{\partial t} = D\nabla^2[f'(\eta) - \varepsilon \nabla^2 \eta] . \qquad \text{(COP)} \tag{3.8}$$

The natural boundary conditions are

$$\frac{\partial \eta}{\partial n} = 0 \quad \text{and} \quad \frac{\partial}{\partial n}[f'(\eta) - \varepsilon \nabla^2 \eta] = 0 \quad (\text{i.e. } \boldsymbol{J} \cdot \boldsymbol{n} = 0) \quad \text{on } \partial\Omega. \tag{3.9}$$

These ensure, as before, that F is a Lyapunov functional, and also that the total amount of order parameter is conserved:

$$\frac{d}{dt}\int_{\Omega} \eta(\boldsymbol{x})\, d^3\boldsymbol{x} = 0 \,. \tag{3.10}$$

The simplest solutions of the equations (3.3) and (3.8) are the constant solutions:

$$\eta(\boldsymbol{x}) = \eta_0 \tag{3.11}$$

with, in the case of equation (3.3) only, the additional condition

$$f'(\eta_0) = 0 \tag{3.12}$$

where the prime denotes a derivative. If, in addition, the condition

$$f''(\eta_0) > 0 \tag{3.13}$$

is satisfied, then it follows from the Lyapunov property of F that the constant solution is stable against small perturbations; but if

$$f''(\eta_0) < 0 \,, \tag{3.14}$$

then the constant solution is unstable in regions Ω large enough for the operator $\varepsilon\nabla^2 - f''(\eta_0)$, or $(f''(\eta_0) - \varepsilon\nabla^2)\,\nabla^2$, to have a positive eigenvalue.

The key to this theory is the function f. At high temperatures it is strictly convex, so that all constant solutions of (3.3) or (3.8) are stable; but these are just the temperatures for which there is no phase transition. At low temperatures, on the other hand, the function f is assumed to have the general character shown in Fig. 3. The dotted line is tangent to the curve at the two points P and Q; these points correspond to two phases which can coexist, such as the ones shown at P and Q in the phase diagram of Fig. 1.

The points of inflection X, Y of the graph in Fig. 3 are also important, because according to equation (3.14) any constant solution of the equations in the form (3.11) will be unstable in large enough regions Ω if η_0 lies between η_X and η_Y. This instability is at the heart of the theory given by Cahn and Hilliard to describe the way an alloy separates into two

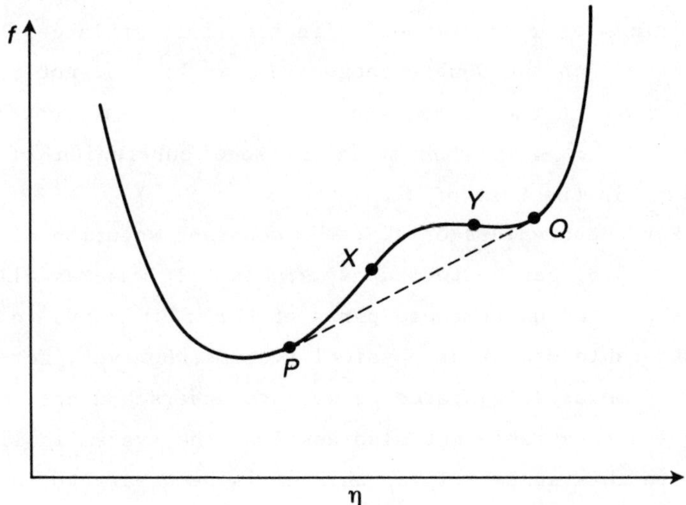

finely intermingled phases after a deep quench — a process known as *spinodal decomposition*. However, Cahn and Hilliard were only able to analyse the very early stages of this process, in which linearization of equation (3.8) about the constant solution is justified. Very little is known with certainty about the behaviour of this equation for medium and late times. A non-rigorous analysis of the behaviour after a shallow quench has, however, been given by Coutsias and Neu [15]; this work establishes the connection between the Cahn-Hilliard equation and the Lifshitz-Slyozov theory described in Section 2 of this article. Even for the simpler NCOP case, equation (3.3), rigorous results are scarce; for this equation some progress has been made for the one-dimensional case (Carr *et al.* [34]) by looking at the behaviour of the equations in the limit of very small ε, and for the three-dimensional case by Caginalp [31].

If the value of the constant η_0 in the solution (3.11) lies between η_P and η_X, or η_Y and η_Q, then this solution is stable under small perturbations, but (if Ω is large enough) not under large ones, since F is not an absolute minimum; the absolute minimum of F corresponds (in the limit of large Ω) to the point at η_0 on the double tangent PQ in Fig. 3, not to the point at η_0 on the curve, and is achieved by a function of $\boldsymbol{x}$ for which $\eta(\boldsymbol{x}) \simeq \eta_P$ when $\boldsymbol{x}$ is in some sub-region of Ω and $\eta(\boldsymbol{x}) \simeq \eta_Q$ in the rest of Ω.

For these values of η_0, the constant solution of equation (3.3) or (3.8) can be termed *metastable*. It was Maxwell [7] who first suggested using these parts of the $f(\eta)$ curve to model the metastable states of physical systems. However, for a proper theory of metastable states we want to understand not only which states are metastable but also how long the system is likely to remain in this state. To do this it is necessary to incorporate fluctuations into the theory. This is usually done by adding a stochastic term to equation (3.3) or (3.8) as proposed originally by Cook [8]. However, the treatment of fluctuations as they apply to these metastable states, given by Langer in an important paper [9], does not work directly from these modified equations. Rather, it uses an *ad hoc* extension of the theory of equilibrium fluctuations, which does not distinguish between the two dynamical models, COP and NCOP.

Even if a rigorous treatment of metastable states in the fluctuating Cahn-Hilliard or Ginzburg-Landau models were available, one might question how far this treatment was applicable to metastable states in real systems. Fluctuations are caused by the motion of individual molecules, and it is not obvious that this motion is correctly represented by grafting stochastic terms on to the macroscopic equations (3.3) or (3.8). A more convincing method would be to derive one of these equations directly from a microscopic model.

One of the difficulties in such derivations, which arises even before we begin to consider dynamics, is that it is not perfectly clear how the function f is related to microscopic quantities. For values of η such that a single-phase state is possible (i.e. values not between η_P and η_Q in Fig. 3) there is no difficulty: $f(\eta)$ is simply the thermodynamic free energy per unit volume, which can be calculated from the partition function formalism of statistical mechanics. However, general principles of thermodynamics require the thermodynamic free energy to be a convex function: a phase transition manifests itself as a straight line in the graph of thermodynamic free energy (see, for example, Griffiths [11]), which must therefore follow the dotted line from P to Q in Fig. 3, rather than the graph of $f(\eta)$.

The one case where a well-defined non-convex free energy function such as $f(\eta)$ in Fig. 3 finds a natural place in rigorous equilibrium statistical mechanics is the mean-field theory (also known, for the liquid-vapour transition, as the van der Waals-Maxwell theory; for ferromagnets, as Curie-Weiss theory; for mixtures, as regular solution theory; and so on). In the limiting case where the phase transition is brought about by an attractive interaction which is very weak and of very long range, mean-field theory is exact (Lebowitz and Penrose [12]) and shows that the thermodynamic free energy function is the convex envelope of a function $f(\eta)$ which at low enough temperatures can have the non-convex character illustrated in Fig. 3. The formula for this last function is

$$f(\eta) = f_0(\eta) - \tfrac{1}{2}\alpha\eta^2$$

where $f_0(\eta)$, a convex function, is the free energy density in the absence of the weak long-range attractive interaction, and α is a positive constant depending on the strength of this interaction.

For short-range interactions, such as the nearest-neighbour interactions popular in studies of lattice spin systems such as the Ising model, it is not so clear how the function $f(\eta)$ is to be defined in microscopic terms. A possible starting point for progress in this area might be to provide such a clarification.

A different approach to these problems which has been tried recently is to consider microscopic models which do not correspond closely to any real physical system but instead are chosen so that their macroscopic kinetic equations can be rigorously derived in a suitable limiting case. In this spirit, de Masi *et al.* [13] have considered a kinetic Ising model in which two processes are taking place simultaneously: a 'stirring' process whose elementary step is to interchange the states of two randomly chosen nearest-neighbour sites, and a 'relaxation' process whose elementary step is to choose a site at random and then, with a probability which depends on the mean value of the order parameter near the chosen site, to reverse the state of that site. In a limit which involves rescaling both the space and time variables, and in which the stirring process goes much faster than the relaxation process, they derive the time-dependent Ginzburg-Landau equation (3.3) as a rigorous consequence of their microscopic assumptions. The application of these results to systems for which equation (3.3) predicts the existence of metastable states is given in a sequel (de Masi *et al.* [14]) to the paper cited; the result is that, in the limit where equation (3.3) becomes rigorously valid, there is indeed a metastable state, and its lifetime is infinite in this limit. The next step along this path might be to find a microscopic kinetic model leading to the Cahn-Hilliard equation.

4. CLUSTER KINETICS

In Section 2 we mentioned the picture used in the Lifshitz-Slyozov theory to describe the late stages of evolution, with COP dynamics, after a shallow quench to a point such as P_1 in Fig. 2. In this picture most parts of the evolving system are thought of as being in a state similar to the nearby equilibrium state P, with spherical inclusions (grains) of the opposite phase, Q in Fig. 2. As it stands this picture is deterministic and makes no allowances for fluctuations; hence it can tell us nothing about the lifetimes of metastable states. The picture can, however, be refined if we do not assume that grains of the same size always behave in the same way.

A system of equations based on this more refined picture was formulated by Becker and Döring in 1935. To describe the Becker-Döring equations conveniently, let us specialize to the case of an alloy consisting of two types of atom, *A* and *B* say, in which atoms of type *A* are in the majority. This could correspond to the point P_1 in Fig. 2 if we take η to be the concentration of the rarer type of atom, i.e. type *B*. The equilibrium phase *P* then consists mainly of *A* atoms, but phase *Q* consists mainly of *B* atoms. In this theory, the grains of the Lifshitz-Slyozov theory are represented by clusters of *B* atoms. The method of defining a 'cluster' is subject to some variation, but the simplest is that a cluster is a maximal connected set of *B* atoms (where two *B* atoms are said to be connected if there is a chain of *B* atoms connecting them, each link in the chain being a pair of *B* atoms which are nearest neighbours in the underlying lattice). The clusters need not be spherical, though it is to be expected that they will be approximately so at temperatures low enough for surface tension between the *A*-rich and *B*-rich phases to have an important influence on their shape.

The size of a cluster is measured by giving the number of *B* atoms in it, rather than its volume as in the Lifshitz-Slyozov description. It is assumed that clusters grow and shrink in the smallest possible steps only; in the COP case we are considering here, this means that a cluster (of size other than 1) can change its size only by absorbing or emitting a single *B* atom (i.e. a cluster of size 1, or *monomer*).

Let us write J_l for the net rate (per unit volume per unit time) at which clusters of size l are being converted to clusters of size $l+1$; then the net rate of change of c_l, the number of l-sized clusters per unit volume, is given, for $l \geqslant 2$, by

$$\frac{dc_l}{dt} = J_{l-1} - J_l \qquad (l \geqslant 2) . \tag{4.1}$$

The rate of change of c_1, the number of monomers per unit volume, can then be determined from the condition that the total number of B atoms is conserved:

$$\frac{d}{dt} \Sigma_l \, l c_l = 0 . \tag{4.2}$$

To use equation (4.1) we need an expression for J_l. Under the assumptions we are using here, J_l can be written as the difference of two terms, one representing the rate per unit volume at which clusters of size l are combining with monomers to form clusters of size $l+1$, and the other representing the rate per unit volume at which clusters of size $l+1$ are breaking up to give a cluster of size l and a monomer. Making the assumption, taken over from chemical kinetics, that each of these rates is proportional to the concentrations of the relevant types of cluster present at the beginning of the step, we obtain

$$J_l = a_l c_1 c_l - b_{l+1} c_{l+1} \qquad (l \geqslant 1) \tag{4.3}$$

where the a_l's and b_l's are constants.

The system of equations (4.1) — (4.3) was formulated by Becker and Döring [17] to describe nucleation in a supersaturated vapour, whose theory is closely analogous to the alloy case considered here. It is believed (Abraham [19], Penrose *et al.* [20]) to give a satisfactory representation of the nucleation and growth of clusters provided the overall concentration of B atoms,

$$\rho = \Sigma \, l c_l , \tag{4.4}$$

is not too large. In their original paper, Becker and Döring did not use the conservation condition (4.2); instead they made the simpler assumption that c_1 is a constant. This is a good approximation for metastable states, which were Becker and Döring's main interest, but not for the phenomena envisaged in the Lifshitz-Slyozov theory, for which the depletion of the monomers as the clusters get larger and larger is important. (The

Lifshitz-Slyozov equations (2.1) and (2.2) can be regarded as an approximation to the Becker-Döring equations (4.1) and (4.3), in which $c_l(t)$ has been replaced by $f(s,t)\,ds/dl$ and $J_l(t)$ by $vf(s,t)\,ds/dl$.)

To study the Becker-Döring equations mathematically, we need information about the coefficients a_l and b_l; this can be obtained from statistical mechanics. First of all, the equilibrium solution of the equations is

$$c_l = Q_l\, c_1^{\,l} \tag{4.5}$$

where

$$Q_l = \frac{a_1 a_2 \cdots a_{l-1}}{b_2 b_3 \cdots b_l}\,. \tag{4.6}$$

Using methods of equilibrium statistical mechanics the quantity Q_l can be expressed in terms of whatever microscopic model is being used to represent the molecules constituting the system (see, for example, Lebowitz and Penrose [16]). For large l, and temperatures well below T_c, it is believed (Perini *et al.* [22]) that Q_l has the asymptotic form

$$Q_l \sim \text{const.}\; z_0^{-l} \exp(-\sigma l^{2/3}) \tag{4.7}$$

where z_0 and σ are positive constants. Formula (4.7) has the important consequence that if c_l is given by (4.5) for all l, then the series (4.4) for the total density of B atoms becomes

$$\sum_{l=1}^{\infty} l\, Q_l\, c_1^{\,l} \tag{4.8}$$

which cannot converge to a value greater than ρ_0, where

$$\rho_0 := \sum_{l=1}^{\infty} l\, Q_l\, z_0^{\,l}\,. \tag{4.9}$$

If the density of B atoms is greater than ρ_0, then the Becker-Döring equations have no equilibrium solution.

The question now arises of describing the long-time behaviour of the solutions of these equations. Does the solution cease to exist after a certain time, indicating the formation of an infinite cluster as in the theory of gelation (see for example Hendriks *et al.* [23]), or does the solution exist for all times but never reach equilibrium? This question is answered in the paper by Ball *et al.* [18], which shows that if a_l/l is bounded as $l \to \infty$ then the equations (4.1) to (4.3) have a solution for all positive times t; on the other hand if $a_l/l \to \infty$ as $l \to \infty$ then no solution exists even for small t. For the alloy system, a_l is believed to be proportional to $l^{1/3}$ (Lifshitz and Slyozov [1], Penrose *et al.* [26]) or at most to $l^{2/3}$ (Abraham [19]) so that a solution does exist and the equations do not lead us to expect the formation of an infinite cluster. (Even so, computer simulations (Heermann [25]) show infinite clusters being formed if the concentration of B atoms is high enough; this discrepancy illustrates the limitations of the Becker-Döring equations as a model at such concentrations.)

The paper of Ball *et al.* [18] also contains results about the behaviour of the solution for large t. If the overall contration of B atoms, as given by (4.4), is less than ρ_0 as given by (4.9) then we have

$$\lim_{t\to\infty} c_l(t) = Q_l z^l \tag{4.10}$$

where z is defined by

$$\rho = \sum_{l=1}^{\infty} l Q_l z^l . \tag{4.11}$$

On the other hand if the overall concentration is greater than ρ_0 we have, in place of (4.10),

$$\lim_{t\to\infty} c_l(t) = Q_l z_0^l \tag{4.12}$$

where z_0 is defined in (4.7). In this case the convergence to the limit is nonuniform in l, so that

$$\sum_{l=1}^{\infty} \lim_{t\to\infty} l c_l(t) < \lim_{t\to\infty} \sum_{l=1}^{\infty} l c_l(t) \tag{4.13}$$

since the left-hand-side is equal to ρ_0 and the right-hand-side to ρ. The difference between the two sides corresponds to the contribution to the sum in (4.4) from the B atoms contained in a group of clusters whose average size grows without limit as t increases but is finite for every finite t.

In the original version of the Becker-Döring equations, c_1 was taken to be constant instead of being determined by equation (4.2). Under this assumption, if $c_1 > z_0$, equations (4.1) and (4.3) have a steady-state solution in which J_l has a positive value independent of l. This solution is

$$J_l = J := 1 \Big/ \sum_{l=1}^{\infty} \left(a_l Q_l c_1^{\,l+1}\right)^{-1} \tag{4.14}$$

$$c_l = Q_l c_1^{\,l} J \sum_{r=l}^{\infty} \left(a_r Q_r c_1^{\,r+1}\right)^{-1} \tag{4.15}$$

the infinite series being convergent because $c_1 > z_0$. The constant J in (4.14) gives the rate per unit volume at which large clusters are acquiring B atoms from the surrounding monomers; it is called the nucleation rate.

Formulas for the nucleation rate arising out of Becker and Döring's expression work well for interpreting experimental data on metastability and nucleation (see for example Sinha *et al.* [24]). Unfortunately, however, the solution (4.14-15) upon which they are based cannot be true for arbitrarily large l, since if it were then the series in formula (4.4) for the total concentration of B atoms would diverge. This can be seen most easily from equation (4.3), which implies, when $J_l = J = \text{const.}$, that

$$c_l \geqslant 1/J c_1 a_l \tag{4.16}$$

and hence that $lc_l \to \infty$ as $l \to \infty$. Out of the outstanding problems in the study of the Becker-Döring equations (4.1)-(4.3) is to derive from them a description of the metastable states and nucleation rate which avoids this difficulty.

Another interesting problem arising from the Becker-Döring equations is to find a systematic way of deriving from them an approximation corresponding to the Lifshitz-Slyozov equations (2.1)-(2.3). A start in this direction was made by Penrose *et al.*[26], but what one would really like, in place of the *ad hoc* methods used there (and in the Becker-Döring theory of nucleation) is a systematic method for constructing approximate large-time solutions of the equations, with proper error estimates and proofs of convergence.

This article has concentrated on mathematical aspects of the kinetics of phase transitions rather than the physical or computational aspects. For further information about the latter the reader is referred to Gunton and Droz [27], Gunton *et al.* [28], or Koch [29].

REFERENCES

1. I.M. Lifshitz and V.V. Slyozov, "The kinetics of precipitation from supersaturated solid solutions", *J. Phys. Chem. Solids* **19**, (1961), pp.35-50.
2. I.M. Lifshitz, "Kinetics of ordering during second-order phase transitions", *Soviet Physics JETP* 15, (1962), pp.939-942.
3. M.K. Phani, Joel L. Lebowitz, M.H. Kalos and O. Penrose, "Kinetics of an order-disorder transition", *Phys. Rev. Lett.* **45**, (1980), pp.366-369.
4. J. Fröhlich, "The statistical mechanics of surfaces", *Springer Lecture Notes in Physics* **216**, (1984), pp.31-57.
5. J.W. Cahn and J.E. Hilliard, "Free energy of a nonuniform system I: Interfacial free energy", *J. Chem. Phys.* **28**, (1958), pp.258-267.
6. J.W. Cahn and J.E. Hilliard, "Free energy of a nonuniform system III: Nucleation in a two component incompressible fluid", *J. Chem. Phys.* **31**, (1959), pp.688-699.
7. J.C. Maxwell, (1874), *Scientific Papers*, vol. 2, (ed. Niven, W.D.), Dover, New York, 1965, p.425.
8. H.E. Cook, "Brownian motion in spinodal decomposition", *Acta. Met.* **18**, (1970), pp.297-306.
9. J.S. Langer, "Theory of the condensation point", *Ann. Phys. NY* **41**, (1967), pp.108-157.
10. O. Penrose and A. Buhagiar, "Kinetics of nucleation in a lattice gas model: microscopic theory and simulation compared", *J. Stat. Phys.* **30**, (1983), pp.219-241.

11. R.B. Griffiths, in *Phase Transitions and Critical Phenomena*, vol. I, (eds. Domb, C. and Green, M.S.) Academic Press, 1972, pp.7-109 (particularly p.35).
12. J.L. Lebowitz and O. Penrose, "Rigorous treatment of the van der Waals-Maxwell theory of the liquid-vapour transition", J. Math. Phys. **7**, (1966), pp.98-113.
13. A. de Masi, P. Ferrari and J.L. Lebowitz, "Reaction-diffusion equations for interacting particle systems", *J. Stat. Phys.* **44**, (1986), pp.589-644.
14. A. de Masi, E. Presutti and M.E. Vares, "Escape from the unstable equilibrium in a random process with infinitely many interacting particles", *J. Stat. Phys.* **44**, (1986), pp.645-696.
15. E.A. Coutsias and J.C. Neu, "The aging of nuclei in a binary mixture", *Physica* **12D**, (1984), pp.295-302.
16. J.L. Lebowitz and O. Penrose, "Cluster and percolation inequalities for lattice systems with interactions", *J. Stat. Phys.* **16**, (1977), pp.321-337.
17. R. Becker and W. Döring, "Kinetische behandlung der keimbildung in übersättigten dämpfen", *Ann. der Phys.* (*Leipzig*), **24**, (1935), pp.719-752.
18. J.M. Ball, J. Carr and O. Penrose, "The Becker-Döring cluster equations: basic properties and asymptotic behaviour of solutions", *Commun. Math. Phys.* **104**, (1986), pp.657-692.
19. F.F. Abraham, *Homogeneous Nucleation Theory*, Academic Press, New York, 1974.
20. O. Penrose, J. Lebowitz, J. Marro, M. Kalos and J. Tobochnik, "Kinetics of a first-order phase transition: computer simulations and theory", *J. Stat. Phys.* **34**, (1984), pp.399-426.
21. S.M. Allen and J.W. Cahn, "A microscopic theory for antiphase boundary motion and its application to antiphase domain coarsening", *Acta. Metall.* **27**, (1979), pp.1085-1095.
22. A. Perini, G. Jacucci and G. Martin, "Cluster free energy in the simple-cubic Ising model", *Phys. Rev.* **B29**. (1984), pp.2689-2697.
23. E.M. Hendriks, M.H. Ernst and R.M. Ziff, "Coagulation equations with gelation", *J. Stat. Phys.* **31**, (1983), pp.519-563.
24. D.N. Sinha, J.S. Semura and L.C. Brodie, "Homogeneous nucleation in ^{4}He: a corresponding-states analysis", *Phys. Rev.* **A26**, (1982), pp.1048-1061.
25. D. Heermann, "Dynamical spinodal: the transition between nucleation and spinodal decomposition", *Z. Phys.* **B55**, (1984), pp.309-315.
26. O. Penrose, J. Lebowitz, J. Marro, M.H. Kalos and A. Sur, "Growth of clusters in a first-order phase transition", *J. Stat. Phys.* **19**, (1978), pp.243-267.
27. J.D. Gunton and M. Droz, "Introduction to the theory of metastable and unstable states", *Springer Lecture Notes in Physics* **183**, (1983).

28. J.D. Gunton, M. San Miguel and P.S. Sahni, in *Phase Transitions and Critical Phenomena*, vol. 8 (eds. Domb, C. and Lebowitz, J.L.) Academic Press, 1983, p.267.
29. S.W. Koch, "Dynamics of first-order phase transitions in equilibrium and nonequilibrium systems", *Springer Lecture Notes in Physics*, **207**, (1984).
30. P.C. Hohenberg and B. Halperin, "Theory of dynamic critical phenomena", *Reviews of Modern Physics* **49**, (1977), pp.435-479.
31. G. Caginalp, "An analysis of a phase field model of a free boundary", *Arch. Rat. Mech. Anal.* **92**, (1986), pp.205-245.
32. J.W. Cahn, "On spinodal decomposition", *Acta. Metall.* **9**, (1961), pp.795-801.
33. J.W. Cahn, "On spinodal decomposition in cubic crystals", *Acta. Metall.* **10**, (1962), pp.179-183,
34. J. Carr, M. Gurtin and M. Slemrod, "Structured phase transitions on a finite interval", *Arch. Rat. Mech. Anal.* **86**, (1984), pp.317-351.

25

ON 1- AND 3-DIMENSIONAL MODELS IN 'NON-CONVEX' ELASTICITY

MARIO PITTERI

1. INTRODUCTION

Recently various authors have analyzed problems in elasticity where certain usually accepted conditions fail for good physical reasons. This is the case of the conditions of ellipticity in elastostatics and of hyperbolicity in elastodynamics, for instance. In [3] Ericksen proposes a one-dimensional equilibrium theory of bars for which the stress is not a monotone function of the strain. Again for one-dimensional bars, James [11] shows that minimizers of an energy functional based on a suitable non-convex stored energy density of the kind proposed by Ericksen [3] are a good model for certain equilibrium configurations of polymers in which two phases, that is, two states of strain, can coexist. The analysis works well for hard as well as soft loading devices, and it allows us to handle body forces and inhomogeneities in the response of the material. Furthermore the coexistence of different phases at equilibrium requires that we consider weak rather than strong relative minimizers according to the calculus of variations.

In the theory above the problem seems to be not the existence, but rather the severe lack of uniqueness of equilibrium solutions, in particular for a homogeneous bar and no body force. According to James [12], there is also lack of uniqueness for the Riemann problem in the elastodynamics of the aforementioned bar, and Slemrod [22], among others, proposes a criterion to select one in a certain class of weak solutions of the equations of motion for a body for which stress is not a monotone function of strain.

The three-dimensional analysis of phenomena like twinning and phase transitions in crystalline solids by means of nonlinear elasticity is a largely unexplored territory. Parry [15], Gurtin [9] and James [13],[14], analyze conditions for piecewise homogeneous deformations to be minimizers of various sorts for an energy functional based on a non-convex stored energy density. The analysis includes the usual kind of loads, in particular dead loads.

In these analyses the material symmetry group of the constitutive equations is a subgroup of the orthogonal group, in agreement with the generally accepted definition of crystalline solids within continuum mechanics proposed by Coleman and Noll [2]. To describe phenomena like twinning and martensitic phase transitions, Ericksen [4] proposed a group which is a conjugate of the group $GL(3,\mathbb{Z})$ of unimodular matrices of integers. This much larger invariance is shown in [17] to be equivalent to the generally accepted one in a neighbourhood of any given configuration. In addition, according to Ericksen [5]-[7] and Pitteri [18]-[20], that invariance provides a description of twinned equilibrium configurations for zero load which encompasses many of the configurations that are experimentally observed. For such piecewise homogeneous natural states the stability analysis can be essentially reduced to kinematics, and this fact explains why the class of piecewise homogeneous deformations constitutes also an attractive first choice when we want to analyze equilibria for a loaded crystal.

On the other hand, having as large a material symmetry group as $GL(3,\mathbb{Z})$ generates difficulties when we consider equilibrium configurations that are not natural states, for a loaded crystal. Indeed, according to Fonseca [8], in the case of a dead traction boundary-value problem various usually accepted conditions fail. Such is the case of ellipticity, which cannot hold everywhere in the domain of the constitutive equations. Also, growth conditions for the stored energy density and sequential weak lower semicontinuity for the stored energy functional do not hold. These conditions are important if we have to apply the direct method of the calculus of variations, as is shown by Ball and Marsden [1], for instance. More, in the case of a pure traction boundary-value problem with body force and dead load, the total energy functional is not even bounded below, except in the trivial case of vanishing body force and traction.

In this paper we analyze a three-dimensional equilibrium problem which is the most simple-minded three-dimensional analogue of the one-dimensional traction problem considered by Ericksen [3] and James [11]. We look for piecewise homogeneous equilibrium configurations of a crystalline solid which in the reference configuration κ has a cylindrical shape and is subjected to a dead load on the bases and to no body force.

We first assume the body to be homogeneous with respect to κ and hyperelastic. We show that, irrespective of additional constitutive assumptions, *no nonhomogeneous piecewise homogeneous deformation* in a suitable neighbourhood of a standard twin *is possible* for nonvanishing load unless any plane of discontinuity is almost *parallel* to the axis of the cylinder. This result follows from the equilibrium equations and boundary conditions alone, so it should follow from any reasonable definition of minimizer of the total energy. On the other hand, experimental observations indicate that twinning planes need neither be almost parallel nor orthogonal to the axis of the specimen, so we may interpret our result by saying that it is not possible to obtain a standard twin from a homogeneous cylindrical crystal by only dead-loading its bases. The aforementioned result rests upon kinematic conditions of compatibility on singular surfaces, and from the condition that the lateral surface of the cylinder be free. Neither condition is effective in the one-dimensional case, and this explains why in this case non-homogeneous piecewise homogeneous deformations can be minimizers of the total energy for non-zero load also.

We show that the aforementioned result of non-existence is unaffected by adding a surface energy of the kind introduced by Parry [16] to the bulk stored energy: that energy is associated with singular surfaces, and depends on the jump of the deformation gradient across the surface.

Secondly, we assume κ to be obtained from a homogeneous reference configuration $\bar{\kappa}$ by applying a standard twinning operation to the shape the crystal occupies in $\bar{\kappa}$. So, now our cylindrical specimen is already twinned in the reference configuration $\bar{\kappa}$, which is a minimizer of the total energy for zero load. We look for a continuous, piecewise homogeneous deformation from $\bar{\kappa}$ which depends smoothly on the load, for sufficiently small load. To simplify the calculations, we assume the cylinder to be a parallelepiped, and look for two-dimensional solutions, that is, displacements which depend on only one of the two coordinates describing the rectangular sections of the parallelepiped. Denoting by $\boldsymbol{C}$ the right Cauchy-Green deformation tensor with respect to $\bar{\kappa}$, and by $\bar{\boldsymbol{C}}_1$ and $\bar{\boldsymbol{C}}_2$ the values of $\boldsymbol{C}$ corresponding to the aforementioned twinned configurations, we assume the stored energy density to have the form

$$\hat{W}(\boldsymbol{F}) = \tilde{W}(C_{LM}) = W(x_1, x_2, x_3) ,$$

where

$$W(x_1,x_2,x_3) = k(x_1 - x_{10})^2/2 + l(x_2 - x_{20})^2(x_2 + x_{20})^2 + m(x_3)^2 \qquad (3.8)_r$$

in suitable neighbourhoods of $\bar{\boldsymbol{C}}_1$ and $\bar{\boldsymbol{C}}_2$. Here k, l, m are positive constants and, in the vector space of symmetric 2×2 tensors, x_2 is a coordinate along the line through $\bar{\boldsymbol{C}}_1$ and $\bar{\boldsymbol{C}}_2$, whereas x_1 and x_3 are coordinates in the plane orthogonal to that line. The choice of $\bar{\boldsymbol{C}}_1$, $\bar{\boldsymbol{C}}_2$ and the constants x_{10} and x_{20}

implies that $\bar{C}_1$ and $\bar{C}_2$ are isolated local minima for W. We regard $(3.8)_r$ as a reasonable local approximation for a stored energy function which is invariant under the transformation that maps the configuration corresponding to $\bar{C}_1$ into the one corresponding to $\bar{C}_2$. For the choice $(3.8)_r$ of W we prove that, to have the continuous deformation we mentioned above, we need the twinning plane to be almost parallel to the axis of the cylinder. This result suggests that, to deal with dead-loaded, twinned crystalline bodies, we should use a class of deformations which is larger than the one of piecewise homogeneous deformations. That result is rather strong because it does not rest on the fact that $GL(3,\mathbb{Z})$ is a large group, since only one of its elements is involved in the proof.

We show that all the aforementioned results on nonexistence remain true if we enlarge the class of deformations to the special nonhomogeneous ones considered by Hagan and Serrin [10] and Silhavy [21], among others. For deformations in that class the displacement is constant along any reference plane which is orthogonal to a given vector $\boldsymbol{v}$.

2. PIECEWISE HOMOGENEOUS DEFORMATIONS

Let us consider a body $\mathcal{B}$ of homogeneous hyperelastic material; let $\kappa:\mathcal{B}\to\mathbb{R}^3$ be a homogeneous reference configuration of $\mathcal{B}$; let

$$x = \hat{x}_\kappa(x) \tag{2.1}$$

be the representation of an admissible deformation of $\mathcal{B}$ with respect to κ; let $F = \operatorname{grad}\hat{x}_\kappa$ denote the deformation gradient with respect to κ, and assume the stored energy density $W = \hat{W}(F)$ per unit mass to have a material symmetry group G which is a conjugate of the group $G = GL(3,\mathbb{Z})$. This implies that it is possible to choose in κ a basis E_a, $a = 1,2,3$, such that the representation of G with respect to that basis is the group G. This group and its relevance to the constitutive equations of elastic crystals have been analyzed by Ericksen [4].

We shall consider piecewise homogeneous equilibrium configurations of a part C of $\mathcal{B}$ which in κ occupies a straight cylinder C^*. More explicitly, letting S^* being a two-dimensional planar, simply connected regular domain immersed in $\mathbb{R}^3$; letting $\boldsymbol{A}_1$ being any one of the two unit vectors orthogonal to S^*, and letting $\boldsymbol{A}_r$, $r=2,3$, be two unit vectors such that $\boldsymbol{A}_r$, $r=1,2,3$ form an orthonormal basis, we assume that, for some real number $L>0$ and some origin O

$$C^* := \kappa(C) = \{y \in \mathbb{R}^3 \,|\, y = \mathrm{O}P^* + \lambda \boldsymbol{A}_1 , \; P^* \in S^* \text{ and } \lambda \in [0,L]\}. \tag{2.2}$$

Let us partition C^* by means of a finite number, say $n-1$, of parallel planes π_i, $i=1,\cdots,n-1$, and at first let us assume that each plane intersects the lateral surface of C^* only. In addition let us select the unit vector $\boldsymbol{N}$ orthogonal to the planes π_i which forms an acute angle with $\boldsymbol{A}_1$. We order the planes according to increasing values of λ for their intersection with any given straight line which is parallel to $\boldsymbol{A}_1$ and intersects S^*. For $r=2,\cdots,n-1$ we denote by C_r^* the part of C^* comprised between π_{r-1} and π_r, and it is quite obvious how this definition can be extended to the cases $r=1$ or $r=n$.

For any partition of C^* as above, we shall consider continuous deformations where F assumes a constant value F_r over each set C_r^*. Then, by a standard theorem presented by Truesdell and Toupin [23], for instance, the compatibility conditions

$$F_{r+1} - F_r = \alpha_r \otimes \boldsymbol{N}, \quad r = 1,\cdots,n-1 \tag{2.3}$$

hold for some choice of the amplitude α_r. If the deformation above is stable or metastable under dead load and no body forces according to any reasonable definition of stability or metastability, it must satisfy the standard equilibrium equations and jump conditions

$$\operatorname{div} \boldsymbol{T}_\kappa = 0 \text{ in } \bigcup_{i=1}^{n} \mathring{C}_i^*, \quad \boldsymbol{T}_\kappa \boldsymbol{\nu} = \tilde{\boldsymbol{f}} \text{ on } \partial C^*, \tag{2.4}$$

and

$$[T_K]N = 0 \quad \text{on} \quad \pi_i, \quad i = 1, \cdots, n-1. \tag{2.5}$$

Here T_K is the Piola stress tensor; v is the outer normal to ∂C^*; $\tilde{f}$ is the prescribed dead traction on ∂C^*; div denotes the divergence with respect to the reference coordinates X, and [ ] denotes the jump on a singular surface. Since B is hyperelastic, T_K is constant on each C_r^* hence $(2.4)_1$ holds trivially for the aforementioned piecewise homogeneous deformation, and we are left with $(2.4)_2$ and (2.5). By (2.4), (2.5) and the divergence theorem

$$\int_{\partial C^*} x \otimes \tilde{f}\, dS = \int_{C^*} T_K^T dV \quad \text{and} \quad \int_{\partial C^*} \tilde{f}\, dS = 0\,. \tag{2.6}$$

Let us now assume that $\tilde{f}$ vanishes on the lateral surface of C^*, and assumes the constant values $-f$ and f on the two bases S^* and $\bar{S}^* := S^* + LA_1$, respectively. Notice that $\tilde{f}$ has to be constant on either one of the bases because of $(2.4)_2$, and that the two constant vectors have to be one the negative of the other because of $(2.6)_2$. These results, together with $(2.4)_2$, imply that, on C_r^*,

$$T_K = b_r \otimes A_1 \tag{2.7}$$

for some choice of vectors b_r. But (2.5) and $(2.4)_2$ imply that

$$T_K = f \otimes A_1 \quad \text{throughout} \quad C^*. \tag{2.8}$$

A similar result on T_K being constant over the body in certain coherent, piecewise homogeneous deformations is obtained by James [14], but under different assumptions.

Equality (2.8) and the balance of rotational momentum severely restrict the possible choices of N. Indeed in $C_1^*\,[C_2^*]$, for instance, the symmetry of the Cauchy stress tensor is equivalent to

$$T_K F_1^T = (T_K F_1^T)^T \quad [T_K F_2^T = (T_K F_2^T)^T]. \tag{2.9}$$

These conditions and (2.3) imply that

$$F_1 A_1 \wedge f = 0 \quad \text{and} \quad \text{either} \quad f \wedge \alpha_1 = 0 \quad \text{or} \quad A_1 \cdot N = 0\,. \tag{2.10}$$

Under the geometric assumptions above, condition $(2.10)_3$ cannot hold, hence the analogue of $(2.10)_2$ must hold for any $\boldsymbol{\alpha}_r$. Roughly, if $\boldsymbol{f} \neq \mathbf{0}$, no matter how small it is, we see that the cylinder C^* is mapped onto a cylinder whose axis has the common orientation $\boldsymbol{V}$ of any one of the parallel vectors $\boldsymbol{F}_r \boldsymbol{A}_1$, and that the applied force $\boldsymbol{f}$ and the amplitudes $\boldsymbol{\alpha}_r$ all have the direction of $\boldsymbol{V}$; in particular

$$\boldsymbol{\alpha}_r = \lambda_r \boldsymbol{F}_r \boldsymbol{A}_1 \quad \text{for suitable} \quad \lambda_r \in \mathbb{R} . \tag{2.11}$$

Therefore, for any choice of $\bar{\boldsymbol{F}}_r$ satisfying the analogues of (2.3) for vectors $\bar{\boldsymbol{\alpha}}_r$ such that

$$\bar{\boldsymbol{\alpha}}_r = \bar{\boldsymbol{F}}_r \boldsymbol{B}_r , \quad \text{where} \quad \boldsymbol{B}_r \neq \mathbf{0} \quad \text{and} \quad \boldsymbol{B}_r \cdot \boldsymbol{N} = 0 , \tag{2.12}$$

there is an $\varepsilon > 0$ such that, for non-vanishing load, there are no piecewise homogeneous solutions to the equilibrium equations (2.4),(2.5) such that †

$$\|\bar{\boldsymbol{F}}_1 - \boldsymbol{F}_1\| < \varepsilon \quad \text{and} \quad \|\bar{\boldsymbol{\alpha}}_r - \boldsymbol{\alpha}_r\| < \varepsilon . \tag{2.13}$$

Indeed, since $\boldsymbol{A}_1 \cdot \boldsymbol{N} > 0$, condition (2.12) implies that $\boldsymbol{B}_r \wedge \boldsymbol{A}_1 \neq \mathbf{0}$. By continuity there is an $\varepsilon > 0$ such that, for all $\boldsymbol{F}_1$ and $\boldsymbol{\alpha}_r$ satisfying (2.13), $\boldsymbol{\alpha}_r \wedge \boldsymbol{F}_r \boldsymbol{A}_1 \neq \mathbf{0}$, contradicting (2.11). Equivalently, it is not possible to obtain nonhomogeneous piecewise homogeneous equilibrium configurations of the aforementioned type from a homogeneous cylinder by only applying dead loads to its bases. This conclusion is bothersome in some respects, because twinned configurations belong to the type of configurations we just mentioned. To simplify matters, assume that $\boldsymbol{F}_r = \boldsymbol{F}_1$ for r odd, $\boldsymbol{F}_r = \boldsymbol{F}_2$ for r even, and

$$\boldsymbol{S}\boldsymbol{F}_1 = \boldsymbol{F}_2 = \boldsymbol{Q}\boldsymbol{F}_1 \boldsymbol{H} , \quad \boldsymbol{S} = \mathbf{1} + \tilde{\boldsymbol{a}} \otimes \boldsymbol{n} , \quad \boldsymbol{F}_1^T \boldsymbol{n} = \boldsymbol{N} \tag{2.14}$$

† Equivalently: any piecewise homogeneous configuration whose gradient in C_r^* is $\bar{\boldsymbol{F}}_r$, the $\bar{\boldsymbol{F}}_r$'s satisfying the conditions above, has a neighbourhood in $W^{1,\infty}$ in which there are no piecewise homogeneous solutions to the equilibrium equations for nonvanishing load.

for some choice of

$$H \in G\ ,\ H^2 = 1\ ,\ Q = Q^{-T}\ ,\ \text{and}\ \tilde{a}\ \text{such that}\ \tilde{a}\cdot n = 0\ . \quad (2.15)$$

Solutions of (2.14) and (2.15), also in the case that $(2.15)_2$ does not hold, have been considered by Ericksen [5] – [7], Gurtin [9], James [13], [14] and Pitteri [18] – [20], for instance. According to their treatment of twinning, if F_1 is a locally unique minimum for $W(F)$, then the piecewise homogeneous deformation satisfying (2.14) – (2.15) is a coherent unstressed stable equilibrium configuration, hence it is a solution of the equilibrium equations for vanishing load. This solution corresponds to what crystallographers and metallurgists would call a twinned crystal, and piecewise homogeneous configurations like this one describe well, at least locally, configurations that are very common in metals and minerals. From the discussion above we conclude that, in general, *it is not possible to twin a cylinder of homogeneous crystalline solid by only applying dead loads to its bases*. On the other hand it is known experimentally that a homogeneous crystal can be twinned by applying suitable loads to its boundary. An easy objection is that we are not adopting the right boundary conditions. Nevertheless the assumption that dead loads are applied to the bases of the cylinder does not seem to provide so bad a description of what we do when we pull a thin strip of tin or indium with our hands. These materials twin when so pulled, and in the case of tin the process is accompanied by a noise that can be heard. A possible resolution of the contradiction above is that, in the aforementioned strip, tin or indium are polycrystals rather than single crystals, and the presence of grain boundaries perhaps renders the model we used too rough. That is, we may have to take additional energy terms into account. This resolution may work in some cases, but certainly not always. Indeed in [25] we see examples of mechanical twinning of cylindrical specimens of zirconium, any cylinder being a coarse-grained polycrystal whose grains occupy a whole section of the cylinder.

Admittedly, in four tests the loading consisted of a dead load applied to the bases of the specimen. Hence we should expect our conclusions above to hold within each grain, in contradiction with the observations.

In spite of the aforementioned difficulties, the result above of nonexistence of nonhomogeneous piecewise homogeneous equilibrium configurations under load may be useful to justify certain experimental facts. Forces of a certain strength are required to twin a crystal, that is, a crystal will not twin for arbitrarily small applied loads. Here one-dimensional models, like the one discussed by James [11], seem to suggest the opposite if one applies them to crystals, at least for homogeneous crystals and no body force. Indeed, in this case, weak relative minima composed of two phases exist for arbitrarily small loads. We do not regard it as reasonable to consider strong relative minima, which correspond to only one phase except at zero load, because they do not seem fit to describe twinned configurations. As we mentioned in the introduction, the difference between the one- and the three-dimensional situation described above comes from jump conditions on singular surfaces and from the assumption that the lateral surface of the cylinder is free, and these conditions are both lost in a one-dimensional model.

REMARK 1. The conditions that all the planes π_i be parallel, and that none of them intersects the bases of the cylinder are not strictly necessary for our conclusion on non-existence of twinned equilibrium configurations under load. What is needed is that the planes π_i do not intersect, and partition C^* into domains C^*_r in each one of which (2.7) holds. This is true if each C^*_r has a non-empty part of its boundary in common with $\partial C^* \setminus (S^* \cup \bar{S}^*)$, but this condition can be further relaxed. If it holds, it is easy to see that (2.7), (2.8) and the analogue of (2.10) for $\boldsymbol{F}_r$, α_r and $\boldsymbol{N}_r$ hold. On the other hand, if C^* is a circular cylinder, then (2.7) holds under the sole assumption that the planes π_i do not intersect.

REMARK 2. By applying suitable piecewise constant forces $\boldsymbol{f}$ to the bases of the cylinder, the forces being actually constant if, for instance, the cylinder is circular, we can show that the aforementioned difficulties for the existence of nonhomogeneous piecewise homogeneous equilibrium configurations do not arise when $\boldsymbol{N}$ is orthogonal to $\boldsymbol{A}_1$. Of course, we still have to show that the constitutive equation for the stress tensor of $\mathcal{B}$ is compatible with equations of the form

$$\boldsymbol{T}_{\kappa} = \boldsymbol{f}_r \otimes \boldsymbol{A}_1 \quad \text{throughout } C_r^* \tag{2.16}$$

when the deformation gradients $\boldsymbol{F}_r$ satisfy (2.3). For certain choices of the stored energy density W these equations may not be compatible, as we show in Section 3 below.

In a recent paper Parry [16] provides a description of shear bands in unloaded crystals by adding to the stored energy W we considered above a surface energy which is defined on any singular surface and depends on the jump of $\boldsymbol{F}$ along the surface. One would hope that the problems we mentioned above would disappear for this modified energy. This is not so. Explicitly, in the simplest but not restrictive case of one singular surface Σ which partitions C^* into two regular domains C_1^* and C_2^*, the total energy of the deformation $\boldsymbol{x}$ is

$$E(\boldsymbol{x}) = \int_{C^*} W(\boldsymbol{F})\, dV + \int_{\partial C^*} \boldsymbol{f}\cdot\boldsymbol{x}\, dS + \int_{\Sigma} \tilde{\gamma}(\boldsymbol{F}_2, \boldsymbol{F}_1)\, dS\,. \tag{2.17}$$

Here $\boldsymbol{x}$ is a deformation which is of class C^{∞} in the closure of C_1^* and C_2^* and of class C^0 in C^*, and $\boldsymbol{F}_1[\boldsymbol{F}_2]$ is the continuous extension to Σ of the field $\boldsymbol{F}$ on C_1^* [on C_2^*]. Instead of the function $\tilde{\gamma}(\boldsymbol{F}_2, \boldsymbol{F}_1)$ it is convenient to consider the function $\gamma(\boldsymbol{F}_1, \boldsymbol{N}, \boldsymbol{\alpha}) = \tilde{\gamma}(\boldsymbol{F}_1 + \boldsymbol{\alpha} \otimes \boldsymbol{N}, \boldsymbol{F}_1)$, where $\boldsymbol{\alpha}$ is the amplitude of the discontinuity across Σ, and $\boldsymbol{N}$ is the unit normal field on Σ, positively oriented towards C_2^*. We can write the following first-order necessary conditions for the piecewise smooth deformation $\boldsymbol{x}$ to be a local minimum for the total energy:

$$\operatorname{div}\frac{\partial W}{\partial \boldsymbol{F}} = \mathbf{0} \quad \text{on} \quad \mathring{C}^*\setminus\Sigma, \qquad \frac{\partial W}{\partial \boldsymbol{F}}\,\boldsymbol{\nu} = \boldsymbol{f} \quad \text{on} \quad \partial C^*$$

$$\frac{\partial\gamma}{\partial\boldsymbol{\alpha}} = \mathbf{0}, \quad \frac{\partial\gamma}{\partial \boldsymbol{F}_1}\,\boldsymbol{N} = \mathbf{0} \quad \text{and} \quad \left(\frac{\partial\gamma}{\partial \boldsymbol{F}_1}\right)_{r}{}^{\alpha}{}_{/\alpha} + \left[\frac{\partial W}{\partial F^{r}_{L}}\right] N_L = 0. \tag{2.18}$$

Here $\boldsymbol{\nu}$ is the outer unit normal field to ∂C^*; $r, L = 1,2,3$ and $\alpha = 1,2$; $(\partial\gamma/\partial\boldsymbol{F}_1)_r{}^{\alpha} := (\boldsymbol{a}^{\alpha})_L\, \partial\gamma/\partial(F_1)^{r}{}_{L}$; $\boldsymbol{a}_{\alpha} := \partial\hat{\boldsymbol{x}}/\partial u^{\alpha}$ is the standard basis on the tangent plane to the surface Σ, which is assumed to have a local representation of the form $\boldsymbol{x} = \hat{\boldsymbol{x}}(u^1, u^2)$; $\boldsymbol{a}^{\alpha}$ denotes the dual basis on that same plane, and / denotes covariant differentiation with respect to the surface metric.

Indeed, let $\boldsymbol{\phi}$ be a deformation which is of class C^{∞} in the closure of C_1^* and C_2^* and of class C^0 in C^*, and denote by $\nabla\phi_1$ and $\nabla\phi_2$ the continuous interior limits of $\nabla\phi$ on Σ, from within C_1^* and C_2^*, respectively. Then the deformation $\boldsymbol{x} + \lambda\boldsymbol{\phi}$ has the same regularity as $\boldsymbol{\phi}$ for any $\lambda \in [0,1]$; the function

$$\begin{aligned} E(\lambda) := &\int_{C_1^*} W(\nabla(\boldsymbol{x}+\lambda\boldsymbol{\phi}))\,dV + \int_{C_2^*} W(\nabla(\boldsymbol{x}+\lambda\boldsymbol{\phi}))\,dV \\ &+ \int_{\Sigma} \gamma\left\{\nabla(\boldsymbol{x}+\lambda\boldsymbol{\phi})_1,\ \boldsymbol{N},\ [\nabla(\boldsymbol{x}+\lambda\boldsymbol{\phi})\,\boldsymbol{N}]\right\} dS \\ &+ \int_{\partial C^*} \boldsymbol{f}\cdot(\boldsymbol{x}+\lambda\boldsymbol{\phi})\,dS \end{aligned} \tag{2.19}$$

is differentiable at $\lambda = 0$, and this derivative has to vanish. By means of the divergence theorem

$$\begin{aligned} 0 = \left.\frac{dE(\lambda)}{d\lambda}\right|_{\lambda=0} &= -\int_{C^*} \operatorname{div}\frac{\partial W(\boldsymbol{x})}{\partial \boldsymbol{F}}\cdot\boldsymbol{\phi}\,dV + \int_{\partial C^*} \boldsymbol{\phi}\cdot\frac{\partial W(\boldsymbol{x})}{\partial \boldsymbol{F}}\,\boldsymbol{\nu}\,dS \\ &+ \int_{\Sigma}\left\{-\boldsymbol{\phi}\cdot\left[\frac{\partial W(\boldsymbol{x})}{\partial \boldsymbol{F}_1}\right]\boldsymbol{N} + \frac{\partial\gamma(\boldsymbol{x})}{\partial \boldsymbol{F}_1}\cdot\nabla\boldsymbol{\phi}_1\right. \\ &\left. + \frac{\partial\gamma(\boldsymbol{x})}{\partial\boldsymbol{\alpha}}\cdot[\nabla\boldsymbol{\phi}]\,\boldsymbol{N}\right\} dS + \int_{\partial C^*} \boldsymbol{f}\cdot\boldsymbol{\phi}\,dS, \end{aligned} \tag{2.20}$$

where, as indicated, the derivatives of W and γ are evaluated at the deformation $\boldsymbol{x}$. As is standard in the calculus of variations, for any given point $\boldsymbol{X}$ which is either in the interior of $C^*\setminus\Sigma$ or on $\partial C^*\setminus\Sigma$, we can choose a function ϕ which assumes an arbitrary value at X and has a compact support which is contained in an arbitrarily chosen neighbourhood of $\boldsymbol{X}$. For such points $(2.18)_1$ and $(2.18)_2$ hold, respectively. For points on $\partial C^*\cap\Sigma$, $(2.18)_2$ still holds for the interior limits. This leaves us with only the last two integrals on the right-hand side of (2.20).

It is not difficult to see that we can arbitrarily choose ϕ such that, first of all, ϕ, $\nabla\phi_1$ and $[\nabla\boldsymbol{\Phi}]\,\boldsymbol{N}$ have arbitrary values at any point $\boldsymbol{X}\in\Sigma$ and, secondly, ϕ vanishes outside an arbitrary neighbourhood of $\boldsymbol{X}$. Since $[\nabla\phi]\,\boldsymbol{N}$ is arbitrary, we obtain $(2.18)_3$, and are only left with the fourth integral and the first and second integrands in the the third integral on the right-hand side of (2.20).

Introducing a local representation

$$\boldsymbol{x} = \hat{\boldsymbol{x}}(u^1, u^2)\ ,\quad (u^1, u^2)\in D \tag{2.21}$$

of Σ in a neighbourhood η of $X\in\Sigma$, and the vectors $\boldsymbol{a}_\alpha$ and $\boldsymbol{a}^\alpha$ we mentioned above, we have

$$\frac{\partial\gamma}{\partial(F^r_L)_1}(\phi^{\ r}_{,L})_1 = \frac{\partial\gamma}{\partial(F^r_R)_1}N_R(\phi^{\ r}_{,S})_1 N^S + \frac{\partial\gamma}{\partial(F^r_R)_1}(\boldsymbol{a}^\alpha)_R(\phi^{\ r}_{,S})_1\,(\boldsymbol{a}_\alpha)^S\ , \tag{2.22}$$

where $(\boldsymbol{a}_\alpha)^S = \partial x^S/\partial u^\alpha$. Therefore, for ϕ whose support is contained in η,

$$\int_\Sigma \frac{\partial\gamma}{\partial F_1}\cdot\nabla\phi_1\,dS = \int_D\left\{\frac{\partial\gamma}{\partial(\boldsymbol{F}_1)^r_R}N_R\phi^{\ r}_{,S}N^S + \left(\frac{\partial\gamma}{\partial\boldsymbol{F}_1}\right)^\alpha_r\phi^r\,(\hat{\boldsymbol{x}}(u^1, u^2))_{,\alpha}\right\}dS\,. \tag{2.23}$$

By the arbitrariness of the normal derivative of ϕ_1 we obtain $(2.18)_4$. Furthermore, by Green's formula and the fact that $\phi(\hat{\boldsymbol{x}}(u^1, u^2))$ vanishes on ∂D

$$\int_D\left(\frac{\partial\gamma}{\partial\boldsymbol{F}_1}\right)^\alpha_r\phi^r_{,\alpha}\,dS = -\int_D\phi^r\left(\left(\frac{\partial\gamma}{\partial(\boldsymbol{F}_1)}\right)^\alpha_r\sqrt{a}\right)_{,\alpha}(a)^{-\frac{1}{2}}\,dS \tag{2.24}$$

where a is the determinant of the surface metric,

$$a = \left\| \frac{\partial x}{\partial u^1} \times \frac{\partial x}{\partial u^2} \right\|^2 , \text{ and } dS = \sqrt{a}\, du^1\, du^2 , \tag{2.25}$$

as is easy to verify. Since $(\sqrt{a})_{;\alpha} = \sqrt{a}\,\Gamma^{\beta}_{\alpha\beta}$, $\Gamma^{\alpha}_{\beta\gamma}$ being the Christoffel symbols associated with the surface metric, we see that

$$\int_D \left(\frac{\partial x}{\partial \boldsymbol{F}_1}\right)^{\alpha}_{r} \phi^r{}_{,\alpha}\, dS = - \int_D \phi^r \left(\left(\frac{\partial \gamma}{\partial (\boldsymbol{F}_1)}\right)^{\alpha}_{r}\right)_{,\alpha} dS . \tag{2.26}$$

Then $(2.18)_5$ follows by the arbitrariness of $\boldsymbol{\phi}$.

Going back to (2.18), we easily see that, for piecewise homogeneous deformations and plane Σ, $(2.18)_5$ reduces to the usual continuity of the traction across Σ. Therefore all the conclusions we obtained above without introducing the surface energy density γ remain unaffected by introducing it.

3. DEFORMATION OF A TWINNED CRYSTAL

Let us consider the cylinder C^* in the previous section, but assume now that it is already twinned. That is, assume $\bar{\kappa}$ to be a homogeneous reference configuration, and κ to be obtained from $\bar{\kappa}$ by means of a piecewise homogeneous deformation whose gradient is $\bar{\boldsymbol{F}}_r$ in $\bar{\kappa}\circ\kappa^{-1}(C^*_r)$. For our purposes it is sufficient to analyze the case that $r = 1,2$, when $\bar{\boldsymbol{F}}_1$ and $\bar{\boldsymbol{F}}_2$ correspond to a standard twin from the reference configuration $\bar{\kappa}$. To keep the calculation as simple as possible, we assume S^* to be a rectangle with edges parallel to $\boldsymbol{A}_2$ and $\boldsymbol{A}_3$, respectively; we consider reference vectors $\boldsymbol{E}_R$, $R = 1,2,3$, such that $E_3 = \boldsymbol{A}_3$, and we restrict our attention to deformations that map E_3 to itself and the plane of E_1 and E_2 onto itself. In these circumstances the original three-dimensional problem reduces to a two-dimensional problem. We further assume E_1 and E_2 to be the standard basis $(1,0),(0,1)$ of $\mathbb{R}^2$, respectively and, following Pitteri [18], analyze the twinning mode associated with the following choices of $m \in G$ and of the actual lattice vectors $\bar{\boldsymbol{e}}_a$ for C^*_1;

$$m = H = \begin{bmatrix} 0 & 1 \\ 1 & 0 \end{bmatrix} \tag{3.1}$$

and, for $a, R = 1, 2$,

$$\bar{e}_a \cdot E_R = (\bar{F}_1)_{aR}, \quad \bar{F}_1 = \frac{1}{\sqrt{\alpha^2+\beta^2}} \begin{bmatrix} \alpha^2 & \beta^2 \\ -\alpha\beta & \alpha\beta \end{bmatrix}, \tag{3.2}$$

where $\alpha > \beta > 0$ and, as the notation already indicates, $\bar{F}_1$ [$\bar{F}_2$ below] is the deformation gradient in $\bar{\kappa} \circ \kappa^{-1}(C_1^*)$ [in $\bar{\kappa} \circ \kappa^{-1}(C_2^*)$]. Then the actual normal N to the twinning plane π_1 and the amplitude $\bar{a}$ of the twinning shear $\bar{S} = (1 + \bar{a} \otimes N)$ have components

$$N = (0, -1) \quad \text{and} \quad \bar{a} = (\alpha\beta)^{-1}(\alpha^2 - \beta^2)(1, 0) \tag{3.3}$$

in the standard basis of $\mathbb{R}^2$. In addition

$$\bar{F}_2 = \bar{S}\bar{F}_1 = Q\bar{F}_1 H, \quad \text{where} \quad Q = Q^T = Q^{-1} = 1 - 2N \otimes N. \tag{3.4}$$

It is not difficult to verify that the right Cauchy-Green tensors associated with $\bar{F}_1$ and $\bar{F}_2$ are

$$\bar{C}_1 = \begin{bmatrix} \alpha^2 & 0 \\ 0 & \beta^2 \end{bmatrix} \quad \text{and} \quad \bar{C}_2 = \begin{bmatrix} \beta^2 & 0 \\ 0 & \alpha^2 \end{bmatrix}, \tag{3.5}$$

respectively, and that the map $C \longmapsto H^T C H$ leaves invariant the two off-diagonal elements, and exchanges the two diagonal elements of any symmetric tensor in $\mathbb{R}^2$. In particular

$$\bar{C}_2 = H^T \bar{C}_1 H. \tag{3.6}$$

For the ease of the reader we present a sketchy deduction of these results in an appendix.

We assume B to be hyperelastic with respect to $\bar{\kappa}$, and to have a stored energy density

$$W = \hat{W}(F) \tag{3.7}$$

per unit reference volume which, in suitably small neighbourhoods of $\bar{C}_1$ and $\bar{C}_2$ has the form

$$\hat{W}(F) = \tilde{W}(C_{LM}) = W(x_1, x_2, x_3), \tag{3.8}$$

where

$$W(x_1, x_2, x_3) = k(x_1 - x_{10})^2/2 + \ell(x_2 - x_{20})^2 (x_2 + x_{20})^2 + m(x_3)^2/2.$$

Here k, ℓ, m are positive constants,

$$x_1 := \tfrac{1}{2}(C_{11} + C_{22}) \, , \; x_2 := \tfrac{1}{2}(C_{11} - C_{22}) \quad \text{and} \quad x_3 := C_{12} \qquad (3.9)$$

and

$$x_{10} = \tfrac{1}{2}(\alpha^2 + \beta^2) \quad \text{and} \quad x_{20} = \tfrac{1}{2}(\alpha^2 - \beta^2) > 0 \, . \qquad (3.10)$$

This is a particularly simple example of a stored energy function which has a minimum at $\bar{\boldsymbol{C}}_1$ and $\bar{\boldsymbol{C}}_2$, and is invariant under the transformation $\boldsymbol{C} \mapsto \boldsymbol{H}^T \boldsymbol{C} \boldsymbol{H}$. We shall consider below only $\boldsymbol{C}$'s that fall into the neighbourhoods of $\bar{\boldsymbol{C}}_1$ or $\bar{\boldsymbol{C}}_2$ where (3.8)-(3.10) hold. Therefore in these sets $\bar{\boldsymbol{C}}_1$ and $\bar{\boldsymbol{C}}_2$ are the only two local minima for W.

We first consider the case of deformations from $\kappa(\mathcal{B})$ which are continuous and have constant deformation gradients Φ_1 in C_1^* and Φ_2 in C_2^*, leaving to Remarks 3 and 4 below the extension to slightly more general deformations. Therefore, for some vector $\boldsymbol{\alpha}$

$$\Phi_2 = \Phi_1 \Sigma \, , \quad \text{where} \quad \Sigma = \mathbf{1} + \boldsymbol{\alpha} \otimes \boldsymbol{N} \, .$$

For any deformation gradient field Φ we denote by $\boldsymbol{K}$ the corresponding left Cauchy-Green deformation tensor field. The results of the preceding section do not rest on special constitutive assumptions, so they still hold for deformations from $\kappa(\mathcal{B})$. However now the conditions that the amplitude $\boldsymbol{\alpha}$ of the discontinuity for the displacement be parallel to $\boldsymbol{A}_1$, and that the axis of the cylinder in the present configuration be parallel to the applied dead load $\boldsymbol{f}$, do not contradict experience. Indeed any orientation of $\boldsymbol{\alpha}$, if this vector has a small norm, produces a small deformation from the reference configuration, which is already twinned. Therefore now, at least for f of small magnitude, it becomes reasonable to look for a solution of the equilibrium equations which is a piecewise homogeneous, suitably small deformation from the twinned reference $\kappa(\mathcal{B})$, and which depends smoothly on $\boldsymbol{f}$ itself. In this section we prove that such a solution does not always exist if the twinning plane has a curve in common with the lateral surface of the cylinder C^*, and in particular it does not exist when the stored energy function W is given by (3.8)-(3.10).

To justify the last assertion we notice that the equilibrium equations and standard boundary conditions of Section 2 still hold. Therefore

$$T_\kappa = f \otimes A_1 \quad \text{throughout} \quad C^* . \tag{3.12}$$

The number of equations and of unknowns suggests that the scalar equations in (3.12) should not have a solution unless they are functionally dependent. Indeed the 18 scalar equations included in (3.12), nine for $(3.12)_1$ throughout C_1^* and nine for $(3.12)_1$ throughout C_2^*, involve the twelve scalar fields that are the components of Φ_1 and of α.

A conceivable reason for the eighteen scalar equations (3.12) to be functionally dependent is that the two reference configurations C_1^* and C_2^* are obtained one from the other by material symmetry and an orthogonal transformation, according to (3.4). This relation implies that the constitutive equations for the stress tensor in C_1^* and C_2^* can be obtained one from the other. In particular, the assumptions on the reference configurations κ and $\bar{\kappa}$ imply that $\mathcal{B}$ is a uniform but not homogeneous hyperelastic body. Indeed, denoting by J the determinant of either $\bar{F}_1$ or $\bar{F}_2$, the stored energy function of $\mathcal{B}$ has the form

$$W = \begin{cases} \hat{W}_1(\Phi_1) = \tilde{W}_1(K_1) & \text{on} \quad C_1^* \\ \hat{W}_2(\Phi_2) = \tilde{W}_2(K_2) & \text{on} \quad C_2^* \end{cases} , \tag{3.13}$$

where the densities $\hat{W}_1$ and $\hat{W}_2$ can be expressed as follows in terms of the density $\hat{W}(F)$ we introduced in (3.7):

$$\begin{aligned} &\hat{W}_1(\Phi_1) = J^{-1}\hat{W}(\Phi_1 \bar{F}_1) \\ \text{and} \quad &\\ &\hat{W}_2(\Phi_2) = J^{-1}\hat{W}(\Phi_2 \bar{F}_2) = J^{-1}\hat{W}(\Phi_2 \bar{S}\bar{F}_1) = \hat{W}_1(\Phi_2 \bar{S}). \end{aligned} \tag{3.14}$$

The invariance of $\hat{W}(F)$ under material symmetry, and (3.4), imply the following invariance for $\hat{W}_1$:

$$\begin{aligned} \hat{W}_1(\Phi_1 Q^T \bar{S}) &= J^{-1}\hat{W}(\Phi_1 Q^T \bar{S}\bar{F}_1) = J^{-1}\hat{W}(\Phi_1 \bar{F}_1 H) \\ &= J^{-1}\hat{W}(\Phi_1 \bar{F}_1) = \hat{W}_1(\Phi_1) . \end{aligned} \tag{3.15}$$

Hence, for any symmetric positive definite tensor K

$$\tilde{W}_2(K) = \tilde{W}_1(\bar{S}^T K \bar{S}) = \tilde{W}_1(Q^T K Q). \tag{3.16}$$

This equality provides the aforementioned relation between the constitutive functions in C_1^* and C_2^*. By differentiating it with respect to K we deduce that

$$\frac{\partial \tilde{W}_2}{\partial K}(K) = Q \left. \frac{\partial \tilde{W}_1}{\partial K'}(K')\, Q^T \right|_{K' = Q^T K Q} . \tag{3.17}$$

In particular this equality holds for $\boldsymbol{K} = \mathbf{1}$, when both derivatives vanish. We use this fact when we first differentiate (3.17) with respect to $\boldsymbol{K}$ and then evaluate the result for $K = \mathbf{1}$. Regarding the second derivatives as linear maps of symmetric tensors into symmetric tensors, we see that, for any symmetric tensor γ

$$\frac{\partial^2 \tilde{W}_2}{\partial K\, \partial K}(\mathbf{1})\, [\gamma] = Q \frac{\partial^2 \tilde{W}_1}{\partial K'\, \partial K'}(\mathbf{1})\, [Q \gamma Q^T]\, Q^T . \tag{3.18}$$

We assume either second derivative in (3.18) to be an invertible map of symmetric tensors into symmetric tensors. This assumption, unlike, for instance, rank-1 convexity, is not necessary for $\mathbf{1}$ to be a minimum for either $\tilde{W}_1$ or $\tilde{W}_2$. Nevertheless it holds for the functions $\tilde{W}_1$ and $\tilde{W}_2$ that (3.13) provides when $\hat{W}(F)$ is given by (3.8)-(3.10), as we are going to see.

Let us use (3.12) and $(3.13)_1$ to compute the Cauchy-Green tensor $\boldsymbol{C}$, which appears in (3.8), in terms of $\boldsymbol{K}_1$, whose components shall be simply indicated by K_{rs}. Since $\boldsymbol{C} = \bar{\boldsymbol{F}}_1^T \boldsymbol{K} \bar{\boldsymbol{F}}_1$, $(3.2)_2$ implies that

$$\begin{aligned} C_{11} &= (\alpha^2 + \beta^2)^{-1} \{ \alpha^4 K_{11} - 2\alpha^3 \beta K_{12} + \alpha^2 \beta^2 K_{22} \} \\ C_{12} &= (\alpha^2 + \beta^2)^{-1} \{ \alpha^2 \beta^2 (K_{11} - K_{22}) + \alpha\beta(\alpha^2 - \beta^2) K_{12} \} \\ C_{22} &= (\alpha^2 + \beta^2)^{-1} \{ \beta^4 K_{11} + 2\alpha\beta^3 K_{12} + \alpha^2 \beta^2 K_{22} \} . \end{aligned} \tag{3.19}$$

Therefore $x_i = \hat{x}_i(K_{rs})$, $i = 1, 2, 3$, and the functions $\hat{x}_i$ are linear; explicitly

$$\begin{aligned} x_1 &= \{2(\alpha^2+\beta^2)\}^{-1}\{(\alpha^4+\beta^4)K_{11} - 2\alpha\beta(\alpha^2-\beta^2)K_{12} + 2\alpha^2\beta^2 K_{22}\} \\ x_2 &= \{2(\alpha^2+\beta^2)\}^{-1}\{(\alpha^4-\beta^4)K_{11} - 2\alpha\beta(\alpha^2+\beta^2)K_{12}\} \\ x_3 &= (\alpha^2+\beta^2)^{-1}\{\alpha^2\beta^2(K_{11}-K_{22}) + \alpha\beta(\alpha^2-\beta^2)K_{12}\} \end{aligned} \tag{3.20}$$

To compute the second derivatives of $\tilde{W}_1$ it is convenient to differentiate it by means of the chain rule. Then

$$\mathbb{D}^{rs\rho\sigma}(\boldsymbol{K}) := \frac{\partial^2 \tilde{W}_1}{\partial K_{rs}\,\partial K_{\rho\sigma}}(\boldsymbol{K}) = \sum_{i=1}^{3} \frac{\partial^2 \tilde{W}_1}{\partial x_i^2}(\boldsymbol{K})\,\frac{\partial x_i}{\partial K_{rs}}\,\frac{\partial x_i}{\partial K_{\rho\sigma}}, \tag{3.21}$$

and we can easily check that

$$\frac{\partial^2 \tilde{W}_1}{\partial x_1^2} = k\,, \quad \frac{\partial^2 \tilde{W}_1}{\partial x_2^2} = 2\ell\{3(x_2)^2 - (x_{20})^2\}\,, \quad \frac{\partial^2 \tilde{W}_1}{\partial x_3^2} = m\,. \tag{3.22}$$

The right-hand side of $(3.22)_2$ assumes the value $4\ell(x_{20})^2$ when $\boldsymbol{K}=\mathbf{1}$, hence is positive for $\boldsymbol{K}$ is a suitable neighbourhood of $\mathbf{1}$. In that neighbourhood the second derivative $\mathbb{D}(\boldsymbol{K})$ of $\tilde{W}_1$ is symmetric and positive definite, as we easily deduce from (3.21). Indeed, by the linearity of the functions $\hat{x}_i$,

$$\boldsymbol{K}' \cdot \mathbb{D}(\boldsymbol{K})[\boldsymbol{K}'] = \sum_{i=1}^{3} \frac{\partial^2 \tilde{W}_1}{\partial x_i^2}(\boldsymbol{K})\,(\hat{x}_i(\boldsymbol{K}'))^2\,, \tag{3.23}$$

hence the left-hand side vanishes if and only if all the x_i's do, and, by (3.20), this happens if and only if $\boldsymbol{K}'=\mathbf{0}$. Since $\mathbb{D}(K)$ is positive definite, it is certainly invertible. In particular, so is $\mathbb{D}(\mathbf{1})$.

Let us replace the variables K_{rs} by the equivalent variables $K_1 := K_{11}$, $K_2 := K_{22}$ and $K_3 := K_{12}$, and, for $i = 1,2,3$, let us correspondingly regard the matrices $\partial x_i/\partial K_{rs}$ as three-dimensional vectors. It is lengthy but easy to verify that

$$\left[\frac{\partial x_1}{\partial K_i}\right] = (\alpha^2+\beta^2)^{-1}\,(\tfrac{1}{2}(\alpha^4+\beta^4)\,,\ \alpha^2\beta^2\,,\ -\alpha\beta(\alpha^2-\beta^2)), \tag{3.24}$$

$$\left[\frac{\partial x_2}{\partial K_i}\right] = (\tfrac{1}{2}(\alpha^2-\beta^2)\,,\ 0\,,\ -\alpha\beta)\,, \tag{3.25}$$

$$\left[\frac{\partial x_3}{\partial K_i}\right] = (\alpha^2+\beta^2)^{-1}\ (\alpha^2\beta^2,\ -\alpha^2\beta^2\ ,\ \alpha\beta(\alpha^2-\beta^2)), \tag{3.26}$$

hence, when $K = 1$,

$$\left[\frac{\partial^2\tilde{W}_1}{\partial K_1\partial K_1}\right] = (\alpha^2+\beta^2)^{-2}\ \{\tfrac{1}{4}k(\alpha^4+\beta^4)^2 + \ell(x_{20})^2(\alpha^4-\beta^4)^2 + m\alpha^4\beta^4\} \tag{3.27}$$

$$\left[\frac{\partial^2\tilde{W}_1}{\partial K_1\partial K_2}\right] = (\alpha^2+\beta^2)^{-2}\ \{\tfrac{1}{2}k\alpha^2\beta^2(\alpha^4+\beta^4) - m\alpha^4\beta^4\} \tag{3.28}$$

$$\left[\frac{\partial^2\tilde{W}_1}{\partial K_1\partial K_3}\right] = (\alpha^2+\beta^2)^{-2}\ \{-\tfrac{1}{2}k\alpha\beta(\alpha^2-\beta^2)(\alpha^4+\beta^4) - 2\ell(x_{20})^2\alpha\beta(\alpha^4-\beta^4)(\alpha^2+\beta^2) + m\alpha^3\beta^3(\alpha^2-\beta^2)\} \tag{3.29}$$

$$\left[\frac{\partial^2\tilde{W}_1}{\partial K_2\partial K_2}\right] = (\alpha^2+\beta^2)^{-2}\ (k+m)\ \alpha^4\beta^4 \tag{3.30}$$

$$\left[\frac{\partial^2\tilde{W}_1}{\partial K_2\partial K_3}\right] = -(\alpha^2+\beta^2)^{-2}\ (k+m)\ (\alpha^2-\beta^2)\,\alpha^3\beta^3 \tag{3.31}$$

$$\left[\frac{\partial^2\tilde{W}_1}{\partial K_3\partial K_3}\right] = \alpha^2\beta^2(\alpha^2+\beta^2)^{-2}\{k(\alpha^2-\beta^2)^2 + 4\ell(x_{20})^2(\alpha^2+\beta^2)^2 + m(\alpha^2-\beta^2)^2\}\ . \tag{3.32}$$

The left-hand sides of (3.27)-(3.32) are the independent components of the matrix Ξ of second derivatives of $\tilde{W}_1$ with respect to the K_i's. Let us use the same convention as above about indexing independent components of a symmetric tensor. In particular, let the single index i correspond as above to the pair of indices (r,s). Then, for any symmetric tensor Λ

$$\mathbb{D}^{rs\rho\sigma}\Lambda_{\rho\sigma} = \Xi^{ij}\,\Phi_j^k\Lambda_k\,,\quad \mathbb{D}^{rs\rho\sigma} := \mathbb{D}^{rs\rho\sigma}(1)\,,\quad \Xi^{ij} = \frac{\partial^2\tilde{W}_1}{\partial K_i\partial K_j}\,. \tag{3.33}$$

where (Φ_j^k) is the diagonal matrix whose diagonal entries are 1, 1 and 2, in this order. Therefore $\mathbb{D}(1)$ is an invertible map of symmetric tensors if and only if Ξ is invertible. In a special case we are going to prove that Ξ is invertible, and compute the components of Ξ^{-1}, that we need later. In this case we recover the result that $\mathbb{D}(\mathbf{1})$ is invertible, which we obtained above in more generality. The invertibility of $\mathbb{D}(\mathbf{1})$ and (3.18) easily imply that the second derivative of $\tilde{W}_2$ at $\mathbf{1}$ is also invertible.

To simplify the calculations, further on we choose $k = 1 = m$, and particularize the right-hand sides of (3.27)-(3.32) to this case. The determinant $\mathcal{D}$ of Ξ is given by

$$\begin{aligned}(\alpha^2+\beta^2)^6\mathcal{D} = \alpha^6\beta^6\{\ell(x_{20})^2\,[2\,(\alpha^4+\beta^4)^2 + 8\,\alpha^4\beta^4 \\ -(\alpha^2-\beta^2)^4\,](\alpha^2+\beta^2)^2 \\ +8\,\ell^2(x_{20})^4\,[(\alpha^4-\beta^4)^2-(\alpha^2-\beta^2)^4\,](\alpha^2+\beta^2)^2\}.\end{aligned} \tag{3.34}$$

Since the square brackets are both positive, so is $\mathcal{D}$. Let $(\mathbb{D}^{-1})_{\rho\sigma rs}$ be the components of the inverse of $\mathbb{D}$. We deduce from (3.18) that the inverse of the second derivative of $\tilde{W}_1$ at $\mathbf{1}$ satisfies the equality

$$\left[\frac{\partial^2\tilde{W}_1}{\partial K\,\partial K}\,(\mathbf{1})\right]^{-1}[\boldsymbol{\gamma}] = \boldsymbol{Q}^T\,\mathbb{D}^{-1}\,[\boldsymbol{Q}^T\,\boldsymbol{\gamma}\,\boldsymbol{Q}]\,\boldsymbol{Q} \qquad \forall\,\boldsymbol{\gamma} = \boldsymbol{\gamma}^T\ . \tag{3.35}$$

Let us consider again the equilibrium equations (3.12), and remember that, by (2.10), $\boldsymbol{f}$ is parallel to both $\Phi_1\boldsymbol{A}_1$ and $\Phi_2\boldsymbol{A}_1$. Then

$$\boldsymbol{f}\otimes\boldsymbol{A}_1 = \Phi_1(\lambda\,\boldsymbol{A}_1\otimes\boldsymbol{A}_1) \quad \text{for some real } \lambda\,. \tag{3.36}$$

By standard results presented by Truesdell and Noll [24], for instance, the Piola stress tensor $\boldsymbol{T}_\kappa$ is given by

$$2\,\Phi_1\,\frac{\partial\tilde{W}_1}{\partial K}\,(K_1) \quad \text{and} \quad 2\,\Phi_2\,\frac{\partial\tilde{W}_2}{\partial K}\,(K_2) \tag{3.37}$$

in C_1^* and C_2^*, respectively. By (3.11) and (3.36), the equilibrium equations (3.12) are equivalent to the following system

$$2\,\frac{\partial \tilde{W}_1}{\partial K}(K_1) = \lambda A_1 \otimes A_1$$

and (3.38)

$$2\Sigma\,\frac{\partial \tilde{W}_2}{\partial K'}(K_2) = \lambda A_1 \otimes A_1\,,$$

which in particular requires the reference amplitude α to be parallel to A_1 for $\lambda \neq 0$, in agreement with the results of Section 2. For the class of deformations we are considering, and for the ones we shall consider in Remark 3 below, the variables in (3.38) are K_1, α, N and λ. We prove now that there are no smooth functions $K_1(\lambda)$, $\alpha(\lambda)$ and $N(\lambda)$ such that $K_1(0) = 1$, $\alpha(0) = 0$ and $N(0) = N$, and such that (3.38) hold identically in any open neighbourhood of $\lambda = 0$, unless the twinning plane is parallel to the axis of the cylinder C^*. We assume, by contradiction, that such functions exist, differentiate (3.38) with respect to λ, and set then $\lambda = 0$. By our assumptions on $\tilde{W}(F)$, $K_1 = 1 = K_2$ is a stable solution of (3.38) for $\lambda = 0$, and both derivatives in (3.38) correspondingly vanish. By the way, either derivative constitutes the second, symmetric, Piola-Kirchhoff stress tensor. Therefore (3.38) imply

$$2\,\mathbb{D}^{rs\rho\sigma}\frac{dK_{1\rho\sigma}}{d\lambda}(0) = A_1^r A_1^s$$

and (3.39)

$$2\,\frac{\partial^2 \tilde{W}_2}{\partial K'_{rs}\,\partial K'_{\rho\sigma}}(1)\,\frac{dK_{2\rho\sigma}}{d\lambda}(0) = A_1^r A_1^s\,.$$

The first equation is equivalent to

$$2\,\frac{dK_{1\rho\sigma}}{d\lambda}(0) = (\mathbb{D}^{-1})_{\rho\sigma rs}\,A_1^r A_1^s\,, \tag{3.40}$$

and, by (3.35), the second is equivalent to

$$2\,\frac{dK_{2\rho\sigma}}{d\lambda}(0) = Q^{\mu}_{\rho}\,(\mathbb{D}^{-1})_{\mu\tau rs}\,Q^r_i\,A_1^i A_1^j\,Q^s_j\,Q^{\tau}_{\sigma}\,. \tag{3.41}$$

On the other hand

$$\frac{dK_{2\rho\sigma}}{d\lambda}(0) = 2\,\frac{d(\alpha_{(\rho}N_{\sigma)})}{d\lambda}(0) + \frac{dK_{1\rho\sigma}}{d\lambda}(0)\,, \tag{3.42}$$

where $(\rho\sigma)$ denotes symmetrization with respect to the indices ρ and σ. Expressing the second summand in (3.42) by means of (3.40), we write (3.41) in the form

$$2\,\frac{d(\boldsymbol{\alpha}\otimes \boldsymbol{N}+\boldsymbol{N}\otimes\boldsymbol{\alpha})}{d\lambda}(0)=\boldsymbol{Q}^T\,\mathbb{D}^{-1}[\boldsymbol{Q}^T \boldsymbol{A}_1\otimes \boldsymbol{A}_1\boldsymbol{Q}]\,\boldsymbol{Q} - \mathbb{D}^{-1}[\boldsymbol{A}_1\otimes \boldsymbol{A}_1]. \tag{3.43}$$

Notice that $\boldsymbol{\alpha}(0)=\boldsymbol{0}$, hence on the left-hand side of (3.43) only the terms that involve differentiation of $\boldsymbol{\alpha}(\lambda)$ actually contribute, whereas it is irrelevant to allow $\boldsymbol{N}$ to depend on λ. Equation (3.40) can be satisfied in a neighbourhood of $\lambda=0$ and $K_1=1$ by one smooth function $K_1(\lambda)$. This result follows by applying the implicit function theorem to $(3.38)_1$. To analyze (3.43) we notice that, by $(3.33)_1$ and its analogue for $\mathbb{D}^{-1}$, the 3×3 matrix Ψ representing $\mathbb{D}^{-1}$ can be expressed in terms of the inverse Ξ^{-1} of the matrix Ξ that we introduced in (3.33): $\Psi=\boldsymbol{\Phi}^{-1}\,\Xi^{-1}\,\boldsymbol{\Phi}^{-1}$. So, for instance,

$$\begin{aligned}(\mathbb{D}^{-1})_{11\,12} &= \Psi_{13}=\tfrac{1}{2}(\Xi^{-1})_{13}\\ &= 2\mathcal{D}^{-1}(\alpha^2+\beta^2)^{-4}\,\alpha^5\beta^5\,\ell(x_{20})^2(\alpha^2-\beta^2)(\alpha^2+\beta^2)^2>0\,.\end{aligned} \tag{3.44}$$

It is long but easy to compute all the other elements of Ξ^{-1}. All that matters is that *they are all positive*, as the one we computed in (3.44). Since any element of $\mathbb{D}^{-1}$ is proportional by a factor of either 1 or $\frac{1}{2}$ or $\frac{1}{4}$ to a suitable element of Ξ^{-1}, we conclude that *all the components of* $\mathbb{D}^{-1}$ *are positive.* By $(3.4)_{3-5}$

$$\boldsymbol{Q}=\begin{bmatrix}1 & 0\\ 0 & -1\end{bmatrix}. \tag{3.45}$$

Therefore, denoting by a prime differentiation with respect to λ for simplicity, we can easily write (3.43) in components as follows:

$$0=-4(\mathbb{D}^{-1})_{11\,12}(A_1)^1(A_1)^2\ , \tag{3.46}$$

$$(\alpha')_1(0) = (\mathbb{D}^{-1})_{12\,11}(A_1)^1(A_1)^1 + (\mathbb{D}^{-1})_{12\,22}(A_1)^2(A_1)^2 \,, \tag{3.47}$$

$$(\alpha')_2(0) = (\mathbb{D}^{-1})_{22\,12}(A_1)^1(A_1)^2 \,. \tag{3.48}$$

A necessary condition for (3.46) to hold is that either $(A_1)^1$ or $(A_1)^2$ vanish, that is, the plane of discontinuity in the reference configuration κ be either orthogonal or parallel to the axis of the cylinder C^*, respectively. This assumption is already in contradiction with a number of experimental observations, which show that the typical twinning plane is neither orthogonal nor parallel to the axis of the specimen. As we remarked in Section 2, our calculations rest on hypotheses implying that the twinning plane should not be parallel to the axis of the cylinder, so (3.46) can hold only if the twinning plane is orthogonal to that axis, that is, *only if* $\boldsymbol{A}_1 = \pm\,(0,1)$. Henceforth we assume this condition to hold, and notice that we still have to take into account the fact that the left-hand side of $(3.38)_2$ is a symmetric tensor, that is, for any λ

$$\varepsilon_{irs}\,\alpha^r t^s = 0, \quad \text{where} \quad t^s := \frac{\partial \tilde{W}_2}{\partial K_{sj}}(K_2)\,N_j \,. \tag{3.49}$$

We differentiate twice $(3.49)_1$ with respect to λ, and evaluate the result for $\lambda = 0$, remembering that both $\boldsymbol{\alpha}$ and the derivative which appears in $\boldsymbol{t}$, hence $\boldsymbol{t}$ itself, vanish for $\boldsymbol{\lambda} = 0$. As a result, $\boldsymbol{\alpha}'(0)$ has to be parallel to $\boldsymbol{t}'(0)$. Moreover

$$\boldsymbol{t}'(0) = \frac{\partial^2 \tilde{W}_2}{\partial K'\,\partial K'}(1)\left[\frac{dK_2}{d\lambda}(0)\right]N \tag{3.50}$$

But $(3.39)_2$ implies that $2\boldsymbol{t}'(0) = (A_1 \cdot \boldsymbol{N})\,\boldsymbol{A}_1 = \pm\,\boldsymbol{A}_1 = \pm\,(0,1)$, and the parallelism of $\boldsymbol{\alpha}'(0)$ and $\boldsymbol{t}'(0)$ is incompatible with (3.47) and (3.48). This proves our assertion above: for a cylinder of twinned crystal with dead loaded ends and twinning plane which is not parallel to the axis and cuts the lateral surface of the cylinder along an arc, there are no continuous, piecewise homogeneous deformations which, for small load, depend smoothly on the load itself, and start from the identity for zero load.

The results in this section do not strictly depend upon our assuming $k=m=1$. Since the value of ℓ remains arbitrary, we can rescale the energy density by multiplying it by an arbitrary positive factor. This is equivalent to requiring k and m to equal an arbitrarily given positive number k. In addition, by continuity, our conclusions above hold for all m in a suitable neighbourhood of k. More in general, those conclusions hold for all energy densities whose restrictions to suitable neighbourhoods of C_1^* and C_2^* are sufficiently uniformly near the density $\hat{W}(\boldsymbol{F})$ that we introduced in (3.8).

REMARK 3. In this section we assumed that the deformation gradient from C^* takes constant values Φ_1 and Φ_2 in exactly the two regions C_1^* and C_2^*, respectively, these being deformed from the homogeneous reference configuration $\bar{\kappa}$ by a standard twinning deformation. The conclusions above about nonexistence of certain deformations of C^* remain unaffected if we assume that, more in general, $\boldsymbol{\Phi}_1$ and $\boldsymbol{\Phi}_2$ are constant over regions C_3^* and C_4^* which partition C^* but do not necessarily coincide with C_1^* and C_2^*. Indeed, as we pointed out above, the Piola stress tensor $\boldsymbol{T}_\kappa$ has anyway to be constant over C^*. Moreover, for any allowed position of the plane separating C_3^* and C_4^*, there are interior points of C^* where either $W=\hat{W}_1(\Phi_1)$ or $W=\hat{W}_2(\boldsymbol{\Phi}_2)$. This fact allows us to carry through the same arguments as those following (3.36).

REMARK 4. Hagan and Serrin [10] and Silhavy [21], for instance, consider classes of nonhomogeneous deformations which contain the piecewise homogeneous ones when the singular planes are all parallel, and which are still tractable. For our purposes, we can follow Silhavy [21], and consider the case that the displacement u has the form

$$u(X) = F_0 X + \hat{u}(X \cdot N) \qquad \hat{u}: \mathbb{R} \to \mathbb{R}^3 . \tag{3.51}$$

We may hope to avoid the problems in Sections 2 and 3, concerning existence, if we look for equilibrium solutions in this enlarged class of deformations. Unfortunately in this way we are not effectively enlarging the class of equilibrium solutions. Indeed, since the lateral surface of the cylinder C^* is free and $(2.4)_1$ has to hold, we see that (3.51) implies again (2.7), hence (2.8) together with its consequences.

REMARK 5. Following a suggestion of Davini and Parry, we notice that the assumption that the body $\mathcal{B}$ occupies a cylinder C^* in the reference configuration κ is important for the arguments in Sections 2 and 3 to be carried through. For instance, assume each C^*_r in Section 2 to be a section of a cylinder whose axis has the direction $A_1^{(r)}$, the intersecting planes π_r and π_{r+1} not being parallel, in general. Then the Piola stress tensor has the form

$$T_\kappa = b_r \otimes A_1^{(r)} \quad \text{throughout} \quad C^*_r . \tag{3.52}$$

Here

$$b_r = f \frac{\|A^1_{(1)}\| (A_1^{(1)} \cdot N)}{A_1^{(r)} \cdot N} ; \tag{3.53}$$

$A_2^{(r)}$ and $A_3^{(r)}$ generate π_r and, for any r and $i = 1,2,3$, $A^i_{(r)}$ are the duals of $A_i^{(r)}$. The additional freedom we have in this case allows us to satisfy both the jump conditions on the singular planes and the balance of moments. For instance, when there are only two regions,

$$T_\kappa = f \|A^1_{(r)}\| \otimes A_1^{(r)} \quad \text{throughout} \quad C^*_r , \quad r = 1,2 .$$

The balance of moments is equivalent to the conditions

$$F_1 A_1^{(1)} \wedge f = 0 \quad \text{and} \quad F_2 A_1^{(2)} \wedge f = 0 .$$

which can always be satisfied by suitably choosing $A_1^{(2)}$. The jump condition on the only singular plane is equivalent to

$$\|A^1_{(1)}\|(A_1^{(1)}\cdot N) = \|A^1_{(2)}\|(A_1^{(2)}\cdot N),$$

and this equality can be satisfied by suitably choosing $A^1_{(2)}$, which is still unrestricted, or equivalently by choosing the orientation of the end base of C_2^*. Of course, these are only necessary conditions for the equilibrium equations to be satisfied. We still have to show that the equations (3.52) are compatible with the constitutive equations, and by the arguments in Section 3 we do not expect this to be a trivial requirement in general.

The problem analyzed in Section 3 above can be trivially solved if, in addition to the hypothesis that C^* be a cylinder, we further relax the hypothesis that the deformation depending on λ reduces to the identity for $\lambda = 0$. Indeed, let C^* be the image of a straight cylinder under a deformation $\bar{\psi}$ from the reference $\bar{\kappa}$ such that grad $\bar{\psi}$ is $\bar{F}_1$ in $\bar{\kappa}\circ\kappa^{-1}(C_1^*)$ and $\bar{F}_2$ in $\bar{\kappa}\circ\kappa^{-1}(C_2^*)$, according to (3.2)-(3.4). Moreover, consider a homogeneous deformation $\psi(\lambda)$ from $\bar{\kappa}$ which solves the equilibrium equations (3.12), that is, the analogues of $(3.38)_1$ for $\hat{W}(F)$, and which depends smoothly on λ for sufficiently small $|\lambda|$. If the second derivative of W as a function of C is invertible at $\bar{F}_1$, as is true for the function $\hat{W}$ in (3.8), the existence of a smooth $\psi(\lambda)$ such that grad $\psi(0) = \bar{F}_1$ is guaranteed by the implicit function theorem. It should be quite clear that, for small λ, $\psi(\lambda)\circ\bar{\psi}^{-1}$ is a nonhomogeneous, piecewise homogeneous stable equilibrium deformation from κ for the loaded crystal.

ACKNOWLEDGEMENTS

I am grateful to J.L. Ericksen for beneficial and influential discussions at the early stages of this paper. This work is part of the research programme on Thermomechanics of Continua on the Italian M.P.I.

APPENDIX

According to Pitteri [18], let $\boldsymbol{e}_a$, $a = 1, 2, 3$, be any 3 linearly independent vectors, and $\boldsymbol{e}^a$ their duals; let m be an arbitrary element of $G := GL(3, \mathbb{Z})$ such that $m^2 = 1$, and let m satisfy

$$\boldsymbol{Q}\boldsymbol{e}_a \neq m_a^b \boldsymbol{e}_b \tag{A.1}$$

for any orthogonal $\boldsymbol{Q}$. As is shown by Ericksen [4], m is conjugate in $GL(3, \mathbb{R})$ to some orthogonal $\hat{\boldsymbol{Q}}$ such that $\hat{\boldsymbol{Q}}^2 = \mathbf{1}$, hence it has a proper vector $v_1 := (v_1^1, v_1^2, v_1^3) \in \mathbb{R}^3$ corresponding to the proper number $D := \det m$, and two proper vectors $v_r := (v_r^1, v_r^1, v_r^2)$, $r = 2, 3$, corresponding to the proper number $-D$. We can choose these proper vectors to have integral components. For $i = 1$ to 3, consider the linearly independent vectors

$$\boldsymbol{v}_i = v_i^a \boldsymbol{e}_a, \tag{A.2}$$

and denote by $\boldsymbol{v}^i$ the dual vectors of the $\boldsymbol{v}_i$'s. According to Pitteri [16, p.33],

$$m_a^b \boldsymbol{e}_b = \boldsymbol{L}\boldsymbol{e}_a, \quad \text{where} \quad \boldsymbol{L} \doteq D(2\boldsymbol{v}_1 \otimes \boldsymbol{v}^1 - \mathbf{1}).$$

It is easy to check that $\boldsymbol{L} = \boldsymbol{Q}\bar{\boldsymbol{S}}$, where $\boldsymbol{Q}$ and $\bar{\boldsymbol{S}}$ are the tensors we introduced in Section 3, if we choose

$$-\bar{\boldsymbol{a}} := 2(\boldsymbol{v}_1 \|\boldsymbol{v}^1\| - \|\boldsymbol{v}^1\|^{-1}\boldsymbol{v}^1) \quad \text{and} \quad \boldsymbol{N} := \|\boldsymbol{v}^1\|^{-1}\boldsymbol{v}^1 .$$

Also, the twinning plane p, which is the plane orthogonal to $\boldsymbol{N}$, is a rational plane. Equivalently, $\boldsymbol{N}$ is a rational combination of the dual vectors $\boldsymbol{e}^a$. On the contrary, in general the direction of $\bar{\boldsymbol{a}}$ is not rational, that is, $\bar{\boldsymbol{a}}$ is not a rational combination of the vectors $\boldsymbol{e}_a$.

The following results are easy to prove when

$$m = H = \begin{bmatrix} 0 & 1 & 0 \\ 1 & 0 & 0 \\ 0 & 0 & 1 \end{bmatrix} \tag{A.5}$$

and

$$\boldsymbol{e}_a \cdot \boldsymbol{E}_R = (\bar{F}_1)_{aR}, \quad \bar{F}_1 = \frac{1}{\sqrt{\alpha^2 + \beta^2}} \begin{bmatrix} \alpha^2 & \beta^2 & 0 \\ -\alpha\beta & \alpha\beta & 0 \\ 0 & 0 & 1 \end{bmatrix} \tag{A.6}$$

The proper vectors v_r and the vectors $\boldsymbol{v}_r$, $\boldsymbol{v}^1$, $\boldsymbol{a}$ and $\boldsymbol{N}$ are

$$
\begin{aligned}
v_1 &= (1, -1, 0), \quad v_2 = (1, 1, 0), \quad v_3 = (0, 0, 1), \\
\boldsymbol{v}_1 &= (\alpha^2 + \beta^2)^{-\frac{1}{2}} (\alpha^2 - \beta^2, -2\alpha\beta, 0), \\
\boldsymbol{v}_2 &= (\alpha^2 + \beta^2)^{-\frac{1}{2}} (\alpha^2 + \beta^2, 0, 0), \\
\boldsymbol{v}_3 &= (0, 0, 1), \quad \boldsymbol{v}^1 = (2\alpha\beta)^{-1} (\alpha^2 + \beta^2)^{\frac{1}{2}} (0, -1, 0) \\
\bar{\boldsymbol{a}} &= (2\alpha\beta)^{-1} (\alpha^2 - \beta^2)(1, 0, 0), \text{ and} \\
\boldsymbol{N} &= (0, -1, 0).
\end{aligned}
$$

REFERENCES

1. J.M. Ball and J.E. Marsden, "Quasiconvexity at the boundary, positivity of the second variation and elastic stability", *Archive for Rational Mechanics and Analysis* **86** (1984), pp.251-277.
2. B.D. Coleman and W. Noll, "Material symmetry and thermostatic inequalities in finite elastic deformations", *Archive for Rational Mechanics and Analysis* **15** (1964), pp.87-111.
3. J.L. Ericksen, "Equilibrium of bars", *Journal of Elasticity* **5** (1975), pp.191-201.
4. J.L. Ericksen, "On the symmetry of deformable crystals", *Archive for Rational Mechanics and Analysis* **72** (1979), pp.1-13.
5. J.L. Ericksen, "Continuum martensitic transition in thermoelastic solids", *Journal of Thermal Stresses* **4** (1981), pp.107-119.
6. J.L. Ericksen, "Some surface defects in unstressed thermoelastic solids", *Archive for Rational Mechanics and Analysis* **88** (1985), pp.337-345.
7. J.L. Ericksen, "Stable equilibrium configurations of elastic crystals", *Archive for Rational Mechanics and Analysis* **94** (1986), pp.1-14.
8. I. Fonseca, "Variational methods for elastic crystals", *Ph.D. Thesis*, University of Minnesota, Minneapolis, July, 1985.
9. M.E. Gurtin, "Two-phase deformations of elastic solids", *Archive for Rational Mechanics and Analysis* **84** (1983), pp.1-29.
10. R. Hagan and J.B. Serrin, "Dynamic changes of phase in a Van der Waals fluid", in *New Perspectives in Thermodynamics*, (Ed. J. Serrin), Springer-Verlag, Berlin-Heidelberg, 1986.
11. R.D. James, "Coexistent phases in the one-dimensional static theory of elastic bars", *Archive for Rational Mechanics and Analysis* **72** (1979), pp.99-140.
12. R.D. James, "The propagation of phase boundaries in elastic bars", *Archive for Rational Mechanics and Analysis*, **73** (1980), pp.125-158.
13. R.D. James, "Finite deformation by mechanical twinning", *Archive for Rational Mechanics and Analysis* **77** (1981), pp.143-176.
14. R.D. James, "Mechanics of coherent phase transformations in solids", *MRL Report*, Brown University, Division of Engineering, October 1982.
15. G.P. Parry, "Twinning in nonlinearly elastic monatomic crystals", *International Journal of Solids and Structures* **16** (1980), pp.275-281.
16. G.P. Parry, "On shear bands in unloaded crystals", (to appear).

17. M. Pitteri, "Reconciliation of local and global symmetries of crystals", *Journal of Elasticity* **14** (1984), pp.175-190.
18. M. Pitteri, "On the kinematics of mechanical twinning in crystals", *Archive for Rational Mechanics and Analysis* **88** (1965), pp.25-27.
19. M. Pitteri, "On Type 2 twins", *International Journal of Plasticity* **2** (1986), pp.99-106.
20. M. Pitteri, "A contribution to the description of natural states for elastic crystalline solids", Metastability and Incompletely Posed Problems, *IMA Volumes in Mathematics and its Applications*, vol.3, Springer-Verlag, 1987.
21. M. Silhavy, "An admissibility criterion for shocks and propagating phase boundaries via thermodynamics of non-simple materials 1", *Journal of Non-Equilibrium Thermodynamics* **9** (1984), pp.177-186.
22. M. Slemrod, "Admissibility criteria for propagation of phase boundaries in a Van der Waals fluid", *Archive for Rational Mechanics and Analysis* **81** (1983), pp.301-316.
23. C. Truesdell and R.A. Toupin, "The classical field theories", *Handbuch der Physik* **III/1** (1960), (Ed. S. Flugge), Springer-Verlag, Berlin, Heidelberg and New York.
24. C. Truesdell and W. Noll, "The nonlinear field theories of mechanics", *Handbuch der Physik* **III/3** (1965), (Ed. S. Flugge), Springer-Verlag, Berlin, Heidelberg and New York.
25. E.J. Rappenport, "Room temperature deformation processes in zirconium", *Acta Metallurgica* **7** (1959), pp.254-260.

26

AN APPLICATION OF THE METHOD OF COMPENSATED COMPACTNESS TO A PROBLEM IN PHASE TRANSITIONS

V. ROYTBURD* *and* M. SLEMROD†

Abstract. This paper studies weak $*L^{\infty}$ limits as $\varepsilon \to 0+$ of solutions of the equations $v_t + p(w)_x = \varepsilon v_{xx}, w_t - v_x = \varepsilon w_{xx}$, $-\infty < x < \infty, t > 0$ with initial data $v(x,0) = v_0(x)$, $w(x,0) = w_0(x)$. The constitutive relation $p(w)$ is similar to that of a van der Waals fluid *with* Maxwell construction. We show that under certain conditions the weak * limits satisfy the "inviscid" $\varepsilon = 0$ problem. The main tools of the analysis are (1) the "method of compactness compactness" as developed by L. Tartar and F. Murat and (2) fundamental ideas of R. DiPerna for the analysis of 2×2 systems of conservation laws.

INTRODUCTION

The purpose of this paper is to consider the asymptotic behaviour of solutions $(v^{\varepsilon}, w^{\varepsilon})$ to the system

$$\begin{aligned} v_t + p(w)_x &= \varepsilon v_{xx}, \\ w_t - v_x &= \varepsilon w_{xx}, \end{aligned} \qquad (P_{\varepsilon})$$

with initial conditions

---

*Partially supported by NSF Grant No. DMS-8603506. This research was done in part while the author was a visiting member of the Courant Institute of Mathematical Sciences.

†This research was sponsored by the Air Force Office of Scientific Research, Air Force Systems Command, USAF, Contract/Grant No. AFDSR-85-0239. The U.S. Government's right to retain a nonexclusive royalty free licence in and to copyright this paper for government purposes is acknowledged.

$$v(x,0) = v_0(x)\ ,\quad w(x,0) = w_0(x)\ ,$$

as $\varepsilon \to 0+$. Here p has the graph shown in Fig. 1.

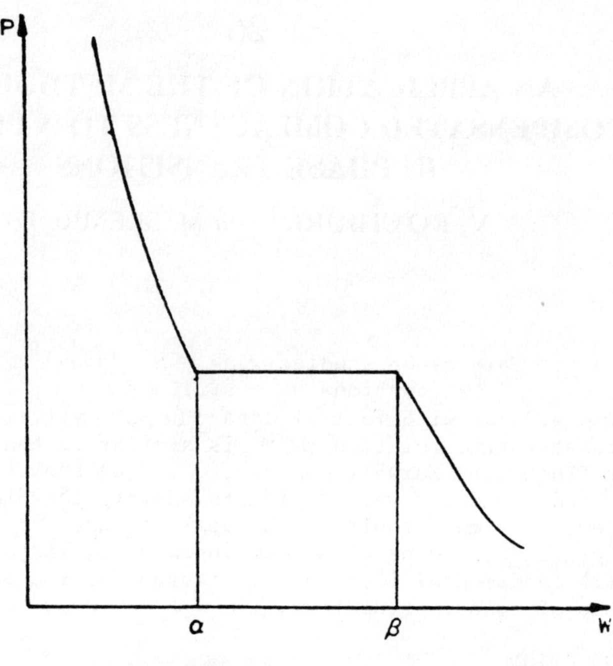

FIG. 1

The motivation for this problem lies in the theory of phase transitions. If $\varepsilon = 0$ equations (P_ε) give the Lagrangian form of the balance of mass and momentum for an elastic fluid (solid) with stress taken as $-p(w)$. Unlike the case of ideal fluid or elastic solid, when $p' < 0$, we consider the nonstandard stress from Fig. 1. One may observe such a p(= pressure) versus w(= specific volume or deformation gradient) diagram in *equilibrium* configurations of materials undergoing first order phase transitions ([8]).

Our goal is to study *nonequilibrium* dynamics of materials exhibiting phase transitions. In classical continuum mechanics these materials are described phenomenologically by nonmonotone constitutive relations (Fig. 1a). It may appear inconsistent to apply the equilibrium constitutive relation for studying the

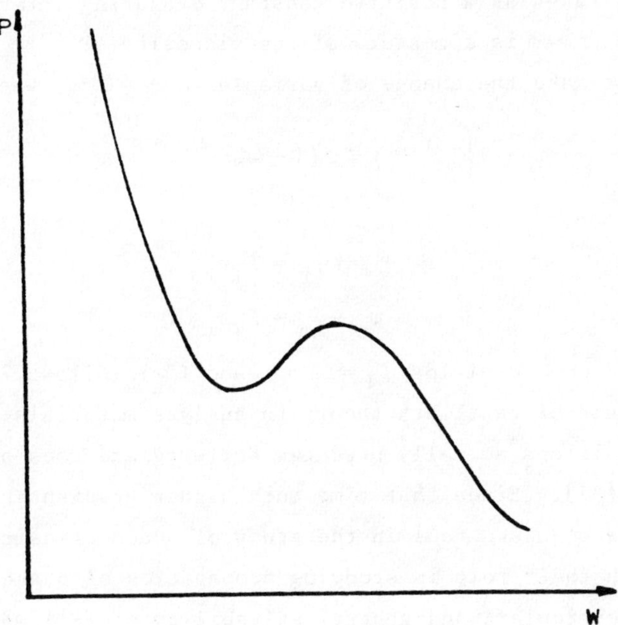

FIG. 1a

nonequilibrium process. While we admit this inconsistency we consider choice of Fig. 1 as a first step to our goal of understanding the limiting behaviour of (P_ε) when p is not strictly monotone. We also note that another motivation for our problem may be found in the recent work of Solomon *et al* [32] on hyperbolic Stefan problems.

While a regularization of the kind given in (P_ε) is standard it does possess the following physical justification. If one considers the balance of mass and momentum in Lagrangian coordinates with specific volume w, velocity u, and stress τ given by Korteweg's theory of capillarity [1], [13], [24], [35],

$$\tau = 2\varepsilon u_x - 4A\varepsilon^2 w_{xx} ,$$

the evolution of u, w is described by the system

$$\begin{aligned} u_t + p(w)_x &= 2\varepsilon u_{xx} - 4A\varepsilon^2 w_{xxx} , \\ w_t - v_x &= 0 . \end{aligned}$$

Here A is taken as a positive constant measuring interfacial energy and $\varepsilon > 0$ is a measure of the viscosity.

If we make the change of variable $v = u - D_2 w_x$ where

$$\left\{ \begin{matrix} D_1 \\ D_2 \end{matrix} \right. = \varepsilon \mp \varepsilon (1 - 4A)^{1/2}$$

then the above system becomes

$$v_t + p(w)_x = D_1 v_{xx} ,$$
$$w_t - v_x = D_2 w_{xx} .$$

The choice $A = 1/4$ yields $D_1 = D_2 = \varepsilon$ and (P_ε) follows.

The use of capillary theory to analyze materials exhibiting phase transitions actually predates Korteweg and goes back to van der Waals [37]. Since that time such higher gradient theories have been a standard tool in the study of phase transitions [3], [36] though their role in studying propagation of phase boundaries, in particular, and general solvability of (P_ε) where p has the shape shown in Fig. 1a is more recent. Basically the only results available so far for the nonmonotone (van der Waals) case are (i) the pioneering paper of R. James [12] who solved special $\varepsilon = 0$ Riemann problems without qualifying admissibility of phase boundaries; (ii) solvability of the $\varepsilon = 0$ Riemann problem where the data are consistent with Rankine-Hugoniot jump conditions (the shock structure problem) [10], [11], [28], [30], [31]; and (iii) solvability of the inviscid $\varepsilon = 0$ Riemann problem for data near the Maxwell line where the shocks are admissible according to the shock structure program (ii) [27], [29]. Hence we view our flat p problem as a logical step towards understanding the connection between Korteweg theory and solvability of the inviscid nonmonotone van der Waals problem for arbitrary initial data.

Our method of analysis is based on the method of *compensated compactness* as developed by L. Tartar and F. Murat [4], [18], [19], [33], [34]. In the context of conservation laws, this method has been first used by Tartar [33] to give a new existence proof for the scalar conservation law

$$u_t + f(u)_x = 0 \tag{0.1}$$

via the limit of vanishing viscosity:

$$u^\varepsilon_t + f(u^\varepsilon)_x = \varepsilon u^\varepsilon_{xx} . \tag{0.2}$$

The main ideas of the method of compensated compactness can be illustrated as follows. Assume that a sequence of solutions u^ε converges to $\bar{u}$ weak $*L^\infty$. The presence of such a sequence u^ε is guaranteed if, for example, one has a uniform in ε L^∞ estimate for solutions of (0.2). If one can show that $f(u^\varepsilon) \to f(\bar{u})$, then $\bar{u}$ is indeed a weak solution of (0.1). The question, when

$$w* \lim f(u^\varepsilon) = f\,(w* \lim u^\varepsilon)$$

is highly nontrivial. In order to address this question Tartar employed a beautiful concept due to L.C. Young.

For each (x,t), the *Young measure* $\gamma_{x,t}$ is a probability measure associated with the weak limit in the following sense. If g is an arbitrary continuous function then

$$g(u^\varepsilon) \to \int g(\lambda)\, d\gamma_{x,t}(\lambda) \equiv \langle g, \gamma \rangle \quad \text{weak}{*}L^\infty$$

as $\varepsilon \to 0+$. It is easily seen that in case when $\gamma_{x,t}$ is a Dirac measure

$$g(u^\varepsilon) \to g(\bar{u}) \quad \text{weak}*L^\infty .$$

Thus, it is of fundamental importance to show that the Young measure associated with a weakly convergent solution sequence reduces to a Dirac mass. This is the case when one can establish some compactness estimates for u^ε, say in variation norm or C^1 norm. Unfortunately aside from the scalar case and the well known result of Glimm [9] for strictly hyperbolic, genuinely nonlinear systems where the oscillation of the initial data is small the compactness estimates are not readily available. However, Tartar realized that entropy type differential restrictions,

$$\partial_t \eta(u^\varepsilon) + \partial_x q(u^\varepsilon) \in \text{compact set of } H^{-1}_{loc} , \tag{0.3}$$

allowed him to infer some information about the support of γ; here η and q form an entropy — entropy flux pair :

$$\nabla\eta \, \nabla f = \nabla q . \tag{0.4}$$

Specifically, Tartar has proved that if (η_1, q_1) and (η_2, q_2) are two entropy-flux pairs satisfying (0.3) then the following functional equation holds:

$$\langle \gamma_{x,t}, \eta_1 q_2 - \eta_2 q_1 \rangle = \langle \gamma_{x,t}, \eta_1 \rangle \langle \gamma_{x,t}, q_2 \rangle - \langle \gamma_{x,t}, \eta_2 \rangle \langle \gamma_{x,t}, q_1 \rangle . \tag{0.5}$$

By using (0.5) Tartar proved that for a scalar conservation law the Young measure reduces to a Dirac mass. He also conjectured that for genuinely nonlinear strictly hyperbolic systems there exists a family of compact entropy fields rich enough to reduce the Young measure to a δ-function.

Subsequently R. DiPerna [6,7] was able to make the breakthrough and apply the compensated compactness method to the case when (0.1) is a 2×2 system of *strictly hyperbolic* conservation laws. Other contributions have been made by M. Schonbek [23], M. Rascle [20], M. Boldrini [2], C. Morawetz [17], D. Serre [25], [26].

Our problem (P_ε) while similar to DiPerna's [6] is different in a very important respect. Since $p'(w) = 0$ in $[\alpha, \beta]$ the limiting $\varepsilon = 0$ problem is <u>not</u> strictly hyperbolic; the eigenvalues of f' are $\lambda_1 = \lambda_2 = 0$ for w in (α, β). So DiPerna's results are not directly applicable. What we have attempted to do is modify DiPerna's arguments to our degenerate case. *Our conjecture is that the Young measure either reduces to a Dirac mass outside the strip* $\alpha \leqslant w \leqslant \beta, -\infty < v < \infty$ *or is supported inside this strip.* In either case we see immediately that

$$\left\{ \begin{matrix} w^\varepsilon \to \bar{w} \\ v^\varepsilon \to \bar{v} \end{matrix} \right\} \text{ implies } p(w^\varepsilon) \to p(\bar{w}) . \tag{0.6}$$

Unfortunately the pathology of the problem has allowed us

only limited success in this program. Our main results show that under some *additional assumptions* on the support of the Young measure, (0.6) does indeed hold.

This paper is organized as follows. In Chapter 1 we collect principal regularity results for solutions of (P_ε). The proofs, based on our paper [21], are given in Appendix A. We note that an alternative derivation of the basic L^∞ estimates from [21] has been subsequently obtained by C. Dafermos in an elegant paper [5].

In Chapter 2 using the energy method we derive L^2 estimates for v_x^ε and w_x^ε. It should be noted since $p'(w) = 0$ for $\alpha < w < \beta$ the energy does not provide good estimates for w_x^ε in the domain $E = \{x, t \mid \alpha < w^\varepsilon(x, t) < \beta\}$. This fact is the source of difficulties which force us in Chapter 4 to construct a special class of entropy-flux pairs so that $\eta_{vw} = 0$ for $\alpha < w < \beta$.

In Chapter 3 we review the compensated compactness method specialized to the situation on hand, and in Chapters 4-5 we construct the entropy-flux pairs. We solve equations (0.4) separately in the three domains $w < \alpha$, $\alpha < w < \beta$, $w > \beta$. In constructions for hyperbolic regions, $w > \beta$, $w < \alpha$, we follow the ideas of P.D. Lax [14]. Some minor technical difficulties are caused by insufficient regularity of η and q. These difficulties are considered in Appendix B.

In Chapter 6 we prove our main results, Theorems 6.1 and 6.3, which in a sense complement each other. We prove that if the support of γ does not contain the corners of the invariant domain or if on the contrary, the support is concentrated in the corners then (0.6) holds, and the viscous solutions $v^\varepsilon, w^\varepsilon$ converge to a weak solution of (P).

1. PRELIMINARIES

As noted in the introduction we consider the Cauchy initial value problem

$$\left.\begin{aligned} v_t + p(w)_x &= 0\,, \\ w_t - v_x &= 0\,, \end{aligned}\right. \qquad -\infty < x < \infty\,, \qquad t > 0 \qquad (P)$$

$$v(x\,,0) = v_0(x)\,, \quad w(x\,,0) = w_0(x)\,,$$

where v_0, $w_0 \in L^\infty(R)$. For the technical convenience we assume $v_0(x)$, $w_0(x) \to a, b$ as $x \to \pm\infty$, and hence subtracting a and b we assume these limits to be zero. In order to analyze (P) we imbed it in the regularized problem

$$\left.\begin{aligned} v_t + p(w)_x &= \varepsilon v_{xx}\,, \\ w_t - v_x &= \varepsilon w_{xx}\,, \end{aligned}\right. \qquad -\infty < w < \infty\,, \qquad t > 0 \quad (P_\varepsilon)$$

with the same initial data as (P). For the regularized problem we have the following result.

THEOREM 1.1 (i) (P_ε) possesses unique continuous weak solutions locally in time. Moreover the weak solutions are Lipschitz continuous in x.

(ii) If p is globally Lipschitz continuous then the above weak solutions are global in t.

(iii) If v_0, w_0 approach zero sufficiently rapidly as $|x| \to \infty$ then the weak solutions $v^\varepsilon, w^\varepsilon$ satisfy

$$v^\varepsilon, v_t^\varepsilon, v_x^\varepsilon, v_{xx}^\varepsilon, w^\varepsilon, w_t^\varepsilon, w_x^\varepsilon, w_{xx}^\varepsilon \quad \text{in} \quad L^2(Q_T)\,.$$

Here $Q_T = (-\infty\,,\infty) \times (0\,,T)$ for any T positive and finite.

(iv) Define the Riemann invariants

$$\left\{\begin{matrix} r \\ s \end{matrix}\right. = v \mp \int_\alpha^w (-p'(\zeta))^{1/2}\, d\zeta\,.$$

If $p'' \geqslant 0$ on $w < \alpha$ and $p'' \leqslant 0$ on $\beta < w$ then there exists a positive invariant D for (P_ε) of the form shown in Figure 2.

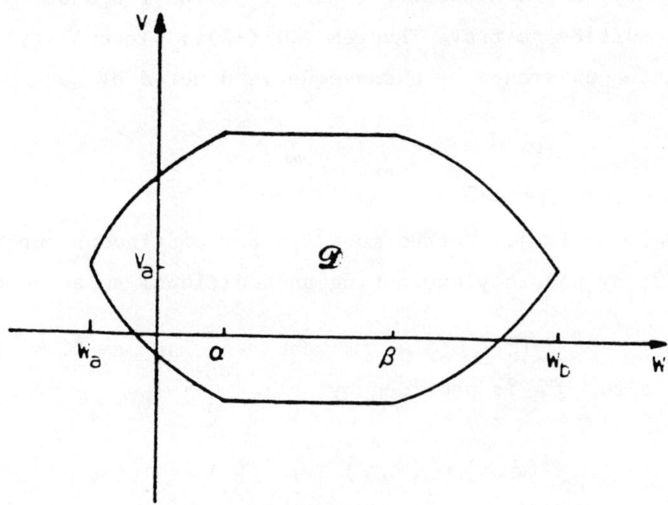

FIG. 2

Region D is described by

$$\begin{aligned} r \leqslant r_a, \; s \geqslant s_a & \quad \text{for} \quad w_a \leqslant w \leqslant \alpha \;, \\ r \geqslant s_a, \; s \leqslant r_a & \quad \text{for} \quad \beta \leqslant w \leqslant w_b \;, \\ s_a \leqslant v \leqslant r_a & \quad \text{for} \quad \alpha \leqslant w \leqslant \beta \;, \end{aligned}$$

where r_a, s_a, w_a, w_b, v_a are constants satisfying

$$\left\{ \begin{matrix} r_a \\ \\ s_a \end{matrix} \right. = v_a \mp \int_{\alpha}^{w_a} (-p'(\zeta)^{1/2} d\zeta = v_a \pm \int_{\alpha}^{w_b} (-p'(\zeta))^{1/2} d\zeta .$$

In particular we know for the weak solutions of (P_ε) that $(v_0(x), w_0(x)) \in D$ for all $x \in R$ implies $(v^\varepsilon(x,t), w^\varepsilon(x,t)) \in D$ for all $x \in R$, $t > 0$.

A proof of this theorem is provided in Appendix A. We note that the existence of the invariant region D implies

$$\|(v^\varepsilon, w^\varepsilon)\|_{L^\infty(Q_T)} \leqslant \text{const.} \qquad (1.1)$$

where const. depends only on v_0, w_0 and is independent of $\varepsilon > 0$.

In what follows we assume (1.1) *independently of whether the hypotheses on* p *in* (iv) *hold.* (We note that the uniform in ε L^∞ bound makes the assumption of p globally Lipschitz unnecessary as the Lipschitz constant on the range of values for w^ε will suffice to prove Theorem 1.1 (ii).) Inequality (1.1) implies the existence of a subsequence denoted by $(v^\varepsilon, w^\varepsilon)$ such that

$$\begin{matrix} v^\varepsilon \to \bar{v} \\ w^\varepsilon \to \bar{w} \end{matrix} \quad \text{weak} * L^\infty \quad \text{as } \varepsilon \to 0,$$

where $L^\infty = L^\infty(Q_T)$. Furthermore for any continuous function $f: R^2 \to R$, by possibly extracting an additional subsequence one obtains

$$f(w^\varepsilon, w^\varepsilon) \to f^* \quad \text{weak} * L^\infty \quad \text{as } \varepsilon \to 0.$$

A formula for f^* is provided by

$$f^*(x,t) = \langle f, \gamma \rangle = \int_{R^2} f(\lambda)\, d\gamma_{x,t}(\lambda), \tag{1.2}$$

where $\gamma_{x,t}$ is the representing Young measure (a probability measure). If $\gamma_{x,t}$ reduces to a Dirac mass supported as $v(x,t)$, $w(x,t)$ then $f^*(x,t) = f(v(x,t), w(x,t))$. In this case hence $\lim f(v^\varepsilon, w^\varepsilon) = f(\text{weak}*\lim(v^\varepsilon, w^\varepsilon))$ and in particular the choice $f(\lambda_1, \lambda_2) = \lambda_1^2 + \lambda_2^2$ shows that $v^\varepsilon \to \bar{v}$, $w^\varepsilon \to \bar{w}$ strongly in $L^2(Q_T)$.

2. ENERGY ESTIMATES

In this section we sharpen the results of Theorem 1.1 (iii). We shall prove the following result.

THEOREM 2.1. Assume v_0, w_0 approach constants (normalized to zero) as $|x| \to \infty$ and

$$\int_{-\infty}^{\infty} \left(\frac{v_0^2}{2} + \sum(w_0)\right) dx \leqslant \text{const.},$$

where

$$\sum(w) = -\int_\alpha^w p(\zeta)\, d\zeta .$$

Then the solutions of (P_ε) satisfy

$$\varepsilon \iint_{Q_T} \left(v_x^\varepsilon\right)^2 dx\, dt \leqslant \text{const.}, \tag{2.1}$$

$$\varepsilon \iint_H \left(w_x^\varepsilon\right)^2 \, dx \, dt \leqslant \text{const.}, \tag{2.2}$$

where

$$H = \{ (x,t) \in Q_T;\ w^\varepsilon(x,t) < \alpha \ \text{ or } \ w^\varepsilon(x,t) > \beta \}$$

and the constants are independent of ε.

PROOF. We do a second regularization and consider the problem

$$\begin{aligned} v_t + p_\delta(w)_x &= \varepsilon v_{xx} \\ w_t - v_x &= \varepsilon w_{xx} \end{aligned} \qquad -\infty < x < \infty \qquad (P_\delta)$$

with initial data $v(x,0) = v_0(x)$, $w(x,0) = w_0(x)$. Here $p_\delta \in C^\infty$ is defined by Proposition 2.2 of [21] and has the shape shown in Fig. 3. For our purposes the exact formula for p_δ is not important.

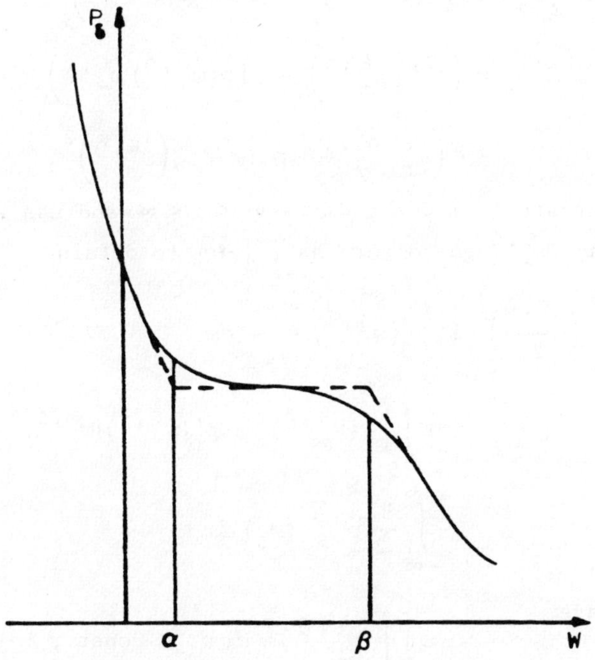

FIG. 3

The relevant facts regarding p_δ are

(a) $p'_\delta < 0$ for all w,

(b) $p_\delta \to p$ uniformly in w,

(c) $\sup_{w \in W_1} |p'_\delta(w) - p'(w)| \leqslant \text{const.}\ \delta$

where $W_1 = \{w;\ |w-\alpha| > \rho\} \cap \{w;\ |w-\beta| > \rho\}$ for $\rho > 0$ small and $0 < \delta < \rho$.

In [21] we have proven that since (b) holds

$$\sup_{0\leqslant t\leqslant T}\ \sup_{x\in R} |(v^{\varepsilon,\delta}, w^{\varepsilon,\delta}) - (v^{\varepsilon}, w^{\varepsilon})| \to 0 \text{ as } \delta \to 0\,,$$

where $v^{\varepsilon,\delta}$, $w^{\varepsilon,\delta}$ denote the solutions of (P_δ) . To obtain our estimates we know that since (P_δ) possesses smooth solutions we can multiply $(P_\delta\, a)$ by $v^{\varepsilon,\delta}$ and $(P_\delta b)$ by $-p_\delta(w^{\varepsilon,\delta})$. Then we have

$$\left(\left(\frac{v^{\varepsilon,\delta}}{2}\right)^2 + \sum(w^{\varepsilon,\delta})\right)_t + \left(v^{\varepsilon,\delta} p(w^{\varepsilon,\delta})\right)_x$$
$$= \left(v^{\varepsilon,\delta} v_x^{\varepsilon,\delta}\right)_x - \varepsilon\left(p(w^{\varepsilon,\delta})\, w_x^{\varepsilon,\delta}\right)_x$$
$$- \varepsilon\left(v_x^{\varepsilon,\delta}\right)^2 + \varepsilon\, p'_\delta(w^{\varepsilon,\delta})\left(w_x^{\varepsilon,\delta}\right)^2.$$

Now integrate from $0 \leqslant t \leqslant T$, $-\infty < x < \infty$ and use the fact that $w_x^{\varepsilon,\delta}$, and $v_x^{\varepsilon,\delta}$ go to zero as $|x| \to \infty$ to obtain

$$\int_{-\infty}^{\infty} \left(\frac{v^{\varepsilon,\delta}}{2}\right)^2 + \sum(w^{\varepsilon,\delta})\Big|_{t=T} dx$$
$$+ \varepsilon \iint_{Q_T} (v_x^{\varepsilon,\delta})^2 - p'_\delta(w^{\varepsilon,\delta})(w_x^{\varepsilon,\delta})^2\, dx\, dt$$
$$= \int_{-\infty}^{\infty} \frac{v_0^2}{2} + \sum(w_0)\, dx$$

so we have

$$\varepsilon \iint_{Q_T} (v_x^{\varepsilon,\delta})^2\, dx\, dt \leqslant \text{const.}, \tag{2.3}$$

$$-\varepsilon \iint_{Q_T} p'_\delta(w^{\varepsilon,\delta})\,(w_x^{\varepsilon,\delta})^2\, dx\, dt \leqslant \text{const.} \tag{2.4}$$

Now set

$$H_\rho = \{(x, t) ; w^{\varepsilon,\delta}(x,t) < \alpha - \rho \text{ or } w^{\varepsilon,\delta}(x,t) > \beta + \rho\} .$$

Then (2.4) implies

$$-\varepsilon \iint_{H_\rho} p'(w^{\varepsilon,\delta})(w_x^{\varepsilon,\delta})^2 \, dx\, dt$$

$$-\varepsilon \iint_{H_\rho} \left(p_\delta'(w^{\varepsilon,\delta}) - p'(w^{\varepsilon,\delta})\right)\left(w_x^{\varepsilon,\delta}\right)^2 dx\, dt \leqslant \text{const.}$$

But on H_ρ, $-p' \geqslant$ const. >0 and (c) holds so that

$$\varepsilon \text{ const. } (1-\delta) \iint_{H_\rho} \left(w_x^{\varepsilon,\delta}\right)^2 dx\, dt \leqslant \text{const.}$$

or

$$\varepsilon \iint_{H_\delta} (w_x^{\varepsilon,\delta})^2 \, dx\, dt \leqslant \text{const.} \tag{2.5}$$

From (2.3) and (2.5) it follows that we can extract subsequences also denoted by $v_x^{\varepsilon,\delta}$, $w_x^{\varepsilon,\delta}$ so that

$$v_x^{\varepsilon,\delta} \to V \text{ weakly in } L^2(Q_T),$$

$$w_x^{\varepsilon,\delta} \to W \text{ weakly in } L^2(H_\rho),$$

as $\delta \to 0+$. But since $w^{\varepsilon,\delta} \to w^\varepsilon$, $v^{\varepsilon,\delta} \to v^\varepsilon$ uniformly in (x, t) we must have $V = v_x^\varepsilon$, $W = w_x^\varepsilon$. Now use the weak lower semicontinuity of the quadratic functionals in (2.3), (2.5) to conclude

$$\varepsilon \iint_{Q_T} (v_x^\varepsilon)^2 \, dx\, dt \leqslant \text{const.}, \tag{2.1}$$

$$\varepsilon \iint_{H_\rho} (w_x^\varepsilon)^2 \, dx\, dt \leqslant \text{const.} \tag{2.6}$$

Finally we let $\rho \to 0+$ observing that

$$\iint_{H_\rho} (w_x^\varepsilon)^2 \, dx\, dt \to \iint_H (w_x^\varepsilon)^2 \, dx\, dt$$

where

$$H = \{(x, t) ; w^\varepsilon(x,t) < \alpha \text{ or } w^\varepsilon(x,t) > \beta\} .$$

Then we obtain

$$\varepsilon \iint_H (w_x^\varepsilon)^2 \, dx\, dt \leq \text{const}.$$

and we have the desired estimates.

3. COMPENSATED COMPACTNESS

Let $\eta(v,w)$ be C and $q(v,w)$ be C^1 except across the lines $w=\alpha$, $w=\beta$ where q is only continuous. If n, q satisfy

$$\eta_w + q_v = 0 \tag{3.1}$$

$$-p'(w)\, \eta_v + q_w = 0 \tag{3.2}$$

for $w \neq \alpha$, $w \neq \beta$ and the jump condition

$$[q_w] = [p'] \, \eta_v, \tag{3.3}$$

across $w=\alpha$, $w=\beta$ then we will call η, q an *entropy-flux pair*.

We use entropy-flux pairs in applying the Div-Curl lemma of Murat and Tartar [18], [19], [33].

DIV-CURL LEMMA

Let $\Omega \subset R^n$ be an open bounded set. Let $\{\phi^\varepsilon\}$, $\{\psi^\varepsilon\}$ be two sequences of vector functions in $(L^2(\Omega))^N$. If $\phi^\varepsilon \to \phi$, $\psi^\varepsilon \to \psi$ in $(L^2(\Omega))^N$ weakly as $\varepsilon \to 0+$ and if $\{\operatorname{div} \phi^\varepsilon\}$, $\{\operatorname{curl} \psi^\varepsilon\}$ lie in a compact subset of $H^{-1}(\Omega)$ then

$$\phi^\varepsilon \cdot \psi^\varepsilon \to \phi \cdot \psi \text{ in the sense of distributions.} \tag{3.4}$$

Now assume for weak solutions $(v^\varepsilon, w^\varepsilon)$ of (P_ε) we can show for two entropy-flux pairs (η_i, q_i) that

$$\eta_i(v^\varepsilon, w^\varepsilon)_t + q_i(v^\varepsilon, w^\varepsilon)_x \qquad (i=1,2)$$

lies in a compact subset of $H^{-1}(\Omega)$ for every bounded open subset $\Omega \subset Q_T$. Then we can identify

$$\phi^\varepsilon(x,t) = \Big(\eta_1(v^\varepsilon(x,t), w^\varepsilon(x,t)),\ q_1(v^\varepsilon(x,t), w^\varepsilon(x,t))\Big),$$

$$\psi^\varepsilon(x,t) = \Big(q_2(v^\varepsilon(x,t), w^\varepsilon(x,t)),\ -\eta_2(v^\varepsilon(x,t), w^\varepsilon(x,t))\Big),$$

$$\operatorname{div} = \left(\frac{\partial}{\partial t}, \frac{\partial}{\partial x}\right), \text{ and } \operatorname{curl} = \left(\frac{-\partial}{\partial x}, \frac{\partial}{\partial t}\right),$$

and set $n=2$, $N=2$ and apply the Div-Curl Lemma to conclude that

$$\Big(\eta_1(v^\varepsilon,w^\varepsilon)\,,q_1(v^\varepsilon,w^\varepsilon)\Big)\cdot\Big(q_2(v^\varepsilon,w^\varepsilon),-\eta_2\,(v^\varepsilon,w^\varepsilon)\Big) \rightharpoonup (\eta_1^*,q_1^*)\cdot(q_2^*\,,-\eta_2^*) \tag{3.5}$$

where for a possible subsequence of $(v^\varepsilon,w^\varepsilon)$ we know

$$\eta_1(v^\varepsilon,w^\varepsilon)\to\eta_1^*\,,\eta_2(v^\varepsilon,w^\varepsilon)\to\eta^*\,, \quad q_1(v^\varepsilon,w^\varepsilon)\to q_1^*\,,q_2(v^\varepsilon,w^\varepsilon)\to q_2^*$$

weakly in $L^2(\Omega)$. Now since

$$\eta_1(v^\varepsilon,w^\varepsilon)\,,\ \eta_2(v^\varepsilon,w^\varepsilon)\,,\ q_1(v^\varepsilon,w^\varepsilon),\ q_2(v^\varepsilon,w^\varepsilon)$$

all possess L^∞ bounds independent of ε they all possess subsequences converging weak $*L^\infty$ to η_1^*, η_2^*, q_1^*, q_2^* respectively. The Young measure allows us to represent these limits as

$$\eta_1^*=\langle\gamma,\eta_1\rangle\,,\ \eta_2^*=\langle\gamma,\eta_2\rangle\,,\ q_1^*=\langle\gamma,q_1\rangle\,,\ q_2^*=\langle\gamma,q_2\rangle\,.$$

Furthermore

$$\Big(\eta_1(v^\varepsilon,w^\varepsilon)\,,q_2(v^\varepsilon,w^\varepsilon)\Big)\cdot\Big(q_2(v^\varepsilon,w^\varepsilon),-\eta_2(v^\varepsilon,w^\varepsilon)\Big)$$

also possesses a convergent subsequence where the weak * limit must be $\langle\gamma,\eta_1q_2-\eta_2q_1\rangle$. We now use (3.5) to conclude Tartar's remarkable identity

$$\langle\gamma,\eta_1q_2-\eta_2q_1\rangle=\langle\gamma,\eta_1\rangle\langle\gamma,q_2\rangle-\langle\gamma,\eta_2\rangle\langle\gamma,q_1\rangle\,. \tag{3.6}$$

where equality holds a.e. in (x,t).

4. CONSTRUCTION OF ENTROPY-FLUX PAIRS

In this section we describe the construction of entropy-flux pairs.

Let $\phi_0(v)\in C_0^\infty(R)$ and define

$$\eta(v,w)=\phi_0(v),\ q(v,w)=0 \quad \text{for } a\leqslant w\leqslant\beta,\ v\in R\,. \tag{4.1}$$

Notice (3.1), (3.2) are trivially satisfied for $w\in(\alpha,\beta)$. For $w<\alpha, w<\beta, \eta, q$ are determined by solutions of the well-posed Cauchy problem (3.1), (3.2) with initial data

$$\eta(v,\alpha) = \phi_0(v), \quad q(v,\alpha) = 0 \tag{4.2}$$

For $w < \alpha$ and

$$\eta(v,\beta) = \phi_0(v), \quad q(v,\beta) = 0 \tag{4.3}$$

for $w < \beta$.

From (3.1) it is clear that η_w is continuous. Furthermore, since η_v is a smooth solution of the wave equation

$$(\eta_v)_{ww} + p'(w)(\eta_v)_{vv} = 0$$

on $w < \alpha, w > \beta$ with initial data

$$\eta_v(v,\alpha) = \phi_0'(v), \quad (\eta_v)_w(v,\alpha) = 0$$

for $w < \alpha$ and

$$\eta_v(v,\beta) = \phi_0'(v), \quad (\eta_v)_w(v,\beta) = 0$$

for $w > \beta$ we see

$$\lim_{w \to \alpha} \eta_v(v,w) = \lim_{w \to \beta} \eta_v(v,w) = \phi_0'(v)$$

and η_v is continuous as well.

Repetition of this argument shows η_v, η_{vw}, η_w, η_{vvw}, etc. are continuous. Also since η_{vw} is continuous in $\alpha < w, w > \beta$ and $\eta_{vw} = \eta_{wv}$ in $\alpha < w, w > \beta$, and

$$\lim_{w \to \alpha -} \eta_{wv}(v,w) = \lim_{w \to \alpha -} \eta_{vw}(v,w)$$

$$= \lim_{w \to \beta +} \eta_{vw}(v,w) = \lim_{w \to \beta +} \eta_{vw}(v,w) = 0$$

and $\eta_{vw} = 0$ for $w \in (\alpha,\beta)$ we see η_{wv} is continuous everywhere as well.

As is readily apparent from (3.2) q_w possesses jumps at $w = \alpha$ and $w = \beta$:

$$[q_w(v,a)] = [p'(\alpha)]\,\phi_0'(v), \quad [q_w(v,\beta)] = [p'(\beta)]\,\phi_0'(v).$$

The presence of these jumps causes some minor technical difficulties in computing $\eta_t + q_x$. Nevertheless we can state the following lemma whose proof is given in Appendix B.

LEMMA 4.1. Let η, q be an entropy-flux pair as described above. Furthermore, assume p is globally Lipschitz so that all the conclusions of Theorem 1.1 hold. Then for $v^\varepsilon, w^\varepsilon$ a solution to (P_ε) we have

$$\eta_t(v^\varepsilon, w^\varepsilon) + q_x(v^\varepsilon, w^\varepsilon) = \varepsilon(\eta_v v_x^\varepsilon)_x + \varepsilon(\eta_w w_x^\varepsilon)_x - \varepsilon\left(\eta_{vv} v_x^{\varepsilon 2} + 2\eta_{vw} v_x^\varepsilon w_x^\varepsilon + \eta_{ww} w_x^{\varepsilon 2}\right) \tag{4.4}$$

where equality holds in the sense of distributions.

From Lemma 4.1 it is fairly straightforward to show that the Div-Curl Lemma will be applicable as we had hoped. Specifically we show the following

LEMMA 4.2. Let $\eta, q, v^\varepsilon, w^\varepsilon$ be as in Lemma 4.1. Then

$$\eta_t(v^\varepsilon, w^\varepsilon) + q_x(v^\varepsilon, w^\varepsilon)$$

lies in a compact subset of $H^{-1}_{\mathrm{loc}}(Q_T)$.

PROOF. Set $I_1^\varepsilon(x,t) = \varepsilon(\eta_v v_x^\varepsilon + \eta_w w_x^\varepsilon)$,

$$I_2^\varepsilon(x,t) = -\varepsilon\left(\eta_{vv} v_x^{\varepsilon 2} + 2\eta_{vw} v_x^\varepsilon w_x^\varepsilon + \eta_{ww} w_x^{\varepsilon 2}\right).$$

As before let

$$H = \{(x,t) \in Q_T;\ w^\varepsilon(x,t) < \alpha \ \text{ or } \ w^\varepsilon(x,t) > \beta\}$$

and set E = complement of H in Q_T. We wish to estimate the L^2 norm of I_2. First note

$$\iint_{Q_T} I_1^\varepsilon(x,t)^2\, dx\, dt = \varepsilon^2 \iint_{Q_T} (\eta_v v_x^\varepsilon + \eta_w w_x^\varepsilon)^2\, dx\, dt$$

$$\leqslant \varepsilon^2 \text{ const.} \left\{ \iint_H w_x^{\varepsilon 2}\, dx\, dt + \iint_{Q_T} v_x^{\varepsilon 2}\, dx\, dt \right\}$$

since $\eta_w = 0$ in E. From (2.1), (2.2) we see

$$\iint_{Q_T} I_1^\varepsilon(x,t)^2\, dx\, dt \leqslant \text{const.} \cdot \varepsilon$$

and hence $I_1^\varepsilon \to 0$ as $\varepsilon \to 0+$ strongly in $L^2(Q_T)$. This yields $I_1^\varepsilon \to 0$ as $\varepsilon \to 0+$ strongly in $H^{-1}_{\mathrm{loc}}(Q_T)$.

Next note

$$\iint_{Q_T} |I_2(x,t)|\, dx\, dt$$
$$= \varepsilon \iint_H |\eta_{vv} v_x^{\varepsilon 2} + 2\,\eta_{vw} v_x^\varepsilon w_x^\varepsilon + \eta_{ww} w_x^{\varepsilon 2}|\, dx\, dt$$
$$+ \varepsilon \iint_E |\eta_{vv} v_x^{\varepsilon 2}|\, dx\, dt \leqslant \text{const.}$$

via (2.1), (2.2). Also we note that as follows from the assumed $L^\infty(Q_T)$ estimates on $(v^\varepsilon, w^\varepsilon)$, $\eta(v^\varepsilon, w^\varepsilon)_t + q(v^\varepsilon, w^\varepsilon)_x$ and hence the right-hand side of (4.4) lies in a bounded subset of $W^{-1,\infty}(Q_T)$. Thus from the above integral estimates we have that the right-hand side of (4.4) is of the form

$$I_1^\varepsilon(x,t)_x + I_2(x,t)$$

where $I_1^\varepsilon(x,t)$ lies in a compact subset $H^{-1}_{\text{loc}}(Q_T)$, $I_2(x,t)$ lies in a bounded subset of $L^1(Q_T)$, and $I_1^\varepsilon(x,t)_x + I_2(x,t)$ lies in a bounded subset of $W^{-1,\infty}(Q_T)$. A lemma of Murat [33] implies the right-hand side of (4.4) lies in a compact subset of $H^{-1}_{\text{loc}}(Q_T)$.

As a consequence of Lemma 4.2 and the Div-Curl Lemma for such $\eta_1, q_1, \eta_2, q_2, v^\varepsilon, w^\varepsilon$ the associated Young measure does indeed satisfy Tartar's equality

$$(3.6) \qquad \langle \gamma, \eta_1 q_2 - \eta_2 q_1 \rangle = \langle \gamma, \eta_1 \rangle \langle \gamma, q_2 \rangle - \langle \gamma, \eta_2 \rangle \langle \gamma, q_1 \rangle .$$

5. LAX ENTROPY CONSTRUCTION

The next step in our analysis is to develope expressions for one parameter families of entropy-flux pairs. We modify a construction of Lax [14] and write exact entropy-flux pairs

$$\eta_k(v,w) = \frac{e^{kr}}{2}\left\{ V_0^r + \frac{V_1^r}{k} + \frac{V_2^r}{k^2} + \cdots \frac{V_N^r}{k^N} + 0\left(\frac{1}{k^N}\right)\right\}$$
$$+ \frac{e^{ks}}{2}\left\{ V_0^s + \frac{V_1^s}{k} + \frac{V_2^s}{k^2} + \cdots \frac{V_N^s}{k^N} + 0\left(\frac{1}{k^N}\right)\right\}$$

$$q_k(v,w) = \frac{e^{kr}}{2}\left\{H_0^r + \frac{H_1^r}{k} + \frac{H_2^r}{k^2} + \cdots \frac{H_N^r}{k^N} + 0\left(\frac{1}{k^N}\right)\right\}$$

$$+ \frac{e^{ks}}{2}\left\{H_0^s + \frac{H_1^s}{k} + \frac{H_2^s}{k^2} + \cdots \frac{H_N^s}{k^N} + 0\left(\frac{1}{k^N}\right)\right\} \tag{5.1}$$

where $r = r(v,w)$, $s = s(v,w)$ are the previously defined Riemann invariants, the coefficients $V_j^s, V_j^r, H_j^s, H_j^r$, $j = 1, \cdots, N$ depend on v, w, and k is a real parameter, and the terms $0\left(\frac{1}{k^N}\right)$ denote error terms which are solved for in the way used by Lax. As our goal is to have entropy-flux pairs which satisfy boundary conditions (4.2), (4.3) we are forced to use both r and s waves in our construction (5.1). Individual r or s waves will not be able to satisfy (4.2), (4.3). This is a serious departure from the arguments of DiPerna [6] and Rascle [20] and will be a source of major difficulties.

For convenience denote

$$\lambda = +\sqrt{-p'(w)}\ .$$

If we substitute (5.1) into (3.1), (3.2) and equate like powers of $\frac{e^{kr}}{k^j}$, $\frac{e^{ks}}{k^j}$ we find

$$\left.\begin{aligned} -\lambda V_0^r + H_0^r &= 0\,, \\ \lambda V_0^s + H_0^s &= 0\,, \end{aligned}\right\} \quad 0(k) \text{ terms}, \tag{5.2}$$

$$\left.\begin{aligned} -\lambda V_1^r + V_{0_w}^r + H_1^r + H_{0_v}^r &= 0\,, \\ \lambda V_1^s + V_{0_w}^s + H_1^s + H_{0_v}^s &= 0\,, \\ \lambda^2\left(V_1^r + V_{0_v}^r\right) - \lambda H_1^r + H_{0_w}^r &= 0\,, \\ \lambda^2\left(V_1^s + V_{0_v}^s\right) + \lambda H_1^s + H_{0_w}^s &= 0\,, \end{aligned}\right\} \quad 0(1) \text{ terms}, \tag{5.3}$$

and so on. From (5.3), (5.3) we find

$$V_{0_w}^r + \lambda V_{0_v}^r + \frac{p''}{4p'} V_0^r = 0\,, \tag{5.4}$$

$$V_{0_w}^s - \lambda V_{0_v}^s + \frac{p''}{4p'} V_0^s = 0\,, \tag{5.5}$$

where the equations are to hold in $w < \alpha, w > \beta$. In order to satisfy the initial conditions (4.2), (4.3) we set

$$\begin{aligned} V_0^r(v,\alpha) &= V_0^r(v,\beta) = V_0^s(v,\alpha) = V_0^s(v,\beta) = \tilde{\phi}_0(v)\,, \\ V_j^r(v,\alpha) &= V_j^r(v,\beta) = V_j^s(v,\alpha) = V_j^s(v,\beta) = 0\,, \end{aligned} \tag{5.6}$$

as well as forcing the error $0(1/k^N)$ terms to zero as $w = \alpha, \beta$, where we now take

$$\phi_0(v) = e^{kv}\,\tilde{\phi}_0(v)\,. \tag{5.7}$$

Equations (5.4)-(5.6) determine V_0^r, V_0^s and hence by (5.2) H_0^r and H_0^s also. Finally note (5.4), (5.5) can be written as

$$(V_0^r)_s + \frac{\lambda_s}{2\lambda} V_0^r = 0\,, \quad (V_0^s)_r + \frac{\lambda_r}{2\lambda} V_0^s = 0\,. \tag{5.8}$$

Hence for data $\tilde{\phi}_0(v)$ with support on $a_2 \leqslant v \leqslant a_1$ V_0^r will have support on the strips $a_2 \leqslant r \leqslant a_1$, $w < \alpha, w > \beta$ and V_0^s will have support on the strips $a_2 \leqslant s \leqslant a_1$, $w < \alpha, w > \beta$. If we examine the equations for V_j^r, V_j^s we see they have these same properties. Note that η_k, being a solution of the wave equation in one space dimension, is supported on the same set. Finally the error terms are supported on the same set as well because they are equal to the difference between η_k and V terms. A picture of the support of η_k is given in Fig. 4.

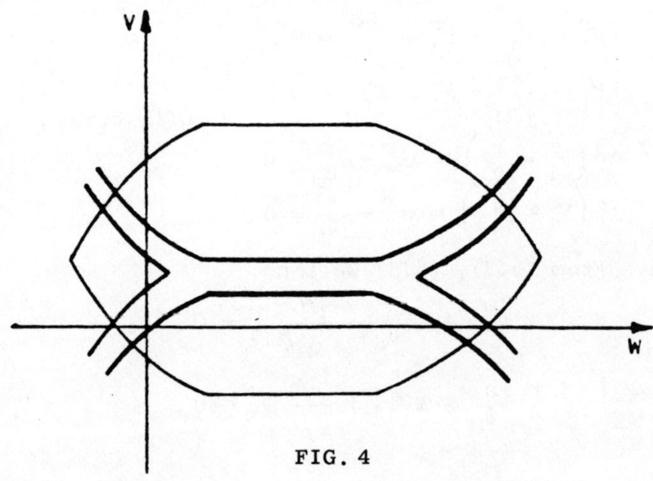

FIG. 4

6. ANALYSIS OF SUPPORT OF γ

Now let R be the smallest characteristic hexagon containing the support of γ. That is R has boundaries given by lines $r = r^+$, $v = v^+$, $s = r^+$ from above and $s = s^-$, $v = s^-$, $r = s^-$ from below. See Fig. 5.

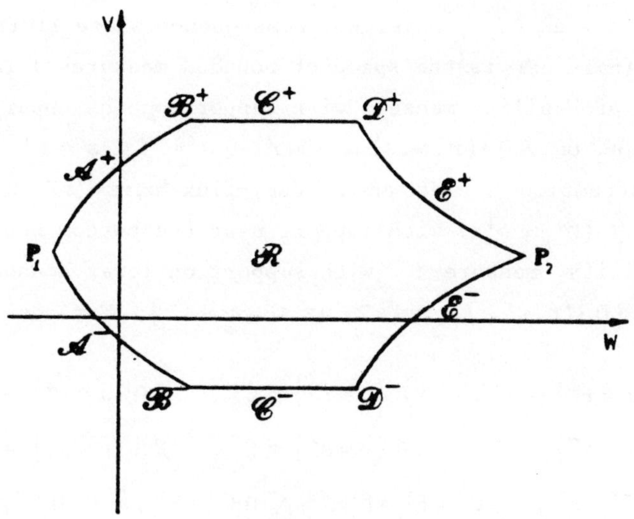

FIG. 5

Let us construct an entropy-flux pair (η_k^+, q_k^+) with data $\tilde{\phi}_0(v) = \Phi(v - r^+)$ where Φ has the graph shown in Fig. 6, and δ

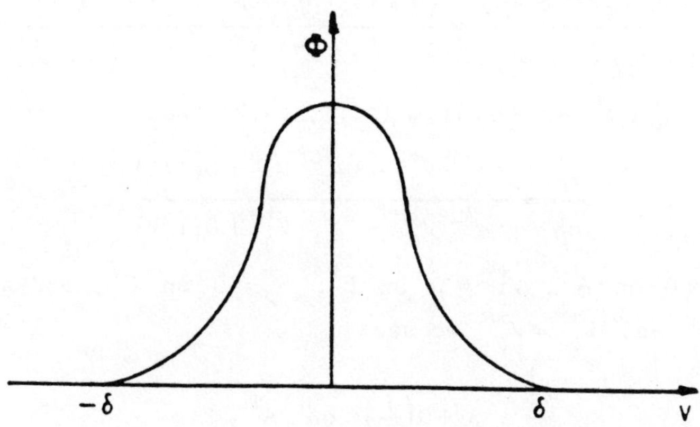

FIG. 6

is small. For h continuous we follow DiPerna and define probability measure

$$\langle \mu_k^+ , h \rangle = \frac{\langle \gamma , \eta_k^+ h \rangle}{\langle \gamma , \eta_k^+ \rangle}$$

so that μ_k^+ has support near the top boundary of R. The sequence $\{\mu_k^+\}$ has a weak $* BM$ convergent subsequence whose limit we denote by μ^+. (Here BM is the space of bounded measures.) The measure μ^+ is a probability measure with support on the upper boundary of R, i.e. on $R \cap \{(r,v,s) : r = r^+ , v = r^+ , s = r^+\}$. A similar construction yields an entropy-flux pair (η_k^- , q_k^-) where $(\eta_k^- , q_k^-) \equiv (\eta_{-k}^+ , q_{-k}^+)$ with support near the bottom boundary and a probability measure μ^- with support on lower boundary of R i.e. on $R \cap \{(r,v,s) : s = s^- , v = s^- , r = s^-\}$. For convenience we set

$$R \cap \{r = r^+\} = A^+ , \quad R \cap \{v = r^+\} = C^+ , \quad R \cap \{s = r^+\} = E^+,$$

$$R \cap \{s = s^-\} = A^- , \quad R \cap \{v = s^-\} = C^- , \quad R \cap \{r = s^-\} = E^-,$$

$$A^+ \cap C^+ = B^+ , \quad C^+ \cap E^+ = D^+ , \quad A^- \cap C^- = B^- , \quad C^- \cap E^- = D^- ,$$

From Tartar's equality (3.6) we know

$$\langle \gamma , \eta_k^+ q_k^- - \eta_k^- q_k^+ \rangle = \langle \gamma , \eta_k^+ \rangle \langle \gamma , q_k^- \rangle - \langle \gamma , \eta_k^- \rangle \langle \gamma , q_k^+ \rangle$$

and hence

$$\frac{\langle \gamma , \eta_k^+ q_k^- - \eta_k^- q_k^+ \rangle}{\langle \gamma , \eta_k^+ \rangle \langle \gamma , \eta_k^- \rangle} = \frac{\langle \gamma , (q_k^- / \eta_k^-)\, \eta_k^- \rangle}{\langle \gamma , \eta_k^- \rangle} - \frac{\langle \gamma , (q_k^+ / \eta_k^+)\, \eta_k^+ \rangle}{\langle \gamma , \eta_k^+ \rangle} . \quad (6.1)$$

Now for η_k^+ , q_k^+ we see from (5.1), (5.2) that

$$\frac{q_k^+}{\eta_k^+} = \frac{\lambda e^{kr} V_0^{+r} - \lambda e^{ks} V_0^{+s} + 0(1/k)}{e^{kr} V_0^{+r} + e^{ks} V_0^{+s} + 0(1/k)} .$$

As $V_0^{+s} = 0$ on A^+, $V_0^{+r} = 0$ on E^+, $q_k^+ = 0$ on C^+, and at B^+ , D^+ $r = s = r^+$ and $V_0^{+r} = V_0^{+s}$ we see

$$\frac{q_k^+}{\eta_k^+} = \lambda + 0\left(\frac{1}{k}\right) \quad \text{on} \quad A^+ ,$$

$$
\begin{aligned}
&= -\lambda + 0\left(\frac{1}{k}\right) \quad \text{on } E^+ , \\
&= 0 + 0\left(\frac{1}{k}\right) \quad \text{at } B^+ \text{ and } D^+ , \qquad (6.2) \\
&= 0 \quad \text{on } C^+ .
\end{aligned}
$$

Similarly we have

$$
\begin{aligned}
\frac{\bar{q}_k}{\bar{\eta}_k} &= -\lambda + 0\left(\frac{1}{k}\right) \quad \text{on } A^- , \\
&= \lambda + 0\left(\frac{1}{k}\right) \quad \text{on } E^- , \\
&= 0\left(\frac{1}{k}\right) \quad \text{at } B^- \text{ and } D^- \qquad (6.3) \\
&= 0 \quad \text{on } C^- .
\end{aligned}
$$

If we define

$$
\sigma = \begin{cases} -\lambda & \text{on } E^+ \text{ and } A^- \\ \lambda & \text{on } A^+ \text{ and } E^- \\ 0 & \text{on } B^+, C^+, D^+, B^-, C^-, D^- \end{cases}
$$

then (6.2), (6.3) can be summarized by the relation

$$
\frac{\overset{+}{q}_k}{\overset{+}{\eta}_k} = \sigma + 0\left(\frac{1}{k}\right), \quad \frac{\bar{q}_k}{\bar{\eta}_k} = \sigma + 0\left(\frac{1}{k}\right) \qquad (6.4)
$$

on the appropriate parts of the boundaries of R.

Now let us make the following assumption.

ASSUMPTION A. There exist open neighbourhoods of the corner points $P_1 = A^- \cap A^+$, $P_2 = E^- \cap E^+$ in which $\gamma \equiv 0$.

Under the above assumption the left hand side of (6.1) is zero if the support of Φ_0 measured by δ is sufficiently small. This is because the support of $\eta_k^+ q_k^- - \eta_k^- q_k^+$ is contained in these open neighbourhoods where $\gamma \equiv 0$.

We now use (6.1)-(6.3) to see

$$
\frac{\langle \gamma, (\sigma + 0\,(1/k))\, \eta_k^- \rangle}{\langle \gamma, \eta_k^- \rangle} = \frac{\langle \gamma, (\sigma + 0\,(1/k))\, \eta_k^+ \rangle}{\langle \gamma, \eta_k^+ \rangle}
$$

and hence taking the limit as $k \to \infty$ we find

$$
\langle \mu^-, \sigma \rangle = \langle \mu^+, \sigma \rangle . \qquad (6.5)
$$

Next let η, q be an entropy-flux pair satisfying boundary conditions (4.2), (4.3). We can apply Tartar's equality to find

$$\begin{aligned} \langle \gamma, \eta q_k^+ - \eta_k^+ q \rangle &= \langle \gamma, \eta \rangle \langle \gamma, q^+ \rangle - \langle \gamma, \eta_k^+ \rangle \langle \gamma, q \rangle \\ \langle \gamma, \eta q_k^- - \eta_k^- q \rangle &= \langle \gamma, \eta \rangle \langle \gamma, q_k^- \rangle - \langle \gamma, \eta_k^- \rangle \langle \gamma, q \rangle . \end{aligned} \tag{6.6}$$

If we employ (6.4) again we see (6.6) shows

$$\begin{aligned} \frac{\langle \gamma, \eta(\sigma + 0(1/k))\, \eta_k^+ - \eta_k^+ q \rangle}{\langle \gamma, \eta_k^+ \rangle} &= \frac{\langle \gamma, \eta \rangle \langle \gamma, (\sigma + 0(1/k)) \eta_k^+ \rangle}{\langle \gamma, \eta_k^+ \rangle} - \langle \gamma, q \rangle \\ \frac{\langle \gamma, \eta(\sigma + 0(1/k))\, \eta_k^- - \eta_k^- q \rangle}{\langle \gamma, \eta_k^- \rangle} &= \frac{\langle \gamma, \eta \rangle \langle \gamma, (\sigma + 0(1/k)\, \eta_k^- \rangle}{\langle \gamma, \eta_k^- \rangle} - \langle \gamma, q \rangle . \end{aligned} \tag{6.7}$$

If we take limits as $k \to \infty$ in (6.7) we find

$$\begin{aligned} \langle \mu^+, \sigma\eta - q \rangle &= \langle \gamma, \eta \rangle \langle \mu^+, \sigma \rangle - \langle \gamma, q \rangle , \\ \langle \mu^-, \sigma\eta - q \rangle &= \langle \gamma, \eta \rangle \langle \mu^-, \sigma \rangle - \langle \gamma, q \rangle , \end{aligned}$$

and hence from (6.5) we have

$$\langle \mu^+, q - \sigma\eta \rangle = \langle \mu^-, q - \sigma\eta \rangle . \tag{6.8}$$

Now from (5.2)-(5.4) we find on

$$\begin{aligned} A^+ : q_k^+ - \sigma\eta_k^+ &= \frac{e^{kr}}{2k} \left(H_1^{+r} - \lambda V_1^{+r} \right) , \\ E^+ : q_k^+ - \sigma\eta_k^+ &= \frac{e^{ks}}{2k} \left(H_1^{+s} + \lambda V_1^{+s} \right) , \\ A^- : q_k^- - \sigma\eta_k^- &= \frac{e^{ks}}{2k} \left(H_1^{-s} + \lambda V_1^{-s} \right) , \\ E^- : q_k^- - \sigma\eta_k^- &= \frac{e^{kr}}{2k} \left(H_1^{-r} - \lambda V_1^{-r} \right) , \end{aligned} \tag{6.9}$$

and

$$\begin{aligned} H_1^{\pm r} - \lambda V_1^{\pm r} &= \frac{p''}{4p'} \, V_0^{\pm r} , \\ H_1^{\pm s} - \lambda V_1^{\pm s} &= \frac{p''}{4p'} \, V_0^{\pm s} . \end{aligned} \tag{6.10}$$

Now choose $\eta = \eta_k^+$, $q = q_k^+$ in (6.8). Then $\langle \mu^-, q_k^+ - \sigma\eta_k^+ \rangle \equiv 0$ since the support of μ^- is strictly contained in the bottom boundary and there is no support in corners P_1 and P_2 because γ has no support there. Hence

$$0 = \langle \mu^+, q^+ - \sigma\eta^+ \rangle = \left\langle \mu^+, \frac{e^{kr}}{2k} \frac{p''}{4p'} V_0^{+r} \right\rangle_{A^+} + \left\langle \mu^+, \frac{e^{ks}}{2k} \frac{p''}{4p'} V_0^{+s} \right\rangle_{E^+} \tag{6.11}$$

where the subscripts A^+, E^+ denote integration over the regions A^+ and E^+ respectively.

A similar argument with $\eta = \eta_k^-$, $q = q_k^-$ yields

$$0 = \left\langle \mu^-, \frac{e^{ks}}{2k} \frac{p''}{4p'} V_0^{-s} \right\rangle_{A^-} + \left\langle \mu^-, \frac{e^{kr}}{2k} \frac{p''}{4p'} V_0^{-r} \right\rangle_{E^-} . \tag{6.12}$$

Note V_0^{-s}, V_0^{-r}, V_0^{+s}, V_0^{+r} have the same sign on A^+, E^+, A^-, E^-. So if p'' is one signed on $w < \alpha, w > \beta$ the only way (6.11), (6.12) can hold is that the supports of μ^- and μ^+ are actually contained in $\alpha \leqslant w \leqslant \beta$. That is the support of γ was contained in $\alpha \leqslant w \leqslant \beta$. We summarize our conclusion in the following theorem.

THEOREM 6.1. Assume there exists a uniform in ε $L^\infty(Q_T)$ bound on $(v^\varepsilon, w^\varepsilon)$. Let R denote the smallest characteristic hexagon containing the support of the associated Young measure γ. If p'' is one signed on $w<\alpha, w>\beta$ and assumption A holds then the support of γ is actually contained in $\alpha \leqslant w \leqslant \beta$.

COROLLARY 6.2. If the hypotheses of Theorem 6.1 hold weak $*$ lim $p(w^\varepsilon) = p(\bar{w})$ as $\varepsilon \to 0^+$ and $(\bar{v}, \bar{w})$ is a solution of (P) is the sense of distributions.

PROOF. We know weak $*$ lim $p(w^\varepsilon) = \langle \gamma, p \rangle$. If supp γ is contained in $\alpha \leqslant w \leqslant \beta$ where $p = p_m$ we have $\langle \gamma, p \rangle = p_m$. Similarly $\alpha \leqslant \bar{w} \leqslant \beta$ so $p(\bar{w}) = p_m =$ weak$*$ lim $p(w^\varepsilon)$.

The above results point out the crucial role of the corner points P_1 and P_2 in determining the support of γ, i.e. excluding open neighbourhoods of P_1 and P_2 shrinks the support of γ

to the region E: $\alpha \leqslant w \leqslant \beta$. The next result considers the complementary issue: what if the support of γ sits only on the corner points P_1 and P_2?

THEOREM 6.3. Assume there exists a uniform in ε $L^\infty(Q_T)$ bound on $(v^\varepsilon, w^\varepsilon)$ and let R be as in Theorem 6.1. If $\gamma_{x,t} = a\delta(P_1) + b\delta(P_2)$ where $a \geqslant 0$, $b \geqslant 0$, $a+b=1$, and $\delta(P_i)$ denotes Dirac masses sitting at P_i, $i=1,2$, then $ab=0$ if $p'(P_1) \neq p'(P_2)$. That is $\gamma_{x,t}$ must be a Dirac mass sitting at either P_1 or P_2.

PROOF. Let $\eta_k^+, q_k^+, \eta_k^-, q_k^-$ be as in the proof of Theorem 6.1. For convenience we delete the subscript k. Then by Tartar's equality we know

$$\langle \gamma, \eta_+ q_- - \eta_- q_+ \rangle = \langle \gamma, \eta_+ \rangle \langle \gamma, q_- \rangle - \langle \gamma, \eta_- \rangle \langle \gamma, q_+ \rangle$$

hence

$$\begin{aligned} a\eta_+(P_1)\, q^-(P_1) - a\eta_-(P_1)\, q^+(P_2) + b\eta_+(P_2)\, q^-(P_2) - b\eta_-(P_2)\, q_+(P_2) \\ = (a\eta_+(P_1) + b\eta_+(P_2))(aq_-(P_1) + bq_-(P_2)) \qquad (6.13) \\ - (a\eta_-(P_1) + b\eta_-(P_2))(aq_+(P_1) + bq_+(P_2)). \end{aligned}$$

As before call $\lambda(P_i) = (-p'(P_i))^{1/2}$ $i=1,2$. From (6.2), (6.3) we know

$$\begin{aligned} q_+(P_1) &= \left(\lambda(P_1) + 0\left(\frac{1}{k}\right)\right)\eta_+(P_1), \\ q_-(P_1) &= \left(-\lambda(P_1) + 0\left(\frac{1}{k}\right)\right)\eta_-(P_1), \\ q_+(P_2) &= \left(-\lambda(P_2) + 0\left(\frac{1}{k}\right)\right)\eta_+(P_2), \\ q_-(P_2) &= \left(\lambda(P_2) + 0\left(\frac{1}{k}\right)\right)\eta_-(P_2). \end{aligned} \qquad (6.14)$$

Also from integrating (5.9) we find

$$\begin{aligned} \eta_+(P_1) &= A\left(\lambda^{-1/2}(P_1) + 0\left(\frac{1}{k}\right)\right), \\ \eta_-(P_1) &= B\left(\lambda^{-1/2}(P_1) + 0\left(\frac{1}{k}\right)\right), \\ \eta_+(P_2) &= A\left(\lambda^{-1/2}(P_2) + 0\left(\frac{1}{k}\right)\right), \\ \eta_-(P_2) &= B\left(\lambda^{-1/2}(P_2) + 0\left(\frac{1}{k}\right)\right), \end{aligned} \qquad (6.15)$$

where $A = e^{kr+} \tilde{\phi}_0(r_+)$, $B = e^{kr-} \tilde{\phi}_0(s_-)$.

If we combine (6.14) and (6.15) we see

$$\begin{aligned} q_+(P_1) &= A\left(\lambda^{1/2}(P_1) + O\left(\frac{1}{k}\right)\right), \\ q_-(P_1) &= -B\left(\lambda^{1/2}(P_1) + O\left(\frac{1}{k}\right)\right), \\ q_+(P_2) &= -A\left(\lambda^{1/2}(P_2) + O\left(\frac{1}{k}\right)\right), \\ q_-(P_2) &= B\left(\lambda^{1/2}(P_2) + O\left(\frac{1}{k}\right)\right). \end{aligned} \tag{6.16}$$

We now substitute (6.15) and (6.16) into (6.13). To leading order in k the left hand side of (6.13) is $2AB(b-a)$ and the right hand side of (6.13) is

$$2AB\left\{a\lambda^{-1/2}(P_1) + b\lambda^{-1/2}(P_2)\right\}\left\{-a\lambda^{1/2}(P_1) + b\lambda^{1/2}(P_2)\right\}.$$

So if we equate the left and right hand sides, divide by AB, and let $k \to \infty$ we find

$$\begin{aligned} b-a = (b-a)(b+a) &+ 2ab\lambda^{-1/2}(P_1)\lambda^{1/2}(P_2) \\ &- 2ab\lambda^{-1/2}(P_2)\lambda^{1/2}(P_1). \end{aligned}$$

Since $a+b = 1$, the quantity $b-a$ cancels and we find

$$0 = ab\left\{\lambda^{-1/2}(P_1)\lambda^{1/2}(P_2) - \lambda^{-1/2}(P_2)\lambda^{1/2}(P_1)\right\}. \tag{6.17}$$

So either $ab = 0$ or the quantity in braces in (6.17) is zero. But this quantity is zero if and only if $\lambda^2(P_1) = \lambda^2(P_2)$ so $p'(P_2) \neq p'(P_1)$ implies $\gamma_{x,t}$ is a Dirac mass.

REMARK. If $p'(z_1) < p'(z_2)$ for $z_1 < \alpha, z_2 > \beta$ then the assumption $p'(P_1) \neq p'(P_2)$ will be satisfied no matter where the characteristic hexagon is located.

REMARK. Of course if $\{(w^\varepsilon, v^\varepsilon)\}$ are such that w^ε lies in one phase for all $\varepsilon > 0$ (say $w^\varepsilon < \alpha$) then (P) is strictly hyperbolic and genuinely nonlinear in this phase and the results of DiPerna [6] apply, i.e. $\gamma_{x,t}$ is a Dirac mass. We have shown in our paper [22] that the two wave construction given here yields this result as well.

APPENDIX 'A'

Proof of Theorem 1.1

In Section 3 of [21] we have proven that for

$$(v_0, w_0) \in L^\infty(R) \times L^\infty(R)$$

(P_ε) has a unique global weak solution when p is Lipschitz continuous. The same proof can be used to show local existence if p is only locally Lipschitz continuous. This establishes (i), (ii). To show (iii) we first approximate p by p_δ as in the proof of Theorem 2.1. For this approximate system we know there exist smooth solutions (see Section 3 of [21]) which satisfy

$$\frac{d}{dt}\int_{-\infty}^{\infty}\left(\tfrac{1}{2}v^{\varepsilon 2} + \sum(w^\varepsilon)\right)dx + \varepsilon\int_{-\infty}^{\infty}\left(v_x^{\varepsilon 2} - p'(w^\varepsilon)\,w_x^{\varepsilon 2}\right)dx = 0 \quad \text{(A.1)}$$

$$\begin{aligned}\varepsilon\frac{d}{dt}\int_{-\infty}^{\infty}\left(v_x^{\varepsilon 2} + w_x^{\varepsilon 2}\right)dx &= -\varepsilon\int_{-\infty}^{\infty}\left(v_x^{\varepsilon 2} + w_t^{\varepsilon 2}\right)dx \\ &+ \int_{-\infty}^{\infty} -\left(v_t^\varepsilon p_x + w_t^\varepsilon v_x^\varepsilon\right)dx \leqslant -\frac{\varepsilon}{2}\int_{-\infty}^{\infty}\left(v_t^{\varepsilon 2} + w_t^{\varepsilon 2}\right)dx \\ &+ C_1(\varepsilon)\int_{-\infty}^{\infty}\left(v_x^{\varepsilon 2} + w_x^{\varepsilon 2}\right)dx\,. \end{aligned} \quad \text{(A.2)}$$

From (A.1), we find

$$\int_{-\infty}^{\infty}\left(\frac{v^{\varepsilon 2}}{2} + \text{const.}\ w^{\varepsilon 2}\right)\Bigg|_{t=T} dx \leqslant \int_{-\infty}^{\infty}\left(\frac{v_0^2}{2} + \sum(w_0)\right)dx\,. \quad \text{(A.3)}$$

Using Gronwall's inequality in (A.2) we see

$$\int_{-\infty}^{\infty}\left(v_x^{\varepsilon 2} + w_x^{\varepsilon 2}\right)\Bigg|_{t=T} dx \leqslant C_2(\varepsilon)\int_{-\infty}^{\infty}\left(v_x^{\varepsilon 2} + w_x^{\varepsilon 2}\right)\Bigg|_{t=0} dx\,. \quad \text{(A.4)}$$

Together with (A.2) this yields:

$$\iint_{Q_T}\left(v_t^{\varepsilon 2} + w_t^{\varepsilon 2}\right)dx\,dt \leqslant C_3(\varepsilon)\int_{-\infty}^{\infty}\left(v_x^{\varepsilon 2} + w_x^{\varepsilon 2}\right)\Bigg|_{t=0} dx\,. \quad \text{(A.5)}$$

Now pass to the limit as $\delta \to 0$ as in the Proof of Theorem 2.1. This shows $v^\varepsilon, v_t^\varepsilon, v_x^\varepsilon, w^\varepsilon, w_t^\varepsilon, w_x^\varepsilon$ are in $L^2(Q_T)$. Use of (P_ε) shows $w_{xx}^\varepsilon, v_{xx}^\varepsilon$ are in $L^2(Q_T)$ as well.

APPENDIX 'B'

In this appendix we give a proof of Lemma 4.1. The basic ingredients of the proof are elementary calculus rules for a special class of non-smooth functions: the chain rule and the product differentiation rule. Probably the necessary results can be found in the literature but we were unsuccessful in our search. The closest ones we were able to find are contained in the two papers by Marcus and Mizel [15], [16]. We have decided to include this appendix for the sake of completeness only.

We will need a chain rule for a class of functions with jumps in first derivatives. First, we prove the following simple lemma.

LEMMA B.1. Let $R(\xi)$ be the "ramp function"

$$R(\xi)=\begin{cases}-\xi & \xi<0\\ 0 & \xi\geqslant 0\end{cases}.$$

If $u(x,t)$ is continuous, and $u_x \in L^1_{\text{loc}}(Q_T)$, then

$$R(u(x,t))_x = R'(u(x,t))\,u_x(x,t) \tag{B.1}$$

where the derivative $R'(\xi)$ is set to be 0 for $\xi=0$, i.e.

$$R(u(x,t))_x = \begin{cases}-u_x\ , & u<0\\ 0\ , & u\geqslant 0\ .\end{cases} \tag{B.2}$$

Equalities (B.1)-(B.2) hold in $L^1_{\text{loc}}(Q_T)$.

PROOF. Let ϕ be a test function. In the sense of distributions,

$$\langle R(u)_x, \phi\rangle = -\int_{Q_T} R(u)\,\phi_x\,dx\,dt = \int_H u\phi_x\,dx\,dt\ , \tag{B.3}$$

where $H=\{(x,t)\,|\,u(x,t)<0\}$. Since u is continuous, H is open.

On the same test function the right hand side of (B.2) yields:

$$-\int_H u_x\,\phi\,dx\,dt\ . \tag{B.4}$$

Now, we note that one can integrate in (B.4) by parts. Indeed,

$$\begin{aligned}\langle u_x\phi,\psi\rangle &= \langle u_x,\phi\psi\rangle = -\langle u,(\phi\psi)_x\rangle\\ &= -\langle u,\phi_x\,\psi\rangle - \langle u,\phi\,\psi_x\rangle = -\langle u\phi_x,\psi\rangle - \langle u\phi,\psi_x\rangle\ .\end{aligned}$$

It follows that in the sense of distributions

$$u_x \phi = (u\phi)_x - u\phi_x . \tag{B.5}$$

Note that (B.5) is a simplest version of the product differentiation rule. Since $u_x\phi,\ u\phi_x \in L^1_{\text{loc}}$, then $(u\phi)_x \in L^1_{\text{loc}}$ as well. Thus, integrating (B.5) over H we get:

$$\int_H u_x \phi \, dx\, dt = -\int_H u\phi_x \, dx\, dt + \int_H (u\phi)_x \, dx\, dt . \tag{B.6}$$

If the last integral in (B.6) vanishes then (B.3) is equal to (B.4) and (B.1) holds in the sense of distributions.

We will show now that

$$\int_H (u\phi)_x \, dx\, dt = 0 . \tag{B.7}$$

By Fubini's theorem

$$\int_H (u\phi)_x \, dx\, dt = \int \Big(\int_{H(t)} (u\phi)_x \, dx \Big) dt,$$

where $H(t) = \{x \mid u(x,t) < 0\}$. Obviously $H(t)$ is open in the line $\{(x,t),\ x \in R\}$, therefore

$$\int_{H(t)} (u\phi)_x \, dx = \int_{\alpha_i}^{\beta_i} (u\phi)_x \, dx ,$$

where $u(\alpha_i) = u(\beta_i) = 0$.

Denote by $\tilde{u}(x)$ the continuation of $u(x,t)\,\phi(x,t)$ by 0 for $x \leqslant \alpha_i$ and $x \geqslant \beta_i$. Then $\tilde{u}_x = (u\phi)_x$ a.e. on $[\alpha_i, \beta_i]$ (everywhere except $x = \alpha_i, \beta_i$). Therefore

$$\int_{\alpha_i}^{\beta_i} \tilde{u}_x \, dx = \int_{\alpha_i}^{\beta_i} (u\phi)_x \, dx .$$

Let $\psi(x)$ be a test function with $\psi(x) \equiv 1$ for $x \in [\alpha_i, \beta_i]$

$$\int_{\alpha_i}^{\beta_i} \tilde{u}_x \, dx = \int_{-\infty}^{\infty} \tilde{u}_x \, dx = \int \tilde{u}_x \psi \, dx = -\int \tilde{u}\, \psi_x \, dx = 0,$$

yielding (B.7).

Since the right hand side of (B.1) belongs to L^1_{loc} and this equality holds in the sense of distributions, it is also true in L^1_{loc}. The lemma is proved.

Another useful chain rule is given by the next lemma.

LEMMA B.2. Let $F(v,w)$ be continuously differentiable. If $v,w \in C(Q_T)$ and $v_x, w_x \in L^1_{loc}(Q_T)$ then

$$F(v,w)_x = F_v w_x + F_w w_x \,. \tag{B.8}$$

Equality (B.8) holds in L^1_{loc}.

PROOF. Let $J_\varepsilon v, J_\varepsilon w$ be mollifications of v and w. Then

$$\begin{aligned} &J_\varepsilon v \to v\,,\quad J_\varepsilon w \to w \quad \text{in } C\,,\\ &(J_\varepsilon v)_x \to v_x\,,\quad (J_\varepsilon w)_x \to w_x \quad \text{in } L^1_{loc}\,. \end{aligned} \tag{B.9}$$

If ϕ is a test function, then

$$\langle F(J_\varepsilon v, J_\varepsilon w)_x\,, \phi\rangle = -\langle F(J_\varepsilon v, J_\varepsilon w)\,, \phi_x\rangle \to -\langle F(v,w)\,, \phi_x\rangle\,. \tag{B.10}$$

On the other hand

$$\begin{aligned} \langle F(J_\varepsilon v, J_\varepsilon w)_x\,, \phi\rangle &= \langle F_v(J_\varepsilon v, J_\varepsilon w)(J_\varepsilon w)_x\,, \phi\rangle \\ &\quad + \langle F_w(J_\varepsilon v, J_\varepsilon w)(J_\varepsilon w)_x\,, \phi\rangle\,. \end{aligned} \tag{B.11}$$

By virtue of (B.9), it converges to

$$\langle F_v v_x\,, \phi\rangle + \langle F_w w_x\,, \phi\rangle\,.$$

Together with (B.10) it yields (B.8) in the sense of distributions which holds in L^1_{loc} as well since its right hand side belongs to L^1_{loc}.

Now we are prepared for *Proof of Lemma* 4.1. We will need to justify the following applications of chain rules

$$p(w)_x = p'(w)\,w_x\,, \tag{B.12}$$

$$q(v,w)_x = q_v v_x + q_w w_x\,, \tag{B.13}$$

$$\eta(v,w)_t = \eta_v v_t + \eta_w w_t\,, \tag{B.14}$$

$$[\eta_v(v,w)]_x = \eta_{vv} v_x + \eta_{vw} w_x\,, \tag{B.15}$$

$$[\eta_w(v,w)]_x = \eta_{wv} v_x + \eta_{ww} w_x\,. \tag{B.16}$$

REMARK. In these equations p', q_w, and η_{ww} are set to be 0 on the phase boundaries $w=\alpha$ and $w=\beta$.

Equalities (B.14)-(B.15) follow immediately from Lemma B.2 since $\eta, \eta_v \in C^1$ (see Section 4 below (4.3)). For the proof of (B.12) we notice that, for $w<\beta$,

$$p(w) = p(\alpha) + p'(\alpha)\, R(w-\alpha) + \Phi(w),$$

where $\Phi \in C^1(\Phi'(\alpha) = 0,\ \Phi \equiv 0,\ \alpha \leqslant w < \beta)$. By Lemmas B.1 and B.2.

$$p(w)_x = p'(\alpha)\, R'\, w_x + \Phi'\, w_x = p'(w)\, w_x,$$

(where $p'(\alpha) = 0$). The argument for $w > \alpha$ is similar.

Arguments for (B.13) and (B.16) are identical. We consider (B.13). For $w > \beta$,

$$\begin{aligned} q(v,w) &= q(v,\alpha) + q_w(v,\alpha-)\, R(w-\alpha) + \Psi(v,w) \\ &= \Phi(v)\, R(w-\alpha) + \Psi(v,w), \end{aligned}$$

Since 0 and $\Phi(v)$ are the initial data for q and q_w, Φ, $\Psi \in C^1$. Therefore

$$\begin{aligned} q(v,w)_x &= \Phi'(v)\, R(w-\alpha) + \Psi_v\, v_x + \Phi(v)\, R'\, w_x \\ &\qquad + \Psi_w\, w_x = q_v v_x + q_w w_x, \end{aligned}$$

modulo the product rule for ΦR which is not difficult.

Thus, by (B.13)-(B.14),

$$\eta(v,w)_t + q(v,w)_x = \eta_v w_t + q_v v_x + q_w w_x. \tag{B.17}$$

We know that

$$v_t = \varepsilon v_{xx} - p'\, w_x,\quad w_t = \varepsilon w_{xx} + v_x$$

in $L^2(Q_T)$ and L^1_{loc} (see Theorem 1.1). These equations can be multiplied by bounded functions η_v and η_w so that (B.17) can be continued:

$$= \varepsilon \eta_v v_{xx} + \varepsilon\, \eta_w w_{xx} + (q_w - p'\, \eta_v) w_x + (q_v + \eta_w) v_x. \tag{B.18}$$

In (B.18), $q_w - p'\eta_v = 0$ for $w<\alpha$ or $w>\beta$ while $q_w w_x = p' w_x = 0$ for $w=\alpha$ or $w=\beta$. Thus, $q_w - p'\eta_v = q_v + \eta_w = 0$, and (B.18) equals to

$$= \varepsilon \eta_v v_{xx} + \varepsilon \eta_w w_{xx} .$$

The proof of Lemma 4.1 will be completed when we justify the two cases of the product rule:

$$(\eta_v v_x)_x = \eta_v v_{xx} + \eta_{vv} v_x^2 + \eta_{vw} v_x w_x , \tag{B.19}$$

$$(\eta_w w_x)_x = \eta_w w_{xx} + \eta_{wv} w_x v_x + \eta_{ww} w_x^2 , \tag{B.20}$$

where equalities hold in the sense of distributions.

The proof of (B.19) proceeds in two steps:

(1) We mollify $\eta_v(v(x,t), w(x,t))$ and $v_x(x,t)$, and prove that

$$(J_\varepsilon \eta_v J_\varepsilon v_x)_x \to (\eta_v v_x)_x \quad \text{in } H^{-1} . \tag{B.21}$$

(2) The product rule gives

$$(J_\varepsilon \eta_v J_\varepsilon v_x)_x = (J_\varepsilon \eta_v)_x J_\varepsilon v_x + J_\varepsilon \eta_v (J_\varepsilon v_x)_x .$$

We prove that

$$(J_\varepsilon \eta_v)_x J_\varepsilon v_x + J_\varepsilon \eta_v (J_\varepsilon v_x)_x \to \eta_{vx} v_x + \eta_v v_{xx} \tag{B.22}$$

in the sense of distributions.

STEP (1). First we observe that

$$\int |\eta_v v_x|^2 \, dx\, dt \leqslant \sup |\eta_v|^2 \int |v_x|^2 \, dx\, dt ,$$

where $\sup|\eta_v|$ is taken over the invariant domain in the vw-plane. In order to prove (B.21) we estimate as follows:

$$\|J_\varepsilon \eta_v J_\varepsilon v_x - \eta_v v_x\|_{L^2} \leqslant \|(J_\varepsilon \eta_v - \eta_v) J_\varepsilon v_x\|_{L^2} + \|\eta_v (J_\varepsilon v_x - v_x)\|_{L^2} . \tag{B.23}$$

Note that η_v is continuous, therefore $J_\varepsilon \eta_v \to \eta_v$ uniformly and the first term tends to 0 by the dominated convergence theorem. The second term in (B.23) tends to 0 because $J_\varepsilon v_x \to v_x$ in L^2. This yields the convergence in (B.21).

STEP (2). To begin with we consider convergence of the first term in (B.22):

$$(J_\varepsilon \eta_v)_x \, J_\varepsilon v_x - (\eta_v)_x v_x$$
$$= [(J_\varepsilon \eta_v)_x - \eta_{vx}] J_\varepsilon v_x + \eta_{vx}(J_\varepsilon v_x - v_x) . \quad \text{(B.24)}$$

Let ϕ be a test function, then

$$\int (J_\varepsilon \eta_v - \eta_v)_x \, J_\varepsilon v_x \, \phi \, dx\, dt \to 0$$

because $(J_\varepsilon \eta_v - \eta_v)_x \to 0$ in H^{-1} and $\phi J_\varepsilon v_x \to \phi v_x$ in H^1. Note that $v_{xx} \in L^2$, therefore v_x and $\phi v_x \in H^1_{\text{loc}}$.

Another term in (B.24) is estimated as follows. Note that

$$\| J_\varepsilon v_x - v_x \|_{L^2} \to 0$$

while by the chain rule

$$\eta_{vx} = \eta_{vv} v_x + \eta_{vw} w_x \in L^2 . \quad \text{(B.25)}$$

Therefore, we see

$$\int \eta_{vx} \, (J_\varepsilon v_x - v_x) \phi \, dx\, dt \to 0 .$$

Convergence of the second term in (B.22) is obtained similarly. We have

$$J_\varepsilon \eta_v (J_\varepsilon v_x)_x - \eta_v v_{xx} = (J_\varepsilon \eta_v)(J_\varepsilon v_x)_x$$
$$+ \eta_v [(J_\varepsilon v_x)_x - v_{xx}] .$$

Note that

$$(J_\varepsilon v_x)_x = J_\varepsilon v_{xx} \to v_{xx} \quad \text{in} \quad L^2 .$$

It follows, since $J_\varepsilon \eta_v \to \eta_v$ uniformly, that

$$\| (J_\varepsilon \eta_v - \eta_v)(J_\varepsilon v_x)_x \|_{L^2} \to 0 .$$

Finally, $\eta_v \in C \subset L^2_{\text{loc}}$, therefore

$$\int \phi \eta_v [(J_\varepsilon v_x)_x - v_{xx}] \, dx\, dt \to 0 .$$

Thus, we have proved that

$$(\eta_v v_x)_x = (\eta_v)_x v_x + \eta_v v_{xx} .$$

We observe that $(\eta_v)_x \in L^2$ (see (B.25)) and $v_x \in L^2$ so it follows that

$$(\eta_v)_x v_x = \eta_{vv} v_x^2 + \eta_{vw} v_x w_x \quad \text{in} \quad L^1_{\text{loc}} .$$

This yields (B.19) in the sense of distributions.

The argument for (B.20) is identical.

REFERENCES

1. E.C. Aifantis and J.B. Serrin, "The mechanical theory of fluid interfaces and Maxwell's rule", *J. Colloid and Interface Science* **96**, (1983), pp.517-529.
2. J.L. Boldrini, "Is elasticity the proper asymptotic theory for materials with memory?" *Preprint*.
3. J.W. Cahn and J.E. Hilliard, "Free energy of a nonuniform system", *J. Chemical Physics* **28**, (1958), pp.258-267.
4. B. Dacorogna. *Weak continuity and weak lower semicontinuity of nonlinear functionals*, *Lecture Notes in Math.*, (1982) Springer-Verlag.
5. C. Dafermos, "Estimates for conservation laws with little viscosity", *SIAM J. Math. Analysis*, to appear.
6. R.J. DiPerna, "Convergence of approximate solution to conservation laws", *Arch. Rational Mech. Anal.* **82**, (1983), pp.27-70.
7. R.J. DiPerna, "Convergence of the viscosity method for isentropic gas dynamics", *Comm. Math. Phys.* **91**, (1983), pp.1-30.
8. E. Fermi, *Thermodynamics*, Dover Publications, New York, (1956).
9. J. Glimm, "Solutions in the large for nonlinear hyperbolic systems of equations", *Comm. Pure and Applied Math.* **18**, (1965), pp.95-105.
10. R. Hagan and M. Slemrod, "The viscosity-capillarity criterion for shocks and phase transitions", *Archive for Rational Mechanics and Analysis* **83**, (1984), pp.333-361.
11. R. Hagan and J. Serrin, "One dimensional shock layers in Korteweg fluids", in *Phase Transformations and Material Instabilities in Solids*, (ed. Gurtin, M.E.), Academic Press, New York, (1984), pp.113-128.
12. R. James, "The propagation of phase boundaries in elastic bars, *Arch. Rational Mech. Anal.* **73**, (1980), pp.125-158.
13. D.J. Korteweg, "Sur la forme que prennent les équations du mouvement des fluides si l'on tient compte des forces capillaires par des variation de densité, *Archives Neerlandaises des Sciences Exactes et Naturelles*, (1901).
14. P.D. Lax, "Shock waves and entropy", in *Contributions to Nonlinear Functional Analysis*, (ed. Zarantonello, E.H.) Academic Press, New York, (1971), pp.603-635.
15. M. Marcus and V.J. Mizel, "Absolute continuity on tracks and mappings of Sobolev spaces", *Archive for Rational Mechanics and Analysis* **45**, (1972), pp.294-320.
16. M. Marcus and V.J. Mizel, "Every superposition operator mapping one Sobolev space into another is continuous", *J. Functional Analysis* **33**, (1979), pp.217-229.
17. C. Morawetz, "On a weak solution for transonic flow", *Comm. Pure and Applied Maths.* **XXXVIII**, (1985), pp.797-818.
18. F. Murat, "Compacité par compensation", *Ann. Scuola Norm. Sup. Pisa* **5**, (1978), pp.489-507.

19. F. Murat, "Compacité par compensation: Condition nécessaire et suffisante de continuité faible sous une hypothèse de rang constant", *Ann. Scuola Norm. Sup. Pisa* **8**, (1981)) pp.79-102.
20. M. Rascle, "Un resultat de 'compacité par compensation à coefficients variables'. Application a l'élasticité non-linéare", *C.R. Acad. Sci. Paris*, **302**, (1986), pp.311-314.
21. V. Roytburd and M. Slemrod, "Positively invariant regions for a problem in phase transitions", *Archive for Rational Mechanics and Analysis* **93**, (1986), pp.61-79.
22. V. Roytburd and M. Slemrod, "Dynamic phase transitions and compensated compactness", in *Dynamical Problems in Continuum Physics*, (ed. J. Bona *et al*). The IMA volumes in Mathematics and its Applications, vol.4, Springer-Verlag, (1987) pp.289-304.
23. M.E. Schonbek, "Convergence of solutions to nonlinear dispersive equations", *Communications in Partial Differential Equations*, **7**, (1982), pp.959-1000.
24. J. Serrin, "Phase transitions and interfacial layers for van der Waals fluids", (eds. Canfora, A., Rionero, S., Sbordone, C. and Trombetti, C.) *Proc. SAFA IV Conf. Recent Methods in Nonlinear Analysis and Applications*, Naples, March 21-28, 1980, (1980).
25. D. Serre, "Compacité par compensation et systèmes hyperboliques de lois de conservation", *C.R. Acad. Sci. Paris* 299, (1984), pp.555-558.
26. D. Serre, "La compacité par compensation pour les systèmes hyperboliques nonlinéaires de deux equations à une dimension d'espace", Preprint.
27. M. Shearer, "The Riemann problem for a class of conservation laws of mixed type", *J. Differential Equations* **46**, (1982), pp.426-443.
28. M. Shearer, "Admissibility criteria for shock wave solutions of a system of conservation laws of mixed type", *Proc. Royal Soc. Edinburgh* **93A**, (1983), pp.233-244.
29. M. Shearer, "Nonuniqueness of admissible solutions of Riemann initial value problems for a system of conservation laws of mixed type", *Arch. Rat. Mech. Anal.* **93**, (1986) pp.45-59.
30. M. Slemrod, "Admissibility criteria for propagating phase boundaries in a van der Waals fluid", *Archive for Rational Mechanics and Analysis* **81**, (1983), pp.301-315.
31. M. Slemrod, "Dynamic phase transitions in a van der Waals fluid", *J. Differential Equations* **52**, (1984), pp.1-23.
32. A.D. Solomon, V. Alexiades, D.G. Wilson and J. Drake, "On the formulation of hyperbolic Stefan problems", *Quarterly J. of Applied Mathematics* **XLIII**, (1985), pp.295-304.
33. L. Tartar, "Compensated compactness and applications to partial differential equations", *Research Notes in Math.*, *Nonlinear Analysis and Mechanics*, Heriot-Watt Symposium, vol.**4**, (ed. Knopps, R.J.), Pitman Press, (1979).

34. L. Tartar, "The compensated compactness method applied to systems of conservation laws", in *System of nonlinear partial differential equations*, (ed. Ball, J.M.), *NATO ASI Series C*, Reidel, (1983).
35. C.A. Truesdell and W. Noll, *The nonlinear field theories of mechanics*, (ed. Flugge, S.) *Vol. III/3 of the Encyclopedia of Physics*, Springer, Heidelberg, New York, (1965).
36. B. Widom, "Structure and thermodynamics of interfaces", in *Statistical Mechanics and Statistical Methods in Theory and Application*, (ed. Landman, U.), Plenum, New York, (1977).
37. J.D. van der Waals, *Veshandel. Konik. Akad. Weten. Amsterdam.* **1**(8), (1979), Translation of J.D. van der Waals' *The Thermodynamic Theory of Capillarity under the Hypothesis of a Continuous Variation of Density*, by S. Rowlinson, *J. Stat. Phys.* **20**, pp.197-244.

27

A REVIEW OF SOME NON-CONVEX PROBLEMS

R. SCHIANCHI

Consider the functional

$$F(u) = \int_\Omega f(x, u(x), Du(x))\, dx$$

where $\Omega \subset \mathbb{R}^n$, $u : \Omega \to R$, $Du = (u_{x_1}, \cdots, u_{x_n})$ and $f : \Omega \times R \times \mathbb{R}^n$ is a real function.

The problem of minimizing $F(u)$ in a suitable functional class A of admissible functions has been studied by direct methods of the Calculus of Variations.

These methods essentially rely on an extension of the well known theorem, due to Weierstrass, concerning the existence of the minimum of a real function of one real variable.

Therefore we get existence of a minimum of $F(u)$ if:

I) F is bounded from below (so that the infimum and therefore a minimizing sequence exist).

II) F is sequentially lower semicontinuous (s.l.s.c.) with respect to some kind of convergence in the class A.

III) There exists at least one minimizing sequence which converges, with respect to the same convergence as in (II), to an admissible function.

For example if $A = H_0^{1,p}(\Omega)$, $p > 1$, we get existence under the following assumptions:

i) Ω is an open, bounded set of $\mathbb{R}^n$,

ii) $c_1|z|^p \leqslant f(x,s,z) \leqslant c_2(1+|z|^p)$, $c_1 > 0$,

$f(x,s,z)$ is measurable in x for all (s,z),
is continuous in s for all z and almost every x, and
is convex in z for all y and almost every x.

It has been proved that the last condition, i.e. convexity in z is necessary and sufficient for the s.l.s.c., so that this approach fails if the integrand f is not a convex function in z.

This does not mean that existence fails: in [5], for the one-dimensional case, P. Marcellini considers the functional

$$F(u) = \int_a^b f(x,u'(x))\,dx$$

and proves existence without convexity.

Other authors proved analogous results in one dimension with different methods and in different situations, for example, Aubert and Tahraoui in [1] and Olech in [12]. The proof of Marcellini's theorem seems to me the simplest one in order to extend this result for dimensions $n>1$ and I think it is useful to say a few words about his proof.

To do this we need the following definition

DEFINITION. We call the "convexified function of f" and denote by $f^{**}(z)$ the greatest convex function less than $f(z)$.

For simplicity we consider the case in which f does not depend explicitly on x.

Let us denote by k the set

$$k = \{z \in R : f(z) > f^{**}(z)\}.$$

Consider the problem of minimizing the functional

$$\int_a^b f(u'(x))\,dx$$

in the class A of Lipschitz continuous functions u such that $u(a) = A$ and $u(b) = B$ and assume that $f(z)$ is a continuous function on R such that $\lim_{|z|\to+\infty} f(z)/|z| = +\infty$.

By Jensen's inequality, for each $v \in A$, we have

$$\int_a^b f(v'(x))\,dx \geq \int_a^b f^{**}(v'(x))\,dx \geq (b-a)\, f^{**}\left(\frac{1}{b-a}\int_a^b v'(x)\,dx\right).$$

Therefore

$$\int_a^b f(v'(x))\,dx \geq (b-a)\; f^{**}\left(\frac{B-A}{b-a}\right) \qquad \forall\, v \in A\,.$$

If

$$\frac{B-A}{b-a} \notin k\,, \text{ since } f\left(\frac{B-A}{b-a}\right) = f^{**}\left(\frac{B-A}{b-a}\right)$$

it is easy to check that the linear function

$$u(x) = \frac{B-A}{b-a}\,(x-a) + A$$

is a solution to the non-convex minimum problem.

On the other hand, if

$$\frac{B-A}{b-a} \in k\,,$$

since $f^{**}(z)$ is a linear function on k, one can easily see that the function

$$u(x) = A + \int_a^x a(t)\,dt$$

$$\text{where } \; a(t) = \begin{cases} \inf k & \text{in some } (a,e) \subset (a,b) \\ \sup k\,, & \text{in } \quad (a,b) \setminus (a,e) \end{cases}$$

is a solution to the non-convex minimum problem.

Let us observe that, in both cases, the derivative of the minimizing function $u(x)$ lies outside the open set k.

In the following I show how it is possible to extend this result to dimension $n > 1$, assuming that f^{**} is affine on k and looking for a minimum of the "convexified" problem whose gradient is in the set where $f = f^{**}$. For details I refer to references [8] – [11].

First consider the problem

$$\left.\begin{aligned} &\min_{u \in C^{0,1}} \int_\Omega f(Du(x))\,dx \\ &u = u_0 \qquad\qquad \text{on } \partial\Omega \end{aligned}\right\} \tag{1}$$

where $u:\Omega\subset\mathbb{R}^n \to R$, $n\geq 2$, Ω is a bounded open subset of $\mathbb{R}^n$, and f is a real continuous function satisfying a suitable boundary condition to which I will come back later. The first step in proving existence for problem (1) is to prove existence of a Lipschitz continuous solution of the problem

$$\left.\begin{aligned} &Du(x)\in\partial k \quad \text{for a.e. } x\in\Omega \\ &u=u_0 \qquad\qquad\quad \text{on } \partial\Omega , \end{aligned}\right\} \tag{2}$$

under the assumptions that k is a bounded convex set and u_0 is a $C^{0,1}$-function such that $Du_0(x)\in k$ a.e. This assumption on u_0 may be weakened; in fact we may assume that

(B.C.) for every $y\in\partial\Omega$ there is an affine function
$$\pi(x)=(p,x-y)+u_0(y)\,,\quad p\in k$$
such that $u_0(x)\leq\pi(x)$ for all $x\in\partial\Omega$.

In [8] the existence of a solution of (2) is proved by a constructive method.

The second step in the proof of the existence of a solution of problem (1) requires the following assumptions:

(i) the set $k=\{z\in\mathbb{R}^n\colon f(z)>f^{**}(z)\}$ is bounded, which means that f is convex close to ∞ but not necessarily coercive.

(ii) the convexified function f^{**} of f is affine on k, i.e. there are $n+1$ real numbers m_i, q such that
$$f^{**}(z)=\sum_{i=1}^{n} m_i z_i+q \qquad \forall\, z\in k\,.$$

The set $\tilde{k}$ where the above identity holds is a convex set since it is a level set of the convex function
$$f^{**}(z)-\sum_{i=1}^{n} m_i z_i-q\,.$$
Therefore the convex hull $ch\,k$ of k is contained in $\tilde{k}$. From the first step we can consider a $C^{0,1}$-solution u of the problem

$$\left.\begin{aligned} &Du(x)\in\partial(ch\,k) \quad \text{a.e. in } \Omega \\ &u=u_0 \qquad\qquad\qquad \text{on } \partial\Omega\,. \end{aligned}\right\}$$

Then we prove that u is also a solution of problem (1). In fact, by the divergence theorem, we get

$$\int_\Omega f^{**}(Du(x))\,dx = \sum_{i=1}^n m_i \int_\Omega u_{x_i}(x)\,dx + q \text{ meas}\,\Omega$$
$$= \sum_{i=1}^n m_i \int_{\partial\Omega} (u_0, \nu)\,d\sigma + q \text{ meas}\,\Omega \qquad (3)$$

where ν is the unit outward normal to $\partial\Omega$ and $d\sigma$ is the $n-1$ dimensional area element on $\partial\Omega$.

On the other hand for each $v \in C^{0,1}(\Omega)$ such that $v = u_0$ on $\partial\Omega$, we have

$$\int_\Omega f(Dv(x))\,dx \geqslant \int_\Omega f^{**}(Dv(x))\,dx$$
$$\geqslant \sum_{i=1}^n m_i \int_\Omega v_{x_i}(x)\,dx + q \text{ meas}\,\Omega. \qquad (4)$$

From (3) and (4) it easily follows that

$$\int_\Omega f(Dv(x)\,dx \geqslant \int_\Omega f^{**}(Du(x))\,dx = \int_\Omega f(Du(x))\,dx.$$

REMARK. The affinity condition on f^{**}, as far as I know, cannot be removed. There is a paper by P. Marcellini [6] where he shows the non-existence of solutions for some problems where the affinity condition fails.

However, the assumption on the boundary datum may be removed. In fact, by a different method, in [9], existence of a solution has been proved for problem (1) under the more general "bounded slope condition with constant L_0" (B.S.C.)

DEFINITION. Let Γ be the set $\{(x, u_0(x)) : x \in \partial\Omega\}$. The function u_0 satisfies the B.S.C. with constant L_0 if for each

$$P_0 \equiv (x_0, u(x_0)) \in \Gamma$$

there exist two planes

$$\pi^\pm(x) = \sum_{i=1}^n a_i^\pm (x_i - x_{0,i}) + u_0(x_0), \quad x \in \partial\Omega$$

such that

(i) for each $x \in \partial\Omega$, $\pi^-(x) \leqslant u_0(x) \leqslant \pi^+(x)$,

(ii) $\left(\sum_{i=1}^{n} (a_i^{\pm})^2\right)^{\frac{1}{2}} \leqslant L_0$.

This assumption on u_0 is better than (B.C.), because it does not relate the slope of u_0 and the set k.

The idea of the new proof is to consider the relaxed problem of (1), i.e. the problem

$$\left.\begin{array}{ll} \min\limits_{u \in C^{0,1}} & \displaystyle\int_\Omega f^{**}(Du)\,dx \\ u = u_0 & \text{on } \partial\Omega \end{array}\right\} \qquad (5)$$

and to look for solutions of (5) with gradient outside the open set k.

By a well known theorem due to Hartman and Stampacchia (see [3]), there exists at least one solution $\bar{u}$ of problem (5) and $|D\bar{u}(x)| \leqslant L_0$.

Since the problem is not strictly convex, there might be more than one solution; in this case, for $L \geqslant L_0$, one considers the set M_L of the solutions u of (5) such that $|Du| \leqslant L$ a.e. in Ω.

It is not difficult to prove that $\bar{u}(x) = \sup M_L$ is still a solution of (5).

Now, by using a lemma by M. Crandall and P.L. Lions (Lemma 4 in Appendix 2 in [2]), it is possible to prove that in every point $x \in \Omega$ where $\bar{u}$ is differentiable, the conditions

$$D\bar{u}(x) \in k \qquad |D\bar{u}(x)| < L$$

cannot be verified at the same time. As has been shown in [9], here the proof requires the assumption that f^{**} is affine on the set k.

Therefore, if $B(0,L)$ is the ball centred at 0 of radius $L \geqslant L_0$, since k is bounded it is possible to find L large enough such that $k \subset B(0,L)$. Then in all points where $\bar{u}$ is differentiable we have $D\bar{u} \in B(0,L) \setminus k$ and $\bar{u}$ is also a solution of problem (1). In fact for each $v \in C^{0,1}(\Omega)$, $v = u_0$ on $\partial\Omega$, we get

$$\int_\Omega f(Dv) \geqslant \int_\Omega f^{**}(Dv) \geqslant \int_\Omega f^{**}(D\bar{u}) = \int_\Omega f(D\bar{u}) .$$

Now if we think of a possible generalization of these theorems to the problem

$$\left.\begin{array}{ll} \min\limits_{\substack{u \in C^{0,1} \\ u = u_0}} \displaystyle\int_\Omega f(x, Du)\, dx & \\ & \text{on } \partial\Omega, \end{array}\right\} \tag{6}$$

as before, we can proceed in two different ways because of two different assumptions on the boundary datum u_0.

The first way is to look for solutions of the problem

$$\left.\begin{array}{ll} Du(x) \in \partial k(x) , & \text{for a.e. } x \in \Omega , \\ u = u_0 & \text{on } \partial\Omega , \end{array}\right\} \tag{7}$$

where

$$k(x) = \{z \in \mathbb{R}^n : f(x,z) > f^{**}(x,z)\}$$

is a convex set for a.e. $x \in \Omega$.

In [10] it is proved that there exists a solution of problem (7), under suitable assumptions on the sets $k(x)$, by reducing (7) to a Dirichlet problem for a Hamilton-Jacobi equation of the type

$$\left.\begin{array}{ll} 1_{k(x)}(Du(x)) = 0 & \text{a.e. in } \Omega \\ u = u_0 & \text{on } \partial\Omega . \end{array}\right\}$$

where

$$1_{k(x)}(z) = \begin{cases} 0 & \text{if} \quad z \in \overline{k(x)} \\ +\infty & \text{if} \quad z \notin \overline{k(x)}. \end{cases}$$

Existence theorems for Dirichlet problems for Hamilton-Jacobi equations are given in [4] and are proved under an assumption on the boundary datum u_0 which, for simplicity, I do not write but which reduces to (B.C.) when f is independent of x. In [10]some non-convex functionals for which there is existence of the minimum are exhibited.

The second way to solve problem (6) is to look for solutions of the relaxed problem whose gradient is outside $k(x)$ for a.e. $x \in \Omega$.

Therefore we consider the relaxed problem of (6)

$$\min_{\substack{u \in C^{0,1} \\ u = u_0 \text{ on } \partial\Omega}} \int_\Omega f^{**}(x, Du)\,dx \tag{8}$$

where $f^{**}(x,z)$ denotes the convexified function of $f(x,z)$ with respect to the variable z.

To my knowledge the theorem of Hartman and Stampacchia used in the case where f^{**} is independent of x has not been extended. Therefore we have first to prove existence of solutions for problem (8); then, by proceeding as in the case where f is independent of x, we can prove existence of solutions of problem (7). All this has been proved in [11]; here I wish to sketch the proof of the following theorem:

Every solution of the problem

$$\min\left\{\int_\Omega g(x, Dv)\,dx : v \in H_0^{1,2} + u_0\right\}$$

is in $C^{0,1}_{\text{loc}}(\Omega)$ if $u_0 \in H^{1,2} \cap L^\infty$ and it is in $C^{0,1}(\Omega)$ *if* $u_0 \in C^{1,1}(\bar{\Omega})$.

Here $g(x,z)$ is assumed to be in $C^0(\bar{\Omega} \times \mathbb{R}^n)$, convex in z and, for $|z|$ large enough, $g(x,z) \in C^2(\bar{\Omega} \times \mathbb{R}^n)$ and is strictly convex in z.

Let us observe that the assumptions on $g(x,z)$ depend on the fact that we think of $g(x,z)$ as the convexified function of some function which is only continuous and convex close to ∞.

Proof of the Theorem. The proof of this theorem rely on classical regularity results and I try to give an idea of it through its main steps.

STEP 1. We approximate the problem with more regular problems obtained by a suitable mollification of the function $g(x,z)$. We consider the solutions u_j of the new problems, these existing because of the direct methods.

STEP 2. Under suitable assumptions it is possible to give a local gradient estimate for u_j with constants independent of j. To do that we truncate u_j in such a way that the integrations occurring in the weak form of the Euler equation are just outside a ball, exactly where everything is regular.

STEP 3. By Step 2 there is a subsequence of $\{u_j\}$ converging to some function $u \in C^{0,1}_{\text{loc}}(\Omega)$. Here we also use the polynomial growth of the integrands and we prove that u is a solution of the original problem.

STEP 4. We prove that every solution of the problem is in $C^{0,1}_{\text{loc}}$ by using the fact that $g(x,z)$ is strictly convex outside a ball.

STEP 5. We use the classical barrier technique and a better regularity of the boundary datum u_0 to prove that every solution of the problem is in $C^{0,1}(\Omega)$.

REFERENCES

1. G. Aubert and R. Tahraoui, "Théorèmes d'existence en calcul des variations", *CRAS Paris*, **285** (1977), pp.355-356.
2. M. Crandall and P.L. Lions, "Viscosity solutions of Hamilton-Jacobi equations", *Trans. Amer. Math. Soc.* **277** (1983), pp.1-42.
3. P. Hartman and G. Stampacchia, "On some nonlinear elliptic differential-functional equations", *Acta Math.* **115** (1965), pp.383-421.
4. P.L. Lions, *Generalised Solutions of Hamilton-Jacobi Equations*, Pitman, 1982.
5. P. Marcellini, "Alcune osservazioni sull'esistensa del minimo di integrali del calcolo delle variazioni senza ipotesi di convessità", *Rend. Mat*, **13** (1980), pp.271-281.
6. P. Marcellini, "A relation between existence of minima for non-convex integrals and uniqueness for non strictly convex integrals of the calculus of variations", *Lecture Notes in Maths.* (1982), pp.216-231.
7. P. Marcellini, "Some remarks on uniqueness in the calculus of variations", in *Nonlinear PDE and their Applications*, College de France Seminar, vol.IX, (1983), Pitman.
8. E. Mascolo and R. Schianchi, "Existence theorems for non-convex problems", *J. Math. Pures et Appl.* **62** (1983), pp.349-359.

9. E. Mascolo and R. Schianchi, "Un Théorème d'existence pour des problèmes de calcul des variations non-convexes", *CRAS Paris*, **297** (1983), pp.615-617.
10. E. Mascolo and R. Schianchi, "Non-convex problems of the calculus of variations", *Nonlinear Analysis* **9** (1985), pp.371-379.
11. E. Mascolo and R. Schianchi, "Existence theorems in the calculus of variations", *J. Diff. Eq.* to appear.
12. C. Olech, "Integrals of set-valued functions and linear optimal control problems", (*Colloque sur la Théorie Mathematique du Contrôle Optimal*, CBRM Vander, Louvain, 1970.

28

NONLINEAR GEOMETRIC OPTICS AND CONSERVATION LAWS

MARIA E. SCHONBEK

In this paper we describe some joint work with A. Majda and R. Rosales. The program deals with the numerical analysis of asymptotic expansions arising in the theory of approximation to conservation laws. The expansions are obtained by the method of nonlinear geometric optics. Here we describe some new phenomena revealed by the numerical computations for the approximate equations. Specifically, for large time we have observed the appearance of resonance and for intermediate times we encounterred a modification of the decay profile for the solutions.

We consider the Cauchy problem for a system of n conservation laws

$$\frac{\partial u}{\partial t} + \frac{\partial f}{\partial x}(u) = 0 \qquad x \in \mathbb{R}, \, t > 0, \tag{1}$$

$$u(x,0) = u_0(x)$$

with $f : \mathbb{R}^n \to \mathbb{R}^n$, $u = u(x,t) \in \mathbb{R}^n$. The structural hypotheses are the following:

(1) The system is strictly hyperbolic, i.e. for each u the Jacobian of f has n real and distinct eigenvalues $\lambda_1 < \lambda_2 < \cdots < \lambda_n$, so that

$$\nabla f \cdot r_j = \lambda_j r_j ,$$

where the r_j are the corresponding right eigenvectors.

(2) The eigenvalues are either genuinely nonlinear or linearly degenerate in the sense of Lax:

$$r_j \cdot \nabla\lambda_j \neq 0 \quad \text{or} \quad r_j \cdot \nabla\lambda_j \equiv 0 \ .$$

The main example is gas dynamics.

The program with Majda and Rosales deals with the representation of periodic solutions through the method of nonlinear geometric optics. We first give some background on the large time behaviour of exact solutions. Two main cases are of interest: data with compact support and periodic data. For compactly supported data we recall that:

THEOREM 1. If the system (1) is genuinely nonlinear, the data u_0 has compact support and $TVu_0 \ll 1$, then

$$TVu \leqslant cTVu_0/\sqrt{t} \ .$$

For a proof we refer the reader to Glimm and Lax [3] in the case $n=2$ and Tai Ping Liu [5] for $n \geqslant 3$.

THEOREM 2. For nondegenerate systems

$$|u(\cdot,t) - \sum_{j=1}^{n} N_j(\cdot,t)|_{L^1} \leqslant c\,t^{-1/6}$$

where N_j is an N-wave.

For a proof we refer the reader to P. Lax [4], for $n=1$, R.J. DiPerna [2], $n=2$ and T.P. Liu, $n \geqslant 3$. We recall that N-waves are piecewise linear approximate solutions to the system of conservation laws.

For completeness we describe briefly the method of multiple scale expansions. Consider as initial data a small perturbation of a constant state based on two spatial scales x and x/ε.

$$u^{\varepsilon}(x,0) = \bar{u} + \varepsilon u_1^0(x, x/\varepsilon). \tag{2}$$

The problem is to find the response of the solution to data of the form (2) in the following situations: (i) u_1^0 has compact support, (ii) u_1^0 is periodic. In both cases we make the ansatz

$$u^{\varepsilon}(x,t) = \sum_{j=1}^{n} \varepsilon^{j} u^{j}(x, x/\varepsilon, t, t/\varepsilon). \tag{3}$$

To obtain the exact expression for the waves $u^{j}(x, x/\varepsilon, t, t/\varepsilon)$ the ansatz (3) is substituted into equation (1) and terms of equal order are equated. The first two terms of the expansion are

u^0 = background state $\bar{u}$;

u^1 = superposition of waves which are associated to the background state eigenvectors modulated by scalar functions.

Specifically, the first order term has the form

$$u^1 = \sum_{j=1}^{n} \sigma_j(t, \phi^j/\varepsilon)\, r_j(\bar{u}), \tag{4}$$

where the phase function ϕ^j satisfies the eikonal equation, and the simplest choice of ϕ^j is $\phi^j = x - \lambda j(\bar{u})t$. The main terms of the expansion is the first order term (4). In order to characterize this term the scalar functions σ_j have to be studied. The compact and the periodic case are distinguished by the behaviour of the σ_j. In the compact case the σ_j are governed by a system of decoupled Burgers' equations of the form:

$$\frac{\partial}{\partial t}\sigma_j + \frac{b_j}{2}\frac{\partial}{\partial\theta}\sigma_j^2 = k_j * \bar{\sigma},$$

where the coupling is done through an integral operator with a known kernel k_j. Through the method of multiple scales a complex system has been reduced to a much simpler nontrivial system which is much easier to analyse.

Classically the asymptotic expansions obtained above were known to be valid up to where shocks form, both for the periodic and compact cases. Specifically, before shocks form

$$|u(\cdot,t) - (u^0 + \varepsilon u^1)|_{L^1} \leqslant c\,\varepsilon^2 .$$

The recent work of R. DiPerna and A. Majda [2] has established the rigorous validity of the weakly nonlinear expansions uniformly in time, in the case of compactly supported data. Uniform validity in the case of periodic data is still an open problem.

The present work with A. Majda and R. Rosales is focussed on the case of periodic data for the equations of gas dynamics. Here the first two terms of the asymptotic expansion take the form

$$u^{\omega} = u^{0} + \varepsilon u_{1} = u^{0} + \varepsilon \sum_{j=1}^{3} \sigma_{j}(x, \phi^{j}/\varepsilon)\, r_{j}(\bar{u})$$

$$u^{\varepsilon}(x, 0) = u^{0} + \varepsilon v(x/\varepsilon),$$

where the background state is given by $u^0 = (\rho_0, 0, S_0)$ and v is periodic, with ρ_0 and S_0 given constants. The asymptotic limit equations for σ_j derived by Majda and Rosales take the form

$$\frac{\partial}{\partial t}\sigma_1 + \frac{\alpha}{2}\frac{\partial}{\partial\theta}\sigma_0^2 + \beta\int_0^1 k(\theta+s)\,\sigma_3(t,s)\,ds = 0$$

$$\frac{\partial\sigma_2}{\partial t} = 0$$

$$\frac{\partial\sigma_3}{\partial t} - \frac{\alpha}{}\frac{\partial}{\partial\theta}\sigma_3^2 - \beta\int_0^1 k(\theta+s)\,\sigma_1(t,s)\,ds = 0$$

where α and β are given constants and k is a known kernel. Specifically k is the projection onto even harmonics of the first derivative of the initial entropy field σ_2. Hence the system is reduced to two equations.

Here we address the following question: How does the kernel k influence the large time behaviour of the solutions and the time of shock development? We are looking for an answer from a numerical point of view. To compute the solutions we used Glimm's random choice method in a context of a fractional step. The nonlocal problem was solved explicitly with a Fourier spectral step. We chose initial data in the form of one or two N-waves per each period, for each of the components σ_2 and σ_3. The kernel was taken to be a function with only even harmonics. Specifically we chose as kernels sine functions. The numerical results indicate that if the amplitude of the kernel k is sufficiently large compared to the amplitude of the initial N-waves new phenomena appear. These phenomena depend on the number of N waves per period, chosen for the data. More precisely, for data with two N waves and kernel a sine function with sufficiently large amplitude, the phenomenon of resonance was observed numerically. The waves after decaying for a certain period of time would oscillate around a constant state, instead of decaying to zero, as occurs for solutions of decoupled Burgers' equations.

In the case of data with one N-wave and the same kernel as above the decay profile of the solutions was modified. Specifically, due to the influence of the kernel new smaller shocks will be formed. This process will slow down the decay rate observed for solutions to the Burgers' equation alone. Details of the results described above will appear in a forth-coming paper [7].

REFERENCES

1. R. DiPerna, "Decay and asymptotic behaviour of solutions to nonlinear hyperbolic systems of conservation laws", *Indiana Univ. Math. J.* **24** (1975), pp.1047-1071.
2. R. DiPerna and A. Majda, "The validity of nonlinear geometric optics for weak solutions of conservation laws", to appear.
3. J. Glimm and P. Lax, "Decay of solutions of systems of nonlinear yperbolic conservation laws", *Amer. Math. Soc. Memoirs* **101**, (1970).
4. P. Lax, "Hyperbolic systems of conservation laws II", *Comm. Pure Appl. Math.* **10**, (1957), pp.537-566.
5. T.P. Liu, "Asymptotic behaviour of solutions of general systems of nonlinear hyperbolic conservation laws", *Indiana Univ. J.* **27** (1957), pp.211-253.
6. T.P. Liu, "Decay to N-waves of solutions of general systems of nonlinear hyperbolic conservation laws", *Comm. Pure and Appl. Math.* **30**, (1977), pp.767-769.
7. A. Majda, R. Rosales and M.E. Schonbek, "A system of integro-differential equations arising in resonant asymptotics", to appear.

29

ON THE ADMISSIBILITY OF SHOCKS AND PROPAGATING PHASE BOUNDARIES IN A VAN DER WAALS FLUID

M. ŠILHAVÝ

Abstract. In this paper two simple admissibility inequalities are formulated and discussed which do not exclude the propagation of the liquid-vapour phase boundaries near the Maxwell co-existence line.

1. INTRODUCTION

The purpose of this paper is to discuss the admissibility of weak solutions to the equations of motion of a fluid capable of liquid-vapour phase transitions. A van der Waals fluid at low temperatures is an example of such a fluid. I shall consider isothermal situations only. The results to be presented below can also be re-interpreted in terms of a one-dimensional elastic solid capable of phase transitions.

Weak solutions to the equations of motion contain propagating singular surfaces at which the velocity and the pressure suffer jump discontinuities. The singular surfaces describe either shock waves or propagating phase boundaries separating the vapour and liquid phases of the fluid. The weak form of equations of motion leads to the Rankine-Hugoniot jump conditions relating the values of the quantities at the two sides of the singular surface. However, it is known that the Rankine-Hugoniot conditions admit too many solutions of which only some have physical meaning. (Cf. the discussion of this issue in Dafermos [1].) Additional conditions, termed admissibility criteria, are therefore sought to eliminate physically inadmissible solutions to the Rankine-Hugoniot conditions. The entropy admissibility criterion, the viscosity admissibility criterion and the entropy rate admissibility criterion (see [1]) provide examples of such conditions.

The entropy admissibility criterion is too weak in many respects. On the other hand, the viscosity admissibility criterion and the entropy rate admissibility criterion are too strong since they rule out propagating phase boundaries near the equilibrium co-existence line, see Slemrod [2].

In this paper I shall state and discuss two closely related inequalities which can play the rôle of admissibility criteria for weak solutions to the equations of motion of a fluid capable of phase transitions. They do not exclude propagating phase boundaries. One of these inequalities was derived in Šilhavý [3] for the class of one-dimensional non-simple materials with internal state variables under appropriate stability assumptions. The stability considerations lead to the presence of the lower convex envelope of the Helmholtz free energy in these inequalities. It is precisely this feature which makes them consistent with the occurrence of propagating phase boundaries.

The non-simple materials with internal variables [4-5], which lead to the mentioned admissibility inequality are fully compatible with standard thermodynamics based on the Clausius-Duhem inequality and the balance equations in their usual forms. This distinguishes the admissibility criteria proposed here from the viscosity-capillarity criterion [2, 6-8], which, too, does not preclude the occurrence of propagating phase boundaries, but which is based on the theory of Korteweg's fluids not compatible with standard thermodynamics. Indeed, Dunn and Serrin [9] had to develop a modification of standard thermodynamics to embrace the Korteweg fluids.

Another conceptual difference is that the admissibility inequalities considered in the present paper contain only the constitutive quantities of the elastic fluid in question and the concepts derived therefrom, while the viscosity-capillarity criterion depends also on the knowledge of two additional constitutive constants (or even functions). Hence, additional constitutive information is needed tc determine whether the singular surface is admissible according to the viscosity-capillarity criterion. In this sense the viscosity-capillarity criterion is less universal than the inequalities proposed here.

2. A ONE-DIMENSIONAL COMPRESSIBLE FLUID FLOW

Consider an isothermal fluid flow taking place in a straight tube along the x-axis. Assume that the velocities are constant over the cross sections $x = \text{const}$. In an appropriately chosen homogeneous reference configuration the motion of the fluid can be described by a function $x = x(X, t)$ giving the value of the x-coordinate of the particles which in the reference configuration have the x-coordinate equal to X. Let the density of the fluid in the reference configuration be equal to 1. The deformation gradient $w = x_X > 0$ then coincides with the specific volume; we denote by u the velocity, $u = x_t$.

In the class of continuous and piecewise smooth processes with singular surfaces, the balance of linear momentum and the dissipation inequality, which are postulated in integral form, reduce to the field equations

$$u_t = -p_X, \tag{2.1}$$

$$(\psi + \tfrac{1}{2}u^2)_t \leqslant -(pu)_X \tag{2.2}$$

at points of smoothness, and the jump conditions

$$U[u] - [p] = 0, \tag{2.3}$$

$$U[w] + [u] = 0, \tag{2.4}$$

$$-U[\psi + \tfrac{1}{2}u^2] \leqslant -[pu] \tag{2.5}$$

across singular surfaces. Here p is the pressure, ψ the specific Helmholtz free energy at the temperature of the process, and the square brackets denote the jumps of quantities across the singular surface, i.e.

$$[u] = u_+ - u_- \ , \quad [p] = p_+ - p_- \ , \quad \text{etc.} \ , \tag{2.6}$$

the subscripts + and − denoting the limiting values of the quantities from the right and the left of the singular surface, respectively. U is the velocity of the singular surface in the reference configuration; if U is positive, then the singular surface moves to the right.

The set of balance equations is supplemented by constitutive equations for p and ψ in the form

$$p = p(w) \ , \qquad \psi = \psi(w) \ , \tag{2.7}$$

giving these quantities as functions of the specific volume. It is assumed for simplicity that the specific volume ranges over the interval $(0,\infty)$. The function $p(w)$ is assumed to be twice continuously differentiable. Thermodynamics requires that

$$p(w) = -\psi'(w) \ , \qquad w > 0 \ , \tag{2.8}$$

where the prime denotes differentiation with respect to w. With (2.8) the inequality (2.2) reduces to equality. In contrast, (2.5) still continues to be an inequality in general.

The jump conditions (2.3) and (2.4) yield

$$U^2 = -[p] \,/\, [w] \ . \tag{2.9}$$

A triple (w_-, w_+, U) of real numbers with $U \geqslant 0$ is called a jump if (2.9) holds with $p_+ = p(w_+)$, $p_- = p(w_-)$. We assume here and henceforth without any loss of generality that the velocity U is non-negative. If U is strictly positive, then the values w_+ , $p(w_+)$, etc. have the following invariant meaning: they are the values of the quantities ahead of the singular surface; the values labelled by the subscript "−" are the values behind the singular surface. For a given jump there always exists a process of the fluid for which w_- and w_+ are the limiting values of the specific volume at some point of a singular surface having the velocity U at that point. For instance, the process may be chosen to be a pairwise homogeneous process with constant specific volumes w_- and w_+ in the regions separated by a propagating singular surface given by the equation $X = Ut + \text{const.}$, and with constant velocities u_- and u_+ chosen in accordance with (2.3), (2.4).

An elimination of u_+ and u_- from (2.5) yields

$$-U[\psi] \leqslant Up_+[w] + \tfrac{1}{2}U^3[w]^2 \; ; \tag{2.10}$$

this shows that for $U = 0$ the entropy inequality (2.5) is trivially satisfied while for $U > 0$ (2.10) yields

$$\psi_- \leqslant \psi_+ - p_+(w_- - w_+) + \tfrac{1}{2}U^2(w_- - w_+)^2 \, . \tag{2.11}$$

The entropy inequality thus reduces to the following condition on a jump (w_-, w_+, U):

(E) <u>If</u> $U > 0$, <u>then</u>

$$\psi(w_-) \leqslant \psi(w_+) - p(w_+)(w_- - w_+) + \tfrac{1}{2}U^2(w_- - w_+)^2 \, . \tag{2.12}$$

We shall discuss this condition and its relationship to the other admissibility inequalities in Section 5.

3. STATICALLY COEXISTENT PHASES AND THE LOWER CONVEX ENVELOPE OF THE HELMHOLTZ FREE ENERGY

Consider a homogeneous equilibrium state of the fluid with specific volume w_0. It is a familiar consequence of Gibbsian thermostatics that if this state is stable in the sense that the energy criterion holds, then the free energy is convex at w_0, i.e., the inequality

$$\psi(w) \geqslant \psi(w_0) - p(w_0)(w - w_0) \tag{3.1}$$

holds for all $w > 0$. (3.1) also implies that

$$p'(w_0) \leqslant 0 \, . \tag{3.2}$$

Consider further a pairwise homogeneous equilibrium state

in which a certain portion of the fluid has specific volume w_α^0 and another portion has specific volume $w_\beta^0 > w_\alpha^0$. The jump conditions (2.3), (2.4) at the immobile interface ($U = 0$) assert the equality of pressures

$$p(w_\alpha^0) = p(w_\beta^0) . \tag{3.3}$$

The state of course corresponds to statically coexisting phases w_α^0 and w_β^0 of the fluid.

Gibbsian thermostatics asserts that if the state is stable in the sense of the energy criterion, then the free energy is convex at w_α^0 and w_β^0, i.e., that the inequalities

$$\psi(w) \geqslant \psi(w_\alpha^0) - p(w_\alpha^0)(w - w_\alpha^0) , \tag{3.4}$$

$$\psi(w) \geqslant \psi(w_\beta^0) - p(w_\beta^0)(w - w_\beta^0) \tag{3.5}$$

hold for every $w > 0$; moreover also

$$p'(w_\alpha^0) \leqslant 0 , \qquad p'(w_\beta^0) \leqslant 0 . \tag{3.6}$$

Setting $w = w_\alpha^0$ in (3.5) and $w = w_\beta^0$ in (3.4) and comparing the resulting inequalities yields the equality of chemical potentials of the coexisting phases

$$\psi(w_\alpha^0) + p(w_\alpha^0)\, w_\alpha^0 = \psi(w_\beta^0) + p(w_\beta^0)\, w_\beta^0 . \tag{3.7}$$

A consequence of the above assertions is that if a fluid admits a stable state of two coexistent phases of specific volumes w_α^0 and w_β^0, then modulo a trivial unlikely case to be indicated below, the free energy cannot be convex at every point in the interval (w_α^0, w_β^0). Indeed, it is a consequence of (3.2), (3.3) that if ψ is convex at every $w_0 \in (w_\alpha^0, w_\beta^0)$, then the pressure is constant in the interval $[w_\alpha^0, w_\beta^0]$, a highly unlikely situation. A realistic assumption is that the graph of $p(w)$ exhibits the familiar serpentine form similar to the form of the van der Waals isotherm at low temperatures. A consequence of the non-monotonicity of the pressure is that there are homogeneous equilibrium states of specific volume w in the interval (w_α^0, w_β^0) which are unstable in the Gibbs sense. However, it is consistent to assume that the homogeneous equilibrium states with specific volumes outside the interval (w_α^0, w_β^0) are stable. This in particular implies $p'(w) \leqslant 0$ for all

$$w \in (0, w_\alpha^0) \cup (w_\beta, \infty),$$

cf. (3.2).

After these motivating considerations we now lay down the following assumption: there exist specific volumes w_α^0, w_α, w_β, w_β^0 with $0 < w_\alpha^0 < w_\alpha < w_\beta < w_\beta^0$ such that the following conditions (H1) — (H3) hold:

(H1) If $w \in (0, w_\alpha) \cup (w_\beta, \infty)$ then $p'(w) < 0$; if $w \in (w_\alpha, w_\beta)$ then $p'(w) > 0$.

(H2) There exists a stable pairwise homogeneous equilibrium state of specific volumes w_α^0 and w_β^0.

(H3) Every homogeneous equilibrium state of specific volume w from $(0, w_\alpha^0) \cup (w_\beta^0, \infty)$ is stable.

In (H2) and (H3) the term "stable" is used as a synonym for the convexity of the free energy as in (3.1), (3.4) and (3.5). Condition (H1) postulates the serpentine form of p discussed above. The numbers w_α, w_β are the turning points of the graph of p. The interval $(0, w_\alpha)$ corresponds to a liquid phase of the fluid and the interval (w_β, ∞) corresponds to the vapour phase of the fluid. Condition (H2) postulates the existence of a stable state describing a static coexistence of the liquid phase of specific volume w_α^0 with the vapour phase of specific volume w_β^0. The homogeneous liquid states from the interval (w_α^0, w_α) are metastable in a certain sense; similarly the homogeneous vapour states from the interval (w_β, w_β^0) are metastable. The homogeneous states from the interval (w_α, w_β) are highly unstable.

Let us now define a function $\bar{\psi} = \bar{\psi}(w)$ by

$$\bar{\psi}(w) = \begin{cases} \psi(w) & \text{if } w \in (0, w_\alpha^0) \cup (w_\beta^0, \infty) \\ \psi(w_\alpha^0) - p(w_\alpha^0)(w - w_\alpha^0) & \text{if } w \in [w_\alpha^0, w_\beta^0]. \end{cases} \tag{3.10}$$

$\bar{\psi}$ is continuously differentiable and if we set

$$\bar{p}(w) = -\,\bar{\psi}'(w)\,, \tag{3.11}$$

then

$$\bar{p}(w) = \begin{cases} p(w) & \text{if} \quad w \in (0, w_\alpha^0) \cup (w_\beta^0, \infty) \\ p(w_\alpha^0) = p(w_\beta^0) & \text{if} \quad w \in [w_\alpha^0, w_\beta^0]\,. \end{cases} \tag{3.12}$$

The graph of the function $\bar{p}$ thus differs from the graph of p only in the interval (w_α^0, w_β^0) delimited by the values of the statically coexistent phases. In this interval the serpentine form of p is replaced by a horizontal line joining the point $(w_\alpha^0, p(w_\alpha^0))$ with the point $(w_\beta^0, p(w_\beta^0))$. This is the Maxwell coexistence line.

The stability assumptions (H2) and (H3) imply that $\bar{\psi}$ is convex and satisfies

$$\bar{\psi}(w) \leqslant \psi(w) \tag{3.13}$$

for all $w > 0$; in fact $\bar{\psi}$ is a lower convex envelope of ψ, i.e., the greatest convex function not exceeding ψ. All this is easily verified. The function $\bar{\psi}$ plays a central rôle in the statement of the admissibility criteria in the next section.

4. ADMISSIBILITY INEQUALITIES

We shall discuss the meaning and consequences of the following two conditions (C) and ($\bar{\text{C}}$) on a jump (w_-, w_+, U).

(C) <u>If</u> $U > 0$, <u>then</u>

$$\bar{\psi}(w) \leqslant \psi(w_+) - p(w_+)(w - w_+) + \tfrac{1}{2}U^2(w - w_+)^2 \tag{4.1}$$

<u>for all</u> w <u>from the closed interval with endpoints</u> w_- <u>and</u> w_+.

Observe that the left hand side of (4.1) contains the lower convex envelope $\bar{\psi}$ of the free energy while the right hand side of (4.1) contains the free energy itself. In contrast, condition ($\bar{\text{C}}$), below, contains the lower convex envelope $\bar{\psi}$ on both sides of (4.2).

($\bar{\mathrm{C}}$). <u>If</u> $U>0$, <u>then</u>

$$\bar{\psi}(w) \leqslant \bar{\psi}(w_+) - p(w_+)(w-w_+) + \tfrac{1}{2}U^2(w-w_+)^2 \tag{4.2}$$

<u>for all</u> w <u>from the closed interval with endpoints</u> w_- and w_+.

The structure of the inequalities (4.1) and (4.2) is similar to the structure of the entropy inequality (2.12); however, the substantial difference is that (4.1) and (4.2) are required to hold for all values of the specific volume between w_- and w_+, not just only for $w = w_-$.

It is immediate that ($\bar{\mathrm{C}}$) is stronger that (C), i.e., if a jump (w_-, w_+, U) satisfies ($\bar{\mathrm{C}}$) then it also satisfies (C); this is so because $\bar{\psi}(w_+) \leqslant \psi(w_+)$, cf. (3.13). Also, ($\bar{\mathrm{C}}$) and (C) coincide for jumps with w_+ outside the interval (w_α^0, w_β^0), for then $\bar{\psi}(w_+) = \psi(w_+)$, see (3.10). The weaker condition (C) was derived in Šilhavý [3] within the framework of thermodynamics of non-simple materials with internal state variables compatible with the standard form of the balance equations. The same inequality can also be derived for non-simple materials for which the memory effects are described in terms of history-dependent functionals, c.f. Beevers and Šilhavý [10].

We shall now describe the geometrical meaning of ($\bar{\mathrm{C}}$); the same interpretation applies to (C) provided w_+ is outside the interval (w_α^0, w_β^0). If w_+ is in the interval (w_α^0, w_β^0), then (C) has a more complicated geometrical interpretation which is not stated here.

By using (2.8) and (3.11), inequality (4.2) can be rewritten in the form

$$\int_{w_+}^{w} (p(w_+) - U^2(w'-w_+))\,dw' \leqslant \int_{w_+}^{w} \bar{p}(w')\,dw' . \tag{4.3}$$

A usual convention is used to give meaning to the integrals when $w_+ \geqslant w$. Equation (2.9) shows that the graph of the function $f(w') = p(w_+) - U^2(w' - w_+)$, $w' > 0$, is a chord which joins

$(w_+, p(w_+))$ to $(w_-, p(w_-))$. We shall bear this in mind and consider separately the cases $w_+ < w_-$ and $w_- < w_+$.

The case $w_+ < w_-$. We have $w_+ \leqslant w \leqslant w_-$ and (4.3) says immediately that the area between the limits w_+ and w and below the chord does not exceed the area below the graph of $\bar{p}$ between the same limits for every $w \in [w_+, w_-]$.

The case $w_- < w_+$. Then we have $w_- \leqslant w \leqslant w_+$ and (4.3) can be rewritten as

$$\int_{w}^{w_+} \bar{p}(w')\,dw' \leqslant \int_{w}^{w_+} (p(w_+) - U^2(w' - w_+))\,dw' . \qquad (4.4)$$

This inequality says that the area below the graph of $\bar{p}$ between the limits w and w_+ does not exceed the area below the chord between the same limits for every $w \in [w_-, w_+]$.

The above conditions on areas can be considered to be extensions of the Maxwell equal area rule for determining the values of the specific volume of the statically coexistent phases.

5. CONSEQUENCES OF THE ADMISSIBILITY INEQUALITIES

In discussing the relationships among and the consequences of ($\bar{\mathrm{C}}$), (C) and (E) we always assume that the velocity U of the jump (w_-, w_+, U) in question is strictly positive.

We first note that neither ($\bar{\mathrm{C}}$) nor (C) generally implies (E). However, the possibility of ($\bar{\mathrm{C}}$) or (C) being satisfied but (E) not, is rather limited. Namely if w_- is outside the interval (w_α^0, w_β^0), then $\bar{\psi}(w_-) = \psi(w_-)$ and (C) and hence also ($\bar{\mathrm{C}}$) gives for $w = w_-$ the inequality

$$\psi(w_-) \leqslant \psi(w_+) - p(w_+)(w_- - w_+) + \tfrac{1}{2}U^2(w_- - w_+)^2 \qquad (5.1)$$

which is the entropy inequality. Consequently, ($\bar{\mathrm{C}}$) or (C) do not imply the entropy inequality only in the case $w_- \in (w_\alpha^0, w_\beta^0)$, and we shall see below in Proposition 4 that this possibility is rather exceptional. Proposition 2 indicates other circumstances when ($\bar{\mathrm{C}}$) or (C) imply (E).

We can summarize the relationships among $(\bar{C})$, (C), and (E) discussed in the preceding section and in the paragraph above as follows.

PROPOSITION 1. Consider a jump (w_-, w_+, U). Then

(i) generally
$$(\bar{C}) \Longrightarrow (C);$$
(ii) if $w_+ \notin (w_\alpha^{\hat{0}}, w_\beta^0)$, we have
$$(\bar{C}) \Longleftrightarrow (C);$$
(iii) if $w_- \notin (w_\alpha^0, w_\beta^0)$, then
$$(\bar{C}) \Longrightarrow (C) \Longrightarrow (E).$$

The following proposition shows that the problem of admissibility of jumps is somewhat trivial for jumps which lie in the region where the pressure curve is convex (i.e., where $p''(w) \geqslant 0$).

PROPOSITION 2. Let (w_-, w_+, U) be a jump with $w_+ \notin (w_\alpha^0, w_\beta^0)$ and with $p''(w) \geqslant 0$ for all w from the closed interval with end-points w_- and w_+. Then for such jump conditions $(\bar{C})$, (C), and (E) are equivalent and these three conditions are satisfied if and only if $w_- < w_+$.

The proof of this proposition uses Proposition 3, below. Namely, inequality (5.2) of that proposition and the convexity of p in combination with $(\bar{C})$ or (C) yield $w_- < w_+$. Also (E) is easily seen to yield $w_- < w_+$ in view of the convexity of p. Finally, the inequality $w_- < w_+$ leads to any of the conditions $(\bar{C})$, (C), and (E), again in view of the convexity of p.

By Proposition 2, in the convexity region of p, the conditions $(\bar{C})$, (C), and (E) assert precisely that rarefaction shocks do not occur. We also note without proof that in the region of convexity of p these conditions are equivalent to the viscosity admissibility criterion and to the Lax shock condition. A comparison with the viscosity-capillarity criterion is less straightforward since the viscosity-capillarity criterion depends on the values and signs of the constants involved in it. I refer to

Slemrod [2], Theorems 4.1 and 4.2 and to Hagan and Slemrod [6, Section 3], for comparison.

TThe next three propositions contain some general consequences of $(\bar{C})$ and (C).

PROPOSITION 3. <u>If a jump</u> (w_-, w_+, U) <u>with</u> $w_+ \notin (w_\alpha^0, w_\beta^0)$ <u>satisfies either $(\bar{C})$ or (C) then</u>

$$U^2 \geqslant -p'(w_+). \tag{5.2}$$

(See Šilhavý [3] for a proof.) This shows that a jump satisfying one of the admissibility inequalities $(\bar{C})$ or (C) and with the state ahead of the singular surface outside the interval (w_α^0, w_β^0) must move with a speed which is supersonic with respect to the state ahead of the jump.

PROPOSITION 4. <u>There exists an</u> $\varepsilon_0 > 0$ <u>such that if</u> (w_-, w_+, U) <u>is a jump satisfying either $(\bar{C})$ or (C) and the inequalities</u>

$$|w_- - w_\alpha^0| < \varepsilon_0 , \quad |w_+ - w_\beta^0| < \varepsilon_0 , \tag{5.3}$$

<u>or the inequalities</u>

$$|w_- - w_\beta^0| < \varepsilon_0 , \quad |w_+ - w_\alpha^0| < \varepsilon_0 , \tag{5.4}$$

<u>then</u>

$$w_+ \in (w_\alpha^0, w_\beta^0) \ \underline{\text{and}} \ w_- \notin (w_\alpha^0, w_\beta^0). \tag{5.5}$$

(See Šilhavý [3] for a proof.) That is to say, the jumps with values sufficiently close to the values of the postulated static phase transition have the state ahead of the jump in the unstable or metastable region (w_α^0, w_β^0) and the state behind the jump in the stable region $(0, w_\alpha^0) \cup (w_\beta^0, \infty)$. The jump thus transforms the unstable or metastable state into a stable state. Note also that for such jumps condition $(\bar{C})$ or (C) automatically implies the entropy condition (E), cf. Proposition 1, (iii).

We conclude with a consequence of $(\bar{C})$.

PROPOSITION 5. <u>If a jump</u> (w_-, W_+, U) <u>satisfies $(\bar{C})$ then</u>

(i) <u>if</u> $w_+ < w_-$, we have $w_+ \notin (w_\beta, w_\beta^0)$

<u>and</u>

(ii) <u>if</u> $w_{-} < w_{+}$, <u>we have</u> $w_{+} \notin (w_{\alpha}, w_{\alpha})$.

The proof follows easily from the geometrical interpretation of ($\bar{C}$) explained in the preceding section.

A jump with $w_{\beta} < w_{+} < w_{-}$ is a jump in the vapour region of the fluid. As the statement (i) of the above proposition shows, such a jump contradicts ($\bar{C}$) if the state ahead of the jump is in the metastability region of the vapour. A similar discussion applies to (ii). We also note that there are jumps with $w_{\beta} < w_{+} < w_{-}$, $w_{+} \in (w_{\beta}, w_{\beta}^{0})$ which satisfy (C). This shows that ($\bar{C}$) is strictly stronger than (C).

REFERENCES

1. C. Dafermos, "Discontinuous thermokinetic processes", *Appendix 4B* to C. Truesdell's *Rational Thermodynamics*, Second Edition, Springer-Verlag New York, Berlin, Heidelberg, Tokyo, 1984.
2. M. Slemrod, "Admissibility criteria for propagating phase boundaries in a van der Waals fluid", *Arch. Rational Mech. Anal.* **81** (1983), pp.301-315.
3. M. Šilhavý, "An admissibility inequality for shocks and propagating phase boundaries via thermodynamics of non-simple materials, I and II", *J. Non-Equilib. Thermodyn.* **9** (1984), pp.177-186; 187-200.
4. M. Šilhavý, "Thermostatics of non-simple materials", *Czech. J. Phys. B* **34** (1984), pp.601-621.
5. M. Šilhavý, "Phase transitions in non-simple bodies", *Arch. Rational Mech. Anal.* **88** (1985), pp.135-161.
6. R. Hagan and M. Slemrod, "The viscosity-capillarity criterion for shocks and phase transitions", *Arch. Rational Mech. Anal.* **83** (1983), pp.333-361.
7. R.D. James, "A relation between the jump in temperature across a propagating phase boundary and the stability of solid phases", *J. of Elasticity* **13** (1983), pp.357-378.
8. R. Hagan and J. Serrin, "One-dimensional shock layers in Korteweg fluids", *Mathematics Research Center Lecture Series*, University of Wisconsin Press (1983).
9. J.E. Dunn and J. Serrin, "On the thermomechanics of interstitial working", *Arch. Rational Mech. Anal.* **88** (1985), pp.95-133.
10. C.E. Beevers and M. Šilhavý, (in preparation).

30

UNILATERAL PROBLEMS IN CONTINUUM MECHANICS

FRANCO TOMARELLI

Consider a Signorini-type problem for a deformable body, say: find an equilibrium configuration under a given field of external forces, when the body (or a prescribed part of it) is constrained to lie inside a given box.

Our strategy is to look for an energy minimizer among admissible configurations.

Whenever one does not have any Dirichlet-type condition (i.e. no part of the body is fixed nor has an *a priori* prescribed displacement) and the box is unbounded, *lack of coerciveness* is the common feature which appears for a wide class of deformation energy functionals. That is, there is no nonempty bounded sublevel of deformed confugurations, with respect to the total energy.

EXAMPLE 1. Let Ω, Q and E be subsets of $\mathbb{R}^3$ such that Ω is bounded and open, Q is closed and

$$E \subset \bar{\Omega} \subset Q . \tag{1}$$

We assume that Ω is a natural state for a hyperelastic body whose generic particle is labelled by $\boldsymbol{x}$. E is the portion of the body constrained to lie inside the box Q. A field of dead forces acting on the body is prescribed via the density $L = L(x)$. The position, after deformation, is given by $\psi(x) = x + v(x)$ in

Lagrangian coordinates. $F(\psi)$ is the elastic energy functional associated with such a deformation.

$$J(\psi) = F(\psi) - \langle L, \psi \rangle \tag{2}$$

is the mechanical energy. Then one equilibrium configuration is given by a solution, if any, of the following problem.

PROBLEM 1. Find $\boldsymbol{\phi}(\boldsymbol{x}) = \boldsymbol{x} + \boldsymbol{u}(\boldsymbol{x})$ such that "$\boldsymbol{\phi}(\boldsymbol{x}) \in Q$ if $\boldsymbol{x} \in E$" and $J(\boldsymbol{\phi}) \leqslant J(\boldsymbol{\psi})$ for all $\boldsymbol{\psi}$ such that "$\boldsymbol{\psi}(\boldsymbol{x}) \in Q$ if $\boldsymbol{x} \in E$".

The property enclosed in quotes will be stated precisely in the following, according to the properties of the space of finite energy deformations $\{\boldsymbol{\psi} : F(\boldsymbol{\psi}) < +\infty\}$. Notice that, even in the case of linearized elasticity (when F is the integral over Ω of a positive definite quadratic form in the linearized strain tensor), one has

$$F(\psi) = 0 \quad \text{if} \quad \boldsymbol{\psi}(\boldsymbol{x}) = \boldsymbol{x} + \boldsymbol{a} \times \boldsymbol{x} + \boldsymbol{b} \quad \text{with} \quad \boldsymbol{a}, \boldsymbol{b} \in \mathbb{R}^3 .$$

EXAMPLE 2. Horizontal plate which is free at the boundary, subject to a vertical load g and constrained to lie on or above a given obstacle. Let T and Γ be given subsets of $\mathbb{R}^2$ such that Γ is open, bounded and $T \subset \overline{\Gamma}$. The scalar function $h : \Gamma \to \mathbb{R}$ denotes the vertical displacement and $E(h)$ the associated mechanical energy. Again one equilibrium configuration is given by a solution, if any, of the following

PROBLEM 2. Min $\{E(h) : h \geqslant 0$ on $T\}$.

The previous examples, like many others, explain the interest of an abstract existence theory for noncoercive functionals. On this subject there are the well-known classical results by Fichera [7],[8] and Lions and Stampacchia [11], for the minimization of quadratic functionals on convex subsets of Hilbert spaces. These results have been recently extended (see [3],[4]) by the following statement (which no longer requires a finite-dimensional kernel of the quadratic form).

THEOREM 1. Let K be a nonempty closed convex subset of a real Hilbert space V, L be a (real) linear continuous functional on V and $a : V \times V \to \mathbb{R}$ be a bilinear continuous form such that $a(v,v) \geq 0$ for all $v \in V$.

K^∞ denotes the recession cone of K (see [12]), and $Y = \{v \in V : a(v,v) = 0\}$. Assume

$$\langle L, w \rangle \leq 0 \quad \forall\, w \in Y \cap K^\infty \tag{3}$$

$$Y \cap \ker L \cap K^\infty \quad \text{is a subspace} \tag{4}$$

$$\exists\, \pi_i : V \to V, \quad i = 0, 1, \tag{5}$$

linear and continuous such that $\pi_0(K)$ is bounded, π_1 is compact and $\exists\, \alpha > 0$, such that

$$a(v,v) + \|\pi_1 v\|^2 + \|\pi_0 v\|^2 \geq \alpha \|v\|^2 \quad \forall\, v \in V.$$

Then there is at least one solution of

$$\text{Min}\,\{\tfrac{1}{2}\, a(v,v) - \langle L, v \rangle : v \in K\}.$$

Moreover the condition (3) is a necessary condition for the boundedness from below of $\{\frac{1}{2}\, a(v,v) - \langle L, v \rangle\}$ on K.

REMARK 1. When a is not symmetric, one can study (instead of the minimum problem) the *variational inequality*:

find u in K such that $a(u, u-v) \leq \langle L, u-v \rangle \quad \forall\, v \in K$.

For this problem one has again existence provided the assumptions (3) and (4) are replaced by (see [4])

$\{w \in K^\infty : a(v,w) \leq \langle L, w \rangle \quad \forall\, v \in K\}$ is a subspace.

REMARK 2. Once the structural assumptions and the compactness (5) are assumed there is still a gap between the necessary condition (3) and the sufficient ones (3)-(4). Actually this gap cannot be eliminated without specializing the geometry of the data: consider Problem 2 for a plate in the framework of linearized elasticity, say

$$E(h) = \tfrac{1}{2}\, a(h,h) - \int_\Gamma gh$$

assuming the existence of two positive constants δ, D such that

$$\delta \|\nabla\nabla h\|^2_{L^2(\Gamma)} \leqslant a(h,h) \leqslant D\|\nabla\nabla h\|^2_{L^2(\Gamma)} \qquad \forall h \in H^2(\Gamma).$$

THEOREM 1 entails that a necessary condition for existence is that the centre c of the forces g belongs to $\overline{co}\,T$ (the closed convex hull of T), while the sufficient condition requires that c belongs to the relative interior of $\overline{co}\,T$.† In the intermediate case (c belongs to the relative boundary of $\overline{co}\,T$) † both possibilities may occur, even assuming g is a negative constant: if T is a convex polygon then there is a solution, but if both Γ and T are circles and the centre of Γ belongs to ∂T then there is no equilibrium (see [4],[9]).

Now let us come back to Problem 1. In the framework of linearized elasticity the Signorini problem was solved by Fichera (see [7],[8]) for a convex box Q.

We wish to study Problem 1 when F is neither quadratic nor convex and the set of admissible displacements may not belong to a reflexive space, and is possibly nonconvex (say Q is not convex). Such generality is useful in the framework of nonlinear elasticity, for instance, to deal with

(A) the polyconvex elastic energy functionals introduced by Ball (see [5]), or with

(B) materials reacting elastically to compression but not to traction: the "masonry-like materials" studied by Giaquinta and Giusti [10], and Anzellotti [1]. In this case the suitable finite-energy space is the space of bounded deformation $BD(\Omega)$, which in fact is not reflexive.

---

† **the relative interior** $ri\,A$ **is the interior of** A **in the topology of the affine hull of** A**. The relative boundary** $r\partial A$ **is** $\bar{A}\setminus ri\,A$ **.**

The idea of the direct method in the calculus of variations is to obtain the existence of a minimum of the energy from semi-continuity, coerciveness and suitable compactness assumptions. In joint research, together with Baiocchi, Buttazzo and Gastaldi, we do not assume any coerciveness and look for sharper conditions which prevent the minimizing sequences from being all unbounded.

Given a real Banach space V which is either reflexive or the dual of a separable normed space, and a proper functional $G : V \longrightarrow]-\infty,+\infty]$, we introduce the following definition

$$G_\infty(v) = \inf\left\{\liminf_n \frac{1}{\lambda_n} G(\lambda_n v_n) \;:\; \lambda_n \to +\infty \,,\; v_n \xrightarrow{w^*} v\right\}.$$

G_∞ extends the definition of recession functional G^∞ which is well defined only when G is convex and lower semicontinuous (see [12]).

THEOREM 2 (see [2]). Assume

G is sequentially $w^* - l.s.c.$ (6)

$$\left\{\begin{array}{l}\text{(compactness) for all sequences } \{\lambda_n\},\ \{v_n\} \text{ such that}\\ \lambda_n \in \mathbb{R},\ \ \lambda_n \to +\infty,\ \ v_n \in V,\ \ v_n \text{ is } w^* \text{ convergent}\\ \text{to some } v \text{ and } G(\lambda_n v_n) \text{ is bounded from above, we}\\ \text{have } v_n \to v \text{ strongly in } V.\end{array}\right. \quad (7)$$

(necessary condition) $G_\infty(v) \geqslant 0 \qquad \forall\, v \in V$ (8)

(compatibility) $\forall z$ such that $G_\infty(z) = 0 \;\; \exists\, \mu > 0$:

$$G(v - \mu z) \leqslant G(v) \qquad \forall v \in V\,. \quad (9)$$

Then G has a minimum on V.

Let us apply the abstract Theorem 2 to the Problem 1. Referring to the notation of Example 1, assume that Ω is a hyperelastic body with a nonlinear constitutive law, such that there are $p > 1$, $\alpha > 0$ and $\beta \in \mathbb{R}$ with

$$F(\psi) \geqslant \alpha \int_\Omega |\nabla\psi|^p \, dx + \beta \qquad \forall \psi \in W^{1,p}(\Omega, \mathbb{R}^3)\,, \quad (10)$$

F is sequentially w-$l.s.c.$ on $W^{1,p}(\Omega, \mathbb{R}^3)$, † (11)

$\boldsymbol{L}$ is a real-valued linear continuous functional on $W^{1,p}(\Omega, \mathbb{R}^3)$. (12)

We define precisely the set K of admissible displacements :

$$K = \{\boldsymbol{\psi} \in W^{1,p}(\Omega, \mathbb{R}^3) : \boldsymbol{\psi}^*(x) \in Q \quad \text{q.e. } \boldsymbol{x} \in E\}$$

where $\boldsymbol{\psi}^*$ denotes the quasi-continuous representative of $\boldsymbol{\psi}$ and q.e. means almost everywhere with respect to the p-capacity. We emphasize the following nontrivial property:

K is sequentially weakly closed.

By setting

$$G(\boldsymbol{\psi}) = F(\boldsymbol{\psi}) - \langle L, \boldsymbol{\psi} \rangle + \chi_K(\boldsymbol{\psi}),$$

where

$$\chi_K(\boldsymbol{\psi}) = 0 \quad \text{if} \quad \boldsymbol{\psi} \in K \quad \text{and} \quad \chi_K(\boldsymbol{\psi}) = +\infty \quad \text{if} \quad \boldsymbol{\psi} \notin K$$

we can rewrite Problem 1 in the following way.

PROBLEM 3. Minimize $\{G(\boldsymbol{\psi}) : \boldsymbol{\psi} \in V\}$.

THEOREM 3. Assume (10), (11), (12) and

$$\exists\, \boldsymbol{\psi}_0 \in K \quad \text{such that} \quad F(\boldsymbol{\psi}_0) < +\infty .$$

Let $\boldsymbol{e}$ be a fixed unit vector in $\mathbb{R}^3$ and $Q = \{\boldsymbol{x} \in \mathbb{R}^3 : \boldsymbol{x} \cdot \boldsymbol{e} \geqslant 0\}$. If $\langle L, \boldsymbol{\psi} \rangle < 0 \quad \forall\, \boldsymbol{\psi}$ such that $\boldsymbol{\psi}(\boldsymbol{x}) \equiv \boldsymbol{\gamma} \in \mathbb{R}^3$ with $\boldsymbol{\gamma} \cdot \boldsymbol{e} > 0$ then Problem 3 has a solution.

For the case of more general boxes (Q different from an half-space and possibly nonconvex) we have the following result. Set

$$Q_\infty = \{\boldsymbol{\gamma} \in \mathbb{R}^3 : \exists\, \lambda_n \to +\infty,\ \exists\, \boldsymbol{\gamma}_n \to \boldsymbol{\gamma} \text{ and } \lambda_n \boldsymbol{\gamma}_n \in Q \ \forall n\}.$$

Q_∞ is called the set of sequentially unbounded directions of Q

---

† This is true, for instance, when F is the integral of a polyconvex function of $\nabla\boldsymbol{\psi}$ and satisfies suitable growth assumptions, as was proved by J.M. Ball in [5].

and, if Q is convex closed, it coincides with the recession cone Q^{∞}. Such a definition is related also the one concerning "dangerous directions" introduced by Ciarlet and Necas in [6], but it is more intrinsic, since, for example, it depends only on the box Q and not on the topology of the set of admissible displacements. Also notice that

$$\chi_{Q_{\infty}} = (\chi_Q)_{\infty} \ .$$

THEOREM 4. Assume (10), (11), (12), (13) and

$$\langle L, \boldsymbol{\psi} \rangle \leqslant 0 \qquad \forall \boldsymbol{\psi} \text{ such that } \boldsymbol{\psi}(\boldsymbol{x}) \equiv \boldsymbol{\gamma} \in Q_{\infty} \tag{14}$$

$$\langle L, \boldsymbol{\psi} \rangle = 0 \text{ and } \boldsymbol{\psi}(\boldsymbol{x}) \equiv \boldsymbol{\gamma} \in Q_{\infty} \text{ imply } \boldsymbol{q} + \lambda\gamma \in Q \text{ for all } \boldsymbol{q} \in Q \text{ and all } \lambda \in \mathbb{R}. \tag{15}$$

Then the Problem 3 has a solution.

The Theorems 3 and 4 extend the previous results contained in [6] since Q may have bilateral unbounded directions (for instance Q might be a half-space or an infinite roof gutter).

Finally, I consider the case of a partly deformable box Q: the hyperelastic body Ω is supported by a horizontal (bounded open) membrane M, which is fixed at the boundary. The energy to be minimized is now

$$F(\psi) - \langle \boldsymbol{L}, \boldsymbol{\psi} \rangle + \int_M |\nabla v|^2 - \int_M fv \tag{16}$$

where F and L are as in Problem 3, $v \in H_0^1(M)$ is the vertical displacement of the membrane and f is the given force acting on the membrane.

The main difficulty is the correct definition of the unilateral constraint (for the admissible displacements) which is of quasi-variational type. Precisely, the point is giving a weak sense to the inequality

$$\psi^3(\boldsymbol{x}) \geqslant v(\psi^1(\boldsymbol{x}) , \psi^2(\boldsymbol{x})) \tag{17}$$

when v belongs to $H_0^1(M)$.

If

$$p > 3 \tag{18}$$

$$F(\psi) = \int_\Omega W(\boldsymbol{x}, \nabla\boldsymbol{\psi}(\boldsymbol{x}))\, d\boldsymbol{x} \quad \text{with} \quad W(\boldsymbol{x}, \boldsymbol{F}) = +\infty \quad \text{if}$$
$$\det \boldsymbol{F} \neq 1, \text{ or } W(\boldsymbol{x}, \boldsymbol{F}) \text{ grows to } +\infty \text{ at a suitable rate when } \det \boldsymbol{F} \longrightarrow 0+, \tag{19}$$

then the difficulty can be overcome and the inequality (17) can be defined weakly in every sublevel of the elastic energy in such a way that (see [13]):

(I) if $\boldsymbol{\psi}$ and v are regular and satisfy (17) in the weak sense, then they satisfy it in the classical sense too, and vice-versa;

(II) if $\boldsymbol{\psi}_n$ and v_n satisfy (17) in the weak sense for every n, and $\boldsymbol{\psi}_n$ (respectively v_n) converges to $\boldsymbol{\psi}$ (respectively to v) weakly in $W^{1,p}(\Omega, \mathbb{R}^3)$ (respectively in $H_0^1(M)$), then $\boldsymbol{\psi}$ and v satisfy the constraint in the weak sense.

The abstract Theorem 2 applies again and gives the following result.

THEOREM 5 (see [13]). Assume (10), (11), (12), (18), (19) and $\exists\ \boldsymbol{\psi}_0, v_0$ satisfying (17) and s.t. the related energy (16) is finite, $\langle \boldsymbol{L}, \boldsymbol{e}^3 \rangle < 0$ where $\boldsymbol{e}^3$ is the vertical unit vector.

Then, for every f in $H^{-1}(M)$, the functional (16) has a finite minimum under the weak constraint (17) in the class $W^{1,p}(\Omega, \mathbb{R}^3) \times H_0^1(M)$. This minimum is obtained in a class such that the Lagrangian coordinates of the corresponding deformed configuration of Ω are an almost everywhere injective and orientation preserving mapping.

REFERENCES

1. G. Anzellotti, "A class of convex noncoercive functionals and masonry-like materials", *Ann. Inst. H. Poincaré, Anal. Non Lin.* **2** (1985), pp.261-307.
2. C. Baiocchi, G. Buttazzo, F. Gastaldi and F. Tomarelli, "General existence results for unilateral problems in continuum Mechanics", Publication I.A.N. C.N.R., Pavia, n.508, (1986).
3. C. Baiocchi, F. Gastaldi and F. Tomarelli, "Inéquations variationnelles non coercives", *C.R. Acad. Sci. Paris, Se.I Math.*, **299** (14), (1984), pp.647-650.
4. C. Baiocchi, F. Gastaldi and F. Tomarelli, "Some existence results on noncoercive variational inequalities", *Ann. Sc. Norm. Pisa* (to appear).
5. J.M. Ball, "Convexity conditions and existence theorems in nonlinear elasticity", *Arch. Rat. Mech. Anal.* **63** (1977), pp.337-406.
6. P.G. Ciarlet and J. Neças, "Unilateral problems in nonlinear three-dimensional elasticity", *Arch. Rat. Mech. Anal.* **87** (1985), pp.319-338.
7. G. Fichera, "Problemi elastostatici con vincoli unilaterali: il problema di Signorini con ambique condizioni al contorno", *Atti Acc. Naz. Lincei, Mem., Sez. I*, **7** (1964), pp.71-140.
8. G. Fichera, "Boundary value problems in elasticity with unilateral constraints", *Handbuch der Physik*, **VI** a/2, Springer-Verlag, Berlin (1972), pp.347-389.
9. F. Gastaldi and F. Tomarelli, "Two-plate problems in linearized elasticity", Publication I.A.N. C.N.R., Pavia, n.537, (1986).
10. M. Giaquinta and E. Giusti, "Researches on the equilibrium of masonry structure", *Arch. Rat. Mech. Anal.* **88** (1985), pp.359-392.
11. J.L. Lions and G. Stampacchia, "Variational inequalities", *Comm. P.A.M.* **22** (1967), pp.153-188.
12. R.T. Rockafellar, *Convex Analysis*, Princeton University Press, Princeton (1970).
13. F. Tomarelli, "A quasi-variational problem in nonlinear elasticity", (to appear).

31

SURFACE TENSION EFFECTS IN PHASE TRANSITION

A. VISINTIN

Abstract. The *surface tension* at a liquid-solid interface is responsible for *supercooling* and *superheating* phenomena; in particular it allows one to explain the high supercooling required for *nucleating* a solid phase in a liquid system, and the *Gibbs-Thomson law*. The latter states that the local mean curvature of the solid-liquid interface is proportional to the relative temperature.

In the stationary case $\theta(x)$ can be assumed to be prescribed. Setting $\chi = 1$ in the liquid and $\chi = -1$ in the solid, one is led to study the minimization of the free enthalpy functional

$$\Phi_\theta(v) := -\alpha \int_\Omega \theta v \, dx + \beta \int_\Omega |\nabla v| + \text{boundary terms} \tag{1}$$

(α, β: constants > 0) in the *non-convex* set $X := \{v \in BV(\Omega) : |v| = 1 \text{ a.e. in } \Omega\}$. The *stable* states correspond to the absolute minima of Φ_θ, and also the *metastable* states can be characterized by means of Φ_θ.

The following *non-convex* functional is then considered:

$$\Psi_\theta(v) := \Phi_\theta(v) - \mu \int_\Omega (v^2 - 1)\, dx \qquad (\mu\text{: constant} > 0)\,, \tag{2}$$

defined for $v \in \tilde{X} := \{v \in BV(\Omega) : |v| \leq 1 \text{ a.e. in } \Omega\}$.

It is proved that all the absolute and relative minima of Ψ_θ are elements of X. The absolute minima of Ψ_θ correspond to the stable states; moreover, if μ is suitably related to the maximum supercooling required for nucleation, then the relative minima of Ψ_θ correspond to the metastable states.

The functional Φ_θ is then obtained from the so-called *phase field model* by means of a Γ-limit. These models are also compared with the Landau-Lifshitz theory of ferromagnetism.

1. INTRODUCTION

In this paper an effort is made to model the effects of surface tension in stationary multi-phase systems. The main features of this problem are the connection between field and shape expressed by the *Gibbs-Thomson law*, and the occurrence of *metastable states*.

In a bounded convex domain Ω of $\mathbb{R}^N$ $(N \geq 2)$, we consider a solid-liquid system composed of a homogeneous substance at equilibrium. We assume that the relative temperature $\theta(x)$ is known and investigate the possible stationary phase configurations.

If the thermal conductivity $k(\theta)$ is phase-independent, we have

$$-\nabla\cdot(k(\theta)\nabla\theta) = g \quad \text{in} \quad \mathcal{D}'(\Omega) \quad \left(\nabla := \left(\frac{\partial}{\partial x_1}, \cdots, \frac{\partial}{\partial x_N}\right)\right), \tag{1.1}$$

g being a prescribed heat source intensity; (1.1) is also coupled with boundary conditions. Thus in this case θ is actually determined independently of the phase configuration.

The domain Ω can be decomposed into two sets Ω^+ and Ω^-, corresponding to the liquid and solid phases respectively, and separated by the interface S.

In the usual model of phase transitions, known in the evolution case as the *Stefan problem*, the surface tension effect is neglected; it is then required that

$$\left.\begin{array}{lll} \theta \geq 0 & \text{in} & \Omega^+ \\ \theta = 0 & \text{on} & S \\ \theta \leq 0 & \text{in} & \Omega^-. \end{array}\right\} \tag{1.2}$$

If instead the *surface tension* effect is considered, then *supercooled* and *superheated* states may appear even at equilibrium; they correspond to $\theta < 0$ in the liquid and $\theta > 0$ in the solid, respectively. Moreover the $(N-1)$-dimensional measure of the interface S is finite, since the surface tension tends to minimize it; mushy regions (namely very fine mixtures of solid and liquid) are also excluded in this framework.

Under suitable regularity conditions, we have the classical *Gibbs-Thomson law*

$$\theta = -\rho\kappa \quad \text{on } S\,; \tag{1.3}$$

here ρ is a positive constant and κ denotes the local mean curvature of S, which is assumed positive for an ice ball. Denoting by ω the angle between the outward normal vectors to Ω and to Ω^+ on $\partial\Omega \cap S$ (if the latter is non-empty), because of the surface tension effect we also have the following contact angle condition

$$\cos\omega = h \quad \text{on } \partial\Omega \cap S\,, \tag{1.4}$$

with $h \in [-1,1]$ dependent on the exterior substance.

In Section 2 we describe the physical aspects of this problem; in particular we present the *nucleation* phenomenon, namely the formation of a new phase in an either completely solid or completely liquid system. Here also *metastable states* appear, besides stable ones. Notice that the occurrence of metastable states in the stationary case corresponds to the appearance of *hysteresis* effects in the evolution.

In (1.3) the constant ρ is very small; hence the standard equilibrium condition prescribing a vanishing θ on S is a good approximation of (1.3). The surface tension effect has more remarkable consequences in the interior of the phases, where interfaces are only potentially present; in particular it is responsible for the high supercooling or superheating (even of the order of hundreds of degrees!) required for nucleation.

In Section 3 we introduce a first variational model. Let us define the characteristic function χ of the phase distribution: $\chi(x) = 1$ in Ω^+, $\chi(x) = -1$ in Ω^-. Taking account of the surface tension contribution and assuming a constant pressure, the *free enthalpy* (or Gibbs free energy) is a functional of the form

$$\Phi_\theta(v) := -\alpha \int_\Omega \theta v\,dx + \beta \int_\Omega |\nabla v| + \text{boundary terms}\,, \tag{1.5}$$

with α and β positive constants; here v belongs to the *non-convex* set of characteristic functions with bounded total variation in Ω, i.e., $v \in X := \{v \in BV(\Omega) : |v| = 1 \text{ a.e. in } \Omega\}$.

The condition of extremality of Φ_θ contains the Gibbs-Thomson law (1.3) and the contact angle condition (1.4) in a weak form.

For any $\theta \in L^1(\Omega)$, Φ_θ has one or more absolute minima. Because of the non-convexity of X, Φ_θ may have several sorts of extremals with respect to the strong topology of $L^1(\Omega)$. But only the absolute minima and a subclass of the relative minima correspond to states of either stable or metastable equilibrium, that is to physically admissible stationary states.

We stress that not all the relative minima of Φ_θ are physically acceptable. For instance, for any $\theta \in L^\infty(\Omega)$ both $\chi = 1$ a.e. in Ω (all liquid) and $\chi \equiv -1$ a.e. in Ω (all solid) correspond to relative minima of Φ_θ. Then we give a first characterization of metastable states. However this is not explicit, hence later on we shall look for a more practical characterization.

In Section 4 we consider a *convex regularization*. The bipolar Φ_θ^{**} of Φ_θ has the same form as Φ_θ, but it is defined in $\tilde{X} := \{v \in BV(\Omega) : |v| \leqslant 1 \text{ a.e. in } \Omega\}$. Φ_θ^{**} may have also non-characteristic absolute minima, namely elements of $\tilde{X} \setminus X$; the physical interpretation of such states is not clear. Moreover the convex functional Φ_θ^{**} has no relative (non-absolute) minima; this apparently excludes the possibility of describing the metastable states.

Our problem is essentially *non-convex*. Thus in Section 5 we introduce what might be named a *concave penalization*; in a simplified form, this consists in fixing a positive constant η and setting

$$\Psi_\theta(v) := \Phi_\theta^{**}(v) - \eta \int_\Omega (v^2 - 1)\, dx \qquad \forall x \in \tilde{X}. \tag{1.6}$$

By means of the co-area formula, we can show that all the absolute and relative minima of Ψ_θ in $\tilde{X}$ are elements of X, namely characteristic functions. Therefore they are extremals of Φ_θ in

X, and consequently they fulfil a weak form of the Gibbs-Thomson law (1.3) and of the contact angle condition (1.4).

The set of the *absolute minima* of Ψ_θ in $\tilde{X}$ coincides with that of the absolute minima of Φ_θ in X, which correspond to the states of stable equilibrium, as we said. Instead the relative minima of Ψ_θ in $\tilde{X}$ are a smaller class than those of Φ_θ in X.

If η is suitably related to the maxima supercooling and superheating required for *homogeneous* nucleation, and if the *heterogeneous* nucleation is somehow avoided, then the local relative minima of Ψ_θ in $\tilde{X}$ (namely the *relative minima* corresponding only to variations with compact support in Ω) represent the metastable states.

In the developments of Section 5, we actually consider a more general form of concave penalization, which does not require the maximum supercooling to be equal to the maximum superheating.

We also introduce a modification which accounts for the heterogeneous nucleation at the boundary of Ω.

In Section 6 we show that for "small" temperatures the functional Φ_θ can be obtained from the so-called *phase-field model* by means of a Γ-limit (in the sense of De Giorgi [17]). We also briefly review the domain model for *ferromagnetism*, which has several analogies with the phase field model.

BIBLIOGRAPHIC NOTES

The physical aspects of surface tension in phase transitions were studied, e.g., by Chalmers [16, chapter 3], by Flemings [22, chapter 9], by Woodruff [43, chapter 2], by Langer [30], by Wollkind and Notestine [42] and by Rogers [34]. The stability of the free boundary was considered by Chadam and Ortoleva [15]. Numerical methods were proposed by Smith [35]. The movement of the interface under the effect of the surface tension was modelled by Brakke [3], by means of an approach based on geometric measure theory.

The so-called *phase field model*, here outlined in Section

6, was introduced by van der Waals [40], and then studied from the physical viewpoint by Cahn and Hilliard [10,11,12]. It was applied to multi-phase problems by Fix [20,21], by Caginalp [4, 5], by Caginalp and Fife [6-9] and by Gurtin [25-27].

The *phase field model* is actually an example of a more general theory of phase transitions due to Landau and Ginzburg [28]. Another example is given by the Landau-Lifshitz theory of *ferromagnetism* [29]; here some differences appear since the phase variable, namely the magnetization field $\bar{\boldsymbol{M}}$ is a vector, cf. Section 6. The mathematical aspects of the latter model were studied by the present author in [38].

Another phenomenon involving the surface tension is the evolution of grain boundaries in a polycrystalline solid; the mathematical aspects of this problem were considered by Duchon and Robert [18] and by others (see the references of the latter paper).

The present author studied the surface tension effect for phase transitions in [36]; those first results were announced in [37]. The evolution Stefan problem with surface tension is dealt with in [39].

2. THE PHYSICAL PROBLEM

The Gibbs-Thomson law. We shall denote the surface tension coefficient by σ, the latent heat per unit volume by L, the relative temperature by θ and the equilibrium absolute temperature for a flat stationary interface by τ_E; thus $\tau_E = 273.14 \cdots {}^\circ\mathrm{K}$ for ice-water. For a solid ball of radius r surrounded by liquid, assuming that the temperature is continuous, the *Gibbs-Thomson law* prescribes that at equilibrium on the interface

$$\theta = -\frac{2\sigma\tau_E}{Lr} \quad \text{on} \quad S\,; \tag{2.1}$$

this condition will be derived in Section 3. In a neighbourhood of equilibrium, the following *kinetic law* can be assumed

$$\alpha\left(\frac{1}{r}\right)\frac{dr}{dt}+\frac{2\sigma\tau_E}{Lr}=-\theta \quad \text{on } S, \tag{2.2}$$

with $\alpha: [0,+\infty[\to]0,+\infty[$.

The condition (2.1) has an intuitive interpretation. On the boundary of a convex solid phase the particles have weaker bonds than on a flat interface [16, chapter 1]. Thus the thermal energy required to remove them from the crystal lattice is smaller, and consequently the corresponding equilibrium temperature is smaller than for a flat interface.

The Gibbs-Thomson law (2.1) represents an *unstable equilibrium*. We check this for a constant θ. If $\theta < -2\sigma\tau_E/Lr$ at some instant t_0, then by (2.2) $dr/dt > 0$ for $t > t_0$; hence $1/r$ decreases and still by (2.2) dr/dt increases; therefore $r\to+\infty$ as $t\to+\infty$. Similarly if $\theta > -2\sigma\tau_E/Lr$ at t_0, then $r\to 0$ in a finite time.

In more general geometries, assuming that the interface S is smooth enough, $1/r$ is replaced by the *local mean curvature* $\kappa := \frac{1}{2}(1/r' + 1/r'')$, where r' and r'' are the local principal radii of curvature and are assumed positive (negative, respect.) if the centre of curvature lies in the solid (in the liquid, respect.). Thus in general the *Gibbs-Thomson law* has the form

$$\theta = -\frac{2\sigma\tau_E}{L}\kappa \quad \text{on } S. \tag{2.3}$$

Near equilibrium the following kinetic law can be assumed

$$\alpha(\kappa)v + \frac{2\sigma\tau_E}{L}\kappa = -\theta \ \text{ on } S, \tag{2.4}$$

where v denotes the normal velocity of the interface, here assumed positive for solidification. Hence for an interface of vanishing mean curvature the equilibrium temperature is $\theta = 0$ and melting (solidification, respect.) occurs with a velocity proportional to θ^+ (θ^-, respect.). Also (2.3) corresponds to an unstable equilibrium.

Nucleation. The form of the kinetic law (2.4) has

consequences for the *nucleation* process, namely for the formation of new phases; [16, chapter 3; 22, chapter 9; 43, chapter 2].

First we take into account the so-called *homogeneous nucleation*. We consider a liquid system at a *uniform* negative temperature θ. Due to *fluctuations*, clusters of molecules are continuously formed; their evolution is governed by (2.4) at each point of their surface; hence they grow (shrink, respect.) where the curvature is smaller (larger, respect.) than the critical value $-(L\theta)/(2\sigma\tau_E) > 0$.

The presence of impurities in the bulk and the contact with other materials at the boundary is responsible for another form of nucleation, named *heterogeneous nucleation*, which is related to the surface tension existing between the liquid and the exterior substance and also to geometrical features.

Although it is possible to maintain a completely liquid system at uniform negative temperatures for a long time, these states are *metastable*, namely they will eventually decay.

Similar considerations hold for the nucleation of a new liquid phase in a solid system; here a threshold $\hat{\theta}_c$ appears for the superheating. Thus, for a single phase system,

$$\left.\begin{array}{ll} \text{the solid is } \textit{stable} & \text{if } \theta \leqslant 0 \\ \text{the liquid is } \textit{stable} & \text{if } 0 \geqslant 0. \end{array}\right\} \qquad (2.5)$$

For a prescribed substance with a prescribed content of impurities and for a prescribed time scale, on account of homogeneous nucleation there exist two positive constants θ_c and $\hat{\theta}_c$ such that

$$\left.\begin{array}{ll} \text{the solid is } \textit{metastable} & \text{if } 0 < \theta < \hat{\theta}_c \\ \text{the liquid is } \textit{metastable} & \text{if } -\theta_c < \theta < 0. \end{array}\right\} \qquad (2.6)$$

For a two-phase system, the Gibbs-Thomson law (2.3) holds; hence there is no *a priori* bound for supercooling and superheating "near" the interface S. Instead "far" from S and from $\partial\Omega$, (2.5) and (2.6) are approximately fulfilled. On the fixed boundary $\partial\Omega$ the constants θ_c and $\hat{\theta}_c$ are reduced, because of heterogeneous nucleation. In the next section this picture will be made precise by means of a variational formulation.

3. FIRST VARIATIONAL FORMULATION

Non-Cartesian surfaces of prescribed mean curvature. We assume that Ω is a bounded Lipschitz region of $\mathbb{R}^N$ $(N \geq 2)$; in order to exclude several pathologies related to the presence of "necks", we also assume that Ω is convex; however this restriction could be considerably weakened. For any $v \in L^1(\Omega)$ we set

$$\int_\Omega |\nabla v| := \sup\left\{\int_\Omega v\,\nabla\cdot\eta\,dx : \eta \in C_0^1(\Omega)^N, \quad |\eta| \leq 1 \text{ in } \Omega\right\}$$

$$BV(\Omega) := \left\{v \in L^1(\Omega) : \int_\Omega |\nabla v| < +\infty\right\}.$$

The latter is the space of *functions with bounded total variation* in Ω; it is a Banach space endowed with the norm

$$\|v\|_{BV(\Omega)} = \|v\|_{L^1(\Omega)} + \int_\Omega |\nabla v| ;$$

for any $v \in BV(\Omega)$, ∇v is a vector-valued Radon measure and $\int_\Omega |\nabla v|$ is its total variation. We recall that the trace operator $C^0(\Omega) \to C^0(\partial\Omega) : v \mapsto v|_{\partial\Omega}$ has a unique continuous extension from $BV(\Omega)$ into $L^1(\partial\Omega)$. For any set A, we shall denote by ∂A its essential boundary, namely the set of the $x \in \mathbb{R}^N$ such that for any $r > 0$

$$\mu(A \cap B(x,r)) > 0, \quad \mu((\mathbb{R}^N \setminus A) \cap B(x,r)) > 0,$$

where μ denotes the usual Lebesgue measure on $\mathbb{R}^N$ and $B(x,r)$ the ball of centre x and radius r.

We denote by σ the surface tension coefficient for a liquid-solid interface and by σ_L (σ_S, respect.) the surface tension coefficient for the contact between the liquid phase (the solid phase, respect.) and an exterior substance. Finally we denote by H_m the m-dimensional Hausdorff measure $(m \leq N)$.

Let us consider a solid-liquid system in which the liquid phase (the solid phase, respect.) occupies a Lipschitz-continuous set $\Omega^+ \subset \Omega$ ($\Omega^- := \Omega \setminus \Omega^+$, respect.); the surface tension contribution to the *free enthalpy* of this system is

$$\sigma H_{N-1}(\partial\Omega^+ \cap \partial\Omega^-) + \sigma_L H_{N-1}(\partial\Omega^+ \cap \partial\Omega) + \sigma_S H_{N-1}(\partial\Omega^- \cap \partial\Omega). \quad (3.1)$$

We introduce the characteristic function of Ω^+

$$\chi_{\Omega^+} := \begin{cases} 1 & \text{in } \Omega^+ \\ -1 & \text{in } \Omega^- := \Omega \setminus \Omega^+ \end{cases} \quad (3.2)$$

(some expressions will be simplified by taking $\chi_{\Omega^+} = -1$ instead of $\chi_{\Omega^+} = 0$ in Ω/Ω^+). Obviously the family of Lebesgue measurable subsets of Ω is in one-to-one correspondence with the family of measurable characteristic functions. We want to extend the formula (3.1) to the case in which Ω^+ is just Lebesgue measurable; following the classical approach of Caccioppoli and De Giorgi [23] we replace $H_{N-1}(\partial\Omega^+ \cap \partial\Omega^-)$ with $\frac{1}{2}\int_\Omega |\nabla\chi_{\Omega^+}|$. We set

$$X := \{v \in BV(\Omega) : |v| = 1 \text{ a.e. in } \Omega\}. \quad (3.3)$$

For any subset A of $L^1(\Omega)$ we define the indicator function of A:

$$I_A(v) := \begin{cases} 0 & \text{if } v \in A \\ +\infty & \text{if } v \in L^1(\Omega) \setminus A. \end{cases}$$

We introduce the convex functional

$$\Lambda(v) := \frac{\sigma}{2}\int_\Omega |\nabla v| + \frac{\sigma_L - \sigma_S}{2}\int_{\partial\Omega} v\, dH_{N-1} + \frac{\sigma_L + \sigma_S}{2} H_{N-1}(\partial\Omega) \qquad \forall v \in L^1(\Omega). \quad (3.4)$$

Then we fix a $\theta \in L^1(\Omega)$ and define the non-convex functional

$$\Phi_\theta(v) := -\frac{L}{2\tau_E}\int_\Omega \theta v \, dx + \Lambda(v) + I_X(v) \qquad \forall v \in L^1(\Omega). \quad (3.5)$$

If $\chi \in X$ corresponds to the liquid-solid configuration, then $\Phi_\theta(\chi)$ represents either the *free enthalpy* of the system at constant pressure or the free energy at constant volume, but for an inessential additive function of θ [16, 22, 43].

REMARKS. (i) For any $\chi \in X$, setting $\Omega^\pm := \{x \in \Omega \,|\, \chi(x) = \pm 1\}$, we have

$$\frac{\sigma_L-\sigma_S}{2}\int_{\partial\Omega}\chi\,dH_{N-1}+\frac{\sigma_L+\sigma_S}{2}H_{N-1}(\partial\Omega)=\int_{\partial\Omega}(\sigma_L\chi^+ +\sigma_S\chi^-)dH_{N-1}=$$

$$=\sigma_L H_{N-1}(\partial\Omega^+\cap\partial\Omega)+\sigma_S H_{N-1}(\partial\Omega^-\cap\partial\Omega);\quad(3.6)$$

this represents the surface tension contribution of the exterior boundary to the free enthalpy.

(ii) In (3.5) $L^1(\Omega)$ can be equivalently replaced by any $L^p(\Omega)$, with $1<p<\infty$; indeed, as Ω is bounded, the strong topologies of the L^p-spaces are equivalent in X , for any $p<\infty$.

LEMMA 1. The bounded subsets of $BV(\Omega)$ are relatively compact in $L^1(\Omega)$ (with respect to the strong topology).

PROOF. See [23], p.17. □

LEMMA 2. The functional Λ is lower semi-continuous if and only if

$$|\sigma_L-\sigma_S|\leqslant\sigma\,.\quad(3.7)$$

PROOF. See [31]. □

THEOREM 1. If (3.7) holds, then for any $\theta\in L^1(\Omega)$, Φ_θ has an absolute minimum.

PROOF. It is sufficient to use the direct method of the calculus of variations and to apply the two previous lemmata. □

REMARK. By Lemma 2, the condition (3.7) is also necessary for Φ_θ to have an absolute minimum for any θ. This fact has remarkable physical consequences, since the absolute minima of Φ_θ correspond to physically acceptable stationary states. For instance for a water-ice system in contact with its vapour $\sigma_L\ \ \sigma<\sigma_S$, and consequently at any temperature $\Phi_\theta(\chi)$ is reduced by the presence of a water layer on $\partial\Omega$ [16, p.85; 43, p.33]. If $\theta<0$ then this layer is very thin, since the term

$$-\frac{L}{2\tau_L}\int_\Omega\theta\chi\,dx$$

forces $\chi=-1$ in the bulk. Of course in our macroscopic model we want to neglect such a layer, hence we replace σ_S with $\sigma+\sigma_L$, so that (3.7) is fulfilled; according to this modified model the ice phase appears on the boundary at any negative temperature, and superheating effects are still excluded there. Similarly, if for the same substance we had $\sigma_S+\sigma<\sigma_L$, then we should replace σ_L with $\sigma+\sigma_S$. In both cases, the modification corresponds to replacing Φ_θ with its lower semi-continuous envelope.

Henceforth we shall always assume that (3.7) holds.

DEFINITION 1. Any $\chi \in X$ is called an extremal of Φ_θ if and only if there exists a $\rho>0$ such that $\Phi_\theta(\cdot)-\Phi_\theta(\chi)$ is either non-negative or non-positive semidefinite in

$$\{v \in X:\ \|\chi-v\|_{L^1(\Omega)} \leq \rho\}\ .$$

PROPOSITION 1. Assume that $\theta \in L^1(\Omega)$ and let $\chi \in X$ be an extremal of Φ_θ. Set $\Omega^{\pm} := \{x \in \Omega:\ \chi(x) = \pm 1\}$ and $S := \partial\Omega^+ \cap \partial\Omega^-$. For any $x \in S$, if S is of class C^2 in a neighbourhood of x and if θ is continuous at x, then, denoting by κ the local mean curvature of S,

$$\kappa(x) = -\frac{L}{2\sigma\tau_F}\,\theta(x)\ .$$

Moreover, if (3.7) holds, and if S is of class C^2 up to $\partial\Omega$ then, denoting by ω the angle between the exterior normal vectors to Ω and to Ω^+ on $\partial\Omega \cap S$,

$$\cos\omega = \frac{\sigma_L-\sigma_S}{\sigma} \quad \text{on } \partial\Omega \cap S\ .$$

PROOF. By a standard procedure [23], S is locally written in Cartesian form and the first variation of Φ_θ is set equal to zero. □

REMARKS. (i) If θ is not continuous at $x \in S$, then (3.8) must be replaced by a weaker condition. For instance if θ is bounded in a neighbourhood of x, then instead of (3.8) we have

$$\mathrm{ap}\underline{\lim}\,\theta(x) := \sup\{\lambda \in \mathbb{R}:\ \lim_{r\to 0} \mu(B(x,r)\cap\{\chi<\lambda\})/\mu(B(x,r)) = 0\} \qquad (3.10)$$

$$\leqslant -\frac{2\sigma\tau_E}{L}\kappa(x) \leqslant \operatorname{ap}\overline{\lim}\,\theta(x) :=$$

$$:= \inf\{\lambda \in \mathbb{R} : \lim_{r\to 0} \mu(B(x,r) \cap \{\chi > \lambda\}) \,/\, \mu(B(x,r)) = 0\} \qquad (3.10)$$

where we have set for example, $\{\chi<\lambda\} := \{y \in \Omega : \chi(y) < \lambda\}$.

(ii) Let $\theta \in L^p_{\mathrm{loc}}(\Omega)$, with $p>N$, and χ be an *absolute* minimum of Φ_θ; let S^* be the *reduced interface*, namely the set of the points of S where an approximate normal vector exists. Then, by a classical result of De Giorgi [23], S^* is a $C^{1,(p-N)/2p}$ hypersurface and $H_s(S\setminus S^*) = 0$ for any $s>N-8$.

(iii) After Barozzi, Gonzalez and Tamanini [1], for any $\chi \in BV(\Omega)$ there exists a $\theta \in L^1(\Omega)$ such that $\Phi_\theta(\chi) = \inf \Phi_\theta$. Thus, as any temperature distribution $\theta \in L^1(\Omega)$ can be obtained by a suitable choice of the source term g in (1.1), in principle any phase configuration χ corresponding to an interface of finite $(N-1)$-dimensional measure may be physically admissible.

Supercooling and superheating. The following two results show that even for the *absolute* minima of Φ_θ there is no bound on the supercooling in sufficiently small regions. This is consistent with the physical picture. For the sake of simplicity, here we shall confine ourselves to the case of three space dimensions; however the extension to the general case is straightforward.

PROPOSITION 2. <u>Let</u> (3.7) <u>hold and</u> $N=3$; <u>let also</u> $\theta \in L^1(\Omega)$, $x \in \Omega$, r: <u>constant</u> >0 <u>be such that</u> $B(x,r) \subset \Omega$, $\theta \leqslant 3\sigma\tau_E/Lr$ <u>in</u> $B(x,r)$ <u>and</u> $\theta \leqslant 0$ <u>in</u> $\Omega\setminus B(x,r)$. <u>Then</u> $\hat\chi \equiv -1$ ($\hat\chi \equiv 1$, respect.) <u>a.e. in</u> Ω <u>is an absolute minimum of</u> Φ_θ ($\Phi_{-\theta}$, <u>respect</u>.).

PROOF. For Φ_θ, it is sufficient to compare $\hat\chi$ with $\tilde\chi$, defined by $\tilde\chi = 1$ in $B(x,r)$ and $\tilde\chi = -1$ in $\Omega\setminus B(x,r)$. Indeed

$$\frac{L}{\tau_E}\int_{B(x,r)} \theta\, dx \leqslant \frac{4L\pi}{3\tau_E} r^3 \operatorname*{esssup}_{B(x,r)} \theta \leqslant 4\pi\sigma r^2 = \frac{\sigma}{2}\int_\Omega |\nabla\chi_{B(x,r)}|.$$

□

PROPOSITION 3. Let (3.7) hold, $\theta \in L^1(\Omega)$ and $\chi \in X$ be an absolute minimum of Φ_θ. Then for any smooth compact set R

$$\int_R \theta \, dx \geqslant -\frac{\sigma\tau_E}{L} H_{N-1}(\partial R) \;\; \text{if} \;\; R \in \Omega^+ \tag{3.11}$$

$$\int_R \theta \, dx \leqslant \frac{\sigma\tau_E}{L} H_{N-1}(\partial R) \;\; \text{if} \;\; R \subset \Omega^- . \tag{3.12}$$

In particular for $N=3$, there is no $x \in \Omega$ and no $r>0$ such that

$$B(x,r) \subset \Omega^+ \;\; \text{and} \;\; \theta < -\frac{6\sigma\tau_E}{Lr} \;\; \text{a.e. in} \;\; B(x,r) \tag{3.13}$$

$$B(x,r) \subset \Omega^- \;\; \text{and} \;\; \theta > \frac{6\sigma\tau_E}{Lr} \;\; \text{a.e. in} \;\; B(x,r). \tag{3.14}$$

PROOF. If $R \subset \Omega^+$, set $\tilde{\chi} = -1$ in R and $\tilde{\chi} = \chi$ in $\Omega \setminus R$; then compare $\Phi_\theta(\tilde{\chi})$ with $\Phi_\theta(\chi)$. The procedure for $R \subset \Omega^-$ is similar. □

DEFINITION 2. Any $\chi \in X$ will be called a relative minimum of Φ_θ if it is not an absolute minimum and if there exists a $\delta > 0$ such that $\Phi_\theta(\chi) \leqslant \Phi_\theta(v)$ for any $v \in X$ with $\|\chi - v\|_{L^1(\Omega)} \leqslant \delta$.

The next result implies that for the *relative* minima of Φ_θ there is no limit for supercooling and superheating even in large sets. Therefore, because of (2.6), not all the relative minima of Φ_θ can be interpreted as metastable states. Here we choose special values for σ_L and σ_S, in order to exclude heterogeneous nucleation effects.

PROPOSITION 4. Let (3.7) hold; $\theta = \text{constant} > 0$, $\sigma_L = \sigma$ and $\sigma_S = 0$; set $\hat{\chi} \equiv -1$ a.e. in Ω. Then $\hat{\chi}$ is a relative minimum of Φ_θ.

PROOF. Here we shall assume that $N=3$; however the extension to the general case is easy. For any $v \in X$ we set

$$r_v := \left[\frac{3\pi}{8} \int_\Omega (v+1) \, dx\right]^{1/3} .$$

Possibly taking a larger Ω (and this is possible by the assumptions on σ_L and σ_S), we can assume that $B(x, r_v) \subset \Omega$, for some $x \in \Omega$. Then, setting $\chi_{B(x,r_v)} = 1$ in $B(x,r_v)$ and $\chi_{B(x,r_v)} = -1$ in $\Omega \setminus B(x, r_v)$,

$$\|\hat{\chi}-v\|_{L^1(\Omega)} = \int_\Omega (v+1)\, dx = \frac{8}{3}\pi r_v^3 = \|\hat{\chi}-\chi_{B(x,r_v)}\|_{L^1(\Omega)} ;$$

since balls minimize the perimeter among sets with prescribed volume, we easily get that

$$\Phi_\theta(\chi_{B(x,r_v)}) \leqslant \Phi_\theta(v) .$$

Therefore in order to verify that $\hat{\chi}$ is a relative minimum, it is sufficient to compare $\hat{\chi}$ with the characteristic functions of the form $\chi_{B(x,r)}$. We have

$$\Phi_\theta(\hat{\chi}) - \Phi_\theta(\chi_{B(x,r)}) = \frac{L}{\tau_E} \int_{B(x,r)} \theta\, dx - 4\pi\sigma r^2$$

$$\leqslant \frac{4\pi L r^2}{3\tau_E} \|\theta\|_{L^\infty(\Omega)} - 4\pi\sigma r^2 ;$$

hence

$$\Phi_\theta(\chi_{B(x,r)}) < \Phi_\theta(\hat{\chi}) \quad \text{only if} \quad r > \frac{3\tau_E\sigma}{L\|\theta\|_{L^\infty(\Omega)}} ,$$

that is only if

$$\|\hat{\chi}-\chi_{B(x,r)}\|_{L^1(\Omega)} \left(= \frac{8}{3}\pi r^3\right) > \frac{8}{3}\pi \left(\frac{3\tau_E\sigma}{L\|\theta\|_{L^\infty(\Omega)}}\right)^3 . \ \square$$

Metastability conditions. As we saw, not all the relative minima of Φ_θ correspond to metastable states. Indeed the states which are not located in a sufficiently deep "potential well" will be removed by thermodynamic fluctuations. Here we try to give a precise variational formulation of metastability, on the basis of the physical descriptions of [16, p.68-71; 22, pp.290-295; 43, pp.25-27].

Let us consider a supercooled liquid. The nucleation rate I, is of the form

$$I = C_1 \exp\left[-C_2\, \Xi(\theta)\right] ;$$

here C_1 and C_2 are positive constants; $\Xi(\theta)\,(>0)$ denotes the difference between the free enthalpy of a critical nucleus (i.e., a spherical nucleus having the temperature-dependent critical radius, cf. (2.1)) and that of the same amount of molecules in the liquid state. As C_1 and C_2 are "large", there exist a critical value Ξ_c and a "small" $\varepsilon > 0$ such that

$$\left.\begin{array}{lll} I \ll 1 & \text{if} & \Xi(\theta) < \Xi_c - \varepsilon \\ I \gg 1 & \text{if} & \Xi(\theta) > \Xi_c + \varepsilon \,. \end{array}\right\} \qquad (3.16)$$

Thus Ξ_c is the largest potential barrier which can be overcome by fluctuations, in the prescribed time scale. Hence $\chi \in X$ corresponds to a metastable state if and only if no state with a smaller free enthalpy can be reached by a continuous path without overcoming the potential $\Phi_\theta(\chi) + \Xi_c$, that is

$$\left.\begin{array}{l} \text{for any curve } \tilde{\chi}:[0,1] \to X \text{ continuous with respect} \\ \text{to the strong topology of } L^1(\Omega) \text{ and such that} \\ \tilde{\chi}(0) = \chi, \text{ if } \Phi_\theta(\tilde{\chi}(\cdot)) \leqslant \Phi_\theta(\chi) + \Xi_c \text{, then } \Phi_\theta(\tilde{\chi}(1)) \geqslant \\ \Phi_\theta(\chi). \end{array}\right\} \qquad (3.17)$$

The equation $\Xi(\theta) = \Xi_c$ defines the critical temperature $\theta = \theta_c$. If the supercooling occurs just in the interior of the sample, then by (2.1) and (3.5) we have

$$\Xi(\theta) = \left[\frac{4\pi L\theta}{3\tau_E} r^3 + 4\pi\sigma r^2\right]_{r = -(2\sigma\tau_E)/L\theta} = \frac{16\,\pi\sigma^3\theta_E^2}{3\,L^2\,\theta^2}; \qquad (3.18)$$

hence $\Xi(\theta) = \Xi_c$ if and only if

$$\theta = \left(\frac{16\,\pi\sigma^3\,\tau_E^2}{3L^2\,\Xi_c}\right)^{1/2} =: \theta_c \,. \qquad (3.19)$$

If instead the supercooling occurs near the exterior boundary $\partial\Omega$, then also heterogeneous nucleation may arise; consequently the maximum supercooling θ_c^* may be smaller; θ_c^* depends also on the local shape of $\partial\Omega$.

The characterization (3.17) of metastable states is quite implicit; hence in sections 4 and 5 we shall search for an alternative representation.

4. CONVEX REGULARIZATION

As the problem of minimizing Φ_θ in X is non-convex, it is natural to look at a related convex formulation; this is also the approach used for the standard Stefan problem without surface tension. We set

$$\tilde{X} := \{v \in BV(\Omega) : |v| \leq 1 \text{ a.e. in } \Omega\} ; \qquad (4.1)$$

it is not difficult to check that $\tilde{X}$ is the closed convex hull of X in the strong topology of $L^1(\Omega)$. For any $\theta \in L^1(\Omega)$ we also set

$$\tilde{\Phi}_\theta(v) := -\frac{L}{2\tau_E}\int_\Omega \theta v \, dx + \Lambda(v) + I_{\tilde{X}}(v) \qquad \forall\, v \in L^1(\Omega). \quad (4.2)$$

If (3.7) holds, then by Lemma 2 for any $\theta \in L^1(\Omega)$, $\tilde{\Phi}_\theta$ is a lower semi-continuous convex functional $L^1(\Omega) \to \mathbb{R} \cup \{+\infty\}$; hence there exists a $\chi \in X$ such that

$$\tilde{\Phi}_\theta(\chi) = \inf \tilde{\Phi}_\theta. \qquad (4.3)$$

THEOREM 2. <u>Let</u> (3.7) <u>hold</u>; $\theta \in L^1(\Omega)$ <u>and</u> $\chi \in \tilde{X}$ <u>be such that</u>

$$\tilde{\Phi}_\theta(\chi) = \inf \tilde{\Phi}_\theta. \qquad (4.4)$$

<u>For any</u> $s \in]-1, 1[$ <u>and any</u> $x \in \Omega$ <u>we set</u>

$$\chi_s(x) := \begin{cases} 1 & \text{if} \quad \chi(x) \geq s \\ -1 & \text{if} \quad \chi(x) < s \end{cases}$$

(<u>notice that</u> $\chi_s \in X$). <u>Then</u> $\tilde{\Phi}_\theta(\chi_s) = \tilde{\Phi}_\theta(\chi)$.

PROOF. For any set $A \subset \Omega$ we set

$$\phi_A(x) = \begin{cases} 1 & \text{if} \quad x \in A \\ 0 & \text{if} \quad x \in \Omega \setminus A . \end{cases}$$

We shall also use notations of the form $\{f \geq s\}$ instead of $\{x \in \Omega : f(x) \geq s\}$. We remind the reader of the *co-area formula* [23, p.20]

$$\int_\Omega |\nabla f| = \int_{-\infty}^{+\infty} ds \int_\Omega |\nabla \phi_{\{f \geq s\}}| \qquad \forall\, f \in BV(\Omega) \qquad (4.5)$$

and of another classical formula:

$$\forall\, p \in [1,\infty[\;,\ \forall f \in L^p(\Omega)\,,\ f \geq 0 \text{ a.e. in } \theta\,,\ \forall g \in L^p(\Omega)'\,,$$

$$\int_\Omega fg\,dx = \int_0^{+\infty} ds \int_\Omega \phi_{\{f\geq s\}}\, g\,dx\,, \tag{4.6}$$

whence, if instead $f \geq -C$ for some $C \in \mathbb{R}^+$,

$$\int_\Omega fg\,dx = \int_{-C}^{+\infty} ds \int_\Omega \phi_{\{f\geq s\}}\, g\,dx - C\int_\Omega g\,dx\,. \tag{4.7}$$

For any $v \in \tilde{X}$ and any $s \in [-1,1]$ we set

$$B_v(s) := -\frac{L}{2\tau_E}\int_\Omega \phi_{\{v\geq s\}}\,\theta\,dx + \frac{\sigma}{2}\int_\Omega |\nabla\phi_{\{v\geq s\}}|$$

$$+\frac{\sigma_L-\sigma_S}{2}\int_{\partial\Omega}\phi_{\{v\geq s\}}\,dH_{N-1}\,; \tag{4.8}$$

hence by (4.5) and (4.7)

$$\tilde{\Phi}_\theta(v) = \int_{-1}^{+1} B_v(s)\,ds + \frac{L}{2\tau_E}\int_\Omega \theta\,dx + \sigma_S H_{N-1}(\partial\Omega) \qquad \forall v \in \tilde{X}\,.$$

Let $\tilde{s} \in]-1,1[$ be such that

$$B_\chi(\tilde{s}) \leq \tfrac{1}{2}\int_{-1}^{+1} B_\chi(s)\,ds\,;$$

denoting the symmetric difference by Δ, for any $s \in]-1,1[$, $\mu(\{\chi_{\tilde{s}} \geq s\}\,\Delta\{\chi \geq \tilde{s}\}) = 0$, whence $B_{\chi_{\tilde{s}}}(s) = B_\chi(\tilde{s})$. Hence

$$\int_{-1}^{+1} B_{\chi_{\tilde{s}}}(s)\,ds = 2B_\chi(\tilde{s}) \leq \int_{-1}^{+1} B_\chi(s)\,ds\,; \tag{4.9}$$

therefore

$$0 \leq \int_{-1}^{\pm 1} B_\chi(s)\,ds - \int_{-1}^{+1} B_{\chi_{\tilde{s}}}(s)\,ds = \tilde{\Phi}_\theta(\chi) - \tilde{\Phi}_\theta(\chi_{\tilde{s}})$$

but χ is an absolute minimum of $\tilde{\Phi}_\theta$, whence the thesis. □

COROLLARY 1. Let (3.7) hold. Then for any $\theta \in L^1(\Omega)$,

$$\inf \Phi_\theta = \inf \tilde{\Phi}_\theta . \tag{4.10}$$

COROLLARY 2. Let (3.7) hold. Then for any $\theta \in L^1(\Omega)$, $\tilde{\Phi}_\theta$ coincides with the bipolar of Φ_θ:

$$\tilde{\Phi}_\theta = \Phi_\theta^{**} . \tag{4.11}$$

PROOF. It is sufficient to notice that $\tilde{\Phi}_\theta$ is convex and lower semi-continuous and that all the hyperplanes supporting the epigraph of $\tilde{\Phi}_\theta$ support also the epigraph of Φ_θ, because (4.10) holds for any $\theta \in L^1(\Omega)$. □

REMARK. If the surface tension contribution $\Lambda(v)$ is neglected, then minimization of $\tilde{\Phi}_\theta$ is equivalent to the condition

$$\begin{cases} \chi = 1 & \text{where} \quad \theta > 0 \\ -1 \leqslant \chi \leqslant 1 & \text{where} \quad \theta = 0 \\ \chi = -1 & \text{where} \quad \theta < 0 \end{cases} \quad \text{a.e. in } \Omega , \tag{4.12}$$

as in the standard weak formulation of the Stefan problem.

The present model exhibits two important drawbacks:

(i) In some cases there exist non-characteristic χ (i.e. $\chi \in \tilde{X} \setminus X$) which minimize Φ_θ^{**}; for instance, take $\sigma_L = \sigma_S$, $\theta \equiv 0$ and $\chi \equiv 0$ in Ω. The physical interpretation of such states is not clear. In the weak formulation of the standard Stefan problem without surface tension, the zones where $-1 < \chi < 1$ are usually interpreted as very fine mixtures of liquid and solid (so-called *mushy regions*), with liquid concentration equal to $\frac{1}{2}(\chi + 1)$. But in this case the $(N-1)$-dimensional measure of the liquid-solid interface is infinite; therefore such an interpretation cannot be accepted if surface tension is taken into account.

(ii) The convex functional Φ_θ^{**} has no relative minima; hence it is not clear how the metastable states could be represented here.

5. CONCAVE PENALIZATION

Here we shall consider a further model, in which the *non-convexity* of the problem is re-established. First let us fix any function γ such that

$$\gamma \in C^2([-1,1]), \quad \gamma'' > 0, \quad \gamma(-1) = \gamma(1) = 0; \qquad (5.1)$$

hence $\gamma(y) < 0$ for $-1 < y < 1$. For instance $\gamma(y) = \eta(y^2 - 1)$, with $\eta > 0$. We set (see (3.4))

$$\Psi_\theta(v) := -\frac{L}{2\tau_E} \int_\Omega [\theta v + \gamma(v)]\,dx + \Lambda(v) + I_{\tilde{X}}(v) \quad \forall v \in L^1(\Omega); \quad (5.2)$$

that is, by (4.2) and (4.11),

$$\Psi_\theta(v) = \Phi_\theta^{**}(v) - \frac{L}{2\tau_E} \int_\Omega \gamma(v)\,dx \qquad \forall v \in L^1(\Omega).$$

We still assume that (3.7) holds. For any $\theta \in L^1(\Omega)$, by the Lemmata 1 and 2, the *non-convex* functional $\Psi_\theta : L^1(\Omega) \to \mathbb{R} \cup \{+\infty\}$ has an absolute minimum.

We notice that $\Psi_\theta(v) = \Phi_\theta(v)$ for any $v \in X$; that is the convex functional Φ_θ^{**} and the non-convex one Ψ_θ are different extensions of Φ_θ from X to $\tilde{X}$.

COROLLARY 3 (of Theorem 2). Let (3.7) hold. Then for any $\theta \in L^1(\Omega)$

$$\inf \Psi_\theta = \inf \Phi_\theta^{**} = \inf \Phi_\theta \qquad (5.3)$$

$$\forall v \in \tilde{X} \setminus X, \quad \inf \Psi_\theta < \Psi_\theta(v). \qquad (5.4)$$

Moreover the set of the absolute minima of Ψ_θ is composed only of elements of X (namely characteristic functions), and coincides with the set of the absolute minima of Φ_θ.

The following result says that not only the absolute minima of ψ_θ, but also its relative minima are characteristic functions; hence they correspond to physically admissible stationary solid-liquid configurations.

THEOREM 3. Let (3.7) and (5.1) hold, $\theta \in L^1(\Omega)$ and let χ be either an absolute or a relative minimum of Ψ_θ. Then $\chi \in X$.

PROOF. We take any $\delta > 0$ and fix a

$$-k \in \left]\min_{[-1,1]} \gamma, 0\right[$$

such that, defined s_1 and s_2 by the conditions $-1 < s_1 < s_2 < 1$ and $\gamma(s_1) = \gamma(s_2) = -k$, then $s_2 - s_1 \leq \delta/\mu(\Omega)$; it is clear that such a k exists. We distinguish two cases:

$$\text{(i)} \qquad \mu(\{\gamma(\chi) < -k\}) > 0 \tag{5.5}$$

$$\text{(ii)} \qquad \mu(\{\gamma(\chi) < -k\}) = 0 . \tag{5.6}$$

CASE (i). For any $s \in]s_1, s_2[$ we set

$$\chi_s(x) := \begin{cases} \chi(x) & \text{if} \quad \chi(x) < s_1 \quad \text{or} \quad s_2 < \chi(x) \\ s_1 & \text{if} \quad s_1 \leq \chi(x) < s \\ s_2 & \text{if} \quad s_1 \leq \chi \leq s_2 \ ; \end{cases}$$

hence

$$\gamma(\chi_s) = \gamma(\chi) \qquad \text{if} \quad \chi(x) \leq s_1 \quad \text{or} \quad s_2 \leq \chi(x)$$

$$\gamma(\chi_s) = -k > \gamma(\chi) \qquad \text{if} \quad s_1 > \chi(x) > s_2 \ ;$$

Then by (5.5) we have

$$\int_\Omega \gamma(\chi_s)\, dx > \int_\Omega \gamma(\chi)\, dx . \tag{5.7}$$

For any $v \in \tilde{X}$ we define B_v as in (4.8) and take $\tilde{s} \in]s_1, s_2[$ such that

$$B_\chi(\tilde{s}) \leq \frac{1}{s_2 - s_1} \int_{s_1}^{s_2} B_\chi(s)\, ds .$$

Denoting the symmetric difference by Δ, we have

$$\mu(\{\chi_{\tilde{s}} \geq s\} \Delta \{\chi \geq s\}) = 0 \quad \text{if} \quad s \leq s_1 \quad \text{or} \quad s_2 < s$$

$$\mu(\{\chi_{\tilde{s}} \geq s\} \Delta \{\chi \geq \tilde{s}\}) = 0 \quad \text{if} \quad s_1 < s \leq s_2 \ ;$$

hence

$$B_{\chi_{\tilde{s}}}(s) = B_\chi(s) \qquad \text{if} \qquad s \leq s_1 \quad \text{or} \quad s_2 < s$$

$$B_{\chi_{\tilde{s}}}(s) = B_\chi(\tilde{s}) \qquad \text{if} \qquad s_1 < s \leq s_2 \ ;$$

therefore

$$\Phi_\theta^{**}(\chi_{\tilde{s}}) - \Phi_\theta^{**}(\chi) = \int_{-1}^{+1} [B_{\chi_{\tilde{s}}}(s) - B_\chi(s)]\, ds = \int_{s_1}^{s_2} [B_\chi(\tilde{s}) - B_\chi(s)]\, ds$$

$$= (s_2 - s_1)\, B_\chi(\tilde{s}) - \int_{s_1}^{s_2} B_\chi(s)\, ds \leqslant 0\,. \qquad (5.8)$$

By (5.7) and (5.8)

$$\Psi_\theta(\chi_{\tilde{s}}) - \Psi_\theta(\chi) < 0\,;$$

on the other hand we have

$$\|\chi - \chi_{\tilde{s}}\|_{L^1(\Omega)} \leqslant (s_2 - s_1)\cdot \mu(\Omega) < \delta\,;$$

as $\delta > 0$ is arbitrary, the assumption on χ is contradicted.

CASE (ii). For any $\lambda \in]0, s_2 + 1/s_1 + 1[$ we set

$$\chi_\lambda(x) := \begin{cases} \chi(x) & \text{if } \chi(x) \geqslant s_2 \\ \lambda[\chi(x)+1] - 1 & \text{if } \chi(x) \leqslant s_1 \end{cases}$$

this function is defined a.e. in Ω, as $\mu(\{s_1 < \chi < s_2\}) = 0$ by (5.6). Hence for any $s \in [-1, s_2[$, $\mu(\{\chi_\lambda \geqslant s\} = \mu(\{\lambda(\chi+1) \geqslant s+1\})$; then by the co-area formula (4.5) we have

$$\int_\Omega |\nabla \chi_\lambda| = \int_{-1}^{s_2} ds \int_\Omega \left|\nabla \phi_{\{\chi_\lambda \geqslant s\}}\right| + \int_{s_2}^{1} ds \int_\Omega \left|\nabla \phi_{\{\chi_\lambda \geqslant s\}}\right| =$$

$$= \int_{-1}^{s_2} ds \int_\Omega \left|\nabla \phi_{\{\chi+1 \geqslant (s+1)/\lambda\}}\right| + \int_{s_2}^{1} ds \int_\Omega \left|\nabla \phi_{\{\chi \geqslant s\}}\right| =$$

$$= \lambda \int_0^{s_1+1} d\xi \int_\Omega \left|\nabla \phi_{\{\chi+1 \geqslant \xi\}}\right| + \int_{s_2}^{1} ds \int_\Omega \left|\nabla \phi_{\{\bar{\chi} \geqslant s\}}\right|.$$

Thus the function $f_1 : \lambda \mapsto \Lambda(\chi_\lambda)$ (see (3.4)) is affine in $]0, (s_2+1)/(s_1+1)[$. Now if $\mu(\{-1 < \chi \leqslant s_1\}) > 0$, then

$$f_2 : \lambda \mapsto -\frac{2}{L\tau_E} \int_\Omega [\theta \chi_\lambda + \gamma(\chi_\lambda)]\, dx$$

is strictly concave by (5.1). Hence also $f_1 + f_2 : \lambda \mapsto \Psi_\theta(\chi_\lambda)$ is strictly concave in $]0, (s_2+1)/(s_1+1)[$; therefore it cannot have either a relative or an absolute minimum at the interior point $\lambda = 1$, corresponding to $\chi_1 = \chi$. Thus $\mu(\{-1 < \chi \leqslant s_1\}) = 0$.

Similarly one can show that $\mu(\{s_1 \leqslant \chi < 1\}) = 0$. Therefore after (5.6) $\chi = \pm 1$ a.e. in Ω. □

REMARK. As θ is any function of $L^1(\Omega)$, the strict convexity of γ is a necessary condition for theorem 3 to hold.

COROLLARY 4. Let (3.7) hold and $\theta \in L^1(\Omega)$. Then any relative minimum of Ψ_θ is also a relative minimum of Φ_θ.

PROOF. Notice that $\Phi_\theta = \Psi_\theta$ in X and $\Phi_\theta = +\infty$ in $\tilde{X} \setminus X$. □

REMARK. The converse of Corollary 4 does not hold. As a counter-example, let us take $\theta \in L^\infty(\Omega)$ and set $\hat{\chi} \equiv -1$ a.e. in Ω. By the Proposition 4, $\hat{\chi}$ is a relative minimum of Φ_θ for any $\theta > 0$, whereas it is not a relative minimum of Ψ_θ if $\theta > -\gamma'(-1)$, as it is easy to check.

PROPOSITION 5. The statements of the Proposition 1 hold also for the absolute and relative minima of Ψ_θ.

PROOF. By the Corollaries 3 and 4, the absolute and relative minima of Ψ_θ are also absolute and relative minima of Φ_θ; hence it is sufficient to apply the Proposition 1 itself. □

The following result shows that any relative minimum of Ψ_θ is an absolute minimum of the functional corresponding to a modified temperature.

THEOREM 4. Let (3.7) hold, $\theta \in L^1(\Omega)$ and χ be either an absolute or a relative minimum of Ψ_θ (whence $\chi \in X$, by the theorem 3). Then, setting $\hat{\theta} := \theta + \gamma'(\chi)$ a.e. in Ω,

$$\Psi_{\hat{\theta}}(\chi) = \Phi_{\hat{\theta}}(\chi) = \inf \Phi_{\hat{\theta}} = \inf \Psi_{\hat{\theta}} . \tag{5.9}$$

PROOF. For any $v \in \tilde{X}$ in a suitable neighbourhood of χ

$$\Phi^{**}_{\hat{\theta}}(\chi) - \Phi^{**}_{\hat{\theta}}(v) = -\frac{1}{2\tau_E}\int_\Omega [\theta + \gamma'(\chi)]\,(\chi - v)\,dx + \Lambda(\chi) - \Lambda(v) \leqslant$$

$$\leqslant -\frac{1}{2\tau_E}\int_\Omega [\theta(\chi - v) + \gamma(\chi) - \gamma(v)]\,dx + \Lambda(\chi) - \Lambda(v) =$$

$$= \Psi_\theta(\chi) - \Psi_\theta(v) \leqslant 0\,.$$

As $\Phi^{**}_{\hat{\theta}}$ is convex, this holds for any $v \in \tilde{X}$; thus

$$\Phi_\theta(\chi) = \Phi^{**}_{\hat{\theta}}(\chi) = \inf \Phi^{**}_{\hat{\theta}}\,;$$

then by (5.3) we get (5.9). □

REMARK. By (5.9), the regularity property pointed out in the second remark following the proposition 1 holds also for the relative minima of Ψ_θ.

PROPOSITION 6. <u>Let</u> (3.7) <u>hold</u>, $\theta \in L^1(\Omega)$ <u>and</u> χ <u>be a relative minimum of</u> Ψ_θ <u>(whence</u> $\chi \in X$, <u>by theorem 3)</u>. <u>Then, setting</u> $\Omega^\pm : \{x \in \Omega\colon \chi(x) = \pm 1\}$, <u>for any "smooth" compact set</u> R

$$\int_R \theta\,dx \geqslant -\frac{\sigma\tau_E}{L} H_{N-1}(\partial R) - \gamma'(1)\,H_N(R) \quad \text{if} \quad R \subset \Omega^+ \tag{5.10}$$

$$\int_R \theta\,dx \leqslant \frac{\sigma\tau_E}{L} H_{N-1}(\partial R) - \gamma'(-1)\,H_N(R) \quad \text{if} \quad R \subset \Omega^-. \tag{5.11}$$

<u>In particular, taking</u> $N = 3$, <u>there is no</u> $x \in \Omega$ <u>and no</u> $r > 0$ <u>such that</u>

$$B(x,r) \subset \Omega^+ \quad \underline{\text{and}} \quad \theta < -\frac{6\sigma\tau_E}{Lr} - \gamma'(1) \quad \underline{\text{a.e. in}}\ B(x,r) \tag{5.12}$$

$$B(x,r) \subset \Omega^- \quad \underline{\text{and}} \quad \theta > \frac{6\sigma\tau_E}{Lr} - \gamma'(-1) \quad \underline{\text{a.e. in}}\ B(x,r) \tag{5.13}$$

PROOF. It is sufficient to apply the Proposition 3 and the Theorem 4. □

REMARKS. (i) By the Propositions 3 and 6, we see that the necessary conditions for the temperature in the interior of the phases are less restrictive for the *relative* minima of Ψ_θ than for its *absolute* minima; cf. also (2.5) and (2.6). However, by the Proposition 5, both the absolute and the relative minima of Ψ_θ contain the Gibbs-Thomson law (3.8) and the contact angle condition (3.9) in a weak form, for any function γ.

(ii) Taking $r\to\infty$ in (5.12) and (5.13), namely for points x "far" from the boundary $\partial\Omega$, we get

$$\theta \geq -\gamma'(1) \;\; (<0) \quad \text{in } \Omega^+ \tag{5.14}$$

$$\theta^- \leq \gamma'(-1) \; (>0) \quad \text{in } \Omega^- . \tag{5.15}$$

This induces to regard the *local relative minima* of Ψ_θ (defined as in Definition 2 of Section 3, but with $\chi - v$ having compact support in Ω) as corresponding to the metastable states determined by homogeneous nucleation, if (cf. (2.6))

$$\gamma'(-1) = -\hat{\theta}_c , \quad \gamma'(1) = \theta_c .$$

(iii) Only $\gamma'(-1)$ and $\gamma'(1)$ have a physical meaning, for the rest γ is just any strictly convex function. If $\gamma'(-1)=-\infty$ and $\gamma'(1)=+\infty$, then also the relative minima of Φ_θ coincide with those of Ψ_θ.

(iv) Setting $\sigma=0$ in (5.12) and (5.13), namely neglecting the surface tension, we still get (5.14) and (5.15). However these conditions do not correspond to any relative minimum of the reduced functional $\hat{\Psi}_\theta$ (namely Ψ_θ without the Λ-term, cf. (5.2)); actually the latter functional has no relative minimum in $L^1(\Omega)$.

For instance, if $\theta>0$ then $\hat{\chi}\equiv -1$ a.e. in Ω is not a relative minimum of $\hat{\Psi}_\theta$; for any measurable $A\subset\Omega$, setting $\chi_A=1$ in A and $\chi_A=-1$ in $\Omega\backslash A$, we have $\hat{\Psi}_\theta(\chi_A)<\hat{\Psi}_\theta(\chi)$ and

$$\|\chi-\chi_A\|_{L^1(\Omega)} = 2\mu(A) .$$

This conclusion can be compared with the Proposition 4.

The model considered in this section is essentially aimed

to represent *homogeneous* nucleation. As we said in Section 2, the nucleation thresholds θ_c and $\hat{\theta}_c$ (see (2.6)) can be reduced by *heterogeneous* nucleation at the boundary $\partial\Omega$; as discussed in [16, chapter 3; 22, chapter 9; 43, chapter 2], the new thresholds θ_c^* and $\hat{\theta}_c^*$ depend also on geometrical features, hence they are (non-negative) functions of the point $x \in \partial\Omega$.

We wish to modify the previous model, so that locally on $\partial\Omega$

$$\left.\begin{array}{lll} \text{the solid is } \textit{metastable} & \text{if} & 0<\theta<\hat{\theta}_c^*(x) \\ \text{the liquid is } \textit{metastable} & \text{if} & -\theta_c^*(x)<\theta<0\,. \end{array}\right\} \tag{5.17}$$

To this aim, we assume that $\theta \in BV(\Omega)$, so that it has a trace in $L^1(\partial\Omega)$. Then we set a.e. on $\partial\Omega$

$$\rho(\theta, x) := \begin{cases} \dfrac{\sigma-\sigma_L+\sigma_S}{2} & \text{if } \theta \geqslant \hat{\theta}_c^*(x) \\[2ex] \dfrac{\sigma-\sigma_L+\sigma_S}{2}\cdot\dfrac{2\theta+\theta_c^*(x)-\hat{\theta}_c^*(x)}{\theta_c^*(x)+\hat{\theta}_c^*(x)} & \text{if } -\theta_c^*(x)<\theta<\hat{\theta}_c^*(x) \\[2ex] \dfrac{\sigma-\sigma_L+\sigma_S}{2} & \text{if } \theta \leqslant -\theta_c^*(x) \end{cases} \tag{5.18}$$

and replace the functional Ψ_θ by

$$\hat{\Psi}_\theta(v) := \Psi_\theta(v) + \int_{\partial\Omega} \rho(\theta(x), x)\, v(x)\, dH_{N-1} \qquad \forall v \in L^1(\Omega)\,; \tag{5.19}$$

thus here the boundary coefficient $(\sigma_L-\sigma_S)/2$ is replaced with $[(\sigma_L-\sigma_S)/2]+\rho(\theta(x),x)$ a.e. on $\partial\Omega$. For $-\theta_c^*<\theta<\hat{\theta}_c^*$, the latter lies between $-\sigma$ and σ; but for $\theta \leqslant -\theta_c^*$ ($\theta \geqslant \hat{\theta}_c^*$, respect.) this coefficient is equal to $-\sigma$ (σ, respect.), and this determines the nucleation of a liquid phase (solid phase, respect.) near the boundary, by the mechanism discussed in the remark following Theorem 1 in Section 3.

However this new model does not seem completely satisfactory, since it entails a temperature-dependence of the contact angle between S and $\partial\Omega$.

An open question. So far we have assumed the temperature to be prescribed. This occurs if the thermal conductivity k is phase-independent, cf. (1.1). If instead $k(\theta)$ is replaced by $k(\theta,\chi)$, then θ and χ are coupled. Does this problem have a solution?

6. COMPARISON WITH OTHER MODELS

The phase field model. The physical aspects of this model were studied by van der Waals [40], Cahn and Hilliard [10, 11, 12]. It is based on the use of a *free enthalpy* functional of the following form, for any $\hat{\theta} \in L^2(\Omega)$,

$$\hat{\Phi}_{\hat{\theta}}(\chi) := \int_\Omega \left[\frac{\xi^2}{2}|\nabla\chi|^2 + \frac{1}{8}(\chi^2-1)^2 - 2\hat{\theta}\chi\right] dx \qquad \forall \chi \in H^1(\Omega), \tag{6.1}$$

with no restriction on $|\chi|$. Here χ represents an *order parameter* and the phase variable is actually given by $\operatorname{sign}(\chi)$. The *correlation length* ξ is very small, hence one is led to study the limit as $\xi \to 0$.

In [6,7] Caginalp and Fife assumed also the temperature $\hat{\theta}$ to be very small. By means of a rescaling, they then replaced $\hat{\theta}$ with $\xi\tilde{\theta}$, so that the minimization of $\hat{\Phi}_{\hat{\theta}}$ became equivalent to that of

$$P^{\xi}_{\tilde{\theta}}(\chi) := \xi^{-1}\hat{\Phi}_{\xi\tilde{\theta}}(\chi) = \int_\Omega \left[\frac{\xi}{2}|\nabla\chi|^2 + \frac{1}{8\xi}(\chi^2-1)^2 - \tilde{\theta}\chi\right] dx\,; \tag{6.2}$$

then, under suitable assumptions on $\tilde{\theta}$, they showed that the stationary points of $P^{\xi}_{\tilde{\theta}}$ correspond to a *layered family*, and that the Gibbs-Thomson law (1.3) holds on the curve corresponding to that layer.

Here we shall compare the latter formulation with the one we gave in Section 3. We set

$$f(y) := \tfrac{1}{2}(y^2-1)\,, \qquad F(y) := \int_0^y f(\eta)\,d\eta \qquad \forall y \in \mathbb{R}\,,$$

and notice that

$$P^{\xi}_{\tilde{\theta}}(\chi) = \int_\Omega \left[|\nabla F(\chi)| - \tilde{\theta}\chi\right] dx + \tfrac{1}{2}\int_\Omega \left[\xi^{\frac{1}{2}}|\nabla\chi| - \xi^{-\frac{1}{2}} f(\chi)\right]^2 dx\,. \tag{6.3}$$

Of course we can extend $P^{\xi}_{\tilde\theta}$ setting

$$P^{\xi}_{\tilde\theta}(\chi) = +\infty \quad \text{if} \quad \chi \in L^1(\Omega)\setminus H^1(\Omega).$$

Then by a result of Modica and Mortola [33], the Γ-limit (in the sense of De Giorgi [17]) of $P^{\xi}_{\tilde\theta}$ as $\xi \to 0$, is

$$P_{\tilde\theta}(\chi) := \begin{cases} \int_\Omega |\nabla F(\chi)| - \int_\Omega \tilde\theta\chi\, dx & \text{if } \chi \in X \\ +\infty & \text{if } \chi \in L^1(\Omega)\setminus X\,; \end{cases} \tag{6.5}$$

more precisely

$$P_{\tilde\theta}(\chi) = \Gamma(\mathbb{R}, L^1_{\mathrm{loc}}(\Omega)^-) - \lim_{\substack{\xi\to 0\\ v\to\chi}} P^{\xi}_{\tilde\theta}(v)\,. \tag{6.5}$$

By definition [17] this means that for any $\chi \in L^1(\Omega)$ and any family $\{\chi_\xi\} \subset L^1(\Omega)$ such that $\chi_\xi \to \chi$ strongly in $L^1_{\mathrm{loc}}(\Omega)$,

$$\varliminf_{\xi\to 0} P^{\xi}_{\tilde\theta}(\chi_\xi) \geqslant P_{\tilde\theta}(\chi) \qquad \text{if } \chi \in X \tag{6.6}$$

$$\lim_{\xi\to 0} P^{\xi}_{\tilde\theta}(\chi_\xi) = +\infty \qquad \text{if } \chi \in L^1(\Omega)\setminus X, \tag{6.7}$$

and that for any $\chi \in L^1(\Omega)$ there exists a family $\{\chi_\xi\} \subset L^1(\Omega)$ such that $\chi_\xi \to \chi$ strongly in $L^1_{\mathrm{loc}}(\Omega)$ and

$$\lim_{\xi\to 0} P^{\xi}_{\tilde\theta}(\chi_\xi) = P^{\xi}_{\tilde\theta}(\chi)\,. \tag{6.8}$$

This entails the following result:

PROPOSITION 7. If $\{\chi_\xi\} \subset L^1(\Omega)$ is such that

$$P^{\xi}_{\tilde\theta}(\chi_\xi) = \inf P^{\xi}_{\tilde\theta}, \quad \chi_\xi \to \chi \text{ strongly in } L^1(\Omega), \tag{6.9}$$

then $P_{\tilde\theta}(\chi) = \inf P_{\tilde\theta}$.

Now we notice that

$$\left|\int_{-1}^{1} f(\eta)\,d\eta\right| = \tfrac{2}{3}, \text{ whence } |\nabla F(\chi)| = \tfrac{2}{3}|\nabla\chi|$$

for any $\chi \in X$; we then replace $\tilde\theta$ with $2L\theta/3\sigma\tau_E$, by means of a second rescaling of the temperature; thus we get

$$R_\theta(\chi) := \frac{3\sigma}{4} P_{\tilde{\theta}}(\chi) = \begin{cases} -\frac{L}{2\tau_E}\int_\Omega \theta\chi\, dx + \frac{\sigma}{2}\int_\Omega |\nabla\chi| & \text{if } \chi \in X \\ +\infty & \text{if } \chi \in L^1(\Omega)\setminus X. \end{cases} \tag{6.10}$$

Thus if $\sigma_L = \sigma_S$ then R_θ coincides with the functional Φ_θ defined in Section 3, but for an unessential additive constant.

These developments can be extended to the case of $\sigma_L \neq \sigma_S$ by replacing $\hat{\Phi}_\theta(\chi)$, cf. (6.1), with

$$\hat{\Phi}_{\hat{\theta}}(\chi) + \xi\,\frac{\sigma_L - \sigma_S}{2}\int_{\partial\Omega} v\, dH_{N-1} + \xi\,\frac{\sigma_L + \sigma_S}{2} H_{N-1}(\partial\Omega) \qquad \forall v \in H^1(\Omega) \tag{6.11}$$

The relation between the initial temperature $\hat{\theta}$ and the final one θ is

$$\hat{\theta} = \frac{2L\xi}{3\sigma\tau_E}\,\theta\,; \tag{6.12}$$

hence ξL and $\sigma\tau_E$ are of the same order; that is ξ is of the order of $\sigma\tau_E/L$, the same constant which appears in the Gibbs-Thomson law (2.1).

We also notice that, denoting by Ψ^n_θ the functional obtained by replacing $\gamma(v)$ with $n\gamma(v)$ in (5.2) for any $n \in \mathbb{N}$, we have

$$\Phi_\theta(v) = \Gamma(\mathbb{R}, L^1_{\text{loc}}(\Omega)^-) - \lim_{\substack{n\to\infty \\ v\to u}} \Psi^n_\theta(v)\,. \tag{6.13}$$

The domain model of ferromagnetism. This model was developed by Weiss, Landau and Lifshitz [28,29]. Below the Curie's critical temperature, a ferromagnetic body is magnetically saturated on a microscopic scale; that is the magnetization field $\boldsymbol{M}$ has a prescribed modulus M and variable orientation. If the body is sufficiently large, it breaks up into small uniformly magnetized regions (*Weiss domains*), separated by thin transition layers (*Bloch walls*). Denoting the magnetic field by $\bar{\boldsymbol{H}}$, the *free enthalpy* is equal to

$$P_{\boldsymbol{H}}(\boldsymbol{M}) := \int_\Omega \left[-\boldsymbol{H}\cdot\boldsymbol{M} + \sum_{l,m=1}^{3} a_{lm}\,\frac{\partial\boldsymbol{M}}{\partial x_l}\cdot\frac{\partial\boldsymbol{M}}{x_m} + \phi(\boldsymbol{M}) \right] dx\,; \tag{6.14}$$

here $\{a_{lm}\}$ is a positive definite tensor and ϕ is a convex real function.

The analogy between (6.1) and (6.14) is obvious; however $\boldsymbol{M}$ is a vector, hence the non-convex constraint $|\boldsymbol{M}| = \mathrm{M}$ corresponds to a *connected* manifold.

The mathematical aspects of this model for either the stationary and evolution cases were studied by the present author in [38].

The coefficients a_{lm} are small, hence the Bloch walls correspond to thin layers, across which $\boldsymbol{M}$ changes its orientation. We can also force $\boldsymbol{M}$ to be discontinuous across these walls, by considering a functional of the following form, for any $\boldsymbol{H} \in L^1(\Omega)^3$:

$$\left.\begin{aligned} \Phi_{\boldsymbol{H}}(\boldsymbol{M}) &:= \int_\Omega \left\{ -\boldsymbol{H}\cdot\boldsymbol{M} + a\left[\sum_{i,j=1}^{3}\left(\frac{\partial M_i}{\partial x_j}\right)^2\right]^{1/2} + \phi(\boldsymbol{M})\right\} dx \\ &\qquad\qquad \forall \boldsymbol{M} \in Y := \{\boldsymbol{v} \in BV(\Omega)^3 : |\boldsymbol{v}| = \mathrm{M}\} \\ \Phi_{\boldsymbol{H}}(\boldsymbol{M}) &:= +\infty \qquad\qquad \forall \boldsymbol{M} \in L^1(\Omega)^3 \setminus Y \end{aligned}\right. \tag{6.15}$$

(a being a positive constant).

One can also introduce a *concave penalization*, setting for instance

$$\left.\begin{aligned} \Psi_{\boldsymbol{H}}(\boldsymbol{M}) &:= E_{\boldsymbol{H}}(\boldsymbol{M}) - \eta \int_\Omega (|\boldsymbol{M}|^2 - \mathrm{M}^2)\, dx \\ &\qquad\qquad \forall \boldsymbol{M} \in \tilde{Y} := \{\boldsymbol{v} \in BV(\Omega)^3 : |\boldsymbol{v}| \leq \mathrm{M}\} \\ \Psi_{\boldsymbol{H}}(\boldsymbol{M}) &:= +\infty \qquad\qquad \forall \boldsymbol{M} \in L^1(\Omega)^3 \setminus \tilde{Y}, \end{aligned}\right. \tag{6.16}$$

η being a positive constant.

Open questions. (i) Can the functional $\Phi_{\boldsymbol{H}}$ be obtained as a Γ-limit of $\Psi_{\boldsymbol{H}}$ as $\eta = +\infty$? We notice that it is not evident how a formula like (6.3) might be extended to the case of a vector variable.

(ii) Are the absolute and relative minima of the functional $\Psi_{\boldsymbol{H}}$ confined to the non-convex constraint Y?

ACKNOWLEDGEMENTS

he author is indebted for useful talks to G. Anzellotti, P. Fife, S. Luckhaus, L. Tamanini, and also to the participants at an evening section at the meeting on free boundary problems held at Oberwolfach in September 1985.

Added in proofs. Recently L. Modica studied the asymptotic behaviour of the phase field model (cf. Section 6), using the techniques of [32,33]:

L. Modica — The gradient theory of phase transitions and the minimal interface criterion — Arch. Rat. Mech. Anal. (to appear).

L. Modica — Gradient theory of phase transitions with boundary contact energy — Annales. Inst. H. Poincaré, Analyse non linéaire (to appear).

REFERENCES

1. E. Barozzi, E. Gonzalez and I. Tamanini, "The mean curvature of a set of finite perimeter", *Proceedings A.M.S.* **99** (1987), pp. 1-4.
2. A. Bossavit, A. Damlamian, M. Fremond (Eds.), *Free Boundary Problems: Applications and Theory*, Vol. III, IV, Pitman, Boston (1985).
3. K.A. Brakke, "The motion of a surface by its mean curvature", *Mathematical Notes*, Princeton University Press, Princeton (1978).
4. G. Caginalp, "Phase field models of solidification: free boundary problems as systems of nonlinear parabolic differential equations", in [2], pp. 107-121.
5. G. Caginalp, "An analysis of a phase field model of a free boundary", *Arch. Rat. Mech. Anal.* **92** (1986), pp. 205-246,
6. G. Cafinalp and P.C. Fife, "Elliptic problems involving phase boundaries satisfying a curvature condition", Preprint.
7. G. Caginalp and P.C. Fife, "Phase field models for interfacial boundaries", *Physical Review B* **33** (1986), pp. 7792-7794.
9. G. Caginalp and P.C. Fife, "Higher order phase field models and detailed anisotropy", Preprint.
10. J.W. Cahn and J.E. Hilliard, "Free energy of a nonuniform system. I. Interfacial free energy", *J. Chem. Phys.* **28** (1957), pp. 258-267.
11. J.W. Cahn, "Free energy of a nonuniform system. II. Thermodynamic basis", *J. Chem. Phys.* **30** (1959), pp. 1121-1124.

12. J.W. Cahn and J.E. Hilliard, "Free energy of a nonuniform system. III. Nucleation in a two component incompressible fluid", *J. Chem. Phys.* **31** (1959), pp.688-699.
13. J. Carr, M.E. Gurtin and M. Slemrod, "Structured phase transitions on a finite interval", *Arch. Rational Mech. Anal.* **86** (1984), pp.317-351.
14. J. Carr, M.E. Gurtin and M. Slemrod, "Structured phase transitions under prescribed loads", *J. Elasticity* (to appear).
15. J. Chadam and P. Ortoleva, "The stabilizing effect of surface tension on the development of the free boundary in a planar, one-dimensional, Cauchy-Stefan problem", *IMA J. Appl. Math.* **30** (1983), pp.57-66.
16. B. Chalmers, *Principles of Solidification*, Wiley, New York, (1964).
17. E. De Giorgi, "Γ-convergenza e G-convergenza", *Boll. U.M.I.* **14**-A (1977), pp.213-220.
18. J. Duchon and R. Robert, "Sur une équation d'évolution non linéaire de la chimie des surfaces", *J. Math. Pures et Appl.* **62** (1983), pp.305-325.
19. A. Fasano and M. Primicerio (Eds.), *Free Boundary Problems: Theory and Applications*, Pitman, Boston (1983).
20. G. Fix, "Numerical simulation of free boundary problems using phase field models", (ed. Whiteman, J.R.), in *The Mathematics of the Finite Elemenents and Applications IV*, Academic Press, London (1982).
21. G. Fix, "Phase field models for free boundary problems", in [19], pp.580-589.
22. M.C. Flemings, *Solidification Processing*, McGraw-Hill, New York (1974).
23. E. Giusti, *Minimal Surfaces and Functions of Bounded Variation*, Birkhäuser, Boston (1984).
24. M.E. Gurtin (Ed.), *Phase Transformations and Material Instabilities in Solids*, Academic Press, Orlando (1984).
25. M.E. Gurtin, "The Gradient theory of phase transitions on a finite interval", in [24], pp.99-102.
26. M.E. Gurtin, "On a theory of phase transitions with interfacial energy", *Arch. Rational Mech. Anal.* **87** (1985), pp.187-212.
27. M.E. Gurtin, "On phase transitions with bulk, interfacial, and boundary energy", *Arch. Rational Mech. Anal.* 96 (1986), pp.243-266.
28. L.D. Landau and E.M. Lifshitz, *Statistical Physics*, Pergamon Press, Oxford (1958).
29. L.D. Landau and E.M. Lifshitz, *Electrodynamics of Continuous Media*, Pergamon Press, Oxford (1960).
30. J.S. Langer, "Instabilities and pattern formation in crystal growth", *Rev. Mod. Phys.* **1** (1980), pp.1-28.
31. U. Massari and L. Pepe, "Su di una impostazione parametrica del problema dei capillari", *Ann. Univ. Ferrara* **20** (1974), pp.21-31.

32. L. Modica, "Gamma convergence to minimal surface problems and global solutions of $\Delta u = 2(u^3 - u)$", in *Proceedings of the International Meeting on Recent Methods in Nonlinear Analysis*. (eds. De Giorgi, E., Magenes, E. and Mosco, U.) Pitagora Editrice, Bologna (1978).
33. L. Modica and S. Mortola, "Il limite nella Γ-convergence di una famiglia di funzionali ellittici", *Boll. Un. Mat. Ital.* **14-A** (1977), pp.526-529.
34. J.C.W. Rogers, "The Stefan problem with surface tension", in [19], pp.263-274.
35. J. Smith, "Shape instability and pattern formation in solidification", *J. Comp. Phys.* **39** (1981), pp.112-127.
36. A. Visintin, "Stefan problem with surface tension", *Preprint of I.A.N. of C.N.R.*, Pavia (1984).
37. A. Visintin, "Models for supercooling and superheating effects", in [2], pp.200-207.
38. A. Visintin, "On Landau-Lifshitz equations for ferromagnetism", *Japan J. Appl. Math.* 2 (1985), pp.69-84.
39. A. Visintin, "Phase transitions with surface tension", (in preparation).
40. J.D. van der Waals, "The thermodynamic theory of capillarity under the hypothesis of a continuous variation of density", (in Dutch), *Verhandel. Konink. Akad. Weten. Amsterdam* (Section 1) **1**, (8) (1893).
41. D.G. Wilson, A.D. Solomon and P.T. Boggs (Eds.) *Moving Boundary Problems*, Academic Press, New York (1978).
42. D.J. Wollkind and R.D. Notestine, "A nonlinear stability analysis of the solidification of a pure substance", *IMA J. Appl. Math.* **27** (1981), pp.85-104.
43. P.D. Woodruff, *The Liquid-Solid Interface*, Cambridge University Press, London (1973).

INDEX